THE ROYAL SOCIETY

DECENNIAL INDEX

1971–1980

INDEX OF AUTHORS IN
PROCEEDINGS
PHILOSOPHICAL TRANSACTIONS
AND
BIOGRAPHICAL MEMOIRS

LONDON

THE ROYAL SOCIETY

1981

PUBLISHED BY
THE ROYAL SOCIETY
6 CARLTON HOUSE TERRACE
LONDON, SW1Y 5AG

ISBN 0 85403 181 2

*Printed in Great Britain
by P.B. Offset Ltd. Mitcham, Surrey
from negatives prepared at the Royal Society*

PREFACE

This index is a sequel to that for the period 1961 to 1970 and covers the scientific journals of the Royal Society and the biographical memoirs of Fellows published during the years 1971 to 1980.

The following publications are indexed under the authors' names:

Proceedings A	volume 320, no. 1543 to volume 373, no. 1754.
Proceedings B	volume 176, no. 1045 to volume 211, no. 1182.
Philosophical Transactions A	volume 268 no. 1192 to volume 291, no. 1442, (excluding no. 1193, volume 269).
Philosophical Transactions B	volume 259, no. 833 to volume 291, no. 1049.
Biographical Memoirs	volume 17 to volume 26.

In addition, Anniversary Addresses, Lectures, and Discussions and Symposia are listed chronologically in appendixes, and a final appendix lists alphabetically the subjects of the *Biographical Memoirs*.

The abbreviations *Proc., Trans.,* and *Biogr. Mem.* are used. The year given in parentheses after the reference is that of the publication of the paper, which is not necessarily the same as the year of completion of the volume.

Authors' names and volume numbers are printed in bold type (except in Appendix IV), and page numbers in ordinary type.

R.W.J. KEAY
Executive Secretary

6 CARLTON HOUSE TERRACE
LONDON, SW1Y 5AG

CONTENTS

DECENNIAL INDEX 1971–1980

Aaboe, A. Scientific astronomy in antiquity (Discussion). *Trans.* A **276**, 21–42 (1974).
Aarseth, S.J. Cosmological *N*-body simulations of galaxy mergers (Discussion).
  Trans. A **296**, 351–353 (1980).
Abbott, J.E. *See* Fenton & Abbott.
Abbott, J.E. & Francis, J.R.D. Saltation and suspension trajectories of solid grains in a water stream.
  Trans. A **284**, 225–254 (1977).
Abboud, M.M. *See* Akhtar, Abboud, Barnard, Jordan & Zaman.
Abdallahi Ould M. Sidia, Skaf, R., Castel, J.M. & Ndiaye, A. OCLALAV and its environment: a
  regional international organization for the control of migrant pests (Discussion).
  Trans. B **287**, 269–276 (1979).
Abe, T., Alemá, S. & Miledi, R. On the purification of β-bungarotoxin. *Proc.* B **207**, 487–490 (1980).
Abe, T., Limbrick, A.R. & Miledi, R. Acute muscle denervation induced by β-bungarotoxin.
  Proc. B **194**, 545–553 (1976).
Abe, T. & Miledi, R. Inhibition of β-bungarotoxin action by bee venom phospholipase A_2.
  Proc. B **200**, 225–230 (1978).
Abegg, K. The growth of turbogenerators (Discussion). *Trans.* A **275**, 51–67 (1973).
Abercrombie, M. Concepts in morphogenesis (Discussion). *Proc.* B **199**, 337–344 (1977).
  The Croonian Lecture, 1978. The crawling movement of metazoan cells.
  Proc. B **207**, 129–147 (1980).
Abraham, E.P. Howard Walter Florey (Baron Florey of Adelaide and Marston).
  Biogr. Mem. **17**, 255–302 (1971).
  See also O'Sullivan (J.), Huddleston & Abraham; Pollock & Abraham.
Accad, Y. *See* Pekeris & Accad; Pekeris, Accad & Shkoller.
Accad, Y. & Pekeris, C.L. Solution of the tidal equations for the M_2 and S_2 tides in the world oceans
  from a knowledge of the tidal potential alone. *Trans.* A **290**, 235–266 (1978).
Acevedo, E. *See* Hsiao, Acevedo, Fereres & Henderson.
Acher, R.A. Molecular evolution of biologically active peptides (Discussion). *Proc.* B **210**, 21–43
  (1980).
Acheson, D.J. On the instability of toroidal magnetic fields and differential rotation in stars. With an
  appendix on the axisymmetric diffusive instability of toroidal magnetic fields in a rotating gas,
  by D.J. Acheson & M.P. Gibbons. *Trans.* A **289**, 459–500 (1978).
Acheson, D.J. & Gibbons, M.P. The axisymmetric diffusive instability of toroidal magnetic fields in
  a rotating gas (appendix to a paper by D.J. Acheson). *Trans.* A **289**, 495–500 (1978).
Acheson, E.D. Record linkage and the identification of long-term environmental hazards (Discussion).
  Proc. B **205**, 165–178 (1979).
Ackefors, H. Mercury pollution in Sweden with special reference to conditions in the water habitat
  (Discussion). *Proc.* B **177**, 365–387 (1971).
Ackerman, J.J.H., Bore, P.J., Gadian, D.G., Grove, T.H. & Radda, G.K. N.m.r. studies of metabolism
  in perfused organs (Discussion). *Trans.* B **289**, 425–436 (1980).
Ackroyd, J.A.D. Laminar natural convection boundary layers on near-horizontal plates.
  Proc. A **352**, 249–274 (1976).
Acrivos, A. *See* Chen (H.-S.) & Acrivos; Rocha & Acrivos.
Acton, L.W. & Catura, R.C. Spectroscopic studies of the solar corona at X-ray wavelengths
  (Discussion). *Trans.* A **281**, 383–390 (1976).
Adal, M.N. *See* Wallace (B.G.), Adal & Nicholls.
Adam, J.A. *See* Craik & Adam.
Adamek, P.M. & Wilson, M.R. The evolution of a uranium province in northern Sweden (Discussion).
  Trans. A **291**, 355–368 (1979).
Adams, D. *See* Rappaport & Adams.
Adams, D.R. & Hirst, W. Frictional traction in elastohydrodynamic lubrication.
  Proc. A **332**, 505–525 (1973).

1

Adams, Eveline M., McDonald, I.R. & Singer, K. Collective dynamical properties of molten salts: molecular dynamics calculations on sodium chloride. *Proc.* A **357**, 37—57 (1977).

Adams, J.M. *See* Evans (S.) Adams & Thomas.

Adams, M.C., Bradley, D.J., Sibbett, W. & Taylor, J.R. Synchronously pumped continuous wave dye lasers (Discussion). *Trans.* A **298**, 217—223 (1980).

Adams, M.J. *See* Landsberg & Adams.

Adams, M.W.W. *See* Hall (D.O.), Adams, Morris & Rao.

Addison, R. *See* Hughes (C.P.), Ingham & Addison.

Adefris Bellehu. The Desert Locust Control Organization for Eastern Africa (DLCOEA) and its background. (Discussion). *Trans.* B **287**, 265—268 (1979).

Adie, R.J. The geology of Antarctica: a review (Discussion). *Trans.* B **279**, 123—130 (1977).

Adinolfi, M., *See* Embury, Seller, Adinolfi & Polani; Seller, Embury, Polani & Adinolfi.

Adkins, C.J. *See* Mott (Sir Nevill), Pepper, Pollitt, Wallis & Adkins.

Adler, J., Barry, P.A. & Bernal, M.J.M. Thermal explosion theory for a slab with time-periodic surface temperature variation. *Proc.* A **370**, 73—88 (1980).

Adolph, K.W. & Butler, P.J.G. Assembly of a spherical plant virus (Discussion). *Trans.* B **276**, 113—122 (1976).

Adrian, Lord. Detlev Wulf Bronk. *Biogr. Mem.* **22**, 1—9 (1976).

Adrian, R.H. Conduction velocity and gating current in the squid giant axon. *Proc.* B **189**, 81—86 (1975).
See also Chandler (W.K.), Schneider, Rakowski & Adrian.

Agarwal, K.L. *See* Steiner (D.F.), Patzelt, Chan, Quinn, Tager, Nielsen and others.

Agg, A.R. *See* Thom & Agg.

Ahmad, N.H., Prutton, M. & Whiting, J.S.S. The magnetoelastic coefficients h_3 and h_4 of (100) nickel films. *Proc.* A **328**, 49—65 (1972).

Ahmed, B.M. & Eltayeb, I.A. On the propagation, reflexion, transmission and stability of atmospheric Rossby-gravity waves on a beta-plane in the presence of latitudinally sheared zonal flows. *Trans.* A **298**, 45—85 (1980).

Air, G.M. *See* Laver, Air, Webster, Gerhard, Ward & Dopheide.

Airth, G.R. X-ray image intensifiers: applications and current limitations (Discussion). *Trans.* A **292**, 257—263 (1979).

Aitchison, Joyce. Numerical treatment of a singularity in a free boundary problem. *Proc.* A **330**, 573—580 (1972).

Ajayi, O.O. A note on Taylor's electrohydrodynamic theory. *Proc.* A **364**, 499—507 (1978).

Akeroyd, Anne V. Archaeological and historical evidence for subsidence in southern Britain (Discussion). *Trans.* A **272**, 151—169 (1972).

Akert, K. *See* Pfenninger, Akert, Moor & Sandri.

Akhtar, M., Abboud, M.M., Barnard, G., Jordan, P. & Zaman, Z. Mechanism and stereochemistry of enzymic reactions involved in porphyrin biosynthesis (Discussion). *Trans.* B **273**, 117—136 (1976).

Akhtar, M., Wilton, D.C., Watkinson, I.A. & Rahimtula, A.D. Substrate activation in pyridine nucleotide-linked reactions: illustrations from the steroid field (Discussion). *Proc.* B **180**, 167—177 (1972).

Akman, Y. *See* Brooks (R.R.), Morrison, Reeves, Dudley & Akman.

Aksenova, N.P. *See* Chailakhyan, Aksenova, Konstantinova & Bavrina.

Ala, F. *See* Lehmann (H.), Ala, Hedeyat, Montazemi and others.

Alabaster, J.S. Suspended solids and fisheries (Discussion). *Proc.* B **180**, 395—406 (1972).

Alber, I.E. The effects of randomness on the stability of two-dimensional surface wavetrains. *Proc.* A **363**, 525—546 (1978).

Albertson, Donna G. & Thomson, J.N. The pharynx of *Caenorhabditis elegans*. *Trans.* B **275**, 299—325 (1976).

Alcock, A.J. & Corkum, P.B. Ultra-short pulse generation with CO_2 lasers (Discussion). *Trans.* A **298**, 365—376 (1980).

Aldred, P.J.E. & Hart, M. The electron distribution in silicon.

I. Experiment. *Proc.* A **332**, 223–238 (1973).

II. Theoretical interpretation. *Proc.* A **332**, 239–254 (1973).

Alemá, S. *See* Abe, Alemá & Miledi.

Aler, B. Total energy: power (nuclear) and district heating (Discussion).
Trans. A **272**, 627–637 (1972).

Alexander, H.G.L. A preliminary assessment of the rôle of the terrestrial decapod crustaceans in the Aldabran ecosystem (Discussion). *Trans.* B **286**, 241–246 (1979).

Alexander, J.M. On the theory of rolling. *Proc.* A **326**, 535–563 (1972).
Corrigenda. *Proc.* A **329**, 493–496 (1972).

Alexander, J.M. & Gunasekera, J.S. On the geometrically similar expansion of a hole in a thin infinite plate. *Proc.* A **326**, 361–373 (1972).

Alger, B.E. *See* Nicoll, Alger & Jahr.

Allan, D. *See* Crumpton, Allan, Auger, Green & Maino.

Allan, P. & Bevis, M. The fold-sector dependence of plastic deformation modes in polyethylene single crystals. *Proc.* A **341**, 75–90 (1974).

Allcock, G.R. Invariant Lagrangian theory of the Poisson bracket for systems with constraints.
Proc. A **344**, 175–198 (1975).
The intrinsic properties of rank and nullity of the Lagrange bracket in the one dimensional calculus of variations. *Trans.* A **279**, 487–545 (1975).

Allègre, C.J. *See* Bottinga & Allègre.

Allègre, C.J., Brévart, O., Dupré, B. & Minster, J.-F. Isotopic and chemical effects produced in a continuously differentiating convecting Earth mantle (Discussion). *Trans.* A **297**, 447–477 (1980).

Allègre, C.J., Shimizu, N. & Treuil, M. Comparative chemical history of the Earth, Moon and parent body of achondrite (Discussion). *Trans.* A **285**, 55–67 (1977).

Allen, C.W. Interpretation of extreme ultraviolet emissions from the Sun (Discussion).
Trans. A **270**, 71–75 (1971).

Allen, E.J. *See* Wareing & Allen.

Allen, G. The molecular basis of rubber elasticity (Discussion). *Proc.* A **351**, 381–396 (1976).

Allen, G., Burgess, J., Edwards, S.F. & Walsh, D.J. On the dimensions of intramolecularly crosslinked polymer molecules.

I. The synthesis and chemical characterization of intramolecularly crosslinked polystyrene molecules having a narrow distribution of molecular weight. *Proc.* A **334**, 453–463 (1973).

II. The theoretical prediction of the dimensions in solution of intramolecularly crosslinked polystyrene molecules. *Proc.* A **334**, 465–476 (1973).

III. The measurement of the dimensions of intramolecularly crosslinked polystyrene.
Proc. A **334**, 477–491 (1973).

Allen, G.I., Eccles, Sir John, Nicoll, R.A., Oshima, T. & Rubia, F.J. The ionic mechanisms concerned in generating the i.p.s.ps of hippocampal pyramidal cells. *Proc.* B **198**, 363–384 (1977).

Allen, J.A. Evolution of the deep sea protobranch bivalves (Discussion).
Trans. B **284**, 387–401 (1978).
See also Oliver & Allen.

Allen, J.A. & Turner, Jill F. On the functional morphology of the family Verticordiidae (Bivalvia) with descriptions of new species from the abyssal Atlantic. *Trans.* B **268**, 401–534 (1974).

Allen, J.E. *See* Andrews (J.G.) & Allen; Phelps & Allen; Prewett & Allen.

Allen, J.E., Fang, M.T.C. & Fraser, D.A. Constant frequency oscillations in a bounded thermally produced plasma. *Proc.* A **322**, 63–72 (1971).

Allen, L.H. *See* Goodrich, Allen & Chatterjee.

Allen, N.L. *See* Waters (R.T.), Allibone, Dring & Allen.

Allen, N.P. Leonard Bessemer Pfeil. *Biogr. Mem.* **18**, 477–489 (1972).

Allen, Sir Peter. Francis Arthur Freeth. *Biogr. Mem.* **22**, 105–118 (1976).

Allen, R.V. *See* Wood (M.D.), Allen & Allen.

Allen, R.V., Wood, D.M. & Mortensen, C.E. Some instruments and techniques for measurement of tidal tilt (Discussion). *Trans.* A **274**, 219–222 (1973).

Allen, S.S. *See* Wood (M.D.), Allen & Allen.

Allen, W. Defining new materials for the 1980s (Discussion). *Trans.* A **272**, 579–584 (1972).

Allen, W.R. & Sheppard, P.M. Copper tolerance in some Californian populations of the monkey flower, *Mimulus guttatus. Proc.* B **177**, 177–196 (1971).

Allibone, T.E. Basil Ferdinand Jamieson Schonland. *Biogr. Mem.* **19**, 629–653 (1973). Dennis Gabor. *Biogr. Mem.* **26**, 107–147 (1980).
See also Waters (R.T.), Allibone, Dring & Allen.

Allibone, T.E. & Dring, D. Lightning and the long spark: the significance of leader-stroke velocity. *Proc.* A **357**, 15–35 (1977).

Allison, A.C. New cell antigens induced by viruses and their biological significance (Discussion). *Proc.* B **177**, 23–39 (1971).
See also Davies (P.), Allison & Cardella.

Allsop, A.L., Bleaney, B., Bowden, G.J., Nambudripad, N., Stone, N.J. & Suzuki, H. Properties of $HoVO_4$ below 1 K. III. A nuclear orientation study of the Van Vleck enhanced nuclear antiferromagnet holmium vanadate. *Proc.* A **372**, 19–31 (1980).

Allsopp, H.L. *See* Erlank, Allsopp, Duncan & Bristow.

Allum, D.R. *See* Ellis (T.), McClintock, Bowley & Allum.

Allum, D.R., McClintock, P.V.E., Phillips, A. & Bowley, R.M. The breakdown of superfluidity in liquid ^{4}He: an experimental test of Landau's theory. *Trans.* A **284**, 179–224 (1976).

Almers, W. Observations on intramembrane charge movements in skeletal muscle (Discussion). *Trans.* B **270**, 507–513 (1975).

Almond, D.C. The Sabaloka igneous complex, Sudan. *Trans.* A **287**, 595–633 (1977).

Almond, D.P., Lea, M.J. & Dobbs, E.R. Ultrasonic studies of the electronic structure of hexagonal metal crystals. III. Shear wave attenuation in magnesium, zinc and cadmium. *Proc.* A **343**, 537–560 (1975).

Almond, J.W. *See* Inglis & Almond.

Alper, A. & Liu, J.T.C. On the interactions between large-scale structure and fine-grained turbulence in a free shear flow. II. The development of spatial interactions in the mean. *Proc.* A **359**, 497–523 (1978).

Al-Rabeh, R., Baker, R.C. & Hemp, J. Induction flow-measurement theory for poorly conducting fluids. *Proc.* A **361**, 93–107 (1978).

Altman, Jennifer S., Anselment, E. & Kutsch, W. Postembryonic development of an insect sensory system: ingrowth of axons from hindwing sense organs in *Locusta migratoria. Proc.* B **202**, 497–516 (1978).

Altman, J.S. & Kien, J. Suboesophageal neurons involved in head movements and feeding in locusts. *Proc.* B **205**, 209–227 (1979).

Altmann, S.L. & Bowen, E.J. Charles Alfred Coulson. *Biogr. Mem.* **20**, 75–134 (1974).

Amaldi, U. Proton—proton scattering at the C.E.R.N. Intersecting Storage Rings (Discussion). *Proc.* A **335**, 431–452 (1973).

Amaya, A. Mendez. *See* Szulejko, Amaya, Morgan, Brenton & Beynon.

Amaya, A. Mendez, Brenton, A.G., Szulejko, J.E. & Beynon, J.H. A method for calculating the shapes of peaks resulting from fragmentations of metastable ions in a mass spectrometer. II. Peak shapes arising from a distribution of kinetic energy releases: determination of distribution function. *Proc.* A **373**, 13–25 (1980).

Ambler, R.P. The structure of β-lactamases (Discussion). *Trans.* B **289**, 321–331 (1980).

Ames, D.L. & Turner, D.W. Photoelectron spectroscopic studies of dinitrogen tetroxide and dinitrogen pentoxide. *Proc.* A **348**, 175–186 (1976).

Amoroso, E.C. & Corner, G.W. Herbert McLean Evans. *Biogr. Mem.* **18**, 83–186 (1972).

Amoroso, E.C. & Perry, J.S. The existence during gestation of an immunological buffer zone at the interface between maternal and foetal tissues (Discussion). *Trans.* B **271**, 343–361 (1975).

Amundson, N.R. *See* Rhee, Aris & Amundson.

Amzel, L.M. *See* Poljak, Amzel, Chen, Phizackerley & Saul.

Ananthakrishna, G. *See* Gopal, Sekhar, Ananthakrishna and others.

van Andel, Tj. H. Tectonic evolution and trace element composition of basement rocks of the Mid-Atlantic Ridge: 8° S [abstract] (Discussion). *Trans.* A **268**, 661 (1971).

Anderle, R.J. The Global Positioning System (Discussion). *Trans.* A **294**, 395–406 (1980).

Anders, E. Chemical compositions of the Moon, Earth and eucrite parent body (Discussion). *Trans.* A **285**, 23–40 (1977).

Andersen, P., Hagan, P.J., Phillips, C.G. & Powell, T.P.S. Mapping by microstimulation of overlapping projections from area 4 to motor units of the baboon's hand. *Proc.* B **188**, 31–60 (1975).

Anderson, A.J. *See* Vogel (K.A.) & Anderson.

Anderson, B. Applications of very-long-baseline interferometry to geodesy and geodynamics (Discussion). *Trans.* A **284**, 469–473 (1977).
See also Palmer (H.P.) & Anderson.

Anderson, B. & Lovell, Sir Bernard. Effect of solar power satellite transmissions on radio-astronomical research (Discussion). *Trans.* A **295**, 507–511 (1980).

Anderson, D.E. Diffusion in metamorphic tectonites: lattice-fixed reference frames (Discussion). *Trans.* A **283**, 241–254 (1976).

Anderson, D.T. & Manton, Sidnie M. Studies on the Onychophora. VIII. The relationship between the embryos and the oviduct in the viviparous placental onychophorans *Epiperipatus trinidadensis* Bouvier and *Macroperipatus torquatus* (Kennel) from Trinidad. *Trans.* B **264**, 161–189 (1972).

Anderson, H.R. *See* Cotes, Anderson & Patrick.

Anderson, J.G. Ultra-high voltage transmission (Discussion). *Trans.* A **275**, 139–151 (1973).

Anderson, J.H.B. & Deckker, B.E.L. Explosion near the ground: a solution to the multi-shock problem in cylindrical symmetry by the particle trajectory method. *Proc.* A **358**, 31–46 (1977).

Anderson, J.I.W. The aquacultural revolution (Discussion). *Proc.* B **191**, 169–184 (1975).

Anderson, J.M. Assessment of the effects of pollutants on physiology and behaviour (Discussion). *Proc.* B **177**, 307–320 (1971).

Anderson, J.S. *See* Browne (J.M.) & Anderson; Hutchison (J.L.), Anderson & Rao; McConnell (J.D.M.), Hutchison & Anderson.

Anderson, J.S., Bevan, D.J.M., Cheetham, A.K., Von Dreele, R.B., Hutchison, J.L. & Strähle, J. The structure of germanium niobium oxide, an inherently non-stoichiometric 'block' structure. *Proc.* A **346**, 139–156 (1975).

Anderson, J.S., Hutchison, J.L. & Lincoln, F.J. Dislocations and related defects in niobium oxide structures. *Proc.* A **352**, 303–323 (1977).

Anderson, N. & Arthurs, A.M. Upper and lower bounds for the torsional stiffness of a prismatic bar. *Proc.* A **328**, 295–299 (1972).

Anderson, P.A.V. *See* Bone, Anderson & Pulsford.

Anderson, P.A.V. & Bone, Q. Communication between individuals in salp chains. II. Physiology. *Proc.* B **210**, 559–574 (1980).

Anderson, R.E. *See* Greenwood (W.R.), Hadley, Anderson, Fleck & Schmidt.

Anding, C., Brandt, R.D., Ourisson, G., Pryce, R.J. & Rohmer, M. Some aspects of the biosynthesis of phytosterols and triterpenes (Discussion). *Proc.* B **180**, 115–124 (1972).

Andreeva, L.A., Katasyev, L.A. & Uvarov, D.B. Winds over Heiss Island from artificial luminous cloud observations (Discussion). *Trans.* A **271**, 559–562 (1972).

Andreeva, L.A., Katasyev, L.A., Uvarov, D.B., Nesterov, V.P. & Chasovitin, Yu.K. Wind profiles from artificial luminous cloud observations and the theory of E_s formation based on wind shear (Discussion). *Trans.* A **271**, 623–629 (1972).

Andreieff, P., Bouysse, P., Curry, D., Fletcher, B.N., Hamilton, D., Monciardini, C. & Smith, A.J. The stratigraphy of the post-Palaeozoic sequences in part of the western Channel (Discussion). *Trans.* A **279**, 79–97 (1975).

Andrew, E.R. N.m.r. imaging of intact biological systems (Discussion). *Trans.* B **289**, 471–481 (1980). Concluding remarks to a Discussion. *Trans.* B **289**, 553 (1980).

Andrew, E.R., Hinshaw, W.S., Hutchins, M.G. & Jasinski, A. A nuclear magnetic resonance investigation of solid tetraphosphorus trisulphide. *Proc.* A **364**, 553–567 (1978).

Andrewes, Sir Christopher. Francis Peyton Rous. *Biogr. Mem.* **17**, 643–662 (1971). James Craigie. *Biogr. Mem.* **25**, 233–240 (1979).

Andrews, A.R. *See* Gibbs (C.F.), Pugh & Andrews.

Andrews, D.A. *See* Newton (G.), Andrews & Unsworth.

Andrews, D.L. & Thirunamachandran, T. A quantum electrodynamical theory of differential scattering based on a model with two chromophores.
  I. Differential Rayleigh scattering of circularly polarized light. *Proc.* A **358**, 297—310 (1978).
  II. Differential Raman scattering of circularly polarized light. *Proc.* A **358**, 311—319 (1978).

Andrews, E.H. & Kinloch, A.J. Mechanics of adhesive failure.
  I. *Proc.* A **332**, 385—399 (1973).
  II. *Proc.* A **332**, 401—414 (1973).

Andrews, E.H., Owen, P.J. & Singh, A. Microkinetics of lamellar crystallization in a long chain polymer. *Proc.* A **324**, 79—97 (1971).

Andrews, E.H. & Voigt-Martin, I.G. Deformation of irradiated single crystals of polyethylene. *Proc.* A **327**, 251—267 (1972).

Andrews, E.H. & Walker, B.J. Fatigue fracture in polyethylene. *Proc.* A **325**, 57—79 (1971).

Andrews, J.G. & Allen, J.E. Theory of a double sheath between two plasmas. *Proc.* A **320**, 459—472 (1971).

Andrews, J.N. *See* Parkin (D.W.), Sullivan & Andrews.

Angel, J.R.P., Sandars, P.G.H. & Woodgate, G.K. The hyperfine structure Stark effect. III. The $^2P_{1/2}$ level of aluminium. *Proc.* A **338**, 95—100 (1974).

Änggård, A. *See* Hökfelt, Lundberg, Schultzberg, Johansson, Skirboll, Änggård and others.

Angus, R.B. Pleistocene *Helophorus* (Coleoptera, Hydrophilidae) from Borislav and Starunia in the Western Ukraine, with a reinterpretation of M.Łomnicki's species, description of a new Siberian species, and comparison with British Weichselian faunas. *Trans.* B **265**, 299—326 (1973).

Anhaeusser, C.R. The evolution of the early Precambrian crust of southern Africa (Discussion). *Trans.* A **273**, 359—388 (1973).

Ankara, A.O. *See* Harris (B.) & Ankara.

Anker, D. & Freeman, N.C. On the soliton solutions of the Davey—Stewartson equation for long waves. *Proc.* A **360**, 529—540 (1978).

Annapurna, N. & Gupta, A.S. Exact analysis of unsteady m.h.d. convective diffusion. *Proc.* A **367**, 281—289 (1979).

Ansell, J.D. *See* Snow (M.H.L.) & Ansell.

Anselment, E. *See* Altman (Jennifer S.), Anselment & Kutsch.

Anstis, S.M. The perception of apparent movement (Discussion). *Trans.* B **290**, 153—168 (1980).

Anthony, J. Évocation des travaux français sur *Latimeria* notamment depuis 1972. *Proc.* B **208**, 349—367 (1980).

Appleton, A.D. Superconducting machines — a new era for the electrical power industry (Discussion). *Trans.* A **275**, 77—84 (1973).

Apte, S.K., Rowell, P. & Stewart, W.D.P. Electron donation to ferredoxin in heterocysts of the N_2-fixing alga *Anabaena cylindrica*. *Proc.* B **200**, 1—25 (1978).

Araki, J., Yano, T., Ueda, M. & Noda, M.T. Spontaneously broken symmetry and the cusp catastrophe. *Proc.* A **345**, 413—416 (1975).

Archer, A.A. *See* Harris (P.M.), Thurrell, Healing & Archer.

Archer, A.S., Cundall, R.B., Evans, G.B. & Palmer, T.F. The effect of temperature, pressure, and excitation wavelength on the photoluminescence of acetaldehyde vapour. *Proc.* A **333**, 385—402 (1973).

Archer, A.S., Cundall, R.B. & Palmer, T.F. The role of excited states in the gas-phase photolysis of acetaldehyde. *Proc.* A **334**, 411—426 (1973).

Arculus, R.J. Recent submarine pillow lavas in the Catania area, eastern Sicily (Symposium). *Trans.* A **274**, 153—162 (1973).

Ardran, G.M. The application and limitation of the use of X-rays in medical diagnosis (Discussion). *Trans.* A **292**, 147—156 (1979).

Aréchiga, H., Huberman, A. & Naylor, E. Hormonal modulation of circadian neural activity in *Carcinus maenas* (L.). *Proc.* B **187**, 299—313 (1974).

Argent, G. *See* Ratter, Richards, Argent & Gifford.

Århem, P. *See* Frankenhaeuser & Århem.

Arifon, P. *See* Voge & Arifon.
Aris, R. *See* Rhee, Aris & Amundson.
Armett-Kibel, Christine, Meinertzhagen, I.A. & Dowling, J.E. Cellular and synaptic organization in the lamina of the dragon-fly *Sympetrum rubicundulum. Proc.* B **196**, 385–413 (1977).
Armit, A.P. Example of an existing system in university research. Multipatch and Multiobject design systems (Discussion). *Proc.* A **321**, 235–242 (1971).
Armstrong, C.M. *See* Bezanilla & Amstrong.
Armstrong, G. World coal resources and their future potential (Discussion).
    Trans. A **276**, 439–452 (1974).
Arnold, E.N. Indian Ocean giant tortoises: their systematics and island adaptations (Discussion).
    Trans. B **286**, 127–145 (1979).
  See also Taylor (J.D.), Braithwaite, Peake & Arnold.
Arnold, Susan. *See* Nesheim, Crompton, Arnold & Barnard.
Arnott, H.J. *See* Nicol & Arnott.
Arnott, H.J., Best, A.C.G., Ito, S. & Nicol, J.A.C. Studies on the eyes of catfishes with special reference to the tapetum lucidum. *Proc.* B **186**, 13–36 (1974).
Arnott, H.J., Nicol, J.A.C. & Querfeld, C.W. Tapeta lucida in the eyes of the seatrout (Sciaenidae).
    Proc. B **180**, 247–271 (1972).
Arridge, R.G.C. *See* Diamant, Keller, Baer, Litt & Arridge.
Arrigo, A.P. *See* Moran, Mirault, Arrigo, Goldschmidt-Clermont & Tissières.
Arrowsmith, J.M. *See* Mintz & Arrowsmith.
Arthur, M. *See* Sibuet, Ryan, Arthur, Barnes, Blechsmidt and others.
Arthurs, A.M. *See* Anderson & Arthurs.
Arup, Sir Ove. Future problems facing the designer (Discussion). *Trans.* A **272**, 573–578 (1972).
Asaad, W.N. & Petrini, D. Relativistic calculation of the K–LL Auger spectrum.
    Proc. A **350**, 381–404 (1976).
Asaro, R.J. Adsorption-induced losses in interfacial cohesion (Discussion).
    Trans. A **295**, 151–163 (1980).
Asgar, M.A. *See* Lee (E.W.) & Asgar.
Ash, R., Barrer, R.M., Chio, H.T. & Edge, A.V.J. Measurements of adsorption for membranes *in situ* with the use of time-lags and steady state flows. *Proc.* A **365**, 267–281 (1979).
Ash, R., Barrer, R.M., Clint, J.H., Dolphin, R.J. & Murray, C.L. Isothermal and thermo-osmotic transport of sorbable gases in microporous carbon membranes. *Trans.* A **275**, 255–307 (1973).
Ashburner, M. *See* Lemeunier & Ashburner.
Ashburner, M. & Lemeunier, Francoise. Relationships within the *melanogaster* species subgroup of the genus *Drosophila (Sophophora).* I. Inversion polymorphisms in *Drosophila melanogaster* and *Drosophila simulans. Proc.* B **193**, 137–157 (1976).
Ashby, Sir Eric. The Bernal Lecture, 1971. Science and antiscience. *Proc.* B **178**, 29–42 (1971).
Ashby, M.F. & Verrall, R.A. Micromechanisms of flow and fracture, and their relevance to the rheology of the upper mantle (Discussion). *Trans.* A **288**, 59–95 (1978).
Asherson, G.L. & Zembala, M. Suppression of contact sensitivity by T cells in the mouse. I. Demonstration that suppressor cells act on the effector stage of contact sensitivity; and their induction following *in vitro* exposure to antigen. *Proc.* B **187**, 329–348 (1974).
Ashford, R.W. *See* Killick-Kendrick, Molyneux & Ashford; Molyneux, Killick-Kendrick & Ashford.
Ashkenazi, V., Crane, S.A., Williams, J.W. & Dean, J.D.A. Terrestrial–Doppler adjustment and analysis of the primary triangulation of Great Britain: preliminary report (Discussion).
    Trans. A **294**, 385–394 (1980).
Ashkenazi, V., Dodson, A.H., Sykes, R.M., Dean, J.D.A. & Blanchard, W.F. United Kingdom Doppler campaigns: field operations and instrumentation (Discussion). *Trans.* A **294**, 245–251 (1980).
Ashkenazi, V., McLintock, D.N. & Sykes, R.M. Doppler integration intervals and correlation (Discussion). *Trans.* A **294**, 377–384 (1980).
Ashkenazi, V. & Sykes, R.M. Doppler translocation and orbit relaxation techniques (Discussion).
    Trans. A **294**, 357–364 (1980).
Ashkenazi, V., Sykes, R.M., Gough, R.J. & Williams, J.W. First United Kingdom Doppler campaign:

Trans. A **291**, 423–431 (1979).

Aumento, F., Loncarevic, B.D. & Ross, D.I. Hudson Geotraverse: geology of the Mid-Atlantic ridge at 45° N (Discussion). *Trans.* B **268**, 623–650 (1971).

Avedik, F. Seismic refraction survey in the Western Approaches to the English Channel: preliminary results (Discussion). *Trans.* A **279**, 29–39 (1975).

Aveston, J. & Kelly, A. Tensile first cracking strain and strength of hybrid composites and laminates (Discussion). *Trans.* A **294**, 519–534 (1980).

Avis, S.J. & Isham, C.J. Vacuum solutions for a twisted scalar field. *Proc.* A **363**, 581–596 (1978).

Axe, J.D. Incommensurate structures (Discussion). *Trans.* B **290**, 593–603 (1980).

Axon, H.J., Nasir, M.J. & Knowles, F. Metal and phosphide phases in Luna 24 soil fragments. *Trans.* A **297**, 7–13 (1980).

Azcárraga, A. *See* Rose (G), Widdel, Azcárraga & Sanchez.

Babiker, M. Bound state quantum electrodynamics with polarization sources. *Proc.* A **342**, 113–129 (1975).

Babiker, M. & Barton, G. Quantum electrodynamics between conducting plates.

II. Spin-dependent effects in neutral atoms. *Proc.* A **326**, 255–275 (1972).

III. Relativistic theory and magnetic moment of the free electron. *Proc.* A **326**, 277–288 (1972).

Babiker, M., Power, E.A. & Thirunamachandran, T. Atomic field equations for Maxwell fields interacting with non-relativistic quantal sources. *Proc.* A **332**, 187–197 (1973).

On a generalization of the Power–Zienau–Woolley transformation in quantum electrodynamics and atomic field equations. *Proc.* A **338**, 235–249 (1974).

Bacchus, S. & Kendall, Marion D. Histological changes associated with enlargement and regression of the thymus glands of the red-billed quelea *Quelea quelea* L. (Ploceidae: weaver-birds). *Trans.* B **273**, 65–78 (1975).

Bach, B. & Ubbelohde, A.R. Synthetic metals based on graphite/aluminium halides. *Proc.* A **325**, 437–445 (1971).

Bachelard, H.S., Daniel, P.M., Love, E.R. & Pratt, O.E. The transport of glucose into the brain of the rat *in vivo*. *Proc.* B **183**, 71–82 (1973).

von Backström, J.W. & Jacob, R.E. Uranium in South Africa and South West Africa (Namibia) (Discussion). *Trans.* A **291**, 307–319 (1979).

Bacon, F.T. & Fry, T.M. Review Lecture. The development and practical application of fuel cells. *Proc.* A **334**, 427–452 (1973). Erratum. *Proc.* A **358**, 561 (1978).

Bacon, G.E., Bacon, P.J. & Griffiths, R.K. Stress distribution in the scapula studied by neutron diffraction. *Proc.* B **204**, 355–362 (1979).

Bacon, M. A gravity survey of the western English Channel between Lyme Bay and St Brieuc Bay (Discussion). *Trans.* A **279**, 69–78 (1975).

Bacon, P.J. *See* Bacon (G.E.), Bacon & Griffiths.

Bacon, R. Carbon fibres from mesophase pitch (Discussion). *Trans.* A **294**, 437–442 (1980).

Baddenhausen, H. *See* Wänke, Palme, Baddenhausen, Dreibus, Kruse & Spettel.

Baer, A.J. The Grenville Province in Helikian times: a possible model of evolution (Discussion). *Trans.* A **280**, 499–515 (1976).

Baer, E. *See* Diamant, Keller, Baer, Litt & Arridge.

Baez, M. *See* Palese, Racaniello, Desselberger, Young & Baez.

Bagg, A. *See* Benedek (G.B.), Clark, Serrallach, Young and others.

Bagnold, R.A. The nature of saltation and of 'bed-load' transport in water. *Proc.* A **332**, 473–504 (1973).

Fluid forces on a body in shear-flow; experimental use of 'stationary flow'. *Proc.* A **340**, 147–171 (1974).

An empirical correlation of bedload transport rates in flumes and natural rivers. *Proc.* A **372**, 453–473 (1980).

Baig, M.A. *See* Connerade & Baig; Connerade, Baig, Garton & Newsom; Connerade, Baig, Mansfield & Radtke; Connerade, Mansfield, Newsom, Tracy, Baig & Thimm.

Baig, M.A. & Connerade, J.P. Extensions to the spectrum of doubly excited Mg I in the vacuum ultra-violet. *Proc.* A **364**, 353—366 (1978).

Baig, M.A., Connerade, J.P. & Newsom, G.H. On the validity of the spectator electrons model in the 3d and 4s excitation spectra of Mn I. *Proc.* A **367**, 381—394 (1979).

Bailey, D.K. Volcanism, Earth degassing and replenished lithosphere mantle (Discussion). *Trans.* A **297**, 309—322 (1980).

Bailey, J.E. *See* Birchall, Howard & Bailey.

Bailey, J.E., Curtis, P.T. & Parvizi, A. On the transverse cracking and longitudinal splitting behaviour of glass and carbon fibre reinforced epoxy cross ply laminates and the effect of Poisson and thermally generated strain. *Proc.* A **366**, 599—623 (1969).

Bailey, P., Madin, A.B. & Preston, L.L. Large transformers (Discussion). *Trans.* A **275**, 95—107 (1973).

Bailey, P.L. *See* Gille, Bailey & Russell.

Bain, W.C. *See* Dickinson (P.H.G.), Bain, Thomas, Williams, Jenkins & Twiddy.

Baines, P.G. The generation of internal tides over steep continental slopes. *Trans.* A **277**, 27—58 (1974).

Baird, D.T. Manipulation of the menstrual cycle (Discussion). *Proc.* B **195**, 137—148 (1976).

Baird, D.T., Land, R.B., Scaramuzzi, R.J. & Wheeler, A.G. Functional assessment of the autotransplanted uterus and ovary in the ewe. *Proc.* B **192**, 463—474 (1976).

Baird, M.H.I. *See* Chan (K.W.), Baird & Round.

Baird, P.E.G. Isotope shifts and hyperfine structure in the atomic spectrum of palladium. *Proc.* A **351**, 267—275 (1976).
See also Kuhn, Baird, Brimicombe, Stacey & Stacey.

Baird, P.E.G., Brambley, R.J., Burnett, K., Stacey, D.N., Warrington, D.M. & Woodgate, G.K. Optical isotope shifts and hyperfine structure in λ 553.5 nm of barium. *Proc.* A **365**, 567—582 (1979).

Baird, P.E.G. & Stacey, D.N. Isotope shifts and hyperfine structure in the optical spectrum of platinum. *Proc.* A **341**, 399—406 (1974).

Bak, A.L. & Zeuthen, J. Higher order structure of mitotic chromozomes (Discussion). *Trans.* B **283**, 415—416 (1978).

Bakayev, V.V. *See* Varshavsky, Bakayev, Bakayeva, Chumackov *et al.*

Bakayeva, T.G. *See* Varshavsky, Bakayev, Bakayeva, Chumackov *et al.*

Baker, J. *See* Miller (D.S.), Baker, Bowden, Evans, Holt and others.

Baker, J.M. & Marsh, D. Interactions between Nd^{3+} ions beyond the first neighbour shell in $LaCl_3$ and $LaBr_3$. *Proc.* A **323**, 341—360 (1971).

Baker, J.R. Julian Sorell Huxley. *Biogr. Mem.* **22**, 207—238 (1976).

Baker, Lord. William Henry Glanville. *Biogr. Mem.* **23**, 91—113 (1977).

Baker, P.E., Buckley, F. & Rex, D.C. Cenozoic volcanism in the Antarctic (Discussion). *Trans.* B **279**, 131—142 (1977).

Baker, P.F. *See* Rubinson & Baker.

Baker, P.F. & Glitsch, H.G. Voltage-dependent changes in the permeability of nerve membranes to calcium and other divalent cations (Discussion). *Trans.* B **270**, 389—409 (1975).

Baker, P.F., Knight, D.E. & Whitaker, M.J. The relation between ionized calcium and cortical granule exocytosis in eggs of the sea urchin *Echinus esculentus*. *Proc.* B **207**, 149—161 (1980).

Baker, R.C. *See* Al-Rabeh, Baker & Hemp.

Baker, R.G. Keynote speech (Symposium). *Trans.* A **282**, 192—196 (1976).
Weldability and its implications for materials requirements (Symposium). *Trans.* A **282**, 207—223 (1976).

Baker, R.R. & Parker, G.A. The evolution of bird coloration. *Trans.* B **287**, 63—130 (1979).

Baker, T.F. *See* Lennon & Baker.

Baker, Wilson & Rangaswami, S. Thiruvenkata Rajendra Seshadri. *Biogr. Mem.* **25**, 505—533 (1979).

Baldwin, J.A. Linked record medical information systems (Discussion). *Proc.* B **184**, 403—420 (1973).

Baldwin, J.E., Jung, M., Singh, P., Wan, T., Haber, S., Herchen, S., Kitchin, J., Demain, A.L., Hunt, N.A., Kohsaka, M., Konomi, T. & Yoshida, M. Recent biosynthetic studies on β-lactam antibiotics (Discussion). *Trans.* B **289**, 169—172 (1980).

Baldwin, P. & Roberts, P.H. On resistive instabilities. *Trans.* A **272**, 303–330 (1972).
Baldwin, Sir Peter. Applications in government policy making (Discussion).
 Trans. A **287**, 509–522 (1977).
Ball, D.J. *See* Venables & Ball.
Ball, E. & Glucksman, J. Biological colonization of Motmot, a recently-created tropical island.
 Proc. B **190**, 421–442 (1975).
Ball, E.E. & Cowan, A.N. Ultrastructure of the antennal sensilla of *Acetes* (Crustacea, Decapoda,
 Natantia, Sergestidae). *Trans.* B **277**, 429–456 (1977).
Ballal, D.R. The structure of a premixed turbulent flame. *Proc.* A **367**, 353–380 (1979).
 The influence of laminar burning velocity on the structure and propagation of turbulent flames.
 Proc. A **367**, 485–502 (1979).
 Further development of the three-region model of a premixed turbulent flame.
 I. Turbulent diffusion dominated region 2. *Proc.* A **368**, 267–282 (1979).
 II. Instability dominated region 1. *Proc.* A **368**, 283–293 (1979).
 III. Eddy entrainment, combustion in depth process of region 3. *Proc.* A **368**, 295–304 (1979).
 Ignition and flame quenching of quiescent dust clouds of solid fuels. *Proc.* A **369**, 479–500 (1980).
Ballal, D.R. & Lefebvre, A.H. The structure and propagation of turbulent flames.
 Proc. A **344**, 217–234 (1975).
 Ignition and flame quenching in flowing gaseous mixtures. *Proc.* A **357**, 163–181 (1977).
 Ignition and flame quenching of quiescent fuel mists. *Proc.* A **364**, 277–294 (1978).
Bamford, C.H., Eastmond, G.C. & Fildes, F.J.T. Studies in polymerization.
XX. On free-radical formation by oxidation of molybdenum carbonyl with carbon tetrahalides: carbon
 tetrachloride. *Proc.* A **326**, 431–451 (1972).
XXI. On free-radical formation by oxidation of molybdenum carbonyl with carbon tetrahalides: carbon
 tetrabromide. *Proc.* A **326**, 453–468 (1972).
Bamford, C.H. & Ferrar, A.N. Studies in polymerization. XIX. Initiation by metal chelates in the
 presence of electron donors. *Proc.* A **321**, 425–443 (1971).
Bamford, C.H. & Hughes, E.O. Studies in polymerization.
XXII. Initiation of polymerization by reaction between tetrakis(triphenyl phosphite)nickel(0) and
 halides: methyl methacrylate. *Proc.* A **326**, 469–487 (1972).
XXIII. Initiation of polymerization by reaction between tetrakis(triphenyl phosphite)nickel(0) and
 halides: styrene. *Proc.* A **326**, 489–501 (1972).
Bamford, D. *See* Owens & Bamford.
Ban, L.L. *See* Jenkins (G.M.), Kawamura & Ban.
Bancroft, G.M. *See* Gupta (R.P.), Tse & Bancroft.
Bancroft, J.B. *See* Goodman (R.M.), McDonald, Horne & Bancroft.
Banerjee, K. W.K.B. approximation and scaling. *Proc.* A **363**, 147–151 (1978).
 General anharmonic oscillators. *Proc.* A **364**, 265–275 (1978).
 Rescaling the perturbation series. *Proc.* A **368**, 155–162 (1979).
Banerjee, K., Bhatnagar, S.P., Choudhry, V. & Kanwal, S.S. The anharmonic oscillator.
 Proc. A **360**, 575–586 (1978).
Bangert, V. *See* Mansfield (P.), Morris, Ordidge, Pykett, Bangert & Coupland.
Banks, Joyce E. A mathematical model of a river–shallow sea system used to investigate tide, surge
 and their interaction in the Thames–southern North Sea region. *Trans.* A **275**, 567–609 (1974).
Banks, P. & Helle, K.B. Chromogranins in sympathetic nerves (Discussion).
 Trans. B **261**, 305–310 (1971).
Banks, R. A.c.–d.c. converter plant (Discussion). *Trans.* A **275**, 233–241 (1973).
Banks, R.J. & Swain, C.J. The isostatic compensation of East Africa. *Proc.* A **364**, 331–352 (1978).
Banks, W.B. *See* Smith (D.N.R.) & Banks.
Bannister, L.H. The interactions of intracellular Protista and their host cells, with special reference to
 heterotrophic organisms (Discussion). *Proc.* B **204**, 141–163 (1979).
Bannwarth, H. & Schweiger, H.-G. Regulation of thymidine phosphorylation in nucleate and anucleate
 cells of *Acetabularia*. *Proc.* B **188**, 203–219 (1975).
Bannwarth, H., Ikehara, N. & Schweiger, H.-G. Thymidine phosphorylating enzymes in *Acetabularia:*

evidence for the occurrence of a thymidine kinase. *Proc.* B **198**, 155—176 (1977).

Nucleo—cytoplasmic interactions in the regulation of thymidine phosphorylation in *Acetabularia*. *Proc.* B **198**, 177—190 (1977).

Baños, Guadalupe, Daniel, P.M., Moorhouse, S.R. & Pratt, O.E. The influx of amino acids into the brain of the rat *in vivo:* the essential compared with some non-essential amino acids. *Proc.* B **183**, 59—70 (1973).

Baranova, N.B. & Zel'dovich, B.Ya. Coriolis contribution to the rotatory ether drag. *Proc.* A **368**, 591—592 (1979).

Barat, P., Cullis, C.F. & Pollard, R.T. The cool-flame oxidation of 3-ethylpentane. *Proc.* A **325**, 469—492 (1971).

The cool-flame oxidation of 3-methylpentane. *Proc.* A **329**, 433—452 (1972).

Barber, D.J. *See* Ashworth (J.R.) & Barber.

Barber, J. *See* Heslop-Harrison (J.), Heslop-Harrison & Barber.

Barbetti, M.F. & McElhinny, M.W. The Lake Mungo geomagnetic excursion. *Trans.* A **281**, 515—542 (1976).

Barclay, I. *See* Bennett (M.D.), Smith & Barclay.

Barcroft, H. Ivan de Burgh Daly. *Biogr. Mem.* **21** 197—226 (1975).

Bardeen, J. Reminiscences of early days in solid state physics (Symposium). *Proc.* A **371**, 77—83 (1980).

Barker, E.A. Surgical techniques and priorities (Discussion). *Proc.* B **199**, 69—72 (1977).

Barker, J.L. & McBurney, R.N. Phenobarbitone modulation of postsynaptic GABA receptor function on cultured mammalian neurons. *Proc.* B **206**, 319—327 (1979).

Erratum. *Proc.* B **207**, 507 (1980).

Barker, P.F. & Griffiths, D.H. The evolution of the Scotia Ridge and Scotia Sea (Discussion). *Trans.* A **271**, 151—183 (1972).

Towards a more certain reconstruction of Gondwanaland (Discussion). *Trans.* B **279**, 143—159 (1977).

Barlow, A.J. *See* Phillips (M.C.), Barlow & Lamb.

Barlow, A.J. & Erginsav, A. Viscoelastic retardation of supercooled liquids. *Proc.* A **327**, 175—190 (1972).

Barlow, A.J., Harrison, G., Irving, J.B., Kim, M.G., Lamb, J. & Pursley, W.C. The effect of pressure on the viscoelastic properties of liquids. *Proc.* A **327**, 403—412 (1972).

Barlow, H.B. The absolute efficiency of perceptual decisions (Discussion). *Trans.* B **290**, 71—82 (1980).

Barlow, H.B. & Gaze, R.M. Introductory remarks to a Discussion. *Trans.* B **278**, 243—244 (1977).

Barlow, J.J. *See* Martin (R.) & Barlow.

Barlow, P.W. & Macdonald, P.D.M. An analysis of the mitotic cell cycle in the root meristem of *Zea mays*. *Proc.* B **183**, 385—398 (1973).

Barlow, S.M. *See* Sullivan (F.M.) & Barlow.

Barltrop, D. Geochemical and man-made sources of lead and human health (Discussion). *Trans.* B **288**, 205—211 (1979).

Barnard, B.J.S., Mahony, J.J. & Pritchard, W.G. The excitation of surface waves near a cut-off frequency. *Trans.* A **286**, 87—123 (1977).

Barnard, D. *See* Nesheim, Crompton, Arnold & Barnard.

Barnard, G. *See* Akhtar, Abboud, Barnard, Jordan & Zaman.

Barndorff-Nielsen, O. Exponentially decreasing distributions for the logarithm of particle size. *Proc.* A **353**, 401—419 (1977).

Models for non-Gaussian variation, with applications to turbulence. *Proc.* A **368**, 501—520 (1979).

Barnes, A.J., Hallam, H.E. & Jones, D. Vapour phase infrared studies of alcohols. I. Intramolecular interactions and self-association. *Proc.* A **335**, 97—111 (1973).

Barnes, C.J. Interactions between particles of arbitrary shape. I. Uncharged particles. *Proc.* A **368**, 177—198 (1979).

Barnes, D.J. The structure and formation of growth-ridges in scleractinian coral skeletons. *Proc.* B **182**, 331—350 (1972).

Barnes, D.W.H., Loutit, J.F. & Sansom, Janet M. Histocompatible cells for the resolution of osteopetrosis in microphthalmic mice. *Proc.* B **188**, 501–505 (1975).

Barnes, G.R.G. *See* Parry, Barnes & Craig; Parry, Craig & Barnes.

Barnes, K.J., Dondi, P.H. & Sarkar, S.C. Nonlinear realizations of chiral algebras and phenomenological Lagrangians. *Proc.* A **330**, 389–415 (1972).

Barnes, P., Tabor, D. & Walker, J.C.F. The friction and creep of polycrystalline ice. *Proc.* A **324**, 127–155 (1971).

Barnes, R. *See* Sibuet, Ryan, Arthur, Barnes, Blechsmidt and others.

Barnes, R.D. *See* Ford (C.E.), Evans, Burtenshaw, Clegg, Tuffrey & Barnes.

Barnett, J.J. The Antarctic atmosphere as seen by satellites (Discussion). *Trans.* B **279**, 247–259 (1977).

Satellite measurements of middle atmosphere temperature structure (Discussion). *Trans.* A **296**, 41–57 (1980).

Barnett, P.R.O. Some changes in intertidal sand communities due to thermal pollution (Discussion). *Proc.* B **177**, 353–364 (1971).

Effects of warm water effluents from power stations on marine life (Discussion). *Proc.* B **180**, 497–509 (1972).

Barnsley, M.F. & Robinson, P.D. Bivariational bounds. *Proc.* A **338**, 527–533 (1974).

Barnstable, C.J. *See* Bodmer (W.F.), Jones, Barnstable & Bodmer; Crumpton, Snary, Walsh, Barnstable and others.

Baron, R.L. *See* Gold (T.), Bilson & Baron.

Barouch, E., Perram, J.W. & Smith, E.R. Theory of films.
I. Single dielectric films. *Proc.* A **334**, 49–58 (1973).
II. The triple-layer film. *Proc.* A **334**, 59–70 (1973).

Barr, M.L. & Rossiter, R.J. James Bertram Collip. *Biogr. Mem.* **19**, 235–267 (1973).

Barr, M.W.C. Crustal shortening in the Zambezi Belt (Discussion). *Trans.* A **280**, 555–567 (1976).

Barr, R.R. & Burdekin, F.M. Design against brittle fracture (Symposium). *Trans.* A **282**, 149–165 (1976).

Barrer, R.M. *See* Ash, Barrer, Chio & Edge; Ash, Barrer, Clint, Dolphin & Murray.

Barrer, R.M. & Davies, J.A. Sorption in decationated zeolites. II. Simple paraffins in H-forms of chabazite and zeolite L. *Proc.* A **322**, 1–19 (1971).

Barrer, R.M. & Klinowski, J. Theory of isomorphous replacement in aluminosilicates. *Trans.* A **285**, 637–676 (1977).

Barrer, R.M. & Papadopoulos, R. The sorption of krypton and xenon in zeolites at high pressures and temperatures. I. Chabazite. *Proc.* A **326**, 315–330 (1972).

Barrer, R.M., Papadopoulos, R. & Ramsay, J.D.F. The sorption of krypton and xenon in zeolites at high pressures and temperatures. II. Comparison and analysis. *Proc.* A **326**, 331–345 (1972).

Barrett, A.J. & Domb, C. Virial expansion for a polymer chain: the two parameter approximation. *Proc.* A **367**, 143–174 (1979).

Barrett, L.M. *See* Beynon (W.J.G.), Barrett & Endean.

Barrett, M.J. Predicting the effect of pollution in estuaries (Discussion). *Proc.* B **180**, 511–520 (1972).

Barrett, M.J. & Mollowney, B.M. Pollution problems in relation to the Thames barrier (Discussion). *Trans.* A **272**, 213–221 (1972).

Barrett, T. *See* Mahy, Barrett, Briedis, Brownson & Wolstenholme.

Barrington, E.J.W. Gavin Rylands de Beer. *Biogr. Mem.* **19**, 65–93 (1973).
Francis Gerald William Knowles. *Biogr. Mem.* **21**, 431–446 (1975).
Chemical communication (Discussion). *Proc.* B **199**, 361–375 (1977).

Barrow, J.D. Galaxy formation: the first million years (Discussion). *Trans.* A **296**, 273–288 (1980).

Barrow, R.F. & Clements, R.M. Rotational analysis of the AO^+, $BO^+ \leftarrow X^1 \Sigma^+$ systems of gaseous AgF. *Proc.* A **322**, 243–249 (1971).

Barrow, R.F. & du Parcq, R.P. Rotational analysis of the AO_u^+, $BO_u^+ - XO_g^+$ systems of gaseous Te_2. *Proc.* A **327**, 279–287 (1972).

Barry, C.D., Hill, H.A.O., Sadler, P.J. & Williams, R.J.P. A magnetic resonance study of metal-porphyrin/caffeine complexes. *Proc.* A **334**, 493–504 (1973).

13

Barry, P.A. *See* Adler, Barry & Bernal.

Bartel, W., Duinker, P., Heintze, J., Heinzelmann, G., Heuer, R.D., Mundhenke, R., Rieseberg, H., Schürlein, B., Steffen, P., Olsson, J.E., Wagner, A. & Walenta, A.H. Neutral and radiative decay modes of the new resonances (Discussion). *Proc.* A **355**, 481–492 (1977).

Barton, D.H.R. & Sammes, P.G. Chemical relationships between cephalosporins and penicillins (Discussion). *Proc.* B **179**, 345–355 (1971).

Barton, G. The interaction of an atom with electromagnetic vacuum fluctuations in the presence of a pair of perfectly conducting plates. *Proc.* A **367**, 117–121 (1979).

See also Babiker & Barton; Plaskett & Barton.

Bartrop, Julie. *See* Jolley, Bartrop & Smith.

Barut, A.O. & Duru, I.H. Introduction of internal coordinates into the infinite-component Majorana equation. *Proc.* A **333**, 217–224 (1973).

Basco, N. & Dogra, S.K. Reactions of halogen oxides studied by flash photolysis.

I. The flash photolysis of chlorine dioxide. *Proc.* A **323**, 29–68 (1971).

II. The flash photolysis of chlorine monoxide and of the ClO free radical. *Proc.* A **323**, 401–415 (1971).

III. The production and reactions of BrO and ClO radicals in the halogen-sensitized decomposition of chlorine dioxide. *Proc.* A **323**, 417–429 (1971).

Basco, N. & Morse, R.D. The adiabatic flash photolysis of NO_2 and SO_2. *Proc.* A **321**, 129–139 (1971).

Energy disequilibrium in the flash photolysis of NO_2. *Proc.* A **334**, 553–568 (1973).

Reactions of halogen oxides studied by flash photolysis. IV. Vacuum ultraviolet kinetic spectroscopy studies on chlorine dioxide. *Proc.* A **336**, 495–505 (1974).

Bassett, D.C. & Hodge, A.M. On lamellar organization in certain polyethylene spherulites. *Proc.* A **359**, 121–132 (1978).

Bassett, P.J. *See* Browning (R.), Bassett, El Gomati & Prutton.

Bastin, J.A. *See* Vickers & Bastin.

Bastow, A.W. & Cullis, C.F. Hydrocarbon cool flames and the influence of hydrogen bromide. *Proc.* A **338**, 327–340 (1974).

The effect of hydrogen bromide on the products of the cool-flame combustion of *n*-pentane. *Proc.* A **341**, 195–212 (1974).

Bastow, B.D., Whittle, D.P. & Wood, G.C. The diffusion-controlled growth of solid solution scales on binary alloys. *Proc.* A **356**, 177–214 (1977).

Basu, A.N. *See* Ghosh, Basu & Sengupta.

Batchelor, G.K. Geoffrey Ingram Taylor. *Biogr. Mem.* **22**, 565–633 (1976).

Batchelor, G.K. & O'Brien, R.W. Thermal or electrical conduction through a granular material. *Proc.* A **355**, 313–333 (1977).

Bateman, D.C. *See* McDonnell, Ashworth, Flavill, Carey, Bateman & Jennison.

Bateman, R.H. *See* Craig (R.D.), Bateman, Green & Millington.

Bates, B. *See* Boksenberg, Kirkham, Michelson, Pettini and others.

Bates, B., Bradley, D.J., McBride, D.A., McKeith, C.D., McKeith, N.E., Burton, W.M., Paxton, H.J.B., Shenton, D.B. & Wilson, R. High resolution interferometric studies of the solar magnesium II doublet spectral region (Discussion). *Trans.* A **270**, 47–53 (1971).

Bates, B., Carson, P.P.D., Dufton, P.L., McKeith, C.D., Boksenberg, A., Kirkham, B. & Pettini, M. Stellar studies in the balloon ultraviolet (Discussion). *Trans.* A **279**, 379–390 (1975).

Bates, Sir David. Dependence of rate of ionic recombination on temperature and density of ambient gas. *Proc.* A **369**, 327–334 (1980).

Bates, D.R. Recombination and electrical networks. *Proc.* A **337**, 15–20 (1974).

Ionization and recombination in flames. *Proc.* A **348**, 427–434 (1976).

Corrigendum *Proc.* A **354**, 537 (1977).

Transition state theory for ion–molecule reactions. *Proc.* A **360**, 1–23 (1978).

Bates, D.R. & McKibbin, C.S. Three-body ion-neutral association. *Proc.* A **339**, 13–28 (1974).

Bates, D.R., Malaviya, V. & Young, N.A. Electron-ion recombination in a dense molecular gas. *Proc.* A **320**, 437–458 (1971).

Bates, D.R. & Mendaš, I. Ionic recombination in an ambient gas.
I. Extension of quasi-equilibrium statistical method into nonlinear region.
  Proc. A **359**, 275–285 (1978).
II. Computer experiment with specific allowance for binary recombination.
  Proc. A **359**, 287–301 (1978).
Bates, R.H.T. & Wall, D.J.N. Null field approach to scalar diffraction.
I. General method. *Trans.* A **287**, 45–78 (1977).
II. Approximate methods. *Trans.* A **287**, 79–95 (1977).
III. Inverse methods. *Trans.* A **287**, 97–114 (1977).
Bath, G.T. A comparison of optical and X-ray novae (Discussion). *Proc.* A **366**, 357–365 (1979).
Batley, M., Bramley, R., Poldy, F. & Robinson, K. Photophysics of the lowest triplet state in 2-benzoylpyridine crystals. II. Optically detected e.p.r. in zero and high magnetic fields.
  Proc. A **369**, 187–206 (1979).
Batley, M., Bramley, R. & Robinson, K. Photophysics of the lowest triplet state in 2-benzoylpyridine crystals. I. Optical spectra. *Proc.* A **369**, 175–185 (1979).
Batt, A.M., Eley, D.D. & Norton, P.R. Chemisorption and catalysis of hydrogen on polycrystalline wires of tungsten and nickel (Discussion). *Proc.* A **331**, 377–381 (1972).
Battarbee, R.W. Observations on the recent history of Lough Neagh and its drainage basin.
  Trans. B **281**, 303–345 (1978).
Batte, A.D., Brear, J.M., Holdsworth, S.R., Myers, J. & Reynolds, P.E. The effects of residual elements and deoxidation practice on the mechanical properties and stress relief cracking susceptibility of ½% CrMoV turbine castings (Discussion). *Trans.* A **295**, 253–264 (1980).
Batte, A.D. & Murphy, M.C. Reheat cracking in 2¼CrMo weld metal: the influence of residual elements and microstructure [abstract] (Discussion). *Trans.* A **295**, 293–294 (1980).
Battersby, A.R. & McDonald, E. Biosynthesis of porphyrins and corrins (Discussion).
  Trans. B **273**, 161–180 (1976).
Battey, M.H. *See* Runcorn, Collinson, O'Reilly, Stephenson, Battey, Manson & Readman.
Bauche, J. *See* Jackson (D.A.), Coulombe & Bauche.
Bauche-Arnoult, Claire. Effects of configuration interaction on atomic hyperfine structure.
  Proc. A **322**, 361–376 (1971).
Bauer, P. Theory of waves incoherently scattered (Discussion). *Trans.* A **280**, 167–191 (1975).
Baumber, M.E. *See* Bunn, Moews & Baumber.
Baur, H. *See* Signer, Baur, Etique, Frick & Funk.
Bavly, Sarah. Biological studies of Yemenite and Kurdish Jews in Israel and other groups in southwest Asia. V. Food intake of Yemenite and Kurdish Jews in Israel. *Trans.* B **266**, 121–126 (1973).
  See also Edholm (O.G.), Samueloff, Mourant, Fox and others.
Bavrina, T.V. *See* Chailakhyan, Aksenova, Konstantinova & Bavrina.
Bawden, Sir Frederick. Alfred Alexander Peter Kleczkowski. *Biogr. Mem.* **17**, 431–440 (1971).
Bawden, Sir Frederick & Pirie, N.W. Factors affecting the amount of tobacco mosaic virus nucleic acid in phenol-treated extracts from tobacco leaves. *Proc.* B **182**, 297–318 (1972).
  The inhibition, inactivation and precipitation of tobacco mosaic virus nucleic acid by components of leaf extracts. *Proc.* B **182**, 319–329 (1972).
Bawn, C.E.H. Karl Ziegler. *Biogr. Mem.* **21**, 569–584 (1975).
Baxter, M.S. & Walton, A. Fluctuations of atmospheric carbon-14 concentrations during the past century. *Proc.* A **321**, 105–127 (1971).
Baxter, R.J. Solvable eight-vertex model on an arbitrary planar lattice. *Trans.* A **289**, 315–346 (1978).
Baxter, R.J., Temperley, H.N.V. & Ashley, S.E. Triangular Potts model at its transition temperature, and related models. *Proc.* A **358**, 535–559 (1978).
Baybutt, P., Guest, M.F. & Hillier, I.H. The rôle of d functions in the Si–N bond.
  Proc. A **333**, 225–236 (1973).
Bayes, K.D. *See* Jones (I.T.N.) & Bayes.
Bayley, F.J. *See* Owen, Haynes & Bayley.
Baylis, J.A. Superconducting cables for a.c. and d.c. power transmission (Discussion).
  Trans. A **275**, 205–224 (1973).

15

Bayliss, M.W. *See* Bennett (M.D.), Rao, Smith & Bayliss.

Baym, G. *See* Hoddeson & Baym.

Bayne, B.L., Moore, M.N., Widdows, J., Livingstone, D.R. & Salkeld, P. Measurement of the responses of individuals to environmental stress and pollution: studies with bivalve molluscs (Discussion).
Trans. B **286**, 563–581 (1979).

Bay-Petersen, J.L. *See* Jarman (Heather N.) & Bay-Petersen.

Beach, A. The interrelations of fluid transport, deformation, geochemistry and heat flow in early Proterozoic shear zones in the Lewisian complex (Discussion). *Trans.* A **280**, 569–604 (1976).

Beahan, P., Bevis, M. & Hull, D. Fracture processes in polystyrene. *Proc.* A **343**, 525–535 (1975).

Beale, G.H. The Leeuwenhoek Lecture, 1976. Protoza and genetics. *Proc.* B **196**, 13–27 (1977).

Beale, S.I. The biosynthesis of δ-aminolaevulinic acid in plants (Discussion).
Trans. B **273**, 99–108 (1976).

Bean, W.J. *See* Webster (R.G.), Hinshaw, Bean & Sriram.

Beaton, J.M. Archaeology and the Great Barrier Reef (Discussion). *Trans.* B **284**, 141–147 (1978).

Beattie, A.M., Stoddart, J.C. & March, N.H. Exchange energy as functional of electronic density from Hartree–Fock theory of inhomogeneous electron gas. *Proc.* A **326**, 97–116 (1971).

Beaty, G.N. & Stefani, E. Calcium dependent electrical activity in twitch muscle fibres of the frog.
Proc. B **194**, 141–150 (1976).

Beaudet, A. *See* Sotelo & Beaudet.

Beaugé, L. & Mullins, L.J. Strophanthidin-induced sodium efflux. *Proc.* B **194**, 279–284 (1976).

Beaumont, Eileen H. Cranial morphology of the Loxommatidae (Amphibia: Labyrinthodontia).
Trans. B **280**, 29–101 (1977).

Beaven, G. *See* Edholm (O.G.), Samueloff, Mourant, Fox and others.

Beaven, G.H. Biological studies of Yemenite and Kurdish Jews in Israel and other groups in south-west Asia. X. Haemoglobin studies of Yemenite and Kurdish Jews in Israel.
Trans. B **266**, 185–193 (1973).

Beaven, G.H., Fox, R.H. & Hornabrook, R.W. The occurrence of haemoglobin-J (Tongariki) and of thalassaemia on Karkar Island and the Papua New Guinea mainland (Discussion).
Trans. B **268**, 269–277 (1974).

Beaven, P.A., Miller, M.K., Williams, P.R., Delargy, K.M. & Smith, G.D.W. Trace element detection at the atomic level by atom probe microanalysis [abstract] (Discussion).
Trans. A **295**, 131–132 (1980).

Beazley, Lynda. *See* Keating, Beazley, Feldman & Gaze.

Becker, G.E. *See* Hagstrum & Becker.

Beckett, A. *See* Hawker & Beckett.

Beckett, A.H. River Thames – removable flood barriers (Discussion). *Trans.* A **272**, 259–274 (1972).

Beckinsale, R.D. Hydrogen, oxygen and silicon isotope systematics in lunar material (Discussion).
Trans. A **285**, 417–426 (1977).

Beckinsale, R.D., Bowie, S.H.U. & Durham, J.J. Oxygen isotope abundance measurements in fine size fractions of Luna 16 and 20 soil. *Trans.* A **284**, 131–136 (1977).

Beckmann, J.S. *See* Bishop (J.O.), Beckmann, Campo, Hastie and others.

Beddard, G.S., Fleming, G.R., Gijzeman, O.L.J. & Porter, Sir George. Vibrational energy dependence of radiationless conversion in aromatic vapours. *Proc.* A **340**, 519–533 (1974).

Beddard, G.S., Fleming, G.R., Porter, Sir George & Robbins, R.J. Time-resolved fluorescence from biological systems: tryptophan and simple peptides (Discussion). *Trans.* A **298**, 321–334 (1980).

Beddard, G.S., Porter, Sir George & Weese, G.M. Model systems for photosynthesis. V. Electron transfer between chlorophyll and quinones in a lecithin matrix. *Proc.* A **342**, 317–325 (1975).

Beddell, C.R., Clark, R.B., Hardy, G.W., Lowe, L.A., Ubatuba, F.B., Vane, J.R. & Wilkinson, S. Structural requirements for opioid activity of analogues of the enkephalins.
Proc. B **198**, 249–265 (1977).

Bedford, D.K. Observations of the micrometeoriod flux from Prospero (Discussion).
Proc. A **343**, 277–287 (1975).

Bedwin, O. The particulate basis of the resistance of a parasitoid to the defence reactions of its insect host. *Proc.* B **205**, 267–270 (1979).

16

An insect glycoprotein: a study of the particles responsible for the resistance of a parasitoid's egg to the defence reactions of its insect host. *Proc.* B **205**, 271—286 (1979).

Beer, G. *See* McCance, Abu Rabiyah, Beer, Edholm, Even-Paz, Luff & Samueloff.

Beeré, W. Stresses and deformation at grain boundaries (Discussion). *Trans.* A **288**, 177—196 (1978).

Beilin, L.J. *See* Coles (E.C.), Beilin, Bulpitt, Dollery, Johnson and others.

Belasco, J.G. *See* Fisher (J.), Belasco, Charnas, Khosla & Knowles.

Belcher, J.H. & Swale, E.M.F. *Luffisphaera* gen.nov., an enigmatic scaly micro-organism.
   Proc. B **188**, 495—499 (1975).

Belderson, R.H. *See* Stride, Belderson & Kenyon.

Bell, F.G. *See* Sparks, Williams & Bell.

Bell, G.D. & Lyzenga, G.A. Absolute *gf*-values for nine resonance lines of neutral scandium.
   Proc. A **351**, 581—584 (1976).

Bell, G.D.H. Introductory remarks to a Discussion. *Trans.* B **267**, 3—4 (1973).

Bell, Janet B.G. & Lacy, D. Studies on the structure and function of the mammalian testis. V. Steroid metabolism by isolated interstitium and seminiferous tubules of the human testis.
   Proc. B **186**, 99—120 (1974).

Bell, Janet B.G., Vinson, G.P. & Lacy, D. Studies on the structure and function of the mammalian testis. III. *In vitro* steroidogenesis by the seminiferous tubules of rat testis.
   Proc. B **176**, 433—443 (1971).

Bell, J.B.G. *See* McDougall, Williams, Hyatt, Bell, Tait & Tait.

Bell, P.R. *See* Duckett & Bell; Sheffield & Bell.

Bell, R.P. & Critchlow, J.E. Reaction orders and isotope effects in the reversible addition of water to 1,3-dichloracetone in aqueous dioxan. *Proc.* A **325**, 35—55 (1971).

Bell, R.P. & Tranter, R.L. Rates and hydrogen isotope effects in the ionization of 1,1-dinitroethane.
   Proc. A **337**, 517—527 (1974).

Bellard, M. *See* Oudet, Germond, Bellard, Spadafora & Chambon.

Bellehu Adefris. *See* Adefris Bellehu.

Béné, G.-J., Borcard, B., Hiltbrand, E. & Magnin, P. *In situ* identification of human physiological fluids by nuclear magnetism in the Earth's field (Discussion). *Trans.* B **289**, 501—502 (1980).

Benedek, G.B. Concluding remarks to a Discussion. *Trans.* A **293**, 469—471 (1979).
   See also Cohen (R.J.), Jedziniak & Benedek.

Benedek, G.B., Clark, J.I., Serrallach, E.N., Young, C.Y., Mengel, L., Sauke, T., Bagg, A. & Benedek, K. Light scattering and reversible cataracts in the calf and human lens (Discussion).
   Trans. A **293**, 329—340 (1979).

Benedek, K. *See* Benedek (G.B.), Clark, Serrallach, Young and others.

Bengtsson, B.-E. Biological variables, especially skeletal deformities in fish, for monitoring marine pollution. *Trans.* B **286**, 457—464 (1979).

Benjamin, P.R. *See* Swindale & Benjamin.

Benjamin, P.R., Slade, Carole T. & Soffe, S.R. The morphology of neurosecretory neurones in the pond snail, *Lymnaea stagnalis*, by the injection of Procion Yellow and horseradish peroxidase.
   Trans. B **290**, 449—478 (1980).

Benjamin, T.B. A unified theory of conjugate flows. *Trans.* A **269**, 587—647 (1971).
   The stability of solitary waves. *Proc.* A **328**, 153—183 (1972).
   Bifurcation phenomena in steady flows of a viscous fluid.
   I. Theory. *Proc.* A **359**, 1—26 (1978).
   II. Experiments. *Proc.* A **359**, 27—43 (1978).

Benjamin, T.B., Bona, J.L. & Mahony, J.J. Model equations for long waves in nonlinear dispersive systems. *Trans.* A **272**, 47—78 (1972).

Ben-Menahem, A. *See* Singh (S.J.), Ben-Menahem & Vered.

van Bennekom, A.J., Gieskes, W.W.C. & Tijssen, S.B. Eutrophication of Dutch coastal waters (Discussion). *Proc.* B **189**, 359—374 (1975).

Bennet-Clark, T.A. John Barker. *Biogr. Mem.* **18**, 35—42 (1972).

Bennett, H.F. *See* Wood (D.S.), Oertel, Singh & Bennett.

Bennett, J.M. *See* Franks (A.), Lindsey, Bennett, Speer, Turner & Hunt.

Bennett, M.D. The duration of meiosis. *Proc.* B **178**, 277–299 (1971).
  Nuclear DNA content and minimum generation time in herbaceous plants.
    Proc. B **181**, 109–135 (1972).
  The time and duration of meiosis (Discussion). *Trans.* B **277**, 201–226 (1977).
Bennett, M.D., Chapman, V. & Riley, R. The duration of meiosis in pollen mother cells of wheat, rye
  and *Triticale. Proc.* B **178**, 259–275 (1971).
Bennett, M.D., Dover, G.A. & Riley, R. Meiotic duration in wheat genotypes with or without homo-
  eologous meiotic chromosome pairing. *Proc.* B **187**, 191–207 (1974).
Bennett, M.D., Finch, R.A., Smith, J.B. & Rao, M.K. The time and duration of female meiosis in
  wheat, rye and barley. *Proc.* B **183**, 301–319 (1973).
Bennett, M.D., Rao, K.M., Smith, J.B. & Bayliss, M.W. Cell development in the anther, the ovule,
  and the young seed of *Triticum aestivum* L. var. Chinese Spring. *Trans.* B **266**, 39–81 (1973).
Bennett, M.D. & Smith, J.B. The effects of polyploidy on meiotic duration and pollen development
  in cereal anthers. *Proc.* B **181**, 81–107 (1972).
  Nuclear DNA amounts in angiosperms. *Trans.* B **274**, 227–274 (1976).
Bennett, M.D., Smith, J.B. & Barclay, I. Early seed development in the Triticeae.
    Trans. B **272**, 199–227 (1975).
Bennett, M.D. & Stern, H. The time and duration of female meiosis in *Lilium.*
    Proc. B **188**, 459–475 (1975).
  The time and duration of preleptotene chromosome condensation stage in *Lilium* hybrid cv. Black
  Beauty. *Proc.* B **188**, 477–493 (1975).
Bennett, M.R. Structure and electrical properties of the autonomic neuromuscular junction (Dis-
  cussion). *Trans.* B **265**, 25–34 (1973).
Bennett, Pauline M. *See* O'Brien, Bennett & Hanson.
Bennett, P.M. *See* Richmond (M.H.), Bennett, Choi, Brown, Brunton and others.
Bennetts, D.A. & Hocking, L.M. On nonlinear Ekman and Stewartson layers in a rotating fluid.
    Proc. A **333**, 469–489 (1973).
Benson, C.W. & Penny, M.J. The land birds of Aldabra (Discussion). *Trans.* B **260**, 417–527 (1971).
Benson, P.C. *See* McGeehin, Henderson & Benson.
Bentley, G.A. *See* Finch (J.T.), Lewit-Bentley, Bentley, Roth & Timmins.
Bentley, G.A. & Mason, S.A. Neutron diffraction studies of proteins (Discussion).
    Trans. B **290**, 505–510 (1980).
Bentley, P.D. & Hands, B.A. The condensation of atmospheric gases on cold surfaces.
    Proc. A **359**, 319–343 (1978).
Benyajati, C. *See* Worcel & Benyajati.
Berci, G. Present and future developments in endoscopy (Discussion). *Proc.* B **195**, 235–242 (1977).
Berelson, B. The impact of new technology (Discussion). *Proc.* B **195**, 25–35 (1976).
Berg, E. *See* Carter (W.E.), Berg & Laurila.
Berg, E. & Lutschak, W. Crustal tilt fields and propagation velocities associated with earthquakes
  (Discussion). *Trans.* A **274**, 261–265 (1973).
Berg, P. The viral genome in transformed cells (Discussion). *Proc.* B **177**, 65–76 (1971).
Berge, G. *See* Jensen (S.), Lange, Berge, Palmork & Renberg.
Bergel, F. Carcinogenic hazards in natural and man-made environments (Discussion).
    Proc. B **185**, 165–181 (1974).
  Alexander Haddow. *Biogr. Mem.* **23**, 133–191 (1977).
Berger, J. & Wyatt, F. Some observations of Earth strain tides in California (Discussion).
    Trans. A **274**, 267–277 (1973).
Bergeron, M.B. *See* Schilling, Bergeron & Evans.
van den Bergh, S. The classification of evolving galaxies (Discussion). *Trans.* A **296**, 319–327 (1980).
Bergman, D.J. Comments on the article: 'The two-fluid model of the helium film'.
    Proc. A **333**, 261–263 (1973).
Beringer, J.E., Brewin, N., Johnston, A.W.B., Schulman, H.M. & Hopwood, D.A. The *Rhizobium–*
  legume symbiosis (Discussion). *Proc.* B **204**, 219–233 (1979).
Berlie, J. *See* Cooper (G.A.), Berlie & Merminod; Ryhming, Cooper & Berlie.

Bernal, J.D. Appendix A: Lessons of the war for science (Discussion). *Proc.* A **342**, 555–574 (1975).

Bernal, M.J.M. *See* Adler, Barry & Bernal.

Bernard, R. Evidence of a prevailing semi-diurnal tide between 100 and 130 km from incoherent scatter observations at Nançay [abstract] (Discussion). *Trans.* A **271**, 611 (1972).

Bernstein, H.J. Resonance Raman spectra (Discussion). *Trans.* A **293**, 287–302 (1979).

Bernstein, I.M. *See* Pressouyre & Bernstein.

Bernstein, J., Regev, H., Herbstein, F.H., Main, P., Rizvi, S.H., Sasvari, K. & Turcsanyi, B. The crystal structure of the π-molecular compound pyrene: *p*-benzoquinone. *Proc.* A **347**, 419–434 (1976).

Berridge, M.J. & Prince, W.T. The electrical response of isolated salivary glands during stimulation with 5-hydroxytryptamine and cyclic AMP (Discussion). *Trans.* B **262**, 111–120 (1971).
Corrigendum. *Trans.* B **262**, 458.

Berrondo, M. Chemical relaxation as a transport process. *Proc.* A **328**, 353–369 (1972).

Berry, M. *See* Hollingworth & Berry.

Berry, M.S. & Cottrell, G.A. Ionic basis of different synaptic potentials mediated by an identified dopamine-containing neuron in *Planorbis*. *Proc.* B **203**, 427–444 (1979).

Berry, M.V. The statistical properties of echoes diffracted from rough surfaces.
Trans. A **273**, 611–654 (1973).
Semi-classical mechanics in phase space: a study of Wigner's function.
Trans. A **287**, 237–271 (1977).
See also Buxton (B.F.) & Berry; Nye & Berry.

Berry, M.V. & Lewis, Z.V. On the Weierstrass—Mandelbrot fractal function.
Proc. A **370**, 459–484 (1980).

Berry, M.V. & Mackley, M.R. The six roll mill: unfolding an unstable persistently extensional flow.
Trans. A **287**, 1–16 (1977).

Berry, M.V., Nye, J.F. & Wright, F.J. The elliptic umbilic diffraction catastrophe.
Trans. A **291**, 453–484 (1979).

Berry, M.V. & Tabor, M. Closed orbits and the regular bound spectrum.
Proc. A **349**, 101–123 (1976).
Level clustering in the regular spectrum. *Proc.* A **356**, 375–394 (1977).

Berry, R.J. & Peters, Josephine. Heterogeneous heterozygosities in *Mus musculus* populations.
Proc. B **197**, 485–503 (1977).

Best, A.C. *See* Sutcliffe (R.C.) & Best.

Best, A.C.G. *See* Arnott, Best, Ito & Nicol.

Best, C.H. & Fisher, A.M. David Alymer Scott. *Biogr. Mem.* **18**, 511–524 (1972).

Bethe, H.A. Recollections of solid state theory, 1926–33 (Symposium).
Proc. A **371**, 49–51 (1980).

Betts, Elizabeth. *See* Rainey (R.C.) & Betts; Rainey (R.C.), Betts & Lumley.

Betts, R.F. *See* Murphy (B.R.), Markoff, Chanock, Spring and others.

Beug, H. *See* Hayman (M.J.), Ramsay, Kitchener, Graf, Beug and others.

Bevan, D.J.M. *See* Anderson (J.S.), Bevan, Cheetham, Von Dreele and others.

Bevan, J.W., Kisiel, Z., Legon, A.C., Millen, D.J. & Rogers, S.C. Spectroscopic investigations of hydrogen bonding interactions in the gas phase. IV. The heterodimer $H_2O \cdots HF$: the observation and analysis of its microwave rotational spectrum and the determination of its molecular geometry and electric dipole moment. *Proc.* A **372**, 441–451 (1980).

Bevan, J.W., Legon, A.C. & Millen, D.J. The microwave spectrum and molecular structure of trimethylene sulphoxide; bends, tilts and twists in the methylene groups.
Proc. A **354**, 491–509 (1977).

Bevan, J.W., Legon, A.C., Millen, D.J. & Rogers, S.C. Spectroscopic investigations of hydrogen bonding interactions in the gas phase. II. The determination of the geometry and potential constants of the hydrogen-bonded heterodimer $CH_3CN \cdots BF$ from its microwave rotational spectrum. *Proc.* A **370**, 239–255 (1980).

Bevan, S., Grampp, W. & Miledi, R. Properties of spontaneous potentials at denervated motor endplates of the frog. *Proc.* B **194**, 195–210 (1976).

Bevan, S.J., Katz, Sir Bernard & Miledi, R. Membrane potential fluctuations produced by glutamate

in nerve cells of the squid. *Proc.* B **191**, 561–565 (1975).

Beveridge, A.E. Kauri forests in the New Hebrides (Discussion). *Trans.* B **272**, 369–383 (1975).

Beverley, K.I. *See* Regan, Beverley & Cynader.

Bevis, M. *See* Allan & Bevis; Beahan, Bevis & Hull.

Bevis, M. & Fearon, E.O. The anisotropy of the deformation twinning behaviour of iron–nickel martensite crystals possessing a transformation twin microstructure. *Proc.* A **354**, 9–25 (1977).

Beynon, J.H. *See* Amaya, Brenton, Szulejko & Beynon; Szulejko, Amaya, Morgan, Brénton & Beynon; Terwilliger, Beynon & Cooks.

Beynon, J.H., Cooks, R.G. & Caprioli, R.M. The mass and ion kinetic energy spectra of H_2, HD and D_2. *Proc.* A **327**, 1–11 (1972).

Beynon, J.H., Morgan, R.P. & Brenton, A.G. New methods of identifying organic compounds (Discussion). *Trans.* A **293**, 157–166 (1976).

Beynon, W.J.G. U.R.S.I. and the early history of the ionosphere (Discussion). (With an appendix by M.V. Wilkes.) *Trans.* A **280**, 47–55 (1975).

Beynon, W.J.G., Barrett, L.M. & Endean, R.P.J. Some studies of sporadic E-layer drifts (Discussion). *Trans.* A **271**, 613–620 (1972).

Bezanilla, F. *See* Keynes, Bezanilla, Rojas & Taylor.

Bezanilla, F. & Armstrong, C.M. Kinetic properties and inactivation of the gating currents of sodium channels in squid axon (Discussion). *Trans.* B **270**, 449–458 (1975).

Bézier, P.E. Example of an existing system in the motor industry: the Unisurf system (Discussion). *Proc.* A **321**, 207–218 (1971).

Bhagavantam, S. Chandrasekhara Venkata Raman. *Biogr. Mem.* **17**, 565–592 (1971).

Bhasavanich, D. & Parker, A.B. The dielectric breakdown of gases at low pressure. *Proc.* A **358**, 385–403 (1978).

Bhatia, M.L. & Cahn, R.W. Recrystallization of porous copper. *Proc.* A **362**, 341–360 (1978).

Bhatnagar, P.L. & Prasad, Phoolan. Study of self-similar and steady flows near singularities. II. A case of multiple characteristic velocity. *Proc.* A **322**, 45–62 (1971).

Bhatnagar, S.P. *See* Banerjee, Bhatnagar, Choudhry & Kanwal.

Bibby, B. Rothberg. *See* Nabarro & Bibby.

Bibby, B. Rothberg, Nabarro, F.R.N., McLachlan, D.S. & Stephen, M.J. The magnetic transition in moderately small superconductors, and the influence of elastic strain. *Trans.* A **278**, 311–341 (1975).

Bibby, J.S. Petrofabric analysis. Appendix to a paper by E.W. MacKie. *Trans.* A **276**, 191–194 (1974).

Bičák, J. The motion of a charged black hole in an electromagnetic field. *Proc.* A **371**, 429–438 (1980).

Biggin, S. *See* Dingley & Biggin.

Bignot, G. *See* Auffret (J.P.), Bignot & Blondeau.

Bijkerk, C. Recreation values of forests and parks (Discussion). *Trans.* B **271**, 179–198 (1975).

Bilby, B.A. & Kolbuszewski, M.L. The finite deformation of an inhomogeneity in two-dimensional slow viscous incompressible flow. *Proc.* A **355**, 335–353 (1977).

Bilham, R.G. The location of Earth strain instrumentation (Discussion). *Trans.* A **274**, 429–433 (1973). *See also* King (G.C.P.) & Bilham.

Billett, D.F. *See* Requena, Billett & Haydon.

Billingham, J. *See* Chubb (J.P.), Billingham, Hancock, Dimbylow & Newcombe.

Billingham, R.E. *See* Zakarian, Streilein & Billingham.

Billington, E.W. & Tate, A. The mechanical properties of copper zinc and aluminium tested in compression. *Proc.* A **327**, 23–46 (1972).

Bilson, E. *See* Gold (T.), Bilson & Baron.

Binney, J. The dynamics, shapes and origins of elliptical galaxies (Discussion). *Trans.* A **296**, 329–338 (1980).

Binnie, A.M. The stability of a falling sheet of water. *Proc.* A **326**, 149–163 (1972). Annular hydraulic jumps in a horizontal tube. *Proc.* A **332**, 269–279 (1973).

Resonating waterfalls. (With an appendix by D.F. Mayers). *Proc.* A **339**, 435—449 (1974).

Unstable flow under a sluice-gate. *Proc.* A **367**, 311—319 (1979).

Binnie, A.M. & Roberts, P.H. Thomas Henry Havelock. *Biogr. Mem.* **17**, 327—377 (1971).

Biot, M.A. New variational-Lagrangian thermodynamics of viscous fluid mixtures with thermomolecular diffusion. *Proc.* A **365**, 467—494 (1979).

Birch, B.J. *See* Rogers (C.A.), Burgess, Halberstam & Birch.

Birch, M.C., Cheng, L. & Treherne, J.E. Distribution and environmental synchronization of the marine insect, *Halobates robustus*, in the Galapagos Islands. *Proc.* B **206**, 33—52 (1979).

Birchall, J.D., Howard, A.J. & Bailey, J.E. On the hydration of Portland cement. *Proc.* A **360**, 445—453 (1978).

Birchley, J.C. Droplet combustion in a stream of oxidant. *Proc.* A **347**, 195—212 (1975).

Bird, A.P. The occurrence and transmission of a pattern of DNA methylation in *Xenopus laevis* ribosomal DNA (Discussion). *Trans.* B **283**, 325—327 (1978).

Bird, B.D., Cooke, Elspeth A., Day, P. & Orchard, A.F. Derivation and testing of a molecular orbital description of ligand field spectra. *Trans.* A **276**, 277—339 (1974).

Birdsall, B. *See* Feeney, Roberts, Birdsall, Griffiths and others.

Birdsall, B., Griffiths, D.V., Roberts, G.C.K., Feeney, J. & Burgen, Sir Arnold. [1]H nuclear magnetic resonance studies of *Lactobacillus casei* dihydrofolate reductase: effects of substrate and inhibitor binding on the histidine residues. *Proc.* B **196**, 251—265 (1977).

Birdsall, N.J.M. *See* Lee (A.G.), Birdsall, Metcalfe, Warren & Roberts.

Birdsall, N.J.M., Hulme, E.C. & Burgen, Sir Arnold. The character of the muscarinic receptors in different regions of the rat brain. *Proc.* B **207**, 1—12 (1980).

Birkinshaw, J.H. Harold Raistrick. *Biogr. Mem.* **18**, 489—509 (1972).

Birks, Hilary H. Studies in the vegetational history of Scotland. IV. Pine stumps in Scottish blanket peats. (With an appendix by V.R. Switsur). *Trans.* B **270**, 181—226 (1975).

Birks, H.J.B., Deacon, Joy & Peglar, Sylvia. Pollen maps for the British Isles 5000 years ago. *Proc.* B **189**, 87—105 (1975).

Birks, J. Introduction to aspects of economics and logistics (Discussion). *Trans.* A **290**, 3—19 (1978).

Birnstiel, M.L., Kressmann, A., Schaffner, W., Portmann, R. & Busslinger, M. Aspects of the regulation of histone genes (Discussion). *Trans.* B **283**, 319—324 (1978).

Birrell, N.D. The application of adiabatic regularization to calculations of cosmological interest. *Proc.* A **361**, 513—526 (1978).

Momentum space techniques for curved space—time quantum field theory. *Proc.* A **367**, 123—141 (1979).

Biscoe, P.V., Cohen, Y. & Wallace, J.S. Daily and seasonal changes of water potential in cereals (Discussion). *Trans.* B **273**, 565—580 (1976).

Biscoe, T.J. *See* Caddy & Biscoe.

Bishop, Ann & Miles, Sir Ashley. Muriel Robertson. *Biogr. Mem.* **20**, 317—347 (1974).

Bishop, A.W., Kumapley, N.K. & El-Ruwayih, A. The influence of pore-water tension on the strength of clay. *Trans.* A **278**, 511—554 (1975).

Bishop, A.W. & Skinner, A.E. The influence of high pore-water pressure on the strength of cohesionless soils. *Trans.* A **284**, 91—130 (1977).

Bishop, D. Productivity in the building industry (Discussion). *Trans.* A **272**, 533—563 (1972).

Bishop, J.A. Commentary on three papers on clines for melanism and polymorphism in moths. *Proc.* B **210**, 273—275 (1980).

See also Conroy & Bishop; Murray (N.D.), Bishop & Macnair; Whittle, Clarke, Sheppard & Bishop.

Bishop, J.A., Cook, L.M. & Muggleton, J. The response of two species of moths to industrialization in northwest England.

I. Polymorphisms for melanism. *Trans.* B **281**, 489—515 (1978).

II. Relative fitness of morphs and population size. *Trans.* B **281**, 517—540 (1978).

Bishop, J.O. Analysis of mRNA populations (Discussion). *Trans.* B **283**, 373—374 (1978).

Bishop, J.O., Beckmann, J.S., Campo, M.S., Hastie, N.D., Izquierdo, M. & Perlman, S. DNA—RNA hybridization (Discussion). *Trans.* B **272**, 147—157 (1975).

Bishop, P.O. Stereopsis and the random element in the organization of the striate cortex (Discussion). *Proc.* B **204**, 415–434 (1979).

Bishop, R.E.D. *See* Fawzy & Bishop.

Bishop, R.E.D., Burcher, R.K. & Price, W.G. The uses of functional analysis in ship dynamics. *Proc.* A **332**, 23–35 (1973).

Application of functional analysis to oscillatory ship model testing. *Proc.* A **332**, 37–49 (1973).

Directional stability analysis of a ship allowing for time history effects of the flow. *Proc.* A **335**, 341–354 (1973).

Bishop, R.E.D. & Fawzy, I. Free and forced oscillation of a vertical tube containing a flowing fluid. *Trans.* A **284**, 1–47 (1976).

Bishop, R.E.D. & Mahalingam, S. The response of an oscillatory system to excitation by a transient displacement. *Proc.* A **324**, 63–77 (1971).

Bishop, R.E.D. & Price, W.G. On modal analysis of ship strength. *Proc.* A **341**, 121–134 (1974).

Antisymmetric response of a box-like ship. *Proc.* A **349**, 157–167 (1976).

On the transverse strength of ships with large deck openings. *Proc.* A **349**, 169–182 (1976).

Bishop, R.E.D. & Taylor, R. Eatock. On wave-induced stress in a ship executing symmetric motions. *Trans.* A **275**, 1–32 (1973).

Black, G.P. The role of the extractive sites in geological education and research (Discussion). *Proc.* A **339**, 389–394 (1974).

Black, H.D. The Transit System, 1977: performance, plans and potential (Discussion). *Trans.* A **294**, 217–236 (1980).

Black, T.R.L. Community-based distribution: the distributive potential and economics of a social marketing approach to family planning (Discussion). *Proc.* B **195**, 199–212 (1976).

Blackett, Lord. Anniversary Address 1970: *Proc.* A **321**, 1–14 (1971): *also Proc.* B **177**, 1–14 (1971).

Blackman, G.E. & Cooper, J.P. Photosynthesis and solar energy conversion (Discussion). *Trans.* B **274**, 321–339 (1976).

Blackman, M. George Ingle Finch. *Biogr. Mem.* **18**, 223–239 (1972).

Heat capacity of crystals and the vibrational spectrum (Symposium). *Proc.* A **371**, 116–119 (1980).

Blackman, R.L. Biological approaches to the control of aphids (Discussion). *Trans.* B **274**, 473–488 (1976).

Blackshaw, Susanna E. Dye injection and electrophysiological mapping of giant neurons in the brain of *Archidoris. Proc.* B **192**, 393–419 (1976).

Blackshaw, Susanna E. & Dorsett, D.A. Behavioural correlates of activity in the giant cerebral neurons of *Archidoris. Proc.* B **192**, 421–437 (1976).

Blair, J.A. *See* Lucas (M.L.) & Blair; Lucas (M.L.), Schneider, Haberich & Blair.

Blake, A.J. Photoionization study of mercury by photoelectron spectroscopy. *Proc.* A **325**, 555–560 (1971).

Blakemore, C. Genetic instructions and developmental plasticity in the kitten's visual cortex (Discussion). *Trans.* B **278**, 425–434 (1977).

The development of stereoscopic mechanisms in the visual cortex of the cat (Discussion). *Proc.* B **204**, 477–484 (1979).

Blakesley, D. Village radiology (Discussion). *Proc.* B **209**, 129–130.

Blanchard, W.F. *See* Ashkenazi, Dodson, Sykes, Dean & Blanchard.

Blaney, T.G., Bradley, C.C., Edwards, G.J., Jolliffe, B.W., Knight, D.J.E., Rowley, W.R.C., Shotton, K.C. & Woods, P.T. Measurement of the speed of light.

I. Introduction and frequency measurement of a carbon dioxide laser. *Proc.* A **355**, 61–88 (1977).

II. Wavelength measurements and conclusion. *Proc.* A **355**, 89–114 (1977).

Blankenburgh, J.C. Doppler—European Datum transformation parameters for the North Sea (Discussion). *Trans.* A **294**, 277–288 (1980).

Blaschko, H.K.F. Introductory remarks to a Discussion. *Trans.* B **261**, 275–277 (1971).

Blaustein, M.P. *See* Reuter, Blaustein & Haeusler.

Blaxter, Sir Kenneth. Herbert Davenport Kay. *Biogr. Mem.* **23**, 283–310 (1977).

Blaxter, K.L. Review Lecture. The nutrition of ruminant animals in relation to intensive methods of

agriculture. *Proc.* B **183**, 321–336 (1973).

The options for British farming (Discussion). *Trans.* B **281**, 77–81 (1977).

Bleaney, B. Properties of HoVO$_4$ below 1 K. I. Predictions from nuclear magnetic resonance.
Proc. A **370**, 313–330 (1980).

See also Allsop, Bleaney, Bowden, Nambudripad, Stone & Suzuki.

Bleaney, B., Loftus, K.V. & Rosenberg, H.M. Properties of HoVO$_4$ below 1 K. II. The radio-frequency susceptibility. *Proc.* A **372**, 9–17 (1980).

Bleaney, B., Robinson, F.N.H. & Wells, M.R. Nuclear magnetic resonance in holmium vanadate, HoVO$_4$. *Proc.* A **362**, 179–194 (1978).

Bleaney, B. & Wells, M.R. Radiofrequency studies of TmVO$_4$. *Proc.* A **370**, 131–153 (1980).

Blechsmidt, G. *See* Sibuet, Ryan, Arthur, Barnes, Blechsmidt and others.

Blennerhassett, P.J. On the generation of waves by wind. *Trans.* A **298**, 451–494 (1980).

Blennerhassett, P.J. & Hall, P. Centrifugal instabilities of circumferential flows in finite cylinders: linear theory. *Proc.* A **365**, 191–207 (1979).

Blest, A.D. A new method for the reduced silver impregnation of arthropod central nervous systems.
Proc. B **193**, 191–197 (1976).

The rapid synthesis and destruction of photoreceptor membrane by a dinopid spider: a daily cycle.
Proc. B **200**, 463–483 (1978).

Blest, A.D. & Day, W.A. The rhabdomere organization of some noctural pisaurid spiders in light and darkness. *Trans.* B **283**, 1–23 (1978).

Blest, A.D. & Land, M.F. The physiological optics of *Dinopis subrufus* L. Koch: a fish-lens in a spider.
Proc. B **196**, 197–222 (1977).

Blest, A.D. & Maples, Joanne. Exocytotic shedding and glial uptake of photoreceptor membrane by a salticid spider. *Proc.* B **204**, 105–112 (1979).

Blight, A.R. *See* Roberts (A.) & Blight.

Blight, A.R. & Llinás, R. The non-impulsive stretch-receptor complex of the crab: a study of depolarization–release coupling at a tonic sensorimotor synapse. *Trans.* B **290**, 219–276 (1980).

Bliss, T.V.P. *See* Chung, Bliss & Keating; Chung, Keating & Bliss.

Bloch, F. Memories of electrons in crystals (Symposium). *Proc.* A **371**, 24–27 (1980).

Block, B. & Dratler, J., Jr. Quartz fibre accelerometers (Discussion). *Trans.* A **274**, 231–243 (1973).

Block, H., Ions, W.D., Powell, G., Singh, R.P. & Walker, S.M. The dielectric behaviour of solutions of some rigid polymers under shear. *Proc.* A **352**, 153–167 (1976).

Blombäck, Birger. *See* Partridge & Blombäck.

Blondeau, A. *See* Auffret (J.P.), Bignot & Blondeau.

Blow, D.M. & Smith, J.M. Enzyme substrate and inhibitor interactions (Discussion).
Trans. B **272**, 87–97 (1975).

Blumberg, P.M. *See* Strominger, Blumberg, Suginaka, Umbreit & Wickus.

Blumenthal, T. Interaction of host-coded and virus-coded polypeptides in RNA phage replication (Discussion). *Proc.* B **210**, 321–335 (1980).

Blundell, T.L. & Powell, H.M. The coordination of the dimethylthallium ion in the crystal and molecular structure of 1,10-phenanthroline-dimethylthallium (III) perchlorate.
Proc. A **331**, 161–169 (1972).

Blythe, P.A. *See* Varley, Kazakia & Blythe.

Boag, J.W. Electrostatic imaging in radiology: limitations and prospects (Discussion).
Trans. A **292**, 273–283 (1979).

Boardman, F.D. Future developments in the control of power systems (Discussion).
Trans. A **275**, 243–253 (1973).

Boardman, J. The olive in the Mediterranean: its culture and use (Discussion).
Trans. B **275**, 187–196 (1976).

Boardman, N.K. Energy from the biological conversion of solar energy (Discussion).
Trans. A **295**, 477–489 (1980).

Bock, W.D. *See* Roberts (D.G.), Montadert, Thompson, Auffret and others.

Boddington, T., Gray, P. & Harvey, D.I. Thermal theory of spontaneous ignition: criticality in bodies of arbitrary shape. *Trans.* A **270**, 467–506 (1971).

Boddington, T., Gray, P. & Robinson, C. Thermal explosions and the disappearance of criticality at small activation energies: exact results for the slab. *Proc.* A **368**, 441—461 (1979).

Boddington, T., Gray, P. & Wake, G.C. Criteria for thermal explosions with and without reactant consumption. *Proc.* A **357**, 403—422 (1977).

Boddington, T., Gray, P. & Walker, I.K. Exothermic systems with diminishing reaction rates: temperature evolution, criticality and spontaneous ignition in the sphere.
Proc. A **373**, 287—310 (1980).

Bodmer, J.G. *See* Bodmer (W.F.), Jones, Barnstable & Bodmer.

Bodmer, W.F. Introduction to a Discussion. *Proc.* B **202**, 3—4 (1978).
See also Crumpton, Snary, Walsh, Barnstable and others.

Bodmer, W.F., Jones, E.A., Barnstable, C.J. & Bodmer, J.G. Genetics of HLA: the major human histocompatibility system (Discussion). *Proc.* B **202**, 93—116 (1978).

Boerma, A.H. The world food and agricultural situation (Discussion). *Trans.* B **267**, 5—12 (1973).

Boillot, G. & Musselec, P. Origine de la Manche d'après une carte géologique au 1:1 000 000e (Discussion). *Trans.* A **279**, 21—27 (1975).

Boisseau, M.R., Lorient, M.F., Born, G.V.R. & Michal, F. Change in electrophoretic mobility associated with the shape change of human blood platelets. *Proc.* B **196**, 471—474 (1977).

Bokhari, M. *See* Pontecorvo & Bokhari.

Boksenberg, A. *See* Bates (B.), Carson, Dufton, McKeith and others.

Boksenberg, A., Kirkham, B., Michelson, E., Pettini, M., Bates, B., Carson, P.P.D., Courts, G.R., Dufton, P.L. & McKeith, C.D. Interstellar gas studies in the balloon ultraviolet (Discussion).
Trans. A **279**, 303—316 (1975).

Boland, B.C., Jones, B.B., Wilson, R., Engstrom, S.F.T. & Noci, G. A high resolution solar ultraviolet spectrum between 200 and 220 nm (Discussion). *Trans.* A **270**, 29—46 (1971).

Boland, B.J., Brown, J.M., Carrington, A. & Nelson, A.C. Microwave spectroscopy of nonlinear free radicals. III. High field Zeeman effect in HCO and DCO. *Proc.* A **360**, 507—528 (1978).

Boley, B.A. & Yagoda, H.P. The three-dimensional starting solution for a melting slab.
Proc. A **323**, 89—110 (1971).

Bollmann, W. *See* Pond & Bollmann.

Bolman, P.S.H., Brown, J.M., Carrington, A. Kopp, I. & Ramsay, D.A. A re-investigation of the $\tilde{A}\,^2\Sigma^+-$ $\tilde{X}\,^2\Pi_i$ band system of NCO. *Proc.* A **343**, 17—44 (1975).

Bolman, P.S.H., Brown, J.M., Carrington, A. & Lycett, G.J. Microwave spectroscopy of non-linear free radicals. II. Zeeman effect studies of DCO. *Proc.* A **335**, 113—126 (1973).

Bolton, H.C., Grant, Janine, McWilliam, I.G., Nicholson, A.J.C. & Swingler, D.L. Ionization in flames. II. Mass-spectrometric and mobility analyses for the flame ionization detector.
Proc. A **360**, 265—277 (1978).

Bolton, H.C. & McWilliam, I.G. Ionization in flames: current—voltage relationships for the flame ionization detector. *Proc.* A **321**, 361—380 (1971).

Bolton, T.B. Action of acetylcholine on the longitudinal muscle of guinea-pig ileum: the role of an electrogenic sodium pump (Discussion). *Trans.* B **265**, 107—114 (1973).
On the latency and form of the membrane responses of smooth muscle to the iontophoretic application of acetylcholine or carbachol. *Proc.* B **194**, 99—119 (1976).

Bomchil, G., Hüller, A., Rayment, T., Roser, S.J., Smalley, M.V., Thomas, R.K. & White, J.W. The structure and dynamics of methane adsorbed on graphite (Discussion).
Trans. B **290**, 537—552 (1980).

Bona, J. On the stability theory of solitary waves. *Proc.* A **344**, 363—374 (1975).

Bona, J.L. *See* Benjamin (T.B.), Bona & Mahony.

Bona, J.L. & Smith, R. The initial-value problem for the Korteweg — de Vries equation.
Trans. A **278**, 555—601 (1975).

Bonatti, E., Honnorez, J. & Ferrara, G. Peridotite—gabbro—basalt complex from the equatorial Mid-Atlantic Ridge (Discussion). *Trans.* A **268**, 385—402 (1971).

Bond, G. Observations on the root nodules of *Purshia tridentata*. *Proc.* B **193**, 127—135 (1976).
See also Mian, Bond & Rodriguez-Barrueco.

Bond, G. & Mackintosh, Anne H. Effect of nitrate-nitrogen on the nodule symbioses of *Coriaria* and

Hippophaë. Proc. B **190**, 199–209 (1975).

Diurnal changes in nitrogen fixation in the root nodules of *Casuarina. Proc.* B **192**, 1–12 (1975).

Bondi, Sir Hermann. Indirect utilization of solar energy (Discussion). *Trans.* A **295**, 501–506 (1980).

Bone, Q. *See* Anderson (P.A.V.) & Bone; Mackie & Bone.

Bone, Q., Anderson, P.A.V. & Pulsford, A. The communication between individuals in salp chains. I. Morphology of the system. *Proc.* B **210**, 549–558 (1980).

Bone, Q. & Ryan, K.P. On the structure and innervation of the muscle bands of *Doliolum* (Tunicata: Cyclomyaria). *Proc.* B **187**, 315–327 (1974).

Bonnet, R.M. The solar extreme ultraviolet continuum (Discussion). *Proc.* A **281**, 305–317 (1976).

Bonnett, R. Neovitamin B$_{12}$ (cyano-13-epicobalamin) (Discussion). *Trans.* B **273**, 295–301 (1976).

Bonnett, R., Davies, J.E., Hursthouse, M.B. & Sheldrick, G.M. The structure of bilirubin. *Proc.* B **202**, 249–268 (1978).

Bonnett, R.M. The solar extreme ultraviolet continuum (Discussion). *Proc.* A **281**, 305–317 (1976).

Bonnor, W.B. A new equation of motion for a radiating charged particle. *Proc.* A **337**, 591–598 (1974).

Bonny, A.P. *See* Pennington, Haworth, Bonny & Lishman.

Booker, H.G. Electromagnetic and hydromagnetic waves in a cold magnetoplasma (Discussion). *Trans.* A **280**, 57–93 (1975).

Boote, R.E. Concluding remarks to a Discussion. *Proc.* B **197**, 101–103 (1977).

Booth, B., Croasdale, R. & Walker, G.P.L. A quantitative study of five thousand years of volcanism on Sao Miguel, Azores. *Trans.* A **288**, 271–319 (1978).

Booth, B. & Self, S. Rheological features of the 1971 Mount Etna lavas (Symposium). *Trans.* A **274**, 99–106 (1973).

Booth, B. & Walker, G.P.L. Ash deposits from the new explosion crater, Etna 1971 (Symposium). *Trans.* A **274**, 147–151 (1973).

Booth, P.B. Genetic distances between certain New Guinea populations studied under the International Biological Programme (Discussion). *Trans.* B **268**, 257–267 (1974). *See also* Boyce (A.J.), Harrison, Platt, Hornabrook and others.

Borcard, B. *See* Béné, Borcard, Hiltbrand & Magnin.

Bore, P.J. *See* Ackerman, Bore, Gadian, Grove & Radda.

Born, G.V.R. *See* Boisseau, Lorient, Born & Michal; Cusack & Born; Cusack, Hickman & Born.

Borochov, N. *See* Eisenberg (H.), Borochov, Kam & Voordouw.

Borrell, Patricia M. *See* Borrell (P.), Borrell, Pedley & Grant.

Borrell, P., Borrell, Patricia M., Pedley, M.D. & Grant, K.R. High temperature studies of singlet excited oxygen, $O_2(^1\Sigma_g^+)$ and $O_2(^1\Delta_g)$, with a combined discharge flow/shock tube method. *Proc.* A **367**, 395–410 (1979).

Boss, K.J. Taxonomic concepts and superfluity in bivalve nomenclature (Discussion). *Trans.* A **284**, 417–424 (1978).

Bostrom, R.C. Arrangement of convection in the Earth by lunar gravity (Discussion). *Trans.* A **274**, 397–407 (1973).

Bostrom, R.C. & Sherif, M.A. On the incorporation of waste material in the solid part of the Earth. *Proc.* A **324**, 353–367 (1971).

Botchan, M. *See* Sambrook, Botchan, Hu, Mitchison & Stringer.

Botstein, D. *See* King (J.), Botstein, Casjens, Earnshaw, Harrison & Lenk.

Bott, M.H.P. Problems of passive margins from the viewpoint of the geodynamics project: a review (Discussion). *Trans.* A **294**, 5–16 (1980).

Bottcher, C. Theory of dissociative recombination and related processes. *Proc.* A **340**, 301–322 (1974).

Bottcher, C., Cravens, T.C. & Dalgarno, A. Collision broadening and relaxation of the resonance lines of lithium and sodium in helium gas. *Proc.* A **346**, 157–170 (1975).

Bottcher, C. & Dalgarno, A. A constructive model potential method for atomic interactions. *Proc.* A **340**, 187–198 (1974).

Böttiger, L.E. The Stockholm County Council's medical information system (Discussion). *Proc.* B **184**, 379–385 (1973).

Bottinga, Y. & Allègre, C.J. Partial melting under spreading ridges (Discussion).
  Trans. A **288**, 501–525 (1978).
Boucher, C. Activities of the Institut Géographique National (I.G.N.) in the field of Doppler satellite geodesy (Discussion). *Trans.* A **294**, 261–270 (1980).
Boucher, E.A. Capillary phenomena. IV. Thermodynamics of rotationally-symmetric fluid bodies in a gravitational field. *Proc.* A **358**, 519–533 (1978).
  Comment from the general discussion (Discussion). *Proc.* A **361**, 179 (1978).
Boucher, E.A. & Evans, M.J.B. Pendent drop profiles and related capillary phenomena.
  Proc. A **346**, 349–374 (1975).
Boucher, E.A., Evans, M.J.B. & Kent, H.J. Capillary phenomena. II. Equilibrium and stability of rotationally symmetric fluid bodies. *Proc.* A **349**, 81–100 (1976).
Boucher, E.A. & Kent, H.J. Capillary phenomena. III. Properties of rotationally symmetric fluid bodies with one asymptote – holms. *Proc.* A **356**, 61–75 (1977).
Bouffard, F.A. *See* Salzmann, Ratcliffe, Bouffard & Christensen.
Bougault, H., Joron, J.L. & Treuil, M. The primordial chondritic nature and large-scale heterogeneities in the mantle: evidence from high and low partition coefficient elements in oceanic basalts (Discussion). *Trans.* A **297**, 203–213 (1980).
Bouley, J.P. *See* Jonard, Briand, Bouley, Witz & Hirth.
Bouloy, Michele. *See* Krug, Bouloy & Plotch.
Boulpaep, E.L. *See* Giebisch, Boulpaep & Whittembury.
Boulter, D., Ramshaw, J.A.M., Thompson, E.W., Richardson, M. & Brown, R.H. A phylogeny of higher plants based on the amino acid sequences of cytochrome *c* and its biological implications. *Proc.* B **181**, 441–455 (1972).
Boulton, G.S. *See* Morland & Boulton.
Boulton, M.G. *See* Curtis (N.A.C.), Orr & Boulton.
Bourn, D. *See* Coe (M.J.), Bourn & Swingland.
Bourn, D. & Coe, M. The size, structure and distribution of the giant tortoise population of Aldabra.
  Trans. B **282**, 139–175 (1978).
Bourn, D. & Coe, M.J. Features of tortoise mortality and decomposition on Aldabra (Discussion).
  Trans. B **286**, 189–193 (1979).
Boustead, I. The effect of biphenyl on the thermoluminescence of squalane.
  Proc. A **328**, 389–399 (1972).
Bouysse, P. *See* Andreieff, Bouysse, Curry, Fletcher and others.
Bouysse, P., Horn, R., Lefort, J.P. & Le Lann, F. Tectonique et structures post-paléozoïques en Manche occidentale (Discussion). *Trans.* A **279**, 41–54 (1975).
Bovée, W.M.M.J., Creyghton, J.H.N., Getreuer, K.W., Korbee, D., Lobregt, S., Smidt, J., Wind, R.A., Lindeman, J., Smid, L. & Posthuma, H. N.m.r. relaxation and images of human breast tumours *in vitro* (Discussion). *Trans.* B **289**, 535–536 (1980).
Bowater, I.C., Brown, J.M. & Carrington, A. Microwave spectroscopy of nonlinear free radicals. I. General theory and application to the Zeeman effect in HCO. *Proc.* A **333**, 265–288 (1973).
Bowden, G.J. A unifying theory for effective field models in magnetism.
  Proc. A **353**, 563–574 (1977).
  See also Allsop, Bleaney, Bowden, Nambudripad, Stone & Suzuki.
Bowden, Lord. Effects of World War II on education in science (Discussion).
  Proc. A **342**, 499–503 (1975).
Bowden, M. *See* Miller (D.S.), Baker, Bowden, Evans, Holt and others.
Bowen, A.J. The tidal régime of the River Thames; long-term trends and their possible causes (Discussion). *Trans.* A **272**, 187–199 (1972).
Bowen, D.K. *See* Duesbery, Vítek & Bowen.
Bowen, E.J. *See* Altmann (S.L.) & Bowen.
Bower, D.I. The magnetostriction coefficients of nickel. *Proc.* A **326**, 87–96 (1971).
Bower, D.R. A sensitive water-level tiltmeter (Discussion). *Trans.* A **274**, 223–226 (1973).
Bowes, D.R., Langer, A.M. & Rohl, A.N. Nature and range of mineral dusts in the environment (Discussion). *Trans.* A **286**, 593–610 (1977).

Bowie, S.H.U. Natural sources of nuclear fuel (Discussion). *Trans.* A **276**, 495–505 (1974).
Modern methods in the search for metalliferous ores in Britain (Discussion).
Proc. A **339**, 299–311 (1974).
Introductory remarks to a Discussion. *Trans.* A **291**, 255 (1979).
The mode of occurrence and distribution of uranium deposits (Discussion).
Trans. A **291**, 289–300 (1979).
Introductory remarks to a Discussion. *Trans.* B **288**, 3 (1979).
See also Beckinsale, Bowie & Durham.
Bowles, J. *See* Woodgate (B.E.), Knight, Uribe, Sheather, Bowles & Nettleship.
Bowles, J.F.W. *See* Simpson (P.R.) & Bowles.
Bowley, R.M. *See* Allum, McClintock, Phillips & Bowley; Cehelnik, Cundall, Timmons & Bowley; Ellis (T.), McClintock, Bowley & Allum.
Bowman, Patricia. *See* Grüneberg, Cattanach, McLaren, Wolfe & Bowman.
Bowrey, M. & Purnell, J.H. The pyrolysis of disilane and rate constants of silene insertion reactions.
Proc. A **321**, 341–359 (1971).
Boxall, C.R. & Simons, J.P. The measurement of rapid rate processes by kinetic spectroscopy during a photolysis flash; vibrational relaxation of NO $\tilde{X}^2\Pi$ from levels $\nu \leqslant 2$.
Proc. A **328**, 515–527 (1972).
Boyce, A.J. *See* Harrison (G.A.), Hiorns & Boyce.
Boyce, A.J., Harrison, G.A., Platt, C.M., Hornabrook, R.W., Serjeantson, S., Kirk, R.L. & Booth, P.B. Migration and genetic diversity in an island population: Karkar, Papua New Guinea.
Proc. B **202**, 269–295 (1978).
Boyce, S.G. The use of bole surface in the estimation of woodland production (Discussion).
Trans. B **271**, 139–148 (1975).
Boycott, B.B. *See* Fisher (S.K.) & Boycott; Wässle, Boycott & Peichl; Wässle, Peichl & Boycott.
Boycott, B.B., Dowling, J.E., Fisher, S.K., Kolb, H. & Laties, A.M. Interplexiform cells of the mammalian retina and their comparison with catecholamine-containing retinal cells.
Proc. B **191**, 353–368 (1975).
Boycott, B.B., Peichl, L. & Wässle, H. Morphological types of horizontal cell in the retina of the domestic cat. *Proc.* B **203**, 229–245 (1978).
Boyd, D.W. *See* Newell (N.D.) & Boyd.
Boyd, P.D.W., Davies, J.E. & Gerloch, M. Antiferromagnetic exchange in triclinic crystals of tetra-μ-benzoato-*bis*(4-methylquinoline)dicobalt II. *Proc.* A **360**, 191–210 (1978).
Boyd, P.D.W., Gerloch, M., Harding, J.H. & Woolley, R.G. Magnetic exchange and anisotropic susceptibilities in high-spin binuclear cobalt II species. *Proc.* A **360**, 161–189 (1978).
Boyd, R.L.F. The Bakerian Lecture, 1978. Cosmic exploration by X-rays.
Proc. A **366**, 1–21 (1979).
Boyd, W.G.C. High-frequency scattering in a certain stratified medium: the three-part problem.
Proc. A **356**, 315–343 (1977).
Boyer, J.S. Photosynthesis at low water potentials (Discussion). *Trans.* B **273**, 501–512 (1976).
Boyle, J. Elizabeth. *See* Muscatine, Boyle & Smith; Trench, Boyle & Smith.
Boyle, J. Elizabeth & Smith, D.C. Biochemical interactions between the symbionts of *Convoluta roscoffensis*. *Proc.* B **189**, 121–135 (1975).
Boyle, L.L. & Green, Kerie F. The representation groups and projective representations of the point groups and their applications. *Trans.* A **288**, 237–269 (1978).
Boylett, F.D.A. & Maclean, I.G. The propagation of electric discharges across the surface of an electrolyte. *Proc.* A **324**, 469–489 (1971).
Boyling, J.B. An axiomatic approach to classical thermodynamics. *Proc.* A **329**, 35–70 (1972).
Bozler, E. Smooth muscle physiology, past and future (Discussion). *Trans.* B **265**, 3–6 (1973).
Brabban, D.H. & Glencross, W.M. Crystal spectrometry of active regions on the Sun.
Proc. A **334**, 231–239 (1973).
Brace, R.C. The functional anatomy of the mantle complex and columellar muscle of tectibranch molluscs (Gastropoda: Opisthobranchia), and its bearing on the evolution of opisthobranch organization. *Trans.* B **277**, 1–56 (1977).

Brachet, J.L. Review Lecture. Nucleocytoplasmic interactions in morphogenesis.
Proc. B **178**, 227–243 (1971).

Bradbury, E.M. Histone interactions, histone modifications and chromatin structure (Discussion).
Trans. B **283**, 291–293 (1978).

Braddick, O.J. Binocular single vision and perceptual processing (Discussion).
Proc. B **204**, 503–512 (1979).

Low-level and high-level processes in apparent motion (Discussion). *Trans.* B **290**, 137–151 (1980).

Brading, Alison F. Ion distribution and ion movements in smooth muscle (Discussion).
Trans. B **265**, 35–46 (1973).

Bradley, C.C. *See* Blaney, Bradley, Edwards, Jolliffe and others.

Bradley, D.J. Improvements of rural domestic water supplies (Discussion).
Proc. B **199**, 37–47 (1977).

Introductory remarks to a Discussion. *Trans.* A **298**, 211–215 (1980).

See also Adams (M.C.), Bradley, Sibbett & Taylor; Bates (B.), Bradley, McBride and others.

Bradley, D.J., Hutchinson, M.H.R. & Koetser, H. Interactions of picosecond laser pulses with organic molecules. II. Two-photon absorption cross-sections. *Proc.* A **329**, 105–119 (1972).

Bradley, D.J., Hutchinson, M.H.R., Koetser, H., Morrow, T., New, G.H.C. & Petty, M.S. Interactions of picosecond laser pulses with organic molecules. I. Two-photon fluorescence quenching and singlet states excitation in Rhodamine dyes. *Proc.* A **328**, 97–121 (1972).

Bradley, D.J., Jones, K.W. & Sibbett, W. Picosecond and femtosecond streak cameras: present and future designs (Discussion). *Trans.* A **298**, 281–285 (1980).

Bradley, J.N. A general mechanism for the high-temperature pyrolysis of alkanes. The pyrolysis of isobutane. *Proc.* A **337**, 199–216 (1974).

Bradley, J.N. & Metcalfe, E. The structure of electron cyclotron resonance absorption signals in gases.
Proc. A **361**, 367–377 (1978).

Bradley, S.G. & Stow, C.D. Collisions between liquid drops. *Trans.* A **287**, 635–675 (1978).

Bradley, W.G. *See* Waxman (S.G.), Bradley & Hartwieg.

Bradshaw, A.D. Conservation problems in the future (Discussion). *Proc.* B **197**, 77–96 (1977).

Braginsky, S.I. & Roberts, P.H. Magnetic field generation by baroclinic waves.
Proc. A **347**, 125–140 (1975).

Braithwaite, A.F. The phytogeographical relationships and origin of the New Hebrides fern flora (Discussion). *Trans.* B **272**, 293–313 (1975).

Braithwaite, C.J.R. Petrology of palaeosols and other terrestrial sediments on Aldabra, Western Indian Ocean. *Trans.* B **273**, 1–32 (1975).

The petrology of oolitic phosphorites from Esprit (Aldabra), western Indian Ocean.
Trans. B **288**, 511–540 (1980).

See also Taylor (J.D.), Braithwaite, Peake & Arnold.

Braithwaite, C.J.R., Taylor, J.D. & Kennedy, W.J. The evolution of an atoll: the depositional and erosional history of Aldabra. *Trans.* B **266**, 307–340 (1973).

Brambley, R.J. *See* Baird (P.E.G.), Brambley, Burnett, Stacey, Warrington & Woodgate.

Bramley, R. *See* Batley, Bramley, Poldy & Robinson; Batley, Bramley & Robinson.

Bramson, B.D. The alinement of frames of reference at null infinity for asymptotically flat Einstein–Maxwell manifolds. *Proc.* A **341**, 451–461 (1975).

Relativistic angular momentum for asymptotically flat Einstein–Maxwell manifolds.
Proc. A **341**, 463–490 (1975).

The invariance of spin. *Proc.* A **364**, 383–392 (1978).

Bramwell, Cherrie D. & Whitfield, G.R. Biomechanics of *Pteranodon*. *Trans.* B **267**, 503–581 (1974).

D.M.S. Watson's notes on pterosaurs. *Trans.* B **267**, 587–589 (1974).

Bramwell, M.E. & Harris, Henry. An abnormal membrane glycoprotein associated with malignancy in a wide range of different tumours. *Proc.* B **201**, 87–106 (1978).

Some further information about the abnormal membrane glycoprotein associated with malignancy.
Proc. B **203**, 93–99 (1978).

Branch, G.D. High temperature mechanical properties as design parameters (Symposium).
Trans. A **282**, 181–192 (1976).

Brand, A.R. *See* Taylor (A.C.) & Brand.

Brander, K.M. *See* Farrow & Brander.

Brandt, R.D. *See* Anding, Brandt, Ourisson, Pryce & Rohmer.

Branton, D. Freeze-etching studies of membrane structure (Discussion).
Trans. B 261, 133–138 (1971).

Braud, J. *See* Tchalenko & Braud.

Braun, E. The contribution of the Göttingen school to solid state physics: 1920–40 (Symposium).
Proc. A 371, 104–111 (1980).

Bray, R.S. *See* Sinden, Canning, Bray & Smalley.

Brazier-Smith, P.R., Brook, M., Latham, J., Saunders, C.P.R. & Smith, M.H. The vibration of electrified water drops. *Proc.* A 322, 523–534 (1971).

Brazier-Smith, P.R., Jennings, S.G. & Latham, J. An investigation of the behaviour of drops and drop-pairs subjected to strong electrical forces. *Proc.* A 325, 363–376 (1971).
The interaction of falling water drops: coalescence. *Proc.* A 326, 393–408 (1972).

Brear, J.M. *See* Batte, Brear, Holdsworth, Myers & Reynolds.

Brear, J.M. & King, B.L. An assessment of the embrittling effects of certain residual elements in two nuclear pressure vessel steels (A533B, A508) [abstract] (Discussion).
Trans. A 295, 291 (1980).

Breathnach, A.S. *See* Logan (A.G.), Tenyi, Peart, Breathnach & Martin.

van Breemen, C., Farinas, Blanca R., Casteels, R., Gerba, Peggy, Wuytack, F. & Deth, R. Factors controlling cytoplasmic Ca^{2+} concentration (Discussion). *Trans.* B 265, 57–71 (1973).

Bregestovski, P.D., Miledi, R. & Parker, I. Blocking of frog endplate channels by the organic calcium antagonist D600. *Proc.* B 211, 15–24 (1980).

Brehm, Lotte & Moult, J. The crystal structure of *p*-nitrophenyl-β-D-*N*-acetylglucosaminide monohydrate. *Proc.* B 188, 425–435 (1975).

Brenner, H. Dispersion resulting from flow through spatially periodic porous media.
Trans. A 297, 81–133 (1980).

Brenner, S. *See* White (J.G.), Southgate, Thomson & Brenner.

Brenton, A.G. *See* Amaya, Brenton, Szulejko & Beynon; Beynon (J.H.), Morgan & Brenton; Szulejko, Amaya, Morgan, Brenton & Beynon.

Breuer, R.A. & Ehlers, J. Propagation of high-frequency electromagnetic waves through a magnetized plasma in curved space–time. I. *Proc.* A 370, 389–406 (1980).

Breuer, R.A., Ryan, M.P., Jr, & Waller, S. Some properties of spin-weighted spheroidal harmonics.
Proc. A 358, 71–86 (1977).

Brévart, O. *See* Allègre, Brévart, Dupré & Minster.

Brewin, N. *See* Beringer, Brewin, Johnston, Schulman & Hopwood.

Brian, P.W. Review Lecture. Hormones in healthy and diseased plants. *Proc.* B 200, 231–243 (1978).

Briand, J.P. *See* Jonard, Briand, Bouley, Witz & Hirth.

Briden, J.C. Application of palaeomagnetism to Proterozoic tectonics (Discussion).
Trans. A 280, 405–416 (1976).

Briden, J.C., Rex, D.C., Faller, A.M. & Tomblin, J.F. K–Ar geochronology and palaeomagnetism of volcanic rocks in the Lesser Antilles island arc. *Trans.* A 291, 485–528 (1979).

Bridge, M.E. & Lambert, R.M. Associative desorption from adlayers of interacting particles: nitric oxide on Pt, Ni and Ru. *Proc.* A 370, 545–560 (1980).

Bridges, Barbara J., Charlesby, A. & Folland, R. Pulsed n.m.r. studies of the crystallization kinetics of polyethylene from the melt and solution. *Proc.* A 367, 343–351 (1979).

Bridgwater, D., Escher, A. & Watterson, J. Tectonic displacements and thermal activity in two contrasting Proterozoic mobile belts from Greenland (Discussion). *Trans.* A 273, 513–533 (1973).

Bridgwater, D., Watson, Janet & Windley, B.F. The Archaean craton of the North Atlantic region (Discussion). *Trans.* A 273, 493–512 (1973).

Briedis, D.J. *See* Mahy, Barrett, Briedis, Brownson & Wolstenholme.

Briggs, C.C. *See* Lilley (T.H.) & Briggs.

Briggs, D.E.G. The morphology, mode of life, and affinities of *Canadaspis perfecta* (Crustacea: Phyllocarida), Middle Cambrian, Burgess Shale, British Columbia. *Trans.* B 281, 439–486 (1978).

Briggs, D.J. *See* Shotton (F.W.), Goudie, Briggs & Osmaston.
Briggs, G.A.D. *See* Savkoor & Briggs.
Brimicombe, M.W.S.M. *See* Kuhn, Baird, Brimicombe, Stacey & Stacey.
Brimicombe, M.S.W.M., Stacey, D.N., Stacey, V., Hühnermann, H. & Menzel, N. Optical isotope shifts and hyperfine structure in Cd. *Proc.* A **352**, 141–152 (1976).
Brinkmann, W.P. See Pollock (M.D.) & Brinkmann.
Brinkworth, B.J. Results of solar heating experiments (Discussion). *Trans.* A **295**, 361–373 (1980).
Brion, C.E. *See* Foo, Brion & Hasted.
Briscoe, B.J. *See* Klein (J.) & Briscoe.
Briscoe, B.J., Scruton, B. & Willis, F.R. The shear strength of thin lubricant films.
    Proc. A **333**, 99–114 (1973).
Bristeau, P. *See* Lambert (G.), Le Roulley & Bristeau.
Bristow, J.W. *See* Erlank, Allsopp, Duncan & Bristow.
Broadbent, D.E. Erratum to the Biographical Memoir of Frederic Charles Bartlett.
    Biogr. Mem. **17**, 757 (1971).
Broadbent, E.G. Morien Bedford Morgan. *Biogr. Mem.* **26**, 371–410 (1980).
Broadbent, E.G. & Moore, D.W. Acoustic destabilization of vortices. *Trans.* A **290**, 353–371 (1979).
Brockmann, H., Jr. Bacteriochlorophyll *e*: structure and stereochemistry of a new type of chlorophyll from Chlorobiaceae (Discussion). *Trans.* B **273**, 277–285 (1976).
Brook, G.B. & Duckworth, W.E. Future developments in non-ferrous metallurgy (Symposium).
    Trans. A **282**, 413–426 (1976).
Brook, M. *See* Brazier-Smith, Brook, Latham, Saunders & Smith.
Brooker, B.E. *See* Frost (A.J.), Hill & Brooker.
Brookes, C.A. & Green, P. Anisotropy in the scratch hardness of cubic crystals.
    Proc. A **368**, 37–57 (1979).
Brookes, C.A., O'Neill, J.B. & Redfern, B.A.W. Anisotropy in the hardness of single crystals.
    Proc. A **322**, 73–88 (1971).
Brooks, B.W. Viscosity effects in the free-radical polymerization of methyl methacrylate.
    Proc. A **357**, 183–192 (1977).
Brooks, C.J.W. Some aspects of mass spectrometry in research on steroids (Discussion).
    Trans. A **293**, 53–67 (1979).
Brooks, D.E., Gaughwin, M. & Mann, T. Structural and biochemical characteristics of the male accessary organs of reproduction in the hairy-nosed wombat (*Lasiorhinus latifrons*).
    Proc. B **201**, 191–207 (1978).
Brooks, D.E., Levine, Y.K., Requena, J. & Haydon, D.A. Van der Waals forces in oil–water systems from the study of thin lipid films. III. Comparison of experimental results with Hamaker co-efficients calculated from Lifshitz theory. *Proc.* A **347**, 179–194 (1975).
Brooks, D.E., Lutwak-Mann, Cecilia, Mann, T. & Martin, A.W., Jr. Motility and energy-rich phosphorus compounds in spermatozoa of *Octopus dofleini martini*. *Proc.* B **178**, 151–160 (1971).
Brooks, D.E., Mann, T. & Martin, A.W. The occurrence of carnitine and glycerylphosphorylcholine in the octopus spermatophore. *Proc.* B **186**, 79–82 (1974).
Brooks, R.R. *See* Jaffré, Kersten, Brooks & Reeves; Kelly (P.C.), Brooks, Dilli & Jaffré.
Brooks, R.R., McCleave, J.A. & Malaisse, F. Copper and cobalt in African species of *Crotalaria* L.
    Proc. B **197**, 231–236 (1977).
Brooks, R.R., Morrison, R.S., Reeves, R.D., Dudley, T.R. & Akman, Y. Hyperaccumulation of nickel by *Alyssum* Linnaeus (Cruciferae). *Proc.* B **203**, 387–403 (1979).
Brooks, R.R. & Radford, C.C. Nickel accumulation by European species of the genus *Alyssum*.
    Proc. B **200**, 217–224 (1978).
Broom, T. & Gow, R.S. Operational experience in nuclear power stations (Discussion).
    Trans. A **276**, 571–586 (1974).
Brotherhood, J. *See* Edholm (O.G.), Humphrey, Lourie, Tredre & Brotherhood.
Brown, A.C.L. Animal health: present and future (Discussion). *Trans.* B **281**, 181–191 (1977).
Brown, A.P. Late-Devensian and Flandrian vegetational history of Bodmin Moor, Cornwall.
    Trans. B **276**, 251–320 (1977).

Brown, B.F. The contributions of physical metallurgy and of fracture mechanics to containing the problem of stress-corrosion cracking (Symposium). *Trans.* A **282**, 235–245 (1976).

Brown, C. *See* Girdler, Brown, Noy & Styles.

Brown, D.M. & Trimm, D.L. Gas phase interaction with catalytic oxidation reactions. *Proc.* A **326**, 215–227 (1972).

Brown, F.F. & Campbell, I.D. N.m.r. studies of red cells (Discussion). *Trans.* B **289**, 395–406 (1980).

Brown, F.F., Halsey, M.J. & Richards, R.E. Halothane interactions with haemoglobin. *Proc.* B **193**, 387–411 (1976).

Brown, G.C. *See* Simpson (P.R.), Brown, Plant & Ostle.

Brown, G.C. & Hennessy, J. The initiation and thermal diversity of granite magmatism (Discussion). *Trans.* A **288**, 631–643 (1978).

Brown, G.M. Two-stage generation of lunar mare basalts (Discussion). *Trans.* A **285**, 169–176 (1977). Two major igneous events in the evolution of the Moon (Discussion). *Trans.* A **286**, 439–451 (1977).

Brown, G.M., O'Hara, M.J. & Oxburgh, E.R. Introduction to a Discussion. *Trans.* A **288**, 385–386 (1978).

Brown, Hilary F., McNaughton, P.A., Noble, D. & Noble, Susan J. Adrenergic control of cardiac pacemaker currents (Discussion). *Trans.* A **270**, 527–537 (1975).

Brown, J. *See* Gabor & Brown.

Brown, J.C. The interpretation of hard and soft X-rays from solar flares (Discussion). *Trans.* A **281**, 473–490 (1976).

Brown, J.J. Computer-based records in the clinic (*b*) (Discussion). *Proc.* B **184**, 399–402 (1973).

Brown, J.M. *See* Boland (B.J.), Brown, Carrington & Nelson; Bolman, Brown, Carrington, Kopp & Ramsay; Bolman, Brown, Carrington & Lycett; Bowater, Brown & Carrington.

Brown, K.S., Jr, Sheppard, P.M. & Turner, J.R.G. Quaternary refugia in tropical America: evidence from race formation in *Heliconius* butterflies. *Proc.* B **187**, 369–378 (1974).

Brown, N. *See* Richmond (M.H.), Bennett, Choi, Brown, Brunton and others.

Brown, P. Jane, Forsyth, J.B. & Mason, R. Magnetization densities and electronic states in crystals (Discussion). *Trans.* B **290**, 481–495 (1980).

Brown, R. Thomas Archibald Bennet-Clark. *Biogr. Mem.* **23**, 1–18 (1977).

Brown, R.A. & Scriven, L.E. The shapes and stability of captive rotating drops. *Trans.* A **297**, 51–79 (1980). The shape and stability of rotating liquid drops. *Proc.* A **371**, 331–357 (1980).

Brown, R.H. *See* Boulter, Ramshaw, Thompson, Richardson & Brown.

Brown, S.N. & Stewartson, K. The evolution of a small inviscid disturbance to a marginally unstable stratified shear flow; stage two. *Proc.* A **363**, 175–194 (1978).

Brown, S.S. Laboratory support for rural health care (Discussion). *Proc.* B **209**, 119–128.

Brown, Susan N. & Stewartson, K. On finite amplitude Bénard convection in a cylindrical container. *Proc.* A **360**, 455–469 (1978).

Brown, T.R. Saturation transfer in living systems (Discussion). *Trans.* B **289**, 441–444 (1980).

Browne, J.L., Sanford, P.A. & Smyth, D.H. Transport and metabolic processes in the small intestine. *Proc.* B **195**, 307–321 (1977). Transfer and metabolism of citrate, succinate, α-ketoglutarate and pyruvate by hamster small intestine. *Proc.* B **200**, 117–135 (1978).

Browne, J.M. & Anderson, J.S. The oxidation of $Nb_{12}O_{29}$: electron microscopy of a solid-state reaction. *Proc.* A **339**, 463–482 (1974).

Browne, S.G. Rural health and disease in five continents (Discussion). *Proc.* B **199**, 9–15 (1977). Closing remarks to a Discussion. *Proc.* B **199**, 187 (1977). Appropriate technologies for the future (Discussion). *Proc.* B **209**, 183–186 (1980).

Brown-Grant, K. *See* Raisman & Brown-Grant.

Brown-Grant, K., Murray, M.A.F., Raisman, G. & Sood, M.C. Reproductive function in male and female rats following extra- and intra-hypothalamic lesions. *Proc.* B **198**, 267–278 (1977).

Brown-Grant, K. & Raisman, G. Abnormalities in reproductive function associated with the destruction of the suprachiasmatic nuclei in female rats. *Proc.* B **198**, 279–296 (1977).

Browning, K.A. Review Lecture. Local weather forecasting. *Proc.* A **371**, 179—211 (1980).

Browning, R., Bassett, P.J., El Gomati, M.M. & Prutton, M. A digital scanning Auger electron micro-scope incorporating a concentric hemispherical analyser. *Proc.* A **357**, 213—230 (1977).

Brownscombe, J.L. *See* Miller (D.E.), Brownscombe, Carruthers, Pick & Stewart.

Brownson, J.M. *See* Mahy, Barrett, Briedis, Brownson & Wolstenholme.

Brownstein, M.J. Peptidergic pathways in the central nervous system (Discussion).
Proc. B **210**, 79—90 (1980).

Brozel, M.R., Evans, T. & Stephenson, R.F. Partial dissociation of nitrogen aggregates in diamond by high temperature—high pressure treatments. *Proc.* A **361**, 109—127 (1978).

Brueckner, G.E. A.t.m. observations on the X u.v. emission from solar flares (Discussion).
Trans. A **281**, 443—459 (1976).

Brundle, C.R. & Roberts, M.W. Some observations on the surface sensitivity of photoelectron spec-troscopy (Discussion). *Proc.* A **331**, 383—394 (1972).

Bruner, J. *See* Thieffry, Bruner & Personne.

Brunet, P.C.J. *See* Pau, Brunet & Williams.

Brunet, P.C.J. & Coles, Barbara C. Tanned silks. *Proc.* B **187**, 133—170 (1974).

Bruns, Gail A.P. & Ingram, V.M. The erythroid cells and haemoglobins of the chick embryo.
Trans. B **266**, 225—305 (1973).

Brunsden, D. & Jones, D.K.C. The evolution of landslide slopes in Dorset (Discussion).
Trans. A **283**, 605—631 (1976).

Brunton, J. *See* Richmond (M.H.), Bennett, Choi, Brown, Brunton and others.

Bruun, H.H. A time-domain evaluation of the large-scale flow structure in a turbulent jet.
Proc. A **367**, 193—218 (1979).

Bryan, G.W. The effects of heavy metals (other than mercury) on marine and estuarine organisms (Discussion). *Proc.* B **177**, 389—410 (1971).
Bioaccumulation of marine pollutants (Discussion). *Trans.* B **286**, 483—505 (1979).

Bryant, F.J., Goodwin, G.K. & Hagston, W.E. Atomic displacement effects on the cathodolumin-escence of zinc selenide implanted with ytterbium ions. *Proc.* A **337**, 21—47 (1974).

Bryant, F.J., Hagston, W.E. & Radford, C.J. The effects of preferential pairing and its detection using electron radiation damage. *Proc.* A **323**, 127—149 (1971).

Bryce, D.J. *See* Ireland & Bryce.

Bryceson, A.D.M. Rehydration in cholera and other diarrhoeal diseases (Discussion).
Proc. B **199**, 109—114 (1977).

Buchsbaum, G. & Goldstein, J.L. Optimum probabilistic processing in colour perception.
I. Colour discrimination. *Proc.* B **205**, 229—247 (1979).
II. Colour vision as template matching. *Proc.* B **205**, 249—266 (1979).

Buchwald, V.F. The mineralogy of iron meteorites (Discussion). *Trans.* A **286**, 453—491 (1977).

Buchwald, V.T. Energy and energy flux in planetary waves. *Proc.* A **328**, 37—48 (1972).

Buckingham, A.D. Intermolecular forces (Discussion). *Trans.* B **272**, 5—12 (1975).
John Wilfrid Linnett. *Biogr. Mem.* **23**, 311—343 (1977).
Polarizability and hyperpolarizability (Discussion). *Trans.* A **293**, 239—248 (1979).

Buckingham, A.D. & Graham, C. The density dependence of the refractivity of gases.
Proc. A **337**, 275—291 (1974).

Buckingham, A.D. & Raab, R.E. Electric-field-induced differential scattering of right and left circularly polarized light. *Proc.* A **345**, 365—377 (1975).

Buckingham, M.J. On the response of steered vertical line arrays to anistropic noise.
Proc. A **367**, 539—547 (1979).

Buckles, R.G. It is appropriate to create new pharmaceutical products within developing countries (Discussion). *Proc.* B **209**, 165—171 (1980).

Buckley, C.P. & Green, A.E. Small deformations of a nonlinear viscoelastic tube: theory and applica-tion to polypropylene. *Trans.* A **281**, 543—566 (1976).

Buckley, C.P., Lloyd, D.W. & Konopasek, M. On the deformation of slender filaments with planar crimp: theory, numerical solution and applications to tendon collagen and textile materials.
Proc. A **372**, 33—64 (1980). [Abstract] . *Proc.* B **209**, 331 (1980).

Buckley, F. *See* Baker (P.E.), Buckley & Rex.
Bucourt, R., Heymès, R., Lutz, A., Penasse, L. & Perronnet, J. New very efficient antibiotics in the field of cephalosporin derivatives [synopsis] (Discussion). *Trans.* B 289, 361–363 (1980).
Budd, G.M. *See* Fox (R.H.), Budd, Woodward, Hackett & Hendrie.
Budd, G.M., Fox, R.H., Hendrie, A.L. & Hicks, K.E. A field survey of thermal stress in New Guinea villagers (Discussion). *Trans.* B 268, 393–400 (1974).
Budden, K.G. Phase integral methods for studying the effect of the ionosphere on radio propagation (Discussion). *Trans.* A 280, 111–130 (1975).
  Radio caustics and cusps in the ionosphere. *Proc.* A 350, 143–164 (1976).
  Resonance tunnelling of waves in a stratified cold plasma. *Trans.* A 290, 405–433 (1979).
Budden, K.G. & Eve, M. Degenerate modes in the Earth-ionosphere waveguide.
  Proc. A 342, 175–190 (1975).
Budden, K.G. & Smith, M.S. The coalescence of coupling points in the theory of radio waves in the ionosphere. *Proc.* A 341, 1–30 (1974).
  Phase memory and additional memory in W.K.B. solutions for wave propagation in stratified media.
  Proc. A 350, 27–46 (1976).
Budden, K.G. & Terry, P.D. Radio ray tracing in complex space. *Proc.* A 321, 275–301 (1971).
Budden, K.G. & Uscinski, B.J. The scintillation of extended radio sources when the receiver has a finite bandwidth.
  II. Multiple scatter. *Proc.* A 321, 15–40 (1971).
  III. Further methods. *Proc.* A 330, 65–77 (1972).
Buffham, B.A. Model-independent aspects of tracer chromatography theory.
  Proc. A 333, 89–98 (1973).
  Model-independent aspects of perturbation chromatography theory. *Proc.* A 364, 443–455 (1978).
  Erratum. *Proc.* A 369, 575 (1980).
Buffler, R.T. *See* Vail (P.R.), Mitchum, Shipley & Buffler.
Buffoni, Franca, Della Corte, Laura & Hope, D.B. Immunofluorescene histochemistry of porcine tissues using antibodies to pig plasma amine oxidase. *Proc.* B 195, 417–423 (1977).
Bülbring, Edith. *See* Szurszewski & Bülbring.
Bülbring, Edith & Kuriyama, H. The action of catecholamines on guinea-pig taenia coli (Discussion).
  Trans. B 265, 115–121 (1973).
Bülbring, Edith & Szurszewski, J.H. The stimulant action of noradrenaline (α-action) on guinea-pig myometrium compared with that of acetylcholine. *Proc.* B 185, 225–262 (1974).
Bülbring, Edith & Tomita, T. The α-action of catecholamines on the guinea-pig taenia coli in K-free and Na-free solution and in the presence of ouabain. *Proc.* B 197, 255–269 (1977).
  Calcium requirement for the α-action of catecholamines on guinea-pig taenia coli.
  Proc. B 197, 271–284 (1977).
Bulewicz, Elżbieta M. & Padley, P.J. Photometric investigations of the behaviour of chromium additives in premixed $H_2 + O_2 + N_2$ flames. *Proc.* A 323, 377–400 (1971).
Bull, G.M. Advanced technology – research and the future (Discussion).
  Proc. B 184, 473–476 (1973).
Bull, K.R., Marshall, R.M. & Purnell, J.H. Mechanism and rate parameters for the pyrolysis of 2,3-dimethylbutane. *Proc.* A 342, 259–277 (1975).
Bull, R.K. *See* Durrani, Bull & McKeever.
Bullard, Sir Edward. Introductory remarks to a Discussion. *Trans.* A 268, 383 (1971).
  The effect of World War II on,the development of knowledge in the physical sciences (Discussion).
  Proc. A 342, 519–536 (1975).
  William Maurice Ewing. *Biogr. Mem.* 21, 269–311 (1975).
Bullen, E.R. How much cultivation? (Discussion). *Trans.* B 281, 153–161 (1977).
Buller, A.J. & Pope, R. Plasticity in mammalian skeletal muscle (Discussion).
  Trans. B 278, 295–305 (1977).
Bullerwell, W. Geophysical studies relating to the Tertiary volcanic structure of the British Isles (Discussion). *Trans.* A 271, 209–215 (1972).
Bullivant, S. Freeze-etching techniques applied to biological membranes (Discussion).
  Trans. B 268, 5–14 (1974).

Bullock, E., Lea, C. & McLean, M. Benefits of minor additions of yttrium to the oxidation and creep behaviour of a nickel-based composite [abstract] (Discussion). *Trans.* A **295**, 332 (1980).

Bullock, G., Eakins, M.N., Sawyer, B.C. & Slater, T.F. Studies on bile secretion with the aid of the isolated perfused rat liver. I. Inhibitory action of sporidesmin and icterogenin.
Proc. B **186**, 333–356 (1974).

Bullock, Suzanne. *See* Fineran & Bullock.

Bullough, K., Denby, M., Gibbons, W., Hughes, A.R.W., Kaiser, T.R. & Tatnall, A.R.D. E.l.f./v.l.f. emissions observed on Ariel 4 (Discussion). *Proc.* A **343**, 207–226 (1975).

Bullough, R. *See* Little (E.A.), Bullough & Wood; Willis (J.R.), Hayns & Bullough.

Bullough, R., Eyre, B.L. & Krishan, K. Cascade damage effects on the swelling of irradiated materials.
Proc. A **346**, 81–102 (1975).

Bullough, R.K. *See* Dodd (R.K.) & Bullough.

Bulpitt, C.J. *See* Coles (E.C.), Beilin, Bulpitt, Dollery, Johnson and others.

Bunch, T.S. & Davies, P.C.W. Stress tensor and conformal anomalies for massless fields in a Robertson–Walker universe. *Proc.* A **356**, 569–574 (1977).

Covariant point-splitting regularization for a scalar quantum field in a Robertson–Walker universe with spatial curvature. *Proc.* A **357**, 381–394 (1977).

Quantum field theory in de Sitter space: renormalization by point-splitting.
Proc. A **360**, 117–134 (1978).

Bunn, C.W., Moews, P.C. & Baumber, M.E. The crystallography of calf rennin (chymosin).
Proc. B **178**, 245–258 (1971).

Bunting, A.H. Review and prospect: where do we go from here? (Discussion).
Trans. B **278**, 611–614 (1977).

Bunyan, P.J. *See* Stanley (P.I.) & Bunyan.

Burbidge, G.R. The extra-galactic contribution to the primary cosmic-ray flux (Discussion).
Trans. A **277**, 481–487 (1974).

Burcher, R.K. *See* Bishop (R.E.D.), Burcher & Price.

Burchfield, R.W. *See* Hatfield, Fisher, Dunigan, Burchfield and others.

Burdekin, F.M. Keynote speech (Symposium). *Trans.* A **282**, 258–263 (1976).
See also Barr & Burdekin.

Burden, R.S. *See* Taylor (H.F.) & Burden.

Burdett, N.A. & Hayhurst, A.N. Kinetics of formation and removal of atomic halogen ions X^- by $HX + e^- \rightleftharpoons H + X^-$ in atmospheric pressure flames for chlorine, bromine and iodine.
Proc. A **355**, 377–405 (1977).

Determination of the rate coefficients of $A + X \rightleftharpoons A^+ + X^-$ and $AX + M \rightleftharpoons A^+ + X^- + M$ where A is a metal atom, X a halogen atom and M a flame species. *Trans.* A **290**, 299–325 (1979).

Burge, R.E., Fiddy, M.A., Greenaway, A.H. & Ross, G. The phase problem.
Proc. A **350**, 191–212 (1976).

Burgen, A.S.V. *See* Burgen (Sir Arnold); King (R.W.) & Burgen.

Burgen, Sir Arnold. Introductory remarks to a Discussion. *Proc.* B **210**, 3–4 (1980).
See also Burgen (A.S.V.); Birdsall (B.), Griffiths, Roberts, Feeney & Burgen; Birdsall (N.J.M.), Hulme & Burgen; Feeney, Roberts, Birdsall, Griffiths and others.

Burger, M.M., Turner, R.S., Kuhns, W.J. & Weinbaum, G. A possible model for cell–cell recognition via surface macromolecules (Discussion). *Trans.* B **271**, 379–393 (1975).

Burgers, J.M. Addendum to a paper by W.G. Burgers (Symposium). *Proc.* A **371**, 130 (1980).

Burgers, W.G. How my brother and I became interested in dislocations (with an addendum by J.M. Burgers) (Symposium). *Proc.* A **371**, 125–130 (1980).

Burges, N.A. The Meathop Wood and the Pasoh rainforest projects (Discussion).
Trans. B **274**, 283–294 (1976).

Burgess, D.A. *See* Rogers (C.A.), Burgess, Halberstam & Birch.

Burgess, J. *See* Allen (G.), Burgess, Edwards & Walsh.

Burggraf, O.R. *See* Jobe & Burggraf.

Burgis, Mary J., Darlington, Johanna P.E.C., Dunn, I.G., Ganf, G.G., Gwahaba, J.J. & McGowan, L.M. The biomass and distribution of organisms in Lake George, Uganda (Discussion).

Proc. B **184**, 271–298 (1973). Corrigendum. *Proc.* B **185**, 469 (1974).

Burke, D.C. The production of interferon by animal viruses (Discussion).
  Proc. B **177**, 17–22 (1971).

Burkill, J.C. Abram Samoilovitch Besicovitch. *Biogr. Mem.* **17**, 1–16 (1971).
  John Edensor Littlewood. *Biogr. Mem.* **24**, 323–367 (1978).

Burkin, A.R. Review Lecture. The winning of non-ferrous metals, 1974.
  Proc. A **338**, 419–437 (1974).

Burland, J.B. *See* Ward (W.H.) & Burland.

Burleigh, R. Radiocarbon dating of eggshell of giant tortoise from Denis Island, Seychelles (appendix
  to a paper by D.R. Stoddart & J.F. Peake) (Discussion). *Trans.* B **286**, 160–161 (1979).

Burnett, K. *See* Baird (P.E.G.), Brambley, Burnett, Stacey, Warrington & Woodgate.

Burns, B. Delisle, Stean, J.P.B. & Webb, A.C. The effects of sleep on neurons in isolated cerebral cortex. *Proc.* B **206**, 281–291 (1979).

Burns, B. Delisle & Webb, A.C. The effects of stationary retinal patterns upon the behaviour of
  neurons in the cat's visual cortex. *Proc.* B **178**, 63–78 (1971).

  The spontaneous activity of neurones in the cat's cerebral cortex. *Proc.* B **194**, 211–223 (1976).

  The correlation between discharge times of neighbouring neurons in isolated cerebral cortex.
  Proc. B **203**, 347–360 (1979).

Burns, G. *See* Wong (W.H.) & Burns.

Burns, R.G. *See* Vaughan (D.J.) & Burns.

Burns, R.G. & Burns, Virginia M. The mineralogy and crystal chemistry of deep-sea manganese
  nodules, a polymetallic resource of the twenty-first century (Discussion).
  Trans. A **286**, 283–301 (1977).

Burns, Virginia M. *See* Burns (R.G.) & Burns.

Burridge, J.C. *See* Mitchell (R.L.) & Burridge.

Burridge, K. *See* Elliott (A.), Offer & Burridge.

Burridge, R. & Sabina, F.J. The propagation of elastic surface waves guided by ridges.
  Proc. A **330**, 417–441 (1972).

Burrows, C.R. & Stanway, R. A coherent strategy for estimating linearized oil-film coefficients.
  Proc. A **370**, 89–105 (1980).

Burrows, Elsie M. Assessment of pollution effects by the use of algae (Discussion).
  Proc. B **177**, 295–306 (1971).

Burrows, J.P., Cliff, D.I., Harris, G.W., Thrush, B.A. & Wilkinson, J.P.T. Atmospheric reactions of the
  HO_2 radical studied by laser magnetic resonance spectroscopy. *Proc.* A **368**, 463–481 (1979).

Burrows, M. Modes of activation of motoneurons controlling ventilatory movements of the locust
  abdomen. *Trans.* B **269**, 29–48 (1974).

  The tracheal supply to the central nervous system of the locust. *Proc.* B **207**, 63–78 (1980).

  See also Horridge & Burrows.

Burrows, M. & Horridge, G.A. The organization of inputs to motoneurons of the locust metathoracic
  leg. *Trans.* B **269**, 49–94 (1974).

Bursill, L.A. *See* Merritt, Hyde, Bursill & Philp.

Burt, M.G. & Peierls, Sir Rudolf. The momentum of a light wave in a refracting medium.
  Proc. A **333**, 149–156 (1973).

Burt, P.B. Exact, multiple soliton solutions of the double sine Gordon equation.
  Proc. A **359**, 479–495 (1978). (Corrigenda. *Proc.* A **362**, 572 (1978)).

Burtenshaw, M.D. *See* Ford (C.E.), Evans, Burtenshaw, Clegg, Tuffrey & Barnes.

Burton, A.J. & Miller, G.F. The application of integral equation methods to the numerical solution
  of some exterior boundary-value problems (Discussion). *Proc.* A **323**, 201–210 (1971).

Burton, J.D. Physico-chemical limitations in experimental investigations (Discussion).
  Trans. B **286**, 443–456 (1979).

Burton, W.M. *See* Bates (B.), Bradley, McBride and others.

Burton, W.M., Evans, R.G. & Griffin, W.G. Skylark rocket observations of ultraviolet spectra of γ^2
  Velorum and ζ Puppis (Discussion). *Trans.* A **279**, 355–369 (1975).

Burton, W.M., Jordan, Carole, Ridgeley, A. & Wilson, R. The structure of the chromosphere—corona

transition region from limb and disk intensities (Discussion). *Trans.* A **270**, 81–98 (1971).

Busby, R.F. Engineering aspects of manned and remotely controlled vehicles (Discussion). *Trans.* A **290**, 135–152 (1978).

Buschhorn, G. Decay properties of new particles (results from the DASP-collaboration/DORIS) (Discussion). *Proc.* A **355**, 539–554 (1977).

Bush, S.F. & Dyer, P. The experimental and computational determination of complex chemical kinetics mechanisms. *Proc.* A **351**, 33–53 (1976).

Bushnell, G.H.S. The beginning and growth of agriculture in Mexico (Discussion). *Trans.* B **275**, 117–120 (1976).

Busslinger, M. *See* Birnstiel, Kressmann, Schaffner, Portmann & Busslinger.

Butcher, R.J., Dennis, R.B. & Smith, S.D. The tunable spin-flip Raman laser. II. Continuous wave molecular spectroscopy. *Proc.* A **344**, 541–561 (1975).

Butcher, R.J., Willetts, D.V. & Jones, W.J. On the use of a Fabry—Perot etalon for the determination of rotational constants of simple molecules — the pure rotational Raman spectra of oxygen and nitrogen. *Proc.* A **324**, 231–245 (1971).

Butler, G. & Ison, H.C.K. Corrosion, design and materials: general and pitting (Symposium). *Trans.* A **282**, 225–234 (1976).

Butler, P.A. Influence of pesticides on marine ecosystems (Discussion). *Proc.* B **177**, 321–329 (1971).

Butler, P.H. Coupling coefficients and tensor operators for chain of groups. *Trans.* A **277**, 545–585 (1975).

Butler, P.J.G. Assembly of tobacco mosaic virus (Discussion). *Trans.* B **276**, 151–163 (1976). *See also* Adolph & Butler.

Butts, D., Tinker, M.H. & Kernahan, J.A. Time-resolved ion cyclotron resonance. *Proc.* A **364**, 367–381 (1978).

Buxton, B.F. Bloch waves and higher order Laue zone effects in high energy electron diffraction. *Proc.* A **350**, 335–361 (1976).

Buxton, B.F. & Berry, M.V. Bloch wave degeneracies in systematic high energy electron diffraction. *Trans.* A **282**, 485–525 (1976).

Buxton, B.F., Eades, J.A., Steeds, J.W. & Rackham, G.M. The symmetry of electron diffraction zone axis patterns. *Trans.* A **281**, 171–194 (1976).

Buxton, G.V. Nanosecond pulse radiolysis of aqueous solutions containing proton and hydroxyl radical scavengers. *Proc.* A **328**, 9–21 (1972).

Buxton, R.E. *See* Scarlett, Buxton & Faulkner.

Byatt-Smith, J.G.B. & Longuet-Higgins, M.S. On the speed and profile of steep solitary waves. *Proc.* A **350**, 175–189 (1976).

Bystrom, J.W. The application of satellites to international interactive service support communication (Discussion). *Proc.* A **345**, 493–510 (1975).

Caddy, K.W.T. & Biscoe, T.J. Structural and quantitative studies on the normal C3H and Lurcher mutant mouse. *Trans.* B **287**, 167–201 (1979).

Cadigan, F.C. *See* Peters (W.), Garnham, Killick-Kendrick, Rajapaksa and others.

Cadogan, P.H. & Turner, G. ^{40}Ar–^{39}Ar dating of Luna 16 and Luna 20 samples. *Trans.* A **284**, 167–177 (1977).

Caggiano, R., Eley, D.D. & Hey, M.J. The relative wetting of sand by crude oil and water. *Proc.* A **340**, 173–185 (1974).

Cahn, R.W. Deformation and recrystallization textures in metals and quartz (Discussion). *Trans.* A **288**, 159–176 (1978). *See also* Bhatia & Cahn.

Cain, A.J. Variation in the spire index of some coiled gastropod shells, and its evolutionary significance. *Trans.* B **277**, 377–428 (1977).
Introduction to general discussion. *Proc.* B **205**, 599–604 (1979).

Caine, P. *See* Walkden & Caine.

Cairns, John. Leeuwenhoek Lecture, 1978. Bacteria as proper subjects for cancer research.
Proc. B **208**, 121–133 (1980).

Cairns-Smith, A.G. A case for an alien ancestry (Discussion). *Proc.* B **189**, 249–274 (1975).

Calderwood, J.H. & Coffey, W.T. On the theory of the Debye and the far infrared absorption of polar fluids. *Proc.* A **356**, 269–286 (1977).

Calderwood, J.H., Coffey, W.T., Morita, A. & Walker, S. A model for the frequency dependence of the polarizability of a polar molecule. *Proc.* A **352**, 275–288 (1976).

Calderwood, J.H. & Scaife, B.K.P. On the motion of space charge in a dielectric medium.
Trans. A **269**, 217–232 (1971).

Caldwell, D.R. & Eide, S.A. Experiments on the resonance of long-period waves near islands.
Proc. A **348**, 359–378 (1976).

Caldwell, D.R. & Longuet-Higgins, M.S. The experimental generation of double Kelvin waves.
Proc. A **326**, 39–52 (1971).

Caldwell, J.B. Stuctures and materials: progress and prospects (Discussion).
Trans. A **273**, 61–75 (1972).

Caldwell, J.R. *See* Polach, McLean, Caldwell & Thom.

Caligari, P.D.S. & Mather, K. Genotype–environment interaction. III. Interactions in *Drosophila melanogaster. Proc.* B **191**, 387–411 (1975).

Caligari, P.D.S. & Mather, Sir Kenneth. Dominance, allele frequency and selection in a population of *Drosophila melanogaster. Proc.* B **208**, 163–187 (1980).

Callaghan, T.V., Smith, R.I.L. & Walton, D.W.H. The I.B.P. Bipolar Botanical Project (Discussion).
Trans. B **274**, 315–319 (1976).

Callahan, R. *See* Todaro, Callahan, Rapp & De Larco.

Callan, H.G. Review Lecture. Replication of DNA in the chromosomes of eukaryotes.
Proc. B **181**, 19–41 (1972).

Introductory remarks (Discussion). *Trans.* B **283**, 381–382 (1978).

See also Sommerville, Malcolm & Callan.

Callan, H.G. & Perry, P.E. Recombination in male and female meiocytes contrasted (Discussion).
Trans. B **277**, 227–233 (1977).

Callanan, J.E., Weir, R.D. & Staveley, L.A.K. The thermodynamics of mixed crystals of ammonium chloride and ammonium bromide.

I. The heat capacity from 8 K to 476 K of an approximately equimolar mixture.
Proc. A **372**, 489–496 (1980).

II. An analysis of the heat capacity of ammonium chloride, ammonium bromide, and an approximately equimolar solid solution of these salts. *Proc.* A **372**, 497–516 (1980).

Callomon, J.H. & Creutzberg, F. The electronic emission spectrum of ionized nitrous oxide, N_2O^+: $A\ ^2\Sigma^+ - \tilde{X}\ ^2\Pi$. *Trans.* A **277**, 157–189 (1974).

Calnan, J.S. Prostheses for joints including flexural elements (Discussion).
Proc. B **192**, 207–215 (1976).

Calverley, T.E. D.c. transmission systems (Discussion). *Trans.* A **275**, 225–232 (1973).

Calvert, J.M. *See* Skeldon, Calvert & Lees.

Calvert, S.E. Mineralogy of silica phases in deep-sea cherts and porcelanites (Discussion).
Trans. A **286**, 239–252 (1977).

Geochemistry of oceanic ferromanganese deposits (Discussion). *Trans.* A **290**, 43–73 (1978).

Camacho, A. *See* Viñuela, Camacho, Jiménez, Carrascosa, Ramírez & Salas.

Cameron, A. *See* Duckworth (R.F.), Paul & Cameron; Grew & Cameron; Paul (G.R.) & Cameron; Ranger, Ettles & Cameron; Robinson (C.L.) & Cameron; Smith (A.J.) & Cameron; Spikes & Cameron.

Campbell, C.M., Coxeter, H.S.M. & Robertson, E.F. Some families of finite groups having two generators and two relations. *Proc.* A **357**, 423–438 (1977).

Campbell, F.W The physics of visual perception (Discussion). *Trans.* B **290**, 5–9 (1980).

Campbell, I.D. *See* Brown (F.F.) & Campbell.

Campbell, I.D., Dobson, C.M. & Williams, R.J.P. Studies of exchangeable hydrogens in lysozyme by means of Fourier transform proton magnetic resonance. *Proc.* B **189**, 485–502 (1975).

Proton magnetic resonance studies of the tyrosine residues of hen lysozyme-assignment and detection of conformational mobility. *Proc.* B **189**, 503–509 (1975).

Assignment of the ^{1}H n.m.r. spectra of proteins (Discussion). *Proc.* A **345**, 23–40 (1975).

Nuclear magnetic resonance studies on the structure of lysozyme in solution (Discussion). *Proc.* A **345**, 41–59 (1975).

Campbell, J. Allan. *See* Hill (P.), Campbell & Petrie.

Campbell, N. & Kemball, C. James Pickering Kendall. *Biogr. Mem.* **26**, 255–273 (1980).

Campo, M.S. *See* Bishop (J.O.), Beckmann, Campo, Hastie and others.

Campos, L.M.B.C. On the emission of sound by an ionized inhomogeneity. *Proc.* A **359**, 65–91 (1978). Erratum. *Proc.* A **360**, 611 (1978).

Candelas, P. & Deutsch, D. On the vacuum stress induced by uniform acceleration or supporting the ether. *Proc.* A **354**, 79–99 (1977).

Fermion fields in accelerated states. *Proc.* A **362**, 251–262 (1978).

Cann, J.R. Major element variations in ocean-floor basalts (Discussion). *Trans.* A **268**, 495–505 (1971).

Petrology of basement rocks from Palmer Ridge, NE Atlantic (Discussion). *Trans.* A **268**, 605–617 (1971).

See also Tarney, Wood, Saunders, Cann & Varet.

Cann, J.R. & Simkin, T. A bibliography of ocean-floor rocks (Discussion). *Trans.* A **268**, 737–743 (1971).

Cannell, P.A. Edge scattering of aerodynamic sound by a lightly loaded elastic half-plane. *Proc.* A **347**, 213–238 (1975).

Acoustic edge scattering by a heavily loaded elastic half-plane. *Proc.* A **350**, 71–89 (1976).

Canning, Elizabeth U. *See* Sinden, Canning, Bray & Smalley; Sinden, Canning & Spain.

Cannon, Helen L. & Petrie, W.L. A review of recent activity in the United States (Discussion). *Trans.* B **288**, 137–149 (1979).

Caprioli, R.M. *See* Beynon (J.H.), Cooks & Caprioli.

Carabine, M.D., Cullis, C.F. & Groome, I.J. The influence of bromine compounds on the combustion of polyolefins. II. Effects on the oxidative degradation. *Proc.* A **324**, 217–229 (1971).

Card, W.I. The computing approach to clinical diagnosis (Discussion). *Proc.* B **184**, 421–432 (1973).

Cardella, C.J. *See* Davies (P.), Allison & Cardella.

Cardwell, D.S.L. Science and World War I (Discussion). *Proc.* A **342**, 447–456 (1975).

Carey, R., Coleman, J.E. & Viney, I.V.F. Magnetization reversal in high anisotropy multi-domain particles. *Proc.* A **328**, 143–151 (1972).

Carey, W.C. *See* McDonnell, Ashworth, Flavill, Carey, Bateman & Jennison.

Carley, A.F. & Roberts, M.W. An X-ray photoelectron spectroscopic study of the interaction of oxygen and nitric oxide with aluminium. *Proc.* A **363**, 403–424 (1978).

Carluccio, F. *See* de Nettancourt, Devreux, Carluccio and others.

Carmichael, I.S.E., Nicholls, J., Spera, F.J., Wood, B.J. & Nelson, S.A. High-temperature properties of silicate liquids: applications to the equilibration and ascent of basic magma (Discussion). *Trans.* A **286**, 373–431 (1977).

Carnochan, D.J., Dworetsky, M.M., Todd, J.J., Willis, A.J. & Wilson, R. A search for ultraviolet objects (Discussion). *Trans.* A **279**, 479–485 (1975).

Caro, C.G., Fitz-Gerald, J.M. & Schroter, R.C. Atheroma and arterial wall shear. Observation, correlation and proposal of a shear dependent mass transfer mechanism for atherogenesis. (With an Appendix by J.M. Fitz-Gerald). *Proc.* B **177**, 109–159 (1971).

Carpenter, G.F., Coe, M.J., Engel, A.R. & Quenby, J.J. Ariel 5 hard X-ray measurements of galactic and extragalactic X-ray sources (Discussion). *Proc.* A **350**, 521–545 (1976).

Carpenter, R.H.S. Cerebellectomy and the transfer function of the vestibulo-ocular reflex in the decerebrate cat. *Proc.* B **181**, 353–374 (1972).

Carrascosa, J.L. *See* Viñuela, Camacho, Jiménez, Carrascosa, Ramírez & Salas.

Carrey, E.A., Mitchinson, C., Pain, R.H. & Virden, R. Reversible deactivation of β-lactamase by penicillin analogues [synopsis] (Discussion). *Trans.* B **289**, 372–373 (1980).

Carrington, A. Review Lecture. Spectroscopy of molecular ion beams. *Proc.* A **367**, 433–449 (1979).

See also Boland (B.J.), Brown, Carrington & Nelson; Bolman, Brown, Carrington, Kopp & Ramsay; Bolman, Brown, Carrington & Lycett; Bowater, Brown & Carrington.

Carrington, A., Hills, G.J. & Webb, K.R. Neil Kensington Adam. *Biogr. Mem.* **20**, 1–26 (1974).

Carrington, T.R. The development of commercial processes for the production of 6-aminopenicillanic acid (6-APA) (Discussion). *Proc.* B **179**, 321–333 (1971).

Carroll, P.K. & Mitchell, P.I. The absorption spectrum of diatomic phosphorus between 1370 and 600 Å. *Proc.* A **342**, 93–111 (1975).

Carruthers, G.P. *See* Miller (D.E.), Brownscombe, Carruthers, Pick & Stewart.

Carson, P.P.D. *See* Bates (B.), Carson, Dufton, McKeith and others; Boksenberg, Kirkham, Michelson, Pettini and others.

Carson, R.A.J. Low-temperature dielectric relaxation in polyethylene. *Proc.* A **332**, 255–268 (1973).

Carter, B. Theory of black holes with accretion disks [abstract] (Discussion).
    Proc. A **368**, 23–25 (1979).
Rheometric structure theory, convective differentiation and continuum electrodynamics.
    Proc. A **372**, 169–200 (1980).

Carter, B. & Quintana, H. Foundations of general relativistic high-pressure elasticity theory.
    Proc. A **331**, 57–83 (1972).

Carter, E.S. The effect of the drought on British agriculture (Discussion).
    Proc. A **363**, 43–54 (1978).

Carter, G.A. *See* Parkes (E.W.) & Carter.

Carter, J.G. *See* Goreau (T.F.), Goreau, Goreau & Carter.

Carter, J.G. & Tevesz, M.J.S. The shell structure of *Ptychodesma* (Cyrtodontidae; Bivalvia) and its bearing on the evolution of the Pteriomorphia (Discussion). *Trans.* B **284**, 367–374 (1978).

Carter, M.R., Druce, O.J. & Wake, G.C. Phase-plane analysis of criticality for thermal explosions with reactant consumption. *Proc.* A **367**, 411–431 (1979).

Carter, R. Theory of black holes with accretion disks (Discussion). *Proc.* A **368**, 23–25 (1979).

Carter, S.R. *See* O'Nions, Evensen, Hamilton & Carter.

Carter, W.E., Berg, E. & Laurila, S. The University of Hawaii lunar ranging experiment geodetic–geophysics support programme (Discussion). *Trans.* A **284**, 451–456 (1977).

Cartwright, Dame Mary L. & Hayman, W.K. Edward Foyle Collingwood.
    Biogr. Mem. **17**, 139–192 (1971).

Cartwright, D.E. Tides and waves in the vicinity of Saint Helena. (With an appendix by J.S. Driver).
    Trans. A **270**, 603–646 (1971).
Oceanographic applications of ranging to artificial satellites (Discussion).
    Trans. A **284**, 537–546 (1977).

Cartwright, D.E., Edden, Anne C., Spencer, R. & Vassie, J.M. The tides of the northeast Atlantic Ocean. *Trans.* A **298**, 87–139 (1980).

Cartwright, D.E. & Ursell, F. Joseph Proudman. *Biogr. Mem.* **22**, 319–333 (1976).

Cartwright, D.E. & Young, Catherine M. Seiches and tidal ringing in the sea near Sheland.
    Proc. A **338**, 111–128 (1974).

Cartwright, S.J. & Coulson, A.F.W. Active site of staphylococcal β-lactamase [synopsis] (Discussion).
    Trans. B **289**, 370–372 (1980).

Case, S.T. *See* Daneholt, Case, Lamb, Nelson & Wieslander.

Casjens, S. *See* King (J.), Botstein, Casjens, Earnshaw, Harrison & Lenk.

Cass, D.T. & Mark, R.F. Re-innervation of axolotl limbs. I. Motor nerves.
    Proc. B **190**, 45–58 (1975).

Cassell, A.C. An alternative method for finite element analysis: a combinatorial approach to the flexibility method for structural continua and to analogous methods in field analysis.
    Proc. A **352**, 73–89 (1976).

Cassell, A.C., Henderson, J.C. de C. & Ramachandran, K. Cycle bases of minimal measure for the structural analysis of skeletal structures by the flexibility method. *Proc.* A **350**, 61–70 (1976).

Cassels, J.W.S. Louis Joel Mordell. *Biogr. Mem.* **19**, 493–520 (1973).

Cassels, J.W.S. & Fröhlich, A. Hans Arnold Heilbronn. *Biogr. Mem.* **22**, 119–135 (1976).

Cassen, P. *See* Schubert, Young & Cassen.

Cassingena, R. & Tournier, P. SV40 specific 'repressor' in infected and transformed cells (Discussion).
Proc. B 177, 77–85 (1971).

Castagné, J.L. *See* Lemblé, Pineau, Castagné, Dumoulin & Guttmann.

Casteels, R. *See* van Breemen, Farinas, Casteels, Gerba, Wuytack & Deth.

Casteels, R., Droogmans, G. & Hendrickx, H. Active ion transport and resting potential in smooth muscle cells (Discussion). *Trans.* B 265, 47–56 (1973).

Castel, J.M. *See* Abdallahi, Skaf, Castel & Ndiaye.

Castle, J.E. & Epler, D. Chemical shifts in photo-excited Auger spectra.
Proc. A 339, 49–72 (1974).

Castro, J.E. Orchidectomy and the immune response.
 I. Effect of orchidectomy on lymphoid tissues of mice. *Proc.* B 185, 425–436 (1974).
 II. Response of orchidectomized mice to antigens. *Proc.* B 185, 437–451 (1974).
 III. The effect of orchidectomy on tumour induction and transplantation in mice.
Proc. B 186, 387–398 (1974).

Cate, Jr, T.R. *See* Murphy (B.R.), Markoff, Chanock, Spring and others.

Catlow, C.R.A. Point defect and electronic properties of uranium dioxide.
Proc. A 353, 533–561 (1977).
 Fission gas diffusion in uranium dioxide. *Proc.* A 364, 473–497 (1978).

Catt, J.A. *See* Mitchell (G.F.), Catt, Weir, McMillan, Margerel & Whatley; West (R.G.), Dickson, Catt, Weir & Sparks.

Cattabeni, F. *See* Fiecchi, Kienle, Scala, Galli and others.

Cattanach, B.M. *See* Grüneberg, Cattanach, McLaren, Wolfe & Bowman.

Cattermole, K.W. Long term developments in switching (Discussion). *Trans.* A 289, 65–77 (1978).
 Concluding remarks to a Discussion. *Trans.* A 289, 227–228 (1978).

Catura, R.C. *See* Acton & Catura.

Caudrey, P.J., Dodd, R.K. & Gibbon, J.D. A new hierarchy of Korteweg–de Vries equations.
Proc. A 351, 407–422 (1976).

Caughey, T.K. *See* Irvine & Caughey.

Caulfield, M.P. *See* Lampen, Nielsen, Izui & Caulfield.

Caveney, S. Cuticle reflectivity and optical activity in scarab beetles: the rôle of uric acid.
Proc. B 178, 205–225 (1971).

Cavenor, M.C. The behaviour of electric probes in gaseous detonation. *Proc.* A 322, 469–481 (1971).

Cawse, P.A. *See* Peirson & Cawse.

Cawthron, E.R. The interaction of atomic particles with solid surfaces at intermediate energies. IV. The energy and angular distributions of particles scattered with electric charge from polycrystalline platinum at bombarding energies of 500–5000 eV. *Proc.* A 341, 213–231 (1974).

Cazenave, Anny. *See* Lambeck & Cazenave.

Cazenave, Anny, Daillet, Sylviane & Lambeck, Kurt. Tidal studies from the perturbations in satellite orbits (Discussion). *Trans.* A 284, 595–606 (1977).

Cebeci, T. Calculation of unsteady two-dimensional laminar and turbulent boundary layers with fluctuations in external velocity. *Proc.* A 355, 225–238 (1977).

Cehelnik, E.D., Cundall, R.B., Timmons, C.J. & Bowley, R.M. Spectroscopic studies of *trans*-1,6-diphenyl-1,3,5-hexatriene in ordered liquid crystal solutions. *Proc.* A 335, 387–405 (1973).

Cekirge, H.M. & Varley, E. Large amplitude waves in bounded media. I. Reflexion and transmission of large amplitude shockless pulses at an interface. *Trans.* A 273, 261–313 (1973).

Cena, K. & Monteith, J.L. Transfer processes in animal coats.
 I. Radiative transfer. *Proc.* B 188, 377–393 (1975).
 II. Conduction and convection. *Proc.* B 188, 395–411 (1975).
 III. Water vapour diffusion. *Proc.* B 188, 413–423 (1975).

Chadwick, P. Thermo-mechanics of rubberlike materials. *Trans.* A 276, 371–403 (1974).

Chadwick, P. & Jarvis, D.A. Surface waves in a pre-stressed elastic body.
Proc. A 366, 517–536 (1979).

Chailakhyan, M.Kh., Aksenova, N.P., Konstantinova, T.N. & Bavrina, T.V. The callus model of plant flowering. *Proc.* B 190, 333–340 (1975).

Chain, Sir Ernst. Thirty years of penicillin therapy (Discussion). *Proc.* B **179**, 293–319 (1971).
Concluding remarks to a Discussion. *Proc.* B **191**, 195–198 (1975).

Chalker, B.E. & Taylor, D.L. Light-enhanced calcification, and the role of oxidative phosphorylation in calcification of the coral *Acropora cervicornis. Proc.* B **190**, 323–331 (1975).
Rhythmic variations in calcification and photosynthesis associated with the coral *Acropora cervicornis* (Lamarck). *Proc.* B **201**, 179–189 (1978).

Challis, E.J. The approach of industry to the assessment of environmental hazards (Discussion). *Proc.* B **185**, 183–197 (1974).

Challis, L.J., de Goër, A.M., Guckelsberger, K. & Slack, G.A. An investigation of the ground state of Cr^{2+} in MgO based on thermal conductivity measurements. *Proc.* A **330**, 29–58 (1972).

Chalmers, I.D., Duffy, H. & Tedford, D.J. The mechanism of spark breakdown in nitrogen, oxygen and sulphur hexafluoride. *Proc.* A **329**, 171–191 (1972).

Chaloner, C.P., Drummond, J.R., Houghton, J.T., Jarnot, R.F. & Roscoe, H.K. Infra-red measurements of stratospheric composition. I. The balloon instrument and water vapour measurements. *Proc.* A **364**, 145–159 (1978).

Chamberlain, A.C., Clough, W.S., Heard, M.J., Newton, D., Stott, A.N.B. & Wells, A.C. Uptake of lead by inhalation of motor exhaust. *Proc.* B **192**, 77–110 (1975).

Chamberlain, A.C., Heard, M.J., Little, P. & Wiffen, R.D. The dispersion of lead from motor exhausts (Discussion). *Trans.* A **290**, 577–589 (1979).

Chambers, C. *See* Turner (Judith), Hewetson, Hibbert, Lowry & Chambers.

Chambon, P. *See* Oudet, Germond, Bellard, Spadafora & Chambon.

Chan, D. & Richmond, P. Van der Waals forces for mica and quartz: calculations from complete dielectric data. *Proc.* A **353**, 163–176 (1977).

Chan, K.W., Baird, M.H.I. & Round, G.F. Behaviour of beds of dense particles in a horizontally oscillating liquid. *Proc.* A **330**, 537–559 (1972).

Chan, S.J. *See* Steiner (D.F.), Patzelt, Chan, Quinn, Tager, Nielsen and others.

Chance, B., D'Ambrosia, C., Leigh, Jr, J.S. & McDonald, G. *In vivo* and freeze-trapped assays of the energy state of brain and skeletal tissues [abstract] (Discussion). *Trans.* B **289**, 457 (1980).

Chandler, R.J. The history and stability of two Lias clay slopes in the upper Gwash valley, Rutland (Discussion). *Trans.* A **283**, 463–491 (1976).

Chandler, R.J., Kellaway, G.A., Skempton, A.W. & Wyatt, R.J. Valley slope sections in Jurassic strata near Bath, Somerset (Discussion). *Trans.* A **283**, 527–556 (1976).

Chandler, T.R.D. *See* Colclough, Quinn & Chandler; Quinn (T.J.), Colclough & Chandler.

Chandler, W.K., Schneider, M.F., Rakowski, R.F. & Adrian, R.H. Charge movements in skeletal muscle (Discussion). *Trans.* B **270**, 501–505 (1975).

Chandrasekhar, S. On the equations governing the perturbations of the Schwarzschild black hole.
Proc. A **343**, 289–298 (1975).
On a transformation of Teukolsky's equation and the electromagnetic perturbations of the Kerr black hole. *Proc.* A **348**, 39–55 (1976).
The solution of Maxwell's equations in Kerr geometry. *Proc.* A **349**, 1–8 (1976).
The solution of Dirac's equation in Kerr geometry. *Proc.* A **349**, 571–575 (1976).
Errata. *Proc.* A **350**, 565 (1976).
The Kerr metric and stationary axisymmetric gravitational fields. *Proc.* A **358**, 405–420 (1978).
The gravitational perturbations of the Kerr black hole.
I. The perturbations in the quantities which vanish in the stationary state.
Proc. A **358**, 421–439 (1978).
II. The perturbations in the quantities which are finite in the stationary state.
Proc. A **358**, 441–465 (1978).
III. Further amplifications. *Proc.* A **365**, 425–451 (1979).
IV. The completion of the solution. *Proc.* A **372**, 475–484 (1980).
On the equations governing the perturbations of the Reissner–Nordström black hole.
Proc. A **365**, 453–465 (1979).
On one-dimensional potential barriers having equal reflexion and transmission coefficients.
Proc. A **369**, 425–433 (1980).

Chandrasekhar, S. & Detweiler, S. The quasi-normal modes of the Schwarzschild black hole.
Proc. A **344**, 441–452 (1975).
On the equations governing the axisymmetric perturbations of the Kerr black hole.
Proc. A **345**, 145–167 (1975).
On the equations governing the gravitational perturbations of the Kerr black hole.
Proc. A **350**, 165–174 (1976).
On the reflexion and transmission of neutrino waves by a Kerr black hole.
Proc. A **352**, 325–338 (1977).
Chandrasekhar, S. & Xanthopoulos, B.C. On the metric perturbations of the Reissner–Nordström black hole. *Proc.* A **367**, 1–14 (1979).
Chang, T.S. *See* Hassard, Chang & Ludford.
Chang, T-T. The rice cultures (Discussion). *Trans.* B **275**, 143–157 (1976).
Changeux, J.P. *See* Mariani, Crepel, Mikoshiba, Changeux & Sotelo.
Channon, R.D. & Hamilton, D. Wave and tidal current sorting of shelf sediments southwest of England [abstract] (Discussion). *Trans.* A **279**, 219–220 (1975).
Chanock, R.M. *See* Murphy (B.R.), Markoff, Chanock, Spring and others.
Chao, E.C.T. Basis for interpretation regarding the ages of the Serenitatis, Imbrium and Orientale events (Discussion). *Trans.* A **285**, 115–126 (1977).
Chapman, D.S. *See* Gass, Chapman, Pollack & Thorpe.
Chapman, G.E., Danyluk, S.S. & McLauchlan, K.A. A model for collagen hydration.
Proc. B **178**, 465–476 (1971).
Chapman, G.R. *See* King (B.C.) & Chapman.
Chapman, J.A. *See* Morrison (W.B.) & Chapman.
Chapman, J.A., Grant, I.S., Taylor, G., Mahmud, Khalida, Sardar-ul-Mulk & Shahid, M.A. Endemic goitre in the Gilgit Agency, West Pakistan. (With an appendix on dermatoglyphics and taste-testing.) *Trans.* B **263**, 459–490 (1972).
Chapman, S., Gupta, J.C. & Malin, S.R.C. The sunspot cycle influence on the solar and lunar daily geomagnetic variations. *Proc.* A **324**, 1–15 (1971).
Chapman, V. *See* Bennett (M.D.), Chapman & Riley.
Chapman, W.A., Cross, M.J., Flower, D.A., Peckham, G.E. & Smith, S.D. A spectral analysis of global atmospheric temperature fields observed by the selective chopper radiometer on the Nimbus 4 satellite during the year 1970–1. *Proc.* A **338**, 57–76 (1974).
Chapman, W.A. & Peckham, G.E. Spectral analyses of wave motions in the middle atmosphere (Discussion). *Trans.* A **296**, 59–63 (1980).
Chappell, J. *See* Thom (B.G.) & Chappell.
Charap, J.M. & Tait, W. A gauge theory of the Weyl group. *Proc.* A **340**, 249–262 (1974).
Charles, J.A. Recycling effects on the composition of non-ferrous metals (Discussion).
Trans. A **295**, 57–68 (1980).
Charlesby, A. *See* Bridges, Charlesby & Folland.
Charlesby, A., Folland, R. & Steven, J.H. Analysis of crosslinked and entangled polymer networks as studied by nuclear magnetic resonance. *Proc.* A **355**, 189–207 (1977).
Charlesworth, B. *See* Charlesworth (D.) & Charlesworth.
Charlesworth, D. & Charlesworth, B. The evolutionary genetics of sexual systems in flowering plants (Discussion). *Proc.* B **205**, 513–530 (1979).
Charnas, R.L. *See* Fisher (J.), Belasco, Charnas, Khosla & Knowles.
Charnley, J. Principles and practice in hip replacement (Discussion). *Proc.* B **192**, 191–198 (1976).
Chase, D. *See* Margulis, Chase & To.
Chasen, S.H. The effect of new computer technology on the design of mechanisms (Discussion).
Proc. A **321**, 177–186 (1971).
Chasovitin, Yu.K. *See* Andreeva, Katasyev, Uvarov, Nesterov & Chasovitin.
Chater, K.F. *See* Hopwood & Chater.
Chatt, J. Prospects for new nitrogen fixation processes (Discussion). *Trans.* B **281**, 243–248 (1977).
Chatterjee, A.K. *See* Goodrich, Allen & Chatterjee.
Chatterjee, S. A rhynchosaur from the Upper Triassic Maleri Formation of India.
Trans. B **267**, 209–261 (1974).

Malerisaurus, a new eosuchian reptile from the late Triassic of India.
Trans. B **291**, 163—200 (1980).
See also Jain (S.L.), Kutty, Roy-Chowdhury & Chatterjee.

Chattopadhyay, D. & Nag, B.R. Diffusion of charge carriers in high electric fields.
Proc. A **354**, 367—378 (1977).

Chaudhri, M.M. *See* Tang (T.B.) & Chaudhri.

Chaudhri, M.M. & Field, J.E. The role of rapidly compressed gas pockets in the initiation of condensed explosives. *Proc.* A **340**, 113—128 (1974).

Chayes, F. Silica saturation in Cenozoic basalt (Discussion). *Trans.* A **271**, 285—296 (1972).

Cheeseman, C.I. & Smyth, D.H. Interaction of amino acids, peptides and peptidases in the small intestine. *Proc.* B **190**, 149—163 (1975).

Cheeseman, K.J. *See* Evans (D.C.B.), Nye & Cheeseman.

Chee-Seng, Lim. Isotropic radiation from a steadily pulsating multidimensional distribution.
Proc. A **323**, 555—580 (1971).
Evolutionary isotropic radiation. *Proc.* A **364**, 181—209 (1978).

Chee-Seng, L., Majumdar, S.R. & Westbrook, D.R. Isotropic wave dispersion.
Proc. A **349**, 205—216 (1976).

Cheetham, A.K. *See* Anderson (J.S.), Bevan, Cheetham, Von Dreele and others; Von Dreele & Cheetham.

Chellone, D.S. & Williams, D.P. A characteristic initial value problem in general relativity in the case of a perfect fluid with axial symmetry. *Proc.* A **332**, 549—560 (1973).

Chen, B.L. *See* Poljak, Amzel, Chen, Phizackerley & Saul.

Chen, H.-S. & Acrivos, A. On the effective thermal conductivity of dilute suspensions containing highly conducting slender inclusions. *Proc.* A **349**, 261—276 (1976).

Cheng, L. *See* Birch (M.C.), Cheng & Treherne.

Cheong, W.H. *See* Peters (W.), Garnham, Killick-Kendrick, Rajapaksa and others.

Cherns, D., Hirsch, Sir Peter & Saka, H. Mechanism of climb of dissociated dislocations.
Proc. A **371**, 213—234 (1980).

Chernysheva, V.I. & Murdmaa, I.O. Metamorphosed igneous rocks from the Mid-Indian rift zones [abstract] (Discussion). *Trans.* A **268**, 621 (1971).

Chesters, J.H. Review Lecture. The prevention of metal breakouts through refractories.
Proc. A **333**, 133—148 (1973).

Chew, W.-L. The phanerogamic flora of the New Hebrides and its relationships (Discussion).
Trans. B **272**, 315—328 (1975).

Chia, W. *See* Rigby, Chia, Clayton & Lovett.

Childs, T.H.C. The persistence of roughness between surfaces in static contact.
Proc. A **353**, 35—53 (1977).

Chilver, A.H. The role of structural materials and forms (Discussion). *Trans.* A **269**, 415—423 (1971).

Chio, H.T. *See* Ash, Barrer, Chio & Edge.

Chipperfield, A.R. & Whittam, R. Evidence that ATP is hydrolysed in a one-step reaction of the sodium pump. *Proc.* B **187**, 269—280 (1974).

Chisholm, J.S.R. Multivariate approximants with branch points.
I. Diagonal approximants. *Proc.* A **358**, 351—366 (1978).
II. Off-diagonal approximants. *Proc.* A **362**, 43—56 (1978).

Chisholm, J.S.R. & Graves-Morris, P.R. Generalizations of the theorem of de Montessus to two-variable approximants. *Proc.* A **342**, 341—372 (1975).

Chisholm, J.S.R. & Jones, R. Hughes. Relative scale covariance of N-variable approximants.
Proc. A **344**, 465—470 (1975).

Chisholm, J.S.R. & McEwan, J. Rational approximants defined from power series in N variables.
Proc. A **336**, 421—452 (1974).

Chisholm, J.S.R. & Roberts, D.E. Rotationally covariant approximants derived from double power series. *Proc.* A **351**, 585—591 (1976).

Choi, C.-L. *See* Richmond (M.H.), Bennett, Choi, Brown, Brunton and others.

Choppin, P.W. *See* Lamb (R.A.) & Choppin.

Chopra, M.G. & Srivastava, R.S. Reflexion and diffraction of shocks interacted by yawed wedges. *Proc.* A **330**, 319–330 (1972).

Chopra, S.K. *See* Singh (G.), Joshi, Chopra & Singh.

Choudhry, V. *See* Banerjee, Bhatnagar, Choudhry & Kanwal.

Choukroune, P. Strain patterns in the Pyrenean Chain (Discussion). *Trans.* A **283**, 271–280 (1976).

Chow, H. *See* Ihrig, Rosensteel, Chow & Trainor.

Christensen, B.G. *See* Salzmann, Ratcliffe, Bouffard & Christensen.

Christiansen, R.L. *See* Lipman, Prostka & Christiansen.

Christiansen, R.L. & Lipman, P.W. Cenozoic volcanism and plate-tectonic evolution of the Western United States. II. Late Cenozoic (Discussion). *Trans.* A **271**, 249–284 (1972).

Christie, G.H. *See* Howard (J.G.), Christie & Courtenay.

Christie, K.N. & Stoward, P.J. A quantitative study of the fixation of acid phosphatase by formalde-hyde and its relevance to histochemistry. *Proc.* B **186**, 137–164 (1974).

Christopher, M. *See* Robbins (N.), Olek, Kelly, Takach & Christopher.

Christopherson, Sir Derman. Richard Vynne Southwell. *Biogr. Mem.* **18**, 549–565 (1972).

Chryssostomidis, C. *See* Mandel & Chryssostomidis.

Chu, Shih-I & Dalgarno, A. The rotational excitation of carbon monoxide by hydrogen atom impact. *Proc.* A **342**, 191–207 (1975).

Chubb, I.W. *See* Somogyi, Chubb & Smith.

Chubb, I.W. & Smith, A.D. Isoenzymes of soluble and membrane-bound acetylcholinesterase in bovine splanchnic nerve and adrenal medulla. *Proc.* B **191**, 245–261 (1975).
Release of acetylcholinesterase into the perfusate from the ox adrenal gland. *Proc.* B **191**, 263–269 (1975).

Chubb, J.P., Billingham, J., Hancock, P., Dimbylow, C. & Newcombe, G. The effect of alloying and residual elements on the strength and hot ductility of cast cupro-nickel [abstract] (Discussion). *Trans.* A **295**, 123 (1980).

Chumackov, P.M. *See* Varshavsky, Bakayev, Bakayeva, Chumackov *et al.*

Chung, S.H. *See* Gaze, Keating & Chung.

Chung, S.H., Bliss, T.V.P. & Keating, M.J. The synaptic organization of optic afferents in the amphib-ian tectum. *Proc.* B **187**, 421–447 (1974).

Chung, S.-H. & Cooke, J. Observations on the formation of the brain and of nerve connections following embryonic manipulation of the amphibian neural tube. *Proc.* B **201**, 335–373 (1978).

Chung, S.H., Keating, M.J. & Bliss, T.V.P. Functional synaptic relations during the development of the retino-tectal projection in amphibians. *Proc.* B **187**, 449–459 (1974).

Chwang, A.T. & Wu, T.Y. A note on the helical movement of micro-organisms. *Proc.* B **178**, 327–346 (1971).

Claesson, S. *See* Fisher (M.), Rämme, Claesson & Szwarc; Rämme, Fisher, Claesson & Szwarc.

Claesson, S. & Pedersen, K.O. The Svedberg. *Biogr. Mem.* **18**, 595–627 (1972).

Clapham, A.R. William Harold Pearsall. *Biogr. Mem.* **17**, 511–540 (1971).
Introductory remarks to a Discussion. *Trans.* B **274**, 277–281 (1976).
Introductory remarks to a Discussion. *Proc.* B **194**, 3–6 (1976).
Edward James Salisbury. *Biogr. Mem.* **26**, 503–541 (1980).

Clapham, A.R. & Harley, J.L. William Owen James. *Biogr. Mem.* **25**, 337–364 (1979).

Clark, C.D. *See* Walker (J.), Vermeulen & Clark.

Clark, C.D. & Walker, J. The neutral vacancy in diamond. *Proc.* A **334**, 241–257 (1973).

Clark, C.J. & Dombrowski, N. Aerodynamic instability and disintegration of inviscid liquid sheets. *Proc.* A **329**, 467–478 (1972).

Clark, D. Energy conversion to electricity (Discussion). *Trans.* A **276**, 559–570 (1974).

Clark, Grahame. Domestication and social evolution (Discussion). *Trans.* B **275**, 5–11 (1976).

Clark, J.I. *See* Benedek (G.B.), Clark, Serrallach, Young and others.

Clark, M.G. & Leslie, F.M. A calculation of orientational relaxation in nematic liquid crystals. *Proc.* A **361**, 463–485 (1978).

Clark, R.B. *See* Beddell, Clark, Hardy, Lowe, Ubatuba, Vane & Wilkinson.

Clark, Jr, S.P. & Turekian, K.K. Thermal constraints on the distribution of long-lived radioactive

elements in the Earth (Discussion). *Trans.* A **291**, 269–275 (1979).

Clark, T.A., Courts, G.R. & Jennings, R.E. A measurement of the brightness temperature of the Sun in the range 65 to 180 cm⁻¹ (Discussion). *Trans.* A **270**, 55–58 (1971).

Clarke, B. *See* Murray (J.) & Clarke.

Clarke, B.C. The evolution of genetic diversity (Discussion). *Proc.* B **205**, 453–474 (1979).

Clarke, C.A. *See* Clarke (Sir Cyril).

Clarke, C.A. & Sheppard, P.M. Further studies on the genetics of the mimetic butterfly *Papilio memnon* L. *Trans.* B **263**, 35–70 (1971).

The genetics of the mimetic butterfly *Papilio polytes* L. *Trans.* B **263**, 431–458 (1972).

The genetics of four new forms of the mimetic butterfly *Papilio memnon* L.
Proc. B **184**, 1–14 (1973).

Clarke, D.D. The resistance of potato tissue to the hyphal growth of fungal pathogens (Discussion).
Proc. B **181**, 303–317 (1972).

Clarke, G. General flow chart of the computer program used in this analysis, and the similarity matrices produced in comparing six superfamilies. An appendix to a paper by R.D. Purchon (Discussion). *Trans.* B **284**, 436 (1978).

Clarke, J.F. & McIntosh, A.C. The influence of a flameholder on a plane flame, including its static stability. *Proc.* A **372**, 367–392 (1980).

Clarke, Patricia H. The Leeuwenhoek Lecture, 1979. Experiments in microbial evolution: new enzymes, new metabolic activities. *Proc.* B **207**, 385–404 (1980).

Microbiology and pollution: the biodegradation of natural and synthetic organic compounds (Discussion). *Trans.* B **290**, 355–367 (1980).

Clarke, P.G.H., Ramachandran, V.S. & Whitteridge, D. The development of the binocular depth cells in the secondary visual cortex of the lamb (Discussion). *Proc.* B **204**, 455–465 (1979).

Clarke, R.T. & Newson, M.D. Some detailed water balance studies of research catchments (Discussion). *Proc.* A **363**, 21–42 (1978).

Clarke, Sir Cyril. Philip Macdonald Sheppard. *Biogr. Mem.* **23**, 465–500 (1977).

See also Clarke (C.A.); Whittle, Clarke, Sheppard & Bishop.

Clarke, Sir Cyril & Ford, E.B. Intersexuality in *Lymantria dispar* (L.). A reassessment.
Proc. B **206**, 381–394 (1980).

Clarke, Sir Cyril, Mittwoch, Ursula & Traut, W. Linkage and cytogenetic studies in the swallowtail butterflies *Papilio polyxenes* Fab. and *Papilio machaon* L. and their hybrids.
Proc. B **198**, 385–399 (1977).

Clarke, Sir Cyril & Sheppard, P.M. The genetics of the mimetic butterfly *Hypolimnas bolina* (L.).
Trans. B **272**, 229–265 (1975).

Clarke, Sir Cyril, Sheppard, P.M. & Scali, V. All-female broods in the butterfly *Hypolimnas bolina* (L.). *Proc.* B **189**, 29–37 (1975).

Clarke, T.A., Mason, R. & Tescari, M. Auger spectroscopy and low-energy electron diffraction studies of the chemisorption of unsaturated molecules by the (100) surface of platinum (Discussion).
Proc. A **331**, 321–333 (1972).

Clarke, W.D. The innovation and implementation of appropriate health education (Discussion).
Proc. B **209**, 141–145 (1980).

Clarkson, E.N.K. *See* Miller (J.) & Clarkson.

Clarkson, R. *See* Richards (W.G.), Clarkson & Ganellin.

Clausing, R.E. *See* White (C.L.), Clausing & Heatherly.

Clay, D.B. & McCutcheon, D.B. Development of line pipe steels (Symposium).
Trans. A **282**, 305–318 (1976).

Clayton, Christine E. *See* Rigby, Chia, Clayton & Lovett.

Clegg, E.J., Jeffries, D.J. & Harrison, G.A. Determinants of blood pressure at high and low altitudes in Ethiopia (Discussion). *Proc.* B **194**, 63–82 (1976).

Clegg, E.J., Pawson, I.G., Ashton, E.H. & Flinn, R.M. The growth of children at different altitudes in Ethiopia. *Trans.* B **264**, 403–437 (1972).

Clegg, H.M. *See* Ford (C.E.), Evans, Burtenshaw, Clegg, Tuffrey & Barnes.

Clegg, J.B. *See* Weatherall & Clegg.

Clement, C.F. Solutions of the continuity equation. *Proc.* A **364**, 107–119 (1978).

Clement, C.F. & Wood, M.H. Equations for the growth of a distribution of small physical objects. *Proc.* A **368**, 521–546 (1979).

Moment and Fokker–Planck equations for the growth and decay of small objects. *Proc.* A **371**, 553–567 (1980).

Clements, J.B. *See* Wilkie (N.M.), Eglin, Sanders & Clements.

Clements, R.M. *See* Barrow (R.F.) & Clements.

Cliff, D.I. *See* Burrows (J.P.), Cliff, Harris, Thrush & Wilkinson.

Clifford, K. *See* Cornforth, Clifford, Mallaby & Phillips.

Cline, D. Excitation of new particles by high-energy neutrino and antineutrino beams (Discussion). *Proc.* A **355**, 555–580 (1977).

Clint, Jane M., Wakely, Jennifer & Ockleford, C.D. Differentiated regions of human placental cell surface associated with attachment of chorionic villi, phagocytosis of maternal erythrocytes and syncytiotrophoblast repair. *Proc.* B **204**, 345–353 (1979).

Clint, J.H. *See* Ash, Barrer, Clint, Dolphin & Murray.

Cloet, R.L. Coast erosion and sediment circulation on the coast of eastern England [abstract] (Discussion). *Trans.* A **272**, 171 (1972).

Clough, P.N., Curran, A.H. & Thrush, B.A. The e.p.r. spectrum of vibrationally excited hydroxyl radicals. *Proc.* A **323**, 541–554 (1971).

Clough, W.S. *See* Chamberlain, Clough, Heard, Newton, Stott & Wells.

Clunie, G.J.A. *See* Morton (H.), Hegh & Clunie.

Clutton-Brock, T.H. & Harvey, P.H. Comparison and adaptation (Discussion). *Proc.* B **205**, 547–565 (1979).

Coad, J.P. & Rivière, J.C. Origin of fine structure in the Auger spectrum of sulphur on a nickel surface (Discussion). *Proc.* A **331**, 403–415 (1972).

Coales, J.F. Review Lecture. The control of industrial processes. *Proc.* A **325**, 291–311 (1971).

Coates, D.J. & Hendry, A. The influence of nitrogen on the oxidation resistance of low alloy steels [abstract] (Discussion). *Trans.* A **295**, 336 (1980).

Cobbold, P.R. Fold shapes as functions of progressive strain (Discussion). *Trans.* A **283**, 129–138 (1976).

Cocconi, G. High-energy physics and proton–proton scattering (Discussion). *Proc.* A **335**, 409–420 (1973).

Cockayne, D.J.H. *See* Ray (I.L.F.) & Cockayne.

Coe, M. *See* Bourn & Coe.

Coe, M.J. *See* Bourn & Coe; Carpenter (G.F.), Coe, Engel & Quenby; Quenby, Coe & Engel; Swingland & Coe.

Coe, M.J., Bourn, D. & Swingland, I.R. The biomass, production and carrying capacity of giant tortoises on Aldabra (Discussion). *Trans.* B **286**, 163–176 (1979).

Coffey, J.M., Oates, G. & Whittle, M.J. The future of ultrasonic examination in engineering (Discussion). *Trans.* A **292**, 285–298 (1979).

Coffey, W.T. *See* Calderwood & Coffey; Calderwood, Coffey, Morita & Walker.

Cogan, B.H., Hutson, A.M. & Shaffer, J.C. Preliminary observations on the affinities and composition of the insect fauna of Aldabra (Discussion). *Trans.* B **260**, 315–325 (1971).

Cogdell, R.J. Carotenoids in photosynthesis (Discussion). *Trans.* B **284**, 569–579 (1978).

Cohen, H. *See* Hawthorne, Cohen & Howell.

Cohen, I. *See* Attwell, Cohen & Eisner.

Cohen, M. *See* Heroux & Cohen; Rosenthal, McEachran & Cohen.

Cohen, M. & Owen, W.S. A forward look at some steel developments based on physical metallurgy (Symposium). *Trans.* A **282**, 329–339 (1976).

Cohen, M.D., Ludmer, Z., Thomas, J.M. & Williams, J.O. The role of structural imperfections in the photodimerization of 9-cyanoanthracene. *Proc.* A **324**, 459–468 (1971).

Cohen, R.J., Jedziniak, Judith A. & Benedek, G.B. Study of the aggregation and allosteric control of bovine glutamate dehydrogenase by means of quasi-elastic light scattering spectroscopy (Discussion). *Proc.* A **345**, 73–88 (1975).

Cohen, S. Review Lecture. Immunity to malaria. *Proc.* B **203**, 323—345 (1979).

Cohen, S.M. & Shulman, R.G. ^{13}C n.m.r. studies of gluconeogenesis in rat liver suspensions and perfused mouse livers (Discussion). *Trans.* B **289**, 407—411 (1980).

Cohen, Y. *See* Biscoe (P.V.), Cohen & Wallace.

Cohen-Tannoudji, C. & Reynaud, S. Atoms in strong light-fields: photon antibunching in single atom fluorescence (Discussion). *Trans.* A **293**, 223—237 (1979).

Cokelet, E.D. Steep gravity waves in water of arbitrary uniform depth.
Trans. A **286**, 183—230 (1977).
See also Longuet-Higgins (M.S.) & Cokelet.

Colapinto, N.D. Skin grafting across the H—Y barrier and the effect of biological adjuvants.
Proc. B **189**, 107—119 (1975).

Colclough, A.R. Low frequency acoustic thermometry in the range 4.2-20 K with implications for the value of the gas constant. *Proc.* A **365**, 349—370 (1979).
See also Quinn (T.J.), Colclough & Chandler.

Colclough, A.R., Quinn, T.J. & Chandler, T.R.D. An acoustic redetermination of the gas constant.
Proc. A **368**, 125—139 (1979).

Cole, G. The East Coast and London tidal flood warning systems (Discussion).
Trans. A **272**, 173—178 (1972).

Cole, H.A. Objectives of biological pollution studies. (Discussion). *Proc.* B **177**, 277—278 (1971).
Summing up and consideration of future research needs (Discussion).
Proc. B **189**, 479—483 (1975).
Pollution of the sea and its effects (Discussion). *Proc.* B **205**, 17—30 (1979).
Summing-up: deficiences and future needs (Discussion). *Trans.* B **286**, 625—633 (1979).

Cole, M. 'β-Lactams' as β-lactamase inhibitors (Discussion). *Trans.* B **289**, 207—223 (1980).

Cole, R.J. & Pack, D.C. Some complementary bivariational principles for linear integral equations of Fredholm type. *Proc.* A **347**, 239—252 (1975).

Cole, S.J. *See* Gillett, Cole & Reeves.

Cole, S.J. & Gillett, J.D. The influence of the brain hormone on retention of blood in the mid-gut of the mosquito *Aedes aegypti* (L.).

II. Early elimination following removal of the medial neurosecretory cells of the brain.
Proc. B **202**, 307—311 (1978).

III. The involvement of the ovaries and ecdysone. *Proc.* B **205**, 411—422 (1979).

Coleman, J.E. *See* Carey (R.), Coleman & Viney.

Coleman, P.G., Griffith, T.C. & Heyland, G.R. A time of flight method of investigating the emission of low energy positrons from metal surfaces. *Proc.* A **331**, 561—569 (1973).

Coleman, Jr, P.J. & Russell, C.T. The remanent magnetic field of the Moon (Discussion).
Trans. A **285**, 489—506 (1977).

Coles, Barbara C. *See* Brunet & Coles.

Coles, E.C., Beilin, L.J., Bulpitt, C.J., Dollery, C.T., Johnson, B.F., Mearns, C., Munro-Faure, A.D. & Turner, S.C. Computer-based records in the clinic (*a*) (Discussion). *Proc.* B **184**, 387—397 (1973).

Coles, H.J. *See* Jennings (B.R.) & Coles.

Coley, G.D. & Field, J.E. The role of cavities in the initiation and growth of explosion in liquids.
Proc. A **335**, 67—86 (1973).

Colhoun, E.A. Quaternary fluviatile deposits from the Pieman Dam site, western Tasmania.
Proc. B **207**, 355—384 (1980).

Colhoun, E.A., Dickson, J.H., McCabe, A.M. & Shotton, F.W. A Middle Midlandian freshwater series at Derryvree, Maguiresbridge, County Fermanagh, Northern Ireland.
Proc. B **180**, 273—292 (1972).

Colhoun, E.A. & Goede, A. The late Quaternary deposits of Blakes Opening and the middle Huon Valley, Tasmania. *Trans.* B **286**, 371—395 (1979).

Collar, A.R. Ernest Frederick Relf. *Biogr. Mem.* **17**, 593—616 (1971).
Arthur Fage. *Biogr. Mem.* **24**, 33—53 (1978).

Collett, J.R. Agricultural use of water (Discussion). *Proc.* B **209**, 31—36 (1980).

Collett, M.S. *See* Erikson (R.L.), Collett, Erikson & Purchio.

Collett, T. *See* Lock & Collett.

Collins, A.T. & Rafique, S. Optical studies of the 2.367 eV vibronic absorption system in irradiated type I*b* diamond. *Proc.* A **367**, 81–97 (1979).

Collins, J.G. *See* White (G.K.) & Collins.

Collins, P. & Lacy, D. Studies on the structure and function of the mammalian testis. IV. Steroid metabolism *in vitro* by isolated interstitium and seminiferous tubules of rat testis after heat sterilization. *Proc.* B **186**, 37–51 (1974).

Collins, R.A., Mühl, S. & Dearnaley, G. The effect of rare-earth impurities on the oxidation resistance of chromium [abstract] (Discussion). *Trans.* A **295**, 331 (1980).

Collins, W.M. & Dennis, S.C.R. Steady flow in a curved tube of triangular cross section. *Proc.* A **352**, 189–211 (1976).

Collinson, D.W. *See* Runcorn, Collinson, O'Reilly, Stephenson, Battey, Manson & Readman; Stephenson (A.), Collinson & Runcorn.

Collinson, D.W., Stephenson, A. & Runcorn, S.K. Intensity and origin of the ancient lunar magnetic field (Discussion). *Trans.* A **285**, 241–247 (1977).

Colquhoun, D. & Hawkes, A.G. Relaxation and fluctuations of membrane currents that flow through drug-operated channels. *Proc.* B **199**, 231–262 (1977).

Coltheart, M. The persistences of vision (Discussion). *Trans.* B **290**, 57–69 (1980).

Colver, G.M. & Weinberg, F.J. Quenching magnetically rotated augmented flames and plasma jets in mixtures containing methane, oxygen and nitrogen. *Proc.* A **326**, 375–391 (1972).

Comins, N. & Schutz, B.F. On the ergoregion instability. *Proc.* A **364**, 211–226 (1978).

Comly, J.B. Solar heating and air conditioning (Discussion). *Trans.* A **295**, 415–422 (1980).

Comninou, Maria & Dundurs, J. Reflexion and refraction of elastic waves in presence of separation. *Proc.* A **356**, 509–528 (1977).

Interaction of elastic waves with a unilateral interface. *Proc.* A **368**, 141–154 (1979).

Compton, G.J. *See* Lowenstein & Compton.

Concordia, C. Future developments of large electric generators (Discussion). *Trans.* A **275**, 39–49 (1973).

Concus, P. & Finn, R. The shape of a pendent liquid drop. *Trans.* A **292**, 307–340 (1979).

Conn, D.L.T. The genetics of mimetic colour polymorphism in the large narcissus bulb fly, *Merodon equestris* Fab. (Diptera: Syrphidae). *Trans.* B **264**, 353–402 (1972).

Connerade, J.P. Magnetic effects at high quantum numbers. *Proc.* A **339**, 127–132 (1974).

Centrifugal barrier perturbation of the *n*d series in Ca II. *Proc.* A **347**, 575–579 (1976).

Potential barrier effects in the absorption spectrum of Xe I between 18 and 90 Å. *Proc.* A **347**, 581–584 (1976).

Inter-subshell correlations and two-electron detachment in the 4d spectrum of In I. *Proc.* A **352**, 561–575 (1977).

Inter-subshell correlations and simultaneous ejection of two photoelectrons in the absorption spectrum of Ga I. *Proc.* A **354**, 511–527 (1977).

On the coupling of a Rydberg series of discrete levels to a continuum of finite bandwidth. *Proc.* A **362**, 361–374 (1978).

See also Baig & Connerade; Baig, Connerade & Newsom; Mansfield (M.W.D.) & Connerade; Rose (S.J.), Grant & Connerade.

Connerade, J.P. & Baig, M.A. Single and double excitation spectra involving the 4d subshell of Ag I. *Proc.* A **365**, 253–265 (1979).

Connerade, J.P., Baig, M. Aslam, Mansfield, M.W.D. & Radtke, E. The absorption spectrum of Ag I in the vacuum ultraviolet. *Proc.* A **361**, 379–398 (1978).

Connerade, J.P., Baig, M.A., Garton, W.R.S. & Newsom, G.H. Potential barrier effects on double excitation series of Ca I and Sr I. *Proc.* A **371**, 295–307 (1980).

Connerade, J.P., Drerup, B. & Mansfield, M.W.D. Potential barrier effects beyond the 4f photoionization threshold in Pb I. *Proc.* A **348**, 235–238 (1976).

Connerade, J.P., Garton, W.R.S., Mansfield, M.W.D. & Martin, M.A.P. The Tl I absorption spectrum in the vacuum ultraviolet. *Proc.* A **350**, 47–60 (1976).

Interchannel interactions and series quenching in the 5d and 6s spectra of Pb I. *Proc.* A **357**, 499–512 (1977).

Connerade, J.P. & Mansfield, M.W.D. Structure in the Hg I absorption spectrum from 20 to 120 Å associated with the exciation of the 4f and 5s subshells. *Proc.* A **335**, 87–96 (1973).
Structure in the Zn I absorption spectrum associated with the excitation of the $3s^2$ subshell. *Proc.* A **339**, 533–537 (1974).
Observation of an 'atomic plasmon' resonance in the Ba I absorption spectrum between 10 and 200 Å. *Proc.* A **341**, 267–275 (1974).
Potential barrier minimum and delayed onset of continuous absorption beyond the 3d limit in Kr I. *Proc.* A **343**, 415–419 (1975).
Continuous and discrete structure in the Tl I absorption spectrum from 20 to 150 Å associated with excitation of the 4f subshell. *Proc.* A **344**, 435–440 (1975).
Centrifugal barrier perturbation of the $nf\,^2F$ series in Ba II. *Proc.* A **346**, 565–570 (1975).
Potential barrier effects in the 3d photoionization continuum of Cs I. *Proc.* A **348**, 239–243 (1976).
Absorption spectra due to excitation of a single 3p electron and simultaneous excitation of two electrons in Rb I. *Proc.* A **348**, 539–552 (1976).
A correction to an apparent discrepancy between theory and experiment in 3d subshell absorption spectra. *Proc.* A **352**, 557–560 (1977).
Molecular damping of centrifugal barrier effects in the 3d absorption spectrum of selenium vapour. *Proc.* A **356**, 135–147 (1977).
Connerade, J.P., Mansfield, M.W.D. & Martin, M.A.P. Observation of a 'giant resonance' in the 3p absorption spectrum of Mn I. *Proc.* A **350**, 405–417 (1976).
Connerade, J.P., Mansfield, M.W.D., Newsom, G.H., Tracy, D.H., Baig, M.A. & Thimm, K. A study of 5p excitation in atomic barium. I. The 5p absorption spectra of Ba I, Cs I and related elements. *Trans.* A **290**, 327–352 (1979).
Connerade, J.P., Mansfield, M.W.D. & Thimm, K. The CdI absorption spectrum below 30 Å. *Proc.* A **337**, 293–295 (1974).
Connerade, J.P. & Martin, M.A.P. On the outermost d-subshell absorption spectra of Ge I and Sn I. *Proc.* A **357**, 103–115 (1977).
Connor, J.W., Hastie, R.J. & Taylor, J.B. High mode number stability of an axisymmetric toroidal plasma. *Proc.* A **365**, 1–17 (1979).
Conroy, B.A. & Bishop, J.A. Maintenance of the polymorphism for melanism in the moth *Phigalia pilosaria* in rural north Wales. *Proc.* B **210**, 285–298 (1980).
Cook, Alan H. Review Lecture. The dynamical properties and internal structures of the Earth, the Moon and the planets. *Proc.* A **328**, 301–336 (1972).
Introductory remarks to a Discussion. *Trans.* A **274**, 183–184 (1973).
Theories of lunar libration (Discussion). *Trans.* A **284**, 573–585 (1977).
Concluding remarks to a Discussion. *Trans.* A **288**, 235–236 (1978).
Cook, J.E. & Horder, T.J. The multiple factors determining retinotopic order in the growth of optic fibres into the optic tectum (Discussion). *Trans.* B **278**, 261–276 (1977).
Cook, L.M. *See* Bishop (J.A.), Cook & Muggleton.
Cook, Sir James & Warner, Sir Frederick. Preface to a Discussion. *Proc.* B **185**, 125–126 (1974).
Comments and conclusions to a Discussion. *Proc.* B **185**, 221–224 (1974).
Cooke, Elspeth A. *See* Bird (B.D.), Cooke, Day & Orchard.
Cooke, G.W. Introductory remarks to a Discussion. *Trans.* B **281**, 75–76 (1977).
Waste of fertilizers (Discussion). *Trans.* B **281**, 231–241 (1977).
Cooke, H.J. Sequence organization of the human Y chromosome (Discussion). *Trans.* B **283**, 329 (1978).
Cooke, J. *See* Chung & Cooke.
Cooks, R.G. *See* Beynon (J.H.), Cooks & Caprioli; Terwilliger, Beynon & Cooks.
Cool, R.L. Proton–proton and proton–deuteron elastic scattering experiments at N.A.L. (Discussion). *Proc.* A **335**, 453–460 (1973).
Coombs, J. The potential of higher plants with the phosphopyruvic acid cycle (Discussion). *Proc.* B **179**, 221–235 (1971).
Coombs, P.H. Can communication satellites overcome crippling educational deficits in the poorest nations? (Discussion). *Proc.* A **345**, 591–600 (1975).

Coope, Elizabeth. *See* Sunderland & Coope.

Coope, G.R. Report on the Coleoptera from Wretton. (Appendix to a paper by West (R.G.), Dickson, Catt, Weir & Sparks). *Trans.* B 267, 414—418 (1974).

Fossil coleopteran assemblages as sensitive indicators of climatic changes during the Devensian (Last) cold stage (Discussion). *Trans.* B 280, 313—340 (1977).

Cooper, A. *See* Garvey, Gold, McAdam & Cooper.

Cooper, G.A. *See* Ryhming, Cooper & Berlie.

Cooper, G.A., Berlie, J. & Merminod, A. A novel concept for a rock-breaking machine. II. Excavation techniques and experiments at larger scale. *Proc.* A 373, 353—372 (1980).

Cooper, J.P. *See* Blackman (G.E.) & Cooper.

Cooper, L.H.N. Hildebrand Wolfe Harvey. *Biogr. Mem.* 18, 331—347 (1972).

Coppack, C.P. Natural gas (Discussion). *Trans.* A 276, 463—483 (1974).

Coppola, J.C. *See* Kennard, Isaacs, Motherwell, Coppola and others.

Corfield, A. *See* Phipps, Richardson, Corfield, Gallagher and others.

Corfield, K.G. Into the world of economic broadband systems (Discussion). *Trans.* A 289, 29—41 (1978).

Corkum, P.B. *See* Alcock & Corkum.

Corner, E.D.S. The fate of fossil fuel hydrocarbons in marine animals (Discussion). *Proc.* B 189, 391—413 (1975).

Corner, E.J.H. *Ficus* in the New Hebrides (Discussion). *Trans.* B 272, 343—367 (1975).

The climbing species of *Ficus*: derivation and evolution. *Trans.* B 273, 359—386 (1976).

Ficus glaberrima Bl. and the pedunculate species of *Ficus* subgen. *Urostigma* in Asia and Australasia. *Trans.* B 281, 347—371 (1978).

Ficus dammaropsis and the multibracteate species of *Ficus* sect. *Sycocarpus.* *Trans.* B 281, 373—406 (1978).

Corner, G.W. *See* Amoroso & Corner.

Cornforth, J.W. *See* Cornforth (Sir John); Todd (Lord) & Cornforth.

Cornforth, J.W., Clifford, K., Mallaby, R. & Phillips, G.T. Stereochemistry of isopentenyl pyrophosphate isomerase. *Proc.* B 182, 277—295 (1972).

Cornforth, Sir John. Review Lecture. The imitation of enzymic catalysis. *Proc.* B 203, 101—117 (1978).

See also Cornforth (J.W.).

Cornforth, Sir John & Ross, F.P. Symmetry of squalene epoxidation *in vivo.* *Proc.* B 199, 213—230 (1977).

Cornish, W.D. & Young, L. Ellipsometric investigation of the electro-optic and electrostrictive effects in anodic Ta_2O_5 films. *Proc.* A 335, 39—50 (1973).

Cosslett, V.E. High voltage electron microscopy and its application in biology (Discussion). *Trans.* B 261, 35—44 (1971).

Review Lecture. Perspectives in high voltage electron microscopy. *Proc.* A 338, 1—16 (1974).

Principles and performance of a 600 kV high resolution electron microscope. *Proc.* A 370, 1—16 (1980).

Errata. *Proc.* A 372, 581 (1980).

Costa, J.L., Dobson, C.M., Kirk, K.L., Poulsen, F.M., Valeri, C.R. & Vecchione, J.J. Nuclear magnetic resonance studies of blood platelets (Discussion). *Trans.* B 289, 413—423 (1980).

Costa, M. *See* Furness & Costa.

Costa, Silvia M. de B. & Porter, Sir George. Model systems for photosynthesis. IV. Photosensitization by chlorophyll *a* monolayers at a lipid/water interface. *Proc.* A 341, 167—176 (1974).

Costa, S.M. de B., Froines, J.R., Harris, J.M., Leblanc, R.M., Orger, B.H. & Porter, G. Model systems for photosynthesis. III. Primary photoprocesses of chloroplast pigments in monomolecular arrays on solid surfaces. *Proc.* A 326, 503—519 (1972).

Cotes, J.E. *See* Edwards (R.H.T.), Miller, Hearn & Cotes; Patrick & Cotes.

Cotes, J.E., Anderson, H.R. & Patrick, J.M. Lung function and the response to exercise in New Guineans: role of genetic and environmental factors (Discussion). *Trans.* B 268, 349—361 (1974).

Cotes, J.E., Dabbs, J.M., Hall, A.M., Lakhera, S.C. Saunders, M.J. & Malhotra, M.S. Lung function of

healthy young men in India: contributory roles of genetic and environmental factors.
 Proc. B **191**, 413—425 (1975).
Cotrufo, R. *See* Metafora, Felsani, Cotrufo, Tajana, Del Rio, De Prisco and others; Metafora, Felsani, Cotrufo, Tajana, Di Iorio, Del Rio and others.
Cotter, Rosalind I. *See* Richards (B.M.), Pardon, Lilley, Cotter *et al.*
Cotton, H.C. Materials requirements for offshore structures (Symposium).
 Trans. A **282**, 53—64 (1976).
Cottrell, G.A. *See* Berry (M.S.) & Cottrell.
Cottrell, Sir Alan. Edward Neville da Costa Andrade. *Biogr. Mem.* **18**, 1—20 (1972).
 The task for the educator (Symposium). *Trans.* A **282**, 467—471 (1976).
 Dislocations in metals: the Birmingham school, 1945—55 (Symposium).
 Proc. A **371**, 144—148 (1980).
Couch, R.B. *See* Murphy (B.R.), Markoff, Chanock, Spring and others.
Coulombe, M.-C. *See* Jackson (D.A.) & Coulombe; Jackson (D.A.), Coulombe & Bauche.
Coulson, A.F.W. *See* Cartwright (S.J.) & Coulson.
Coulson, C.A. Samuel Francis Boys. *Biogr. Mem.* **19**, 95—115 (1973).
Coulson, C.A. & Emerson, D. Crystal growth and orientational disorder in bromoform.
 Proc. A **337**, 151—165 (1974).
Coulson, C.A. & Robertson, G.N. A theory of the broadening of the infrared absorption spectra of hydrogen-bonded species.
 I. *Proc.* A **337**, 167—197 (1974).
 II. The coupling of anharmonic ν(XH) and ν(XH..Y) modes. *Proc.* A **342**, 289—315 (1975).
Count, B.M. On the dynamics of wave-power devices. *Proc.* A **363**, 559—579 (1978).
Coupland, R.E. *See* Mansfield (P.), Morris, Ordidge, Pykett, Bangert & Coupland.
Courtenay, Barbara M. *See* Howard (J.G.), Christie & Courtenay.
Courtès, G., Laget, M., Sivan, J.P., Viton, M., Vuillemin, A. & Atkins, H. Recent results in ultra-violet astronomy obtained by a wide field rocket camera and the French S 183 Skylab experiment (Discussion). *Trans.* A **279**, 401—404 (1975).
Courtier, G.M. *See* Smith (J.F.) & Courtier.
Courts, G.R. *See* Boksenberg, Kirkham, Michelson, Pettini and others; Clark (T.A.), Courts & Jennings.
Cowan, A.N. *See* Ball (E.E.) & Cowan.
Coward, M.P. Archaean deformation patterns in southern Africa (Discussion).
 Trans. A **283**, 313—331 (1976).
Coward, M.P., Graham, R.H., James, P.R. & Wakefield, J. A structural interpretation of the northern margin of the Limpopo orogenic belt, southern Africa (Discussion).
 Trans. A **273**, 487—491 (1973).
Cowey, A. & Porter, J. Brain damage and global stereopsis (Discussion).
 Proc. B **204**, 399—407 (1979).
Cowley, R.A. Percolation in antiferromagnetic insulators (Discussion). *Trans.* B **290**, 583—592 (1980).
Cowling, T.G. Sydney Chapman. *Biogr. Mem.* **17**, 53—89 (1971).
Cox, A.B. *See* Nagasawa, Cox & Lett.
Cox, D.R. & Herzberg, Agnes M. On a statistical problem of E.A. Milne.
 Proc. A **331**, 273—283 (1972).
Cox, D.R. & Isham, Valerie. A bivariate point process connected with electronic counters.
 Proc. A **356**, 149—160 (1977).
Cox, J.B., Jones, A.R. & Weinberg, F.J. Heat transfer from augmented flames and plasma jets based on magnetically rotated arcs. *Proc.* A **325**, 269—289 (1971).
Cox, K.G. Komatiites and other high-magnesia lavas: some problems (Discussion).
 Trans. A **288**, 599—609 (1978).
Cox, Nancy J. *See* Murphy (B.R.), Markoff, Chanock, Spring and others.
Cox, R.A. Photochemical oxidation of atmospheric sulphur dioxide (Discussion).
 Trans. A **290**, 543—550 (1979).
Cox, R.G. *See* Torza, Cox & Mason.
Cox, S. *See* Leslie (M.), Jenkin, Hayter, White, Cox & Warner.

Coxeter, H.S.M. *See* Campbell (C.M.), Coxeter & Robertson.

Cragg, B.G. *See* Waite & Cragg.

Craig, A.S. *See* Parry, Barnes & Craig; Parry, Craig & Barnes.

Craig, D.P. Ronald Sydney Nyholm. *Biogr. Mem.* **18**, 445–475 (1972).

Craig, D.P. & Dissado, L.A. Absorption and fluorescence by thin molecular crystals.
  I. Theory. *Proc.* A **325**, 1–21 (1971).
  II. Application to anthracene. *Proc.* A **332**, 419–437 (1973).
  Exciton—phonon coupling in molecular crystals: absorption band profiles in the first singlet system of anthracene. *Proc.* A **363**, 153–173 (1978).

Craig, D.P., Power, E.A. & Thirunamachandran, T. The interaction of optically active molecules. *Proc.* A **322**, 165–179 (1971).
  The dynamic terms in induced circular dichroism. *Proc.* A **348**, 19–38 (1976).

Craig, D.P., Radom, L. & Stiles, P.J. A pairwise additivity scheme for conformational energies of substituted ethanes. *Proc.* A **343**, 1–10 (1975).
  Intramolecular chiral discrimination between *meso* and *d, l* isomers. *Proc.* A **343**, 11–16 (1975).

Craig, D.P. & Schipper, P.E. Electrostatic terms in the interaction of chiral molecules. *Proc.* A **342**, 19–37 (1975).

Craig, I.W. *See* Munro (E.), Siegel, Craig & Sly.

Craig, R. & Offer, G. The location of C-protein in rabbit skeletal muscle. *Proc.* B **192**, 451–461 (1976).

Craig, R.D., Bateman, R.H., Green, B.N. & Millington, D.S. Mass spectrometry instrumentation for chemists and biologists (Discussion). *Trans.* A **293**, 135–155 (1979).

Craik, A.D.D. Second order resonance and subcritical instability. *Proc.* A **343**, 351–362 (1975).
  Evolution in space and time of resonant wave triads. II. A class of exact solutions. *Proc.* A **363**, 257–269 (1978).

Craik, A.D.D. & Adam, J.A. Evolution in space and time of resonant wave triads. I. The 'pump-wave approximation'. *Proc.* A **363**, 243–255 (1978).

Crampin, M. *See* McCarthy & Crampin.

Crampin, M. & McCarthy, P.J. Representations of the Bondi—Metzner—Sachs group. IV. Cantoni representations are induced. *Proc.* A **351**, 55–70 (1976).

Crampin, S. *See* Taylor (D.B.) & Crampin.

Crampin, S. & Willmore, P.L. Small earthquakes observed with local seismometer networks (Discussion). *Trans.* A **274**, 383–387 (1973).

Crane, G.G. *See* Hornabrook, Crane & Stanhope.

Crane, S.A. *See* Ashkenazi, Crane, Williams & Dean.

Crangle, J. & Goodman, G.M. The magnetization of pure iron and nickel. *Proc.* A **321**, 477–491 (1971).

Crapper, G.D., Dombrowski, N. & Jepson, W.P. Wave growth on thin sheets of non-Newtonian liquids. *Proc.* A **342**, 225–236 (1975).

Crapper, G.D., Dombrowski, N. & Pyott, G.A.D. Large amplitude Kelvin—Helmholtz waves on thin liquid sheets. *Proc.* A **342**, 209–224 (1975).

Craven, J.D. & Frank, L.A. Observations of angular distributions of low energy electron intensities over the auroral zones with Ariel 4 (Discussion). *Proc.* A **343**, 167–188 (1975).

Cravens, T.C. *See* Bottcher, Cravens & Dalgarno.

Crawford, A.C. *See* Fettiplace & Crawford.

Crawford, A.C. & McBurney, R.N. The post-synaptic action of some putative excitatory transmitter substances. *Proc.* B **192**, 481–489 (1976).

Crawford, L.V. Introduction: virus—host interactions (Discussion). *Proc.* B **210**, 319–320 (1980).
  See also Lane & Crawford.

Craxford, S.R. & Weatherley, Marie-Louise P.M. Air pollution in towns in the United Kingdom (Discussion). *Trans.* A **269**, 503–513 (1971).

Creed, D., Hales, B.J. & Porter, Sir George. Photochemistry of the plastoquinones. *Proc.* A **334**, 505–521 (1973).

Creed, E.R. Melanism in the two spot ladybird: the nature and intensity of selection. *Proc.* B **190**, 135–148 (1975).

Creed, Kate E. *See* Gillespie (J.S.), Creed & Muir.

Creek, D.M. & Nicholls, R.W. A comprehensive re-analysis of the O_2 (B $^3\Sigma_u^-$ – X $^3\Sigma_g^-$) Schumann–Runge band system. *Proc.* A **341**, 517–536 (1975).

Crepel, F. *See* Mariani, Crepel, Mikoshiba, Changeux & Sotelo.

Cresti, M. *See* de Nettancourt, Devreux, Carluccio and others.

Creutzberg, F. *See* Callomon & Creutzberg.

Crewe, A.V. High resolution scanning microscopy of biological specimens (Discussion).
Trans. B **261**, 61–70 (1971).

Creyghton, J.H.N. *See* Bovée, Creyghton, Getreuer, Korbee, Lobregt and others.

Cribb, A.B. Algae collected on Ingram–Beanley Reef (appendix to a paper by Flood & Scoffin).
(Discussion). *Trans.* A **291**, 69–71 (1978).

Crighton, D.G. Radiation properties of the semi-infinite vortex sheet. *Proc.* A **330**, 185–198 (1972).

Crighton, D.G. & Leppington, F.G. Singular perturbation methods in acoustics: diffraction by a plate of finite thickness. *Proc.* A **335**, 313–339 (1973).

Crighton, D.G. & Scott, J.F. Asymptotic solutions of model equations in nonlinear acoustics.
Trans. A **292**, 101–134 (1979).

Crisp, D.J. The British contribution to the I.B.P. programme on marine productivity (Discussion).
Trans. B **274**, 393–399 (1976).
See also Richardson (C.A.), Crisp & Runham.

Cristofolini, R. Recent trends in the study of Etna (Symposium). *Trans.* A **274**, 17–35 (1973).

Critchley, R. *See* Unwin (S.D.) & Critchley.

Critchlow, J.E. *See* Bell (R.P.) & Critchlow.

Croasdale, R. *See* Booth (B.), Croasdale & Walker.

Crocco, L. Coordinate perturbation and multiple scale in gasdynamics.
Trans. A **272**, 275–301 (1972). Corrigenda. *Ibid.*, p. 699.

Crofts, A.R. The potential of bacterial photosynthesis in recycling of human wastes (Discussion).
Proc. B **179**, 209–219 (1971).

Croker, M.N., Fidler, R.S. & Smith, R.W. The characterization of eutectic structures.
Proc. A **335**, 15–37 (1973).

Crompton, D., Hirst, W. & Howse, M.G.W. The wear of diamond. *Proc.* A **333**, 435–454 (1973).

Crompton, D.W.T. *See* Nesheim, Crompton, Arnold & Barnard; Parshad, Crompton & Nesheim.

Cronin, E.L. Pollution prevention (Discussion). *Proc.* B **177**, 439–450 (1971).

Cross, F.B. *See* Zyznar, Cross & Nicol.

Cross, G.A.M. Antigenic variation in trypanosomes (Discussion). *Proc.* B **202**, 55–72 (1978).

Cross, M.J. *See* Chapman (W.A.), Cross, Flower, Peckham & Smith.

Crosswhite, H.M. *See* Garton, Tomkins & Crosswhite.

Crothers, D.S.F. & Hughes, J.G. Proton–hydrogen close-capture collision spectroscopy.
Proc. A **359**, 345–363 (1978).
Proton–hydrogen collisions: differential and total cross sections for H(2s) and H(2p) production in the 1–7 keV range. *Trans.* A **292**, 539–561 (1979).

Crowe, Dame Sylvia. Forests in relation to landscape and amenity (Discussion).
Trans. B **271**, 199–211 (1975).

Crowther, R.A. Procedures for three-dimensional reconstruction of spherical viruses by Fourier synthesis from electron micrographs (Discussion). *Trans.* B **261**, 221–230 (1971).

Crozaz, G., Poupeau, G., Walker, R.M., Zinner, E. & Morrison, D.A. The record of solar and galactic radiations in the ancient lunar regolith and their implications for the early history of the Sun and Moon (Discussion). *Trans.* A **285**, 587–592 (1977).

Cruft, H.J. Edgar Stedman. *Biogr. Mem.* **22**, 529–553 (1976).

Cruickshank, A.R.I. & Skews, B.W. The functional significance of nectridean tabular horns (Amphibia: Lepospondyli). *Proc.* B **209**, 513–537 (1980).

Crumpton, M.J., Allan, D., Auger, Judy, Green, N.M. & Maino, V.C. Recognition at cell surfaces: phytohaemagglutinin–lymphocyte interaction (Discussion). *Trans.* B **272**, 173–180 (1975).

Crumpton, M.J., Snary, D., Walsh, F.S., Barnstable, C.J., Goodfellow, P.N., Jones, E.A. & Bodmer, W.F. Molecular structure of the gene products of the human HLA system: isolation and

characterization of HLA-A, -B, -C and Ia antigens (Discussion). *Proc.* B **202**, 150—175 (1978).

Crussard, C. Theory of dislocations: pre-war nucleation, post-war crystallization and present growth (Symposium). *Proc.* A **371**, 139—143 (1980).

Culhane, J.L. Studies of the X-ray emission from supernova remnants (Discussion).
  Proc. A **340**, 423—437 (1974).
The structure and spectra of the X-ray sources in clusters of galaxies (Discussion).
  Proc. A **366**, 403—421 (1979).
Hot gas in clusters of galaxies (Discussion). *Trans.* A **296**, 385—397 (1980).

Cull-Candy, S.G., Miledi, R., Nakajima, Y. & Uchitel, O.D. Visualization of satellite cells in living muscle fibres of the frog. *Proc.* B **209**, 563—568 (1980).

Cullen, A.L., Nagenthiram, P. & Williams, A.D. A variational approach to the theory of the open resonator. *Proc.* A **329**, 153—169 (1972).

Cullen, A.L. & Yu, P.K. The accurate measurement of permittivity by means of an open resonator.
  Proc. A **325**, 493—509 (1971).
Complex source-point theory of the electromagnetic open resonator. *Proc.* A **366**, 155—171 (1979).

Cullen, M.J. The jumping mechanism of *Xenopsylla cheopis*. II. The fine structure of the jumping muscle. *Trans.* B **271**, 491—497 (1975).

Cullis, C.F. *See* Barat, Cullis & Pollard; Bastow (A.W.) & Cullis; Carabine, Cullis & Groome.

Cullis, C.F. & Foster, C.D. Application of thermal ignition theory to determination of spontaneous ignition temperature limits of hydrocarbon—oxygen mixtures. *Proc.* A **355**, 153—165 (1977).

Cullis, C.F. & Hirschler, M.M. The formation and destruction of pentenes during the combustion of pentane. *Proc.* A **364**, 75—88 (1978).
Isotopic tracer studies of the further reactions of pentenes in the combustion of pentane.
  Proc. A **364**, 309—329 (1978).

Cullis, C.F., Hucknall, D.J. & Shepherd, J.V. Studies of the reactions of ethynyl radicals with hydrocarbons. *Proc.* A **335**, 525—545 (1973).

Cullis, C.F., Keene, D.E. & Trimm, D.L. Studies of the heterogeneous oxidation of methanol in a pulsed-flow reactor. *Proc.* A **325**, 121—131 (1971).

Cullis, C.F. & Nevell, T.G. The kinetics of the catalytic oxidation over palladium of some alkanes and cycloalkanes. *Proc.* A **349**, 523—534 (1976).

Cummins, H.Z. Brillouin scattering spectroscopy of ferroelectric and ferroelastic phase transitions (Discussion). *Trans.* A **293**, 393—405 (1979).

Cumpsty, N.A. & Marble, F.E. The interaction of entropy fluctuations with turbine blade rows; a mechanism of turbojet engine noise. *Proc.* A **357**, 323—344 (1977).

Cundall, R.B. *See* Archer (A.S.), Cundall, Evans & Palmer; Archer (A.S.), Cundall & Palmer; Cehelnik, Cundall, Timmons & Bowley.

Curle, S.N. Solution of an integral equation of Lighthill. *Proc.* A **364**, 435—441 (1978).
Calculation of the axisymmetric boundary layer on a long thin cylinder.
  Proc. A **372**, 555—564 (1980).

Curran, A.H. *See* Clough (P.N.), Curran & Thrush.

Curran, A.H., Macdonald, R.G., Stone, A.J. & Thrush, B.A. Gas-phase electron paramagnetic resonance spectrum and dipole moment of $NF(^1\Delta)$. *Proc.* A **332**, 355—363 (1973).

Curray, J.R. The IPOD programme on passive continental margins (Discussion).
  Trans. A **294**, 17—33 (1980).

Currey, H.L.F. Non-operative treatment (Discussion). *Proc.* B **192**, 157—161 (1976).

Currey, J.D. Mechanical properties of mother of pearl in tension. *Proc.* B **196**, 443—463 (1977).

Curry, A.S. Review Lecture. Forensic science. *Proc.* B **199**, 189—198 (1977).

Curry, D. A method of classification of samples into groups of similar faunal content and its application to some rocks from the floor of the English Channel (Discussion).
  Trans. A **279**, 99—107 (1975).
See also Andreieff, Bouysse, Curry, Fletcher and others; Smith (A.J.) & Curry.

Curry, D. & Smith, A.J. New discoveries concerning the geology of the central and eastern parts of the English Channel (Discussion). *Trans.* A **279**, 155—167 (1975).

Curtis, C.D. Sedimentary geochemistry: environments and processes dominated by involvement of an

aqueous phase (Discussion). *Trans.* A **286**, 353–372 (1977).

Curtis, G.E. Twistors and multipole moments. *Proc.* A **359**, 133–149 (1978).

Curtis, N.A.C., Orr, D.C. & Boulton, M.G. The action of some β-lactam antibiotics on the penicillin-binding proteins of Gram-negative bacteria [synopsis] (Discussion).
Trans. B **289**, 368–370 (1980).

Curtis, P.D., Houghton, J.T., Peskett, G.D. & Rodgers, C.D. Remote sounding of atmospheric temperature from satellites. V. The pressure modulator radiometer for Nimbus F.
Proc. A **337**, 135–150 (1974).

Curtis, P.T. *See* Bailey (J.E.), Curtis & Parvizi.

Cusack, N.J. & Born, G.V.R. Inhibition of adenosine deaminase and of platelet aggregaticn by 2-azidoadenosine, a photolysable analogue of adenosine. *Proc.* B **193**, 307–311 (1976).
Effects of photolysable 2-azido analogues of adenosine, AMP and ADP on human platelets.
Proc. B **197**, 515–520 (1977).

Cusack, N.J., Hickman, M.E. & Born, G.V.R. Effects of D- and L- enantiomers of adenosine, AMP and ADP and their 2-chloro- and 2-azido- analogues on human platelets.
Proc. B **206**, 139–144 (1979).

Cusatis, C. & Hart, M. The anomalous dispersion corrections for zirconium.
Proc. A **354**, 291–302 (1977).

Cushing, D.H. The monitoring of biological effects: the separation of natural changes from those induced by pollution (Discussion). *Trans.* B **286**, 597–609 (1979).

Cuthbert, A.W. Neurohypophyseal hormones and sodium transport (Discussion).
Trans. B **262**, 103–109 (1971).

Cuthbert, A.W. & Shum, W.K. Effects of vasopressin and aldosterone on amiloride binding in toad bladder epithelial cells. *Proc.* B **189**, 543–575 (1975).

Cutter, M.A., Key, P.Y. & Little, V.I. Non-steady-state stimulated scattering: a convenient solution in terms of convolution integrals. *Proc.* A **352**, 539–555 (1977).

Cynader, M. *See* Hoffmann (K.-P.) & Cynader; Regan, Beverley & Cynader.

Dabbs, J.M. *See* Cotes, Dabbs, Hall, Lakhera, Saunders & Malhotra.

Dahl, H.A. *See* Lømo, Westgaard & Dahl.

Dahlen, F.A. & Smith, M.L. The influence of rotation on the free oscillations of the Earth.
Trans. A **279**, 583–624 (1975).

Dahlkamp, F.J. Uranium occurrences in northern Saskatchewan, Canada, and their mode of origin (Discussion). *Trans.* A **291**, 369–384 (1979).

Dahlström, Annica. Axoplasmic transport (with particular respect to adrenergic neurons) (Discussion).
Trans. B **261**, 325–358 (1971).

Daillet, Sylviane. *See* Cazenave, Daillet & Lambeck.

Dainton, Sir Frederick, Janovský, I. & Salmon, G.A. The radiation chemistry of liquid methanol. I. The oxidizing radical. *Proc.* A **327**, 305–316 (1972).

Dainton, Sir Frederick, May, R., Morrow, T., Salmon, G.A. & Thompson, G.F. The radiation-induced formation of excited states of aromatic hydrocarbons in cyclohexane. IV. Effect of electron scavengers. *Proc.* A **328**, 497–513 (1972).

Dainton, Sir Frederick, Morrow, T., Salmon, G.A. & Thompson, G.F. The radiation-induced formation of excited states of aromatic hydrocarbons in benzene and cyclohexane.
II. Yields of excited singlet and triplet state solute molecules. *Proc.* A **328**, 457–479 (1972).
III. Effect of singlet state quenchers. *Proc.* A **328**, 481–496 (1972).

Dainton, Sir Frederick, Salmon, G.A. & Zucker, U.F. The radiolysis of n-propanol glass containing naphthalene or diphenyl. *Proc.* A **325**, 23–33 (1971).

Dale, D. *See* Venkatesan, Dale, Hodgkin, Nockolds, Moore & O'Connor.

Dale, M.R. Gear-train design (Discussion). *Proc.* A **321**, 163–167 (1971).

Dale, P. Axisymmetric gravitational fields: a nonlinear differential equation that admits a series of exact eigenfunction solutions. *Proc.* A **362**, 463–468 (1978).

Dalgaard, Esper. The spin Hamiltonian for electron spin resonance of radicals in doublet states: calculations in the energy weighted maximum overlap model. *Proc.* A **361**, 487–512 (1978).

Dalgarno, A. Interstellar molecular absorption lines (Discussion). *Trans.* A **279**, 323–329 (1975).
See also Bottcher, Cravens & Dalgarno; Bottcher & Dalgarno; Chu & Dalgarno; Drake (G.W.F.) & Dalgarno.

Dalitz, R.H. Introductory remarks about new particles (Discussion). *Trans.* A **355**, 443–445 (1977).
Charm and the ψ-meson family (Discussion). *Proc.* A **355**, 601–619 (1977).
Glossary for new particles and new quantum numbers (Discussion). *Proc.* A **355**, 629–631 (1977).

Dalziel, K. Dynamic aspects of enzyme specificity (Discussion). *Trans.* B **272**, 109–122 (1975).

Dalziel, R. The Ariel 4 satellite (Discussion). *Proc.* A **343**, 161–165 (1975).

Damadian, R. Field focusing n.m.r. (FONAR) and the formation of chemical images in man (Discussion). *Trans.* B **289**, 489–500 (1980).

D'Ambrosia, C. *See* Chance, D'Ambrosia, Leigh & McDonald.

Damms, Susan M. & Küchemann, D. On a vortex-sheet model for the mixing between two parallel streams. I. Description of the model and experimental evidence. *Proc.* A **339**, 451–461 (1974).

Dance, D.F. & Walker, Isobel C. Threshold electron energy-loss spectra for some unsaturated molecules. *Proc.* A **334**, 259–277 (1973).

Dando, M.R. *See* Maynard & Dando.

Daneholt, B., Case, S.T., Lamb, M.M., Nelson, L. & Wieslander, L. The 75*S* RNA transcription unit in Balbiani ring 2 and its relation to chromosome structure (Discussion).
Trans. B **283**, 383–389 (1978).

Daniel, P.M. *See* Bachelard, Daniel, Love & Pratt; Baños, Daniel, Moorhouse & Pratt.

Daniel, P.M., Love, E.R. & Pratt, O.E. The influence of insulin upon the metabolism of glucose by the brain. *Proc.* B **196**, 85–104 (1977).

Daniel, P.M., Pratt, O.E. & Spargo, E. The mechanism by which glucagon induces the release of amino acids from muscle and its relevance to fasting. *Proc.* B **196**, 347–365 (1977).

Daniel, P.M., Pratt, O.E. & Wilson, Penelope A. The transport of L-leucine into the brain of the rat *in vivo:* saturable and non-saturable components of influx. *Proc.* B **196**, 333–346 (1977).

Daniels, A., Korda, A., Tanswell, P., Williams, A. & Williams, R.J.P. The internal structure of the chromaffin granule. *Proc.* B **187**, 353–361 (1974).

Daniels, P.G. The effect of distant sidewalls on the transition to finite amplitude Bénard convection. *Proc.* A **358**, 173–197 (1977).
Finite amplitude two-dimensional convection in a finite rotating system.
Proc. A **363**, 195–215 (1978).

Danylewych, L.L. & Nicholls, R.W. Intensity measurements on the C_2 ($d^3\Pi_g$–$a^3\Pi_u$) Swan band system.
I. Intercept and partial band methods. *Proc.* A **339**, 197–212 (1974).
II. Interpretation of band intensity measurements from synthetic spectra.
Proc. A **339**, 213–222 (1974).
Intensity measurements and transition probabilities for bands of the CN violet ($B^2\Sigma$–$X^2\Sigma$) band system. *Proc.* A **360**, 557–573 (1978).

Danyluk, S.S. *See* Chapman (G.E.), Danyluk & McLauchlan.

D'Arcy, R.J. *See* Goodall, Hopkins, Tulunay & D'Arcy.

Darlington, C.D. Horace Newton Barber. *Biogr. Mem.* **18**, 21–33 (1972).
Meiosis in perspective (Discussion). *Trans.* B **277**, 185–189 (1977).

Darlington, Johanna P.E.C. *See* Burgis, Darlington, Dunn, Ganf, Gwahaba & McGowan; Moriarty (D.J.W.), Darlington, Dunn, Moriarty & Tevlin.

Darmstadter, J. & Schurr, S.H. The world energy outlook to the mid-1980s: the effect of an alternative supply path in the United States (Discussion). *Trans.* A **276**, 413–430 (1974).

Darwent, J.R., Kalyanasundaram, K. & Porter, Sir George. Model systems for photosynthesis. VII. Chlorophyll *a* photosensitized reduction of methyl viologen by hydroquinones.
Proc. A **373**, 179–187 (1980).

Dass, G.V. & Kabir, P.K. Limits on variances from symmetry in neutral kaon decays.
Proc. A **330**, 331–347 (1972).

Dass, G.V., Kabir, P.K. & Kenny, B.G. The tests of TCP invariance in neutral kaon decays.
Proc. A **325**, 101–119 (1971).

Davenport, A.G. The response of six building shapes to turbulent wind (Discussion).
Trans. A **269**, 385–394 (1971).
Davenport, A.T. & Honeycombe, R.W.K. Precipitation of carbides at γ–α boundaries in alloy steels.
Proc. A **322**, 191–205 (1971).
Davenport, H. & Heilbronn, H. On the density of discriminants of cubic fields. II.
Proc. A **322**, 405–420 (1971).
Davey, A. *See* Freeman (N.C.) & Davey.
Davey, A. & Stewartson, K. On three-dimensional packets of surface waves.
Proc. A **338**, 101–110 (1974).
David, Y. & Reeves, H. Weak interactions in the early Universe: is the Universe open? (Discussion).
Trans. A **296**, 415–429 (1980).
Davidson, C.F. Millimetric waveguide systems (Discussion). *Trans.* A **289**, 123–134 (1978).
Davidson, J.N. William Ogilvy Kermack. (With appendixes by F. Yates and W.M. McCrea.)
Biogr. Mem. **17**, 399–429 (1971).
Davies, A.J.S., Leuchars, E., Wallis, V. & Doenhoff, M.J. A system for lymphocytes in the mouse
(Discussion). *Proc.* B **176**, 369–384 (1971).
Davies, A.K., McKellar, J.F. & Phillips, G.O. Photochemistry of the piperidinoanthraquinones.
Proc. A **323**, 69–87 (1971).
Davies, A.W. Pollution problems arising from the 1975–76 drought (Discussion).
Proc. A **363**, 97–107 (1978).
Davies, C.T.M. *See* Samueloff, Davies & Shvartz.
Davies, D.S. The Fifth Royal Society Technology Lecture. Discontinuities in chemistry and chemical
technology. *Proc.* A **330**, 149–172 (1972).
Industrial microbiology: a view from Whitehall (Discussion). *Trans.* B **290**, 281–290 (1980).
Davies, D.S. & Stammers, Judith R. The effect of World War II on industrial science (Discussion).
Proc. A **342**, 505–518 (1975).
Davies, D.T. The neuroendocrine control of gonadotrophin release in the Japanese quail. III. The role
of the tuberal and anterior hypothalamus in the control of ovarian development and ovulation.
Proc. B **206**, 421–437 (1980).
Davies, D.T. & Follett, B.K. The neuroendocrine control of gonadotrophin release in the Japanese
quail.
I. The role of the tuberal hypothalamus. *Proc.* B **191**, 285–301 (1975).
II. The role of the anterior hypothalamus. *Proc.* B **191**, 303–315 (1975).
Davies, G. Optical properties of electron-irradiated type I*a* diamond. *Proc.* A **336**, 507–523 (1974).
See also de Sa & Davies; Thomaz & Davies.
Davies, G. & Evans, T. Graphitization of diamond at zero pressure and at a high pressure.
Proc. A **328**, 413–427 (1972).
Davies, G. & Hamer, M.F. Optical studies of the 1.945 eV vibronic band in diamond.
Proc. A **348**, 285–298 (1976).
Davies, G. & Nazaré, Maria H. Optical study of the secondary absorption edge in type I*a* diamonds.
Proc. A **365**, 75–94 (1979).
Davies, G., Nazaré, M.H. & Hamer, M.F. The H3 (2.463 eV) vibronic band in diamond: uniaxial
stress effects and the breakdown of mirror symmetry. *Proc.* A **351**, 245–265 (1976).
Davies, G. & Penchina, C.M. The effect of uniaxial stress on the GR1 doublet in diamond.
Proc. A **338**, 359–374 (1974).
Davies, H. *See* Perrin, Rose & Davies.
Davies, J.A. *See* Barrer & Davies.
Davies, J.B. *See* Le Berre, Garms, Davies, Walsh & Philippon.
Davies, J.E. *See* Bonnett (R.), Davies, Hursthouse & Sheldrick; Boyd (P.D.W.), Davies & Gerloch.
Davies, J.M. & Gamble, J.C. Experiments with large enclosed ecosystems (Discussion).
Trans. B **286**, 523–544 (1979).
Davies, P., Allison, A.C. & Cardella, C.J. The relation between connective tissue cells and intercellular
substances, including basement membranes (Discussion). *Trans.* B **271**, 363–377 (1975).
Davies, P.A. *See* Schnurmann & Davies.

Davies, P.B., Russell, D.K., Thrush, B.A. & Radford, H.E. Analysis of the laser magnetic resonance spectra of $NH_2(\tilde{X}\,^2B_1)$. *Proc.* A **353**, 299–318 (1977).

Davies, P.C.W. On the origin of black hole evaporation radiation. *Proc.* A **351**, 129–139 (1976).
 The thermodynamic theory of black holes. *Proc.* A **353**, 499–521 (1977).
 Quantum vacuum stress without regularization in two-dimensional space-time.
  Proc. A **354**, 529–532 (1977).
 See also Bunch & Davies; Fulling & Davies.

Davies, P.C.W. & Fulling, S.A. Quantum vacuum energy in two dimensional space-times.
  Proc. A **354**, 59–77 (1977).
 Radiation from moving mirrors and from black holes. *Proc.* A **356**, 237–257 (1977).

Davies, P.C.W. & Unruh, W.G. Neutrino stress tensor regularization in two-dimensional space—time.
  Proc. A **356**, 259–268 (1977).

Davies, R.C. *See* Wider de Xifra, Sandy, Davies & Neuberger.

Davies, R.D. Evidence from the 21 cm line relating to intergalactic gas (Discussion).
  Trans. A **296**, 407–414 (1980).

Davis, A.M.J. & Leppington, F.G. The scattering of electromagnetic surface waves by circular or elliptic cylinders. *Proc.* A **353**, 55–75 (1977).
 The scattering by a conducting sphere of an electromagnetic surface wave travelling along a duct.
  Proc. A **358**, 243–251 (1977).

Davis, J.F. Uranium in the U.S.A.: genesis and exploration implications (Discussion).
  Trans. A **291**, 301–306 (1979).

Davis, J. Newsom. *See* Ito (Y.), Miledi, Molenaar, Polak and others.

Davis, M.A.F. *See* Atkinson (D.), Davis & Leslie.

Davis, R.C. *See* Edelson & Davis.

Davis, R.J. The Celescope survey and the galactic distribution of interstellar absorption (Discussion).
  Trans. A **279**, 345–354 (1975).

Davy, D.W. Computer-aided press-tool design (Discussion). *Proc.* A **321**, 157–161 (1971).

Dawkins, R. & Krebs, J.R. Arms races between and within species (Discussion).
  Proc. B **205**, 489–511 (1979).

Dawson, J.B. Kimberlites and their relation to the mantle (Discussion).
  Trans. A **271**, 297–311 (1972).

Dawson, J.B., Smith, J.V. & Hervig, R.L. Heterogeneity in upper-mantle lherzolites and harzburgites (Discussion). *Trans.* A **297**, 323–331 (1980).

Dawson, M. Joan, Gadian, D.G. & Wilkie, D.R. Studies of the biochemistry of contracting and relaxing muscle by the use of ^{31}P n.m.r. in conjunction with other techniques (Discussion).
  Trans. B **289**, 445–455 (1980).

Day, J.T. Computer parts description (Discussion). *Proc.* A **321**, 169–175 (1971).

Day, J.B.W. & Rodda, J.C. The effects of the 1975–76 drought on groundwater and aquifers (Discussion). *Proc.* A **363**, 55–68 (1978).

Day, J.H. The effect of dopant elements on the structure and high temperature creep behaviour of tungsten and molybdenum wires [abstract] (Discussion). *Trans.* A **295**, 297 (1980).

Day, M.F. *See* Rees (F. Gwendolen) & Day.

Day, M.J., Dixon-Lewis, G. & Thompson, K. Flame structure and flame reaction kinetics. VI. Structure, mechanism and properties of rich hydrogen+nitrogen+oxygen flames.
  Proc. A **330**, 199–218 (1972).

Day, P. *See* Bird (B.D.), Cooke, Day & Orchard.

Day, R.L. & Spinks, A. Christopher Alwyne Jack Young. *Biogr. Mem.* **25**, 555–573 (1979).

Day, W.A. *See* Blest & Day.

Deacon, Joy. *See* Birks (H.J.B.), Deacon & Peglar.

Deacon, Sir George. Postscript to a Discussion. *Trans.* A **270**, 465 (1971).

Deam, R.T. & Edwards, S.F. The theory of rubber elasticity. *Trans.* A **280**, 317–353 (1976).

Dean, J.D.A. *See* Ashkenazi, Crane, Williams & Dean; Ashkenazi, Dodson, Sykes, Dean & Blanchard.

Dean, P.M. & Wakelin, L.P.G. The docking manoeuvre at a drug receptor: a quantum mechanical study of intercalative attack of ethidium and its carboxylated derivative on a DNA fragment.

Trans. B **287**, 571–604 (1979).
Electrostatic components of drug–receptor recognition.
I. Structural and sequence analogues of DNA polynucleotides. *Proc.* B **209**, 453–471 (1980).
II. The DNA-binding antibiotic actinomycin. *Proc.* B **209**, 473–487 (1980).
Dearnaley, G. *See* Collins (R.A.), Mühl & Dearnaley.
De Beuckeleer, M. *See* Schell, Van Montagu, De Beuckeleer, De Block and others; Van Montagu, Holsters, Zambryski, Hernalsteens, Depicker and others.
Debler, W.R. & Vest, C.M. Observations of a stratified flow by means of holographic interferometry. *Proc.* A **358**, 1–16 (1977).
De Block, M. *See* Schell, Van Montagu, De Beuckeleer, De Block and others.
De Charpal, O. *See* Sibuet, Ryan, Arthur, Barnes, Blechsmidt and others.
Deckker, B.E.L. *See* Anderson (J.H.B.) & Deckker.
Deer, W.A. & Nockolds, S.R. Cecil Edgar Tilley. *Biogr. Mem.* **20**, 381–400 (1974).
De Graciansky, P.C. *See* Sibuet, Ryan, Arthur, Barnes, Blechsmidt and others.
DeKock, R.L., Lloyd, D.R., Hillier, I.H. & Saunders, V.R. Experimental and theoretical study of the electronic structures of sulphuryl fluoride and perchloryl fluoride. *Proc.* A **328**, 401–411 (1972).
De Larco, J.E. *See* Todaro, Callahan, Rapp & De Larco.
Delargy, K.M. *See* Beaven (P.A.), Miller, Williams, Delargy & Smith.
DeLey, J. The recognition of bioenergetic processes (Discussion). *Proc.* B **189**, 235–248 (1975).
Della Corte, Laura. *See* Buffoni, Della Corte & Hope.
Dellimore, J.W. Conductimetric investigation of erythrocyte behaviour during shear flow of concentrated suspensions through a large tube. *Proc.* B **193**, 359–385 (1976).
Del Rio, A. *See* Metafora, Felsani, Cotrufo, Tajana, Del Rio, De Prisco and others; Metafora, Felsani, Cotrufo, Tajana, Di Iorio, Del Rio and others.
DeLucia, M.L., Kelly, J.A. & Mangion, M.M., Moews, P.C. & Knox, J.R. Tertiary and secondary structure analysis of penicillin-binding proteins [synopsis] (Discussion). *Trans.* B **289**, 374–376 (1980).
Demain, A.L. *See* Baldwin (J.E.), Jung, Singh, Wan, Haber and others.
Demling, L. The clinical significance of new instruments for endoscopy of the digestive tract (Discussion). *Proc.* B **195**, 227–233 (1977).
Dempster, J.P. The scientific basis of practical conservation: factors limiting the persistence of populations and communities of animals and plants (Discussion). *Proc.* B **197**, 69–76 (1977).
Denby, M. *See* Bullough (K), Denby, Gibbons, Hughes and others.
Dence, M.R. The contribution of major impact processes to lunar crustal evolution (Discussion). *Trans.* A **285**, 259–265 (1977).
Dene, H., Goodman, M. & Romero-Herrera, A.E. The amino acid sequence of elephant (*Elephas maximus*) myoglobin and the phylogeny of Proboscidea. *Proc.* B **207**, 111–127 (1980).
Denholm, I. A forward view of British shipping (Discussion). *Trans.* A **273**, 5–11 (1972).
Dennett, M.D. *See* Elston & Dennett.
Dennis, M.J. *See* Harris (A.J.), Kuffler & Dennis; Jan (Y.N.), Jan & Dennis; Kuffler, Dennis & Harris.
Dennis, M.J., Harris, A.J. & Kuffler, S.W. Synaptic transmission and its duplication by focally applied acetylcholine in parasympathetic neurons in the heart of the frog. *Proc.* B **177**, 509–539 (1971).
Dennis, R.B. *See* Butcher, Dennis & Smith.
Dennis, R.B., Pidgeon, C.R., Smith, S.D., Wherrett, B.S. & Wood, R.A. Stimulated spin-flip Raman scattering: a magnetically tunable infrared laser. I. *Proc.* A **331**, 203–236 (1972).
Dennis, R.B., Smith, S.D. & Summers, C.J. Oscillatory non-resonant interband Faraday rotation in InSb and PbTe – a new magneto-optical effect. *Proc.* A **321**, 303–320 (1971).
Dennis, S.C.R. *See* Collins (W.M.) & Dennis.
Dennis, S.C.R. & Smith, F.T. Steady flow through a channel with a symmetrical constriction in the form of a step. *Proc.* A **372**, 393–414 (1980).
Denton, E.J. Examples of the use of active transport of salts and water to give buoyancy in the sea (Discussion). *Trans.* B **262**, 277–287 (1971).
Croonian Lecture, 1973. On buoyancy and the lives of modern and fossil cephalopods.

Proc. B **185**, 273–299 (1974).

Trevor Ian Shaw. *Biogr. Mem.* **20**, 359–380 (1974).

Denton, E.J., Gilpin-Brown, J.B. & Wright, P.G. The angular distribution of the light produced by some mesopelagic fish in relation to their camouflage. *Proc.* B **182**, 145–158 (1972).

Denton, E.J. & Land, M.F. Mechanism of reflexion in silvery layers of fish and cephalopods. *Proc.* B **178**, 43–61 (1971).

Depicker, A. *See* Schell, Van Montagu, De Beuckeleer, De Block and others; Van Montagu, Holsters, Zambryski, Hernalsteens, Depicker and others.

De Prisco, P.P. *See* Metafora, Felsani, Cotrufo, Tajana, Del Rio, De Prisco and others; Metafora, Felsani, Cotrufo, Tajana, Di Iorio, Del Rio and others.

De Robertis, E.M. *See* Gurdon, Wyllie & De Robertis.

De Robertis, E.M., Partington, G.A. & Gurdon, J.B. Selective gene expression by somatic nuclei injected into amphibian oocytes (Discussion). *Trans.* B **283**, 375–377 (1978).

DeRosier, D.J. Three-dimensional image reconstruction of helical structures (Discussion). *Trans.* B **261**, 209–210 (1971).

Derrick, G.H. *See* McKenzie (D.R.), McPhedran & Derrick.

Derrick, P.J. *See* Martinussen-Runde, Melrose & Derrick.

Derrick, P.J. & Robertson, A.J.B. The determination of the kinetics of unimolecular ionic dissociations by field ionization mass spectrometry with the use of a sharp edge. *Proc.* A **324**, 491–502 (1971).

De Sanctis, Sofia Candeloro, Grdenić, D., Taylor, N. & Hodgkin, Dorothy Crowfoot. The structure of ferroverdin. II. Rhombohedral ferroverdin crystals. *Proc.* B **184**, 137–148 (1973).

De Sanctis, Sofia Candeloro & Hodgkin, Dorothy Crowfoot. The structure of ferroverdin. I. Monoclinic ferroverdin crystals. *Proc.* B **184**, 121–135 (1973).

Desselberger, U. *See* Palese, Racaniello, Desselberger, Young & Baez.

Destombes, J.-P. *See* Kellaway, Redding, Shephard-Thorn & Destombes.

Destombes, J.-P., Shephard-Thorn, E.R. & Redding, J.H. A buried valley system in the Strait of Dover (with an appendix by M.T. Morzadec-Kerfourn) (Discussion). *Trans.* A **279**, 243–256 (1975).

Deth, R. *See* van Breemen, Farinas, Casteels, Gerba, Wuytack & Deth.

Detweiler, S. On the equations governing the electromagnetic perturbations of the Kerr black hole. *Proc.* A **349**, 217–230 (1976).

On resonant oscillations of a rapidly rotating black hole. *Proc.* A **352**, 381–395 (1977).

See also Chandrasekhar & Detweiler.

Deutsch, D. *See* Candelas & Deutsch.

Deverall, B.J. Phytoalexins and disease resistance (Discussion). *Proc.* B **181**, 233–246 (1972).

De Villiers, J.M. Asymptotic solutions of the Orr—Sommerfeld equation. *Trans.* A **280**, 271–316 (1975).

Devine, C.E. *See* Somlyo, Devine, Somlyo & Rice.

Devine, C.E., Somlyo, Avril V. & Somlyo, A.P. Sarcoplasmic reticulum and mitochondria as cation accumulation sites in smooth muscle (Discussion). *Trans.* B **265**, 17–23 (1973).

Devoy, R.J.N. Flandrian sea level changes and vegetational history of the lower Thames esturary. *Trans.* B **285**, 355–407 (1979).

Devreux, M. *See* de Nettancourt, Devreux, Carluccio and others.

Dewar, M.J.S. & Metiu, H. Ground states of molecules. XXI. MINDO (2) potential surface for ethane. *Proc.* A **330**, 173–184 (1972).

Dewey, J.M. The properties of a blast wave obtained from an analysis of the particle trajectories. *Proc.* A **324**, 275–299 (1971).

Dewhurst, P. On the non-uniqueness of the machining process. *Proc.* A **360**, 587–610 (1978).

De Wilde, M. *See* Schell, Van Montagu, De Beuckeleer, De Block and others.

Dexter, K. The political economy of our arable and grassland production (Discussion). *Trans.* B **281**, 83–92 (1977).

Dhingra, A.K. Alumina fibre FP (Discussion). *Trans.* A **294**, 411–417 (1980).

Metal matrix composites reinforced with Fibre FP (α-Al$_2$O$_3$) (Discussion). *Trans.* A **294**, 559–564 (1980).

Dhir, Indra. *See* Verma, Malik & Dhir.

Diamant, J., Keller, A., Baer, E., Litt, M. & Arridge, R.G.C. Collagen; ultrastructure and its relation to mechanical properties as a function of ageing. *Proc.* B **180**, 293–315 (1972).

Diamond, A.W. The ecology of the sea birds of Aldabra (Discussion). *Trans.* B **260**, 561–571 (1971).
Dynamic ecology of Aldabran seabird communities (Discussion). *Trans.* B **286**, 231–240 (1979).
See also Penny (M.J.) & Diamond.

Diamond, Jared M. Water–solute coupling and ion selectivity in epithelia (Discussion).
Trans. B **262**, 141–151 (1971).

Dickens, F. Edward Charles Dodds. *Biogr. Mem.* **21**, 227–267 (1975).

Dicker, D. Contribution to a Discussion: critical wind speeds for suspension bridges.
Trans. A **269**, 410–413 (1971).

Dickinson, H.G. & Heslop-Harrison, J. Ribosomes, membranes and organelles during meiosis in angiosperms (Discussion). *Trans.* B **277**, 327–342 (1977).

Dickinson, H.G. & Lawson, Janice. Pollen tube growth in the stigma of *Oenothera organensis* following compatible and incompatible intraspecific pollinations (Discussion).
Proc. B **188**, 327–344 (1975).

Dickinson, H.G. & Lewis, D. Cytochemical and ultrastructural differences between intraspecific compatible and incompatible pollinations in *Raphanus*. *Proc.* B **183**, 21–38 (1973).
The formation of the tryphine coating the pollen grains of *Raphanus*, and its properties relating to the self-incompatibility system. *Proc.* B **184**, 149–165 (1973).

Dickinson, P.H.G., Bain, W.C., Thomas, L., Williams, E.R., Jenkins, D.B. & Twiddy, N.D. The determination of the atomic oxygen concentration and associated parameters in the lower ionosphere.
Proc. A **369**, 379–408 (1980).

Dickson, Camilla A. *See* West (R.G.), Dickson, Catt, Weir & Sparks.

Dickson, J.H. Moss remains from Wretton. Appendix to a paper by West (R.G.), Dickson, Catt, Weir & Sparks. *Trans.* B **267**, 418–420 (1974).
See also Colhoun, Dickson, McCabe & Shotton.

Dieminger, W. Trends in early ionospheric research in Germany (Discussion).
Trans. A **280**, 27–34 (1975).

Diesendorf, M.O. *See* Horridge, Ninham & Diesendorf.

Diesendorf, M.O. & Horridge, G.A. Two models of the partially focused clear zone compound eye.
Proc. B **183**, 141–158 (1973).

Diesendorf, M., Stange, G. & Snyder, A.W. A theoretical investigation of radiation mechanisms of insect chemoreception. *Proc.* B **185**, 33–49 (1974). Erratum. *Proc.* B **186**, 399 (1974).

Di Iorio, G. *See* Metafora, Felsani, Cotrufo, Tajana, Di Iorio, Del Rio and others.

Dillamore, I.L., Morris, P.L., Smith, C.J.E. & Hutchinson, W.B. Transition bands and recrystallization in metals. *Proc.* A **329**, 405–420 (1972).

Dilli, S. *See* Kelly (P.C.), Brooks, Dilli & Jaffré.

Dilworth, S.M. *See* Griffin (Beverly E.), Dilworth, Ito & Novak.

Dimbleby, G.W. Climate, soil and man (Discussion). *Trans.* B **275**, 197–208 (1976).

Dimbylow, C. *See* Chubb (J.P.), Billingham, Hancock, Dimbylow & Newcombe.

Dimroth, E. Significance of diagenesis for the origin of Witwatersrand-type uraniferous conglomerates. *Trans.* A **291**, 277–287 (1979).

Dingle, J.T. The secretion of enzymes into the pericellular environment (Discussion).
Trans. B **271**, 315–324 (1975).

Dingley, D.J. & Biggin, S. Grain boundary structure, intergranular fracture and the role of segregants as embrittling agents [abstract] (Discussion). *Trans.* A **295**, 165 (1980).

Dingwall, R.G. Sub-bottom infilled channels in an area of the eastern English Channel (Discussion).
Trans. A **279**, 233–241 (1975).

Dippenaar, R.J. & Honeycombe, R.W.K. The crystallography and nucleation of pearlite.
Proc. A **333**, 455–467 (1973).

DiPrima, R.C. *See* Stuart (J.T.) & DiPrima.

DiPrima, R.C. & Pridor, A. The stability of viscous flow between rotating concentric cylinders with an axial flow. *Proc.* A **366**, 555–573 (1979).

Dirac, P.A.M. A positive-energy relativistic wave equation. *Proc.* A **322**, 435—445 (1971).
A positive-energy relativistic wave equation. II. *Proc.* A **328**, 1—7 (1972).
Long range forces and broken symmetries. *Proc.* A **333**, 403—418 (1973).
Cosmological models and the Large Numbers hypothesis. *Proc.* A **338**, 439—446 (1974).
The Large Numbers hypothesis and the Einstein theory of gravitation.
 Proc. A **365**, 19—30 (1979).
Dissado, L.A. *See* Craig (D.P.) & Dissado.
Ditchburn, R.W. & Drysdale, A.E. The effect of retinal-image movements on vision.
 I. Step-movements and pulse-movements. *Proc.* B **197**, 131—144 (1977).
 II. Oscillatory movements. *Proc.* B **197**, 385—406 (1977).
Ditchburn, R.W. & Rochester, G.D. Samuel Tolansky. *Biogr. Mem.* **20**, 429—455 (1974).
Dixon, A.J., von Engel, A. & Harrison, M.F.A. A measurement of the electron impact ionization cross section of atomic hydrogen in the metastable 2S state. *Proc.* A **343**, 333—349 (1975).
Dixon, H.E., Earnshaw, J.C., Hook, J.R., Hough, J.H., Smith, G.J., Stephenson, W. & Turver, K.E. Computer simulations of cosmic-ray air showers. I. Average characteristics of proton initiated showers. *Proc.* A **339**, 133—155 (1974).
Dixon, H.E. & Turver, K.E. Computer simulations of cosmic-ray air showers. III. Fluctuations in shower development. *Proc.* A **339**, 171—195 (1974).
Dixon, H.E., Turver, K.E. & Waddington, C.J. Computer simulations of cosmic-ray air showers. II. Showers initiated by heavy primary particles. *Proc.* A **339**, 157—170 (1974).
Dixon, R.N. & Field, D. Rotational energy transfer in collisions between orbitally non-degenerate open-shell systems. *Proc.* A **366**, 225—246 (1979).
Rotational energy transfer within the $\tilde{A}\,^2A_1$ state of NH_2; absolute rate coefficients and the influence of spin exchange in collisions with H atoms. *Proc.* A **366**, 247—276 (1979).
Rotationally inelastic collisions of orbitally degenerate molecules; maser action in OH and CH.
 Proc. A **368**, 99—123 (1979).
Dixon, W.G. Dynamics of extended bodies in general relativity. III. Equations of motion.
 Trans. A **277**, 59—119 (1974).
Dixon, W.R., Garcia, A.G. & Kirpekar, S.M. Depletion and recovery of catecholamines in the rat adrenal medulla and its relationship with dopamine β-hydroxylase. *Proc.* B **194**, 403—416 (1976).
Dixon-Lewis, G. Flame structure and flame reaction kinetics. VII. Reactions of traces of heavy water, deuterium and carbon dioxide added to rich hydrogen+nitrogen+oxygen flames.
 Proc. A **330**, 219—245 (1972).
Kinetic mechanism, structure and properties of premixed flames in hydrogen—oxygen—nitrogen mixtures. *Trans.* A **292**, 45—99 (1979).
See also Day (M.J.), Dixon-Lewis & Thompson.
Dixon-Lewis, G., Goldsworthy, F.A. & Greenberg, J.B. Flame structure and flame reaction kinetics. IX. Calculation of properties of multi-radical premixed glames. *Proc.* A **346**, 261—278 (1975).
Dixon-Lewis, G., Isles, G.L. & Walmsley, R. Flame structure and flame reaction kinetics. VIII. Structure, properties and mechanism of a rich hydrogen+nitrogen+oxygen flame at low pressure.
 Proc. A **331**, 571—584 (1973).
Djerassi, C. The manufacture of steroidal contraceptives: technical versus political aspects (Discussion).
 Proc. B **195**, 175—186 (1976).
Dmitriev, L.V. *See* Udintsev, Dmitriev & Vinogradov; Vinogradov, Dmitriev & Udintsev.
Dmitriev, L.V., Vinogradov, A.P. & Udintsev, G.B. Petrology of ultrabasic rocks from rift zones in the Mid-Indian Ocean Ridge (Discussion). *Trans.* A **268**, 403—408 (1971).
Corrigendum. *Trans.* A **269**, 555, 647 (1971).
Dobb, M.G., Johnson, D.J. & Saville, B.P. Structural aspects of high modulus aromatic polyamide fibres (Discussion). *Trans.* A **294**, 483—485 (1980).
Dobbs, E.R. *See* Almond, Lea & Dobbs; Lea (M.J.), Llewellyn, Peck & Dobbs.
Dobbs, E.R., Lea, M.J. & Peck, D.R. Ultrasonic studies of the electronic structure of hexagonal metal crystals. II. Superconducting state in zinc and cadmium. *Proc.* A **334**, 379—396 (1973).
Dobson, C.M. *See* Campbell (I.D.), Dobson & Williams; Costa (J.L.), Dobson, Kirk, Poulsen, Valeri & Vecchione.

Dobson, J.C. *See* Atkins (P.W.) & Dobson.
Dobzhansky, Th., **Levene, H. & Spassky, B.** Effects of selection and migration on geotactic and phototactic behaviour of *Drosophila*. III. *Proc.* B **180**, 21—41 (1972).
Dobzhansky, Th. & Powell, J.R. Rates of dispersal of *Drosophila pseudoobscura* and its relatives. *Proc.* B **187**, 281—298 (1974).
Dockray, G.J. & Gregory, R.A. Relations between neuropeptides and gut hormones (Discussion). *Proc.* B **210**, 151—164 (1980).
Dodd, J.M. *See* Wilson (J.F.), Goos & Dodd.
Dodd, R.K. *See* Caudrey, Dodd & Gibbon.
Dodd, R.K. & Bullough, R.K. Bäcklund transformations for the sine—Gordon equations. *Proc.* A **351**, 499—523 (1976).
Polynomial conserved densities for the sine—Gordon equations. *Proc.* A **352**, 481—503 (1977).
Dodd, R.K. & Gibbon, J.D. The prolongation structure of a higher order Korteweg—de Vries equation. *Proc.* A **358**, 287—296 (1978).
The prolongation structures of a class of nonlinear evolution equations. *Proc.* A **359**, 411—433 (1978).
Dodson, A.H. *See* Ashkenazi, Dodson, Sykes, Dean & Blanchard.
Dodson, Eleanor. Appendix to a paper by Venkatesan, Dale, Hodgkin, Nockolds, Moore & O'Connor. *Proc.* A **323**, 484—487 (1971).
Doenhoff, M.J. *See* Davies (A.J.S.), Leuchars, Wallis & Doenhoff.
Dogra, S.K. *See* Basco & Dogra.
Doherty, L. OAO observations of magnesium II emission in late-type stars (Discussion). *Trans.* A **270**, 189—195 (1971).
Doig, P. & Edington, J.W. The influence of solute depleted zones on the stress-corrosion susceptibility of aged Al-7.2 mass % Mg and Al-4.4 mass % Cu alloys. *Proc.* A **339**, 37—47 (1974).
Doig, P. & Flewitt, P.E.J. The electrode potential within a growing stress corrosion crack. *Proc.* A **357**, 439—452 (1977).
X-ray microanalysis of grain boundary segregation in steels by s.t.e.m. [abstract] (Discussion). *Trans.* A **295**, 137 (1980).
Dolan, A.K. & Edwards, S.F. Theory of the stabilization of colloids by adsorbed polymer. *Proc.* A **337**, 509—516 (1974).
The effect of excluded volume on polymer dispersant action. *Proc.* A **343**, 427—442 (1975).
D'Olier, B. Some aspects of Late Pleistocene—Holocene drainage of the River Thames in the eastern part of the London Basin (Discussion). *Trans.* A **279**, 269—277 (1975).
Doll, Sir Richard. Preface to a Discussion. *Proc.* B **205**, 3—4 (1979).
The pattern of disease in the post-infection era: national trends (Discussion). *Proc.* B **205**, 47—61 (1979).
See also Vessey & Doll.
Dollery, C.T. *See* Coles (E.C.), Beilin, Bulpitt, Dollery, Johnson and others.
Dollfus, A. & Geake, J.E. Polarimetric and photometric studies of lunar samples (Discussion). *Trans.* A **285**, 397—402 (1977).
Dolling, G., Pawley, G.S. & Powell, B.M. Interatomic forces in hexamethylenetetramine. *Proc.* A **333**, 363—384 (1973).
Dolphin, R.J. *See* Ash, Barrer, Clint, Dolphin & Murray.
Domb, C. *See* Barrett (A.J.) & Domb.
de Dombal, F.T. Surgical diagnosis assisted by a computer (Discussion). *Proc.* B **184**, 433—440 (1973).
Dombrowski, N. *See* Clark (C.J.) & Dombrowski; Crapper, Dombrowski & Jepson; Crapper, Dombrowski & Pyott.
Donald, H.P. Animal breeding: contributions to the efficiency of livestock production (Discussion). *Trans.* B **267**, 131—144 (1973).
Donaldson, C.H., Drever, H.I. & Johnston, R. Supercooling on the lunar surface: a review of analogue information (Discussion). *Trans.* A **285**, 207—217 (1977).
Donaldson, I.M.L. & Whitteridge, D. The nature of the boundary between cortical visual areas II and

III in the cat. *Proc.* B **199**, 445–462 (1977).

Donath, F.A. & Wood, D.S. Experimental evaluation of the deformation path concept (Discussion). *Trans.* A **283**, 187–201 (1976).

Dondi, P.H. *See* Barnes (K.J.), Dondi & Sarkar.

Donnelly, R.J. & Roberts, P.H. Stochastic theory of the nucleation of quantized vortices in superfluid helium. *Trans.* A **271**, 41–100 (1971).

Donner, Mireille & Mehrishi, J.N. The lymphocyte surface: differences in the surface chemistry of murine spleen T lymphocytes of varying major histocompatibility haplotypes. *Proc.* B **201**, 271–284 (1978).

Donnet, J.B. *See* Ehrburger & Donnet.

Donovan, D.T. *See* Lloyd (A.J.), Savage, Stride & Donovan.

Dopheide, T.A.A. *See* Laver, Air, Webster, Gerhard, Ward & Dopheide.

Dörffling, K. The possible rôle of xanthoxin in plant growth and development (Discussion). *Trans.* B **284**, 499–507 (1978).

Dormand, J.R. & Woolfson, M.M. The evolution of planetary orbits. *Proc.* A **340**, 349–365 (1974).

Dorsett, D.A. *See* Blackshaw & Dorsett.

Double, D.D., Hellawell, A. & Perry, S.J. The hydration of Portland cement. *Proc.* A **359**, 435–451 (1978).

Dougherty, J.P. Waves in a hot plasma (Discussion). *Trans.* A **280**, 95–110 (1975).

Douglas, Jr, R.G. *See* Murphy (B.R.), Markoff, Chanock, Spring and others.

Douglas, T.D. The Quaternary deposits of western Leicestershire. *Trans.* B **288**, 259–286 (1980).

Dover, G.A. *See* Bennett (M.D.), Dover & Riley.

Dover, G.A. & Riley, R. Inferences from genetical evidence on the course of meiotic chromosome pairing in plants (Discussion). *Trans.* B **277**, 313–326 (1977).

Dowie, R.S., Kemball, C., Kempling, J.C. & Whan, D.A. The use of a combined gas chromatograph–mass spectrometer to study the deuterolysis of neopentane over iron films. *Proc.* A **327**, 491–500 (1972).

Dowker, J.S. *See* Mayes & Dowker.

Dowling, A.P., Williams, J.E. Ffowcs & Goldstein, M.E. Sound production in a moving stream. *Trans.* A **288**, 321–349 (1978).

Dowling, J.E. *See* Armett-Kibel, Meinertzhagen & Dowling; Boycott, Dowling, Fisher, Kolb & Laties; Hedden & Dowling.

Dowling, J.E. & Ehinger, B. The interplexiform cell system. I. Synapses of the dopaminergic neurons of the goldfish retina. *Proc.* B **201**, 7–26 (1978).

Downes, M.J. Some experimental studies on the 1971 lavas from Etna (Symposium). *Trans.* A **274**, 55–62 (1973).

Doyle, B.B., Hulmes, D.J.S., Miller, A., Parry, D.A.D., Piez, K.A. & Woodhead-Galloway, J. A D-periodic narrow filament in collagen. *Proc.* B **186**, 67–74 (1974).

Axially projected collagen structures. *Proc.* B **187**, 37–46 (1974).

Doyle, E.D., Horne, J.G. & Tabor, D. Frictional interactions between chip and rake face in continuous chip formation. *Proc.* A **366**, 173–183 (1979).

Doyle, M.J., Maranci, A., Orowan, E. & Stork, S.T. The fracture of glassy polymers. *Proc.* A **329**, 137–151 (1972).

Drake, G.W.F. & Dalgarno, A. The $1/Z$ expansion study of the 2s2p ^{1}P and ^{3}P autoionizing resonances of the helium isoelectronic sequence. *Proc.* A **320**, 549–560 (1971).

Drake, J.F. The Princeton equipment on board (Discussion). *Proc.* A **340**, 403–409 (1974).

Interstellar molecules (Discussion). *Proc.* A **340**, 457–469 (1974).

Drake, L.D. & Shreve, R.L. Pressure melting and regelation of ice by round wires. *Proc.* A **332**, 51–83 (1973).

Drake, Sir Eric. Oil reserves and production (Discussion). *Trans.* A **276**, 453–462 (1974).

Dran, J.C., Duraud, J.P., Klossa, J., Langevin, Y. & Maurette, M. Microprobe studies of space weathering effects in extraterrestrial dust grains (Discussion). *Trans.* A **285**, 433–439 (1977).

Dratler, J., Jr. *See* Block (B.) & Dratler.

Drazin, P.G. On the instability of an internal gravity wave. *Proc.* A **356**, 411–432 (1977).

Dreibus, G. *See* Wänke, Palme, Baddenhausen, Dreibus, Kruse & Spettel.

Dreibus, G., **Spettel, B. & Wänke, H.** Lithium and halogens in lunar samples (Discussion). *Trans.* A **285**, 49–54 (1977).

Drerup, B. *See* Connerade, Drerup & Mansfield.

Dresch, J. The evaluation and exploitation of the West African Sahel (Discussion). *Trans.* B **278**, 537–542 (1977).

Drever, H.I. *See* Donaldson (C.H.), Drever & Johnston.

Drever, R.W.P., Hough, J., Pugh, J.R., Edelstein, W.A., Ward, H., Ford, G.M. & Robertson, N.A. Gravitational wave detectors [abstract] (Discussion). *Proc.* A **368**, 11–13 (1979).

Drewry, D.J. *See* Robin, Drewry & Meldrum.

Driedonks, R.A., Krijgsman, P.C.J. & Mellema, J.E. A study of the states of aggregation of alfalfa mosaic virus protein (Discussion). *Trans.* B **276**, 131–141 (1976).

Dring, D. *See* Allibone & Dring; Waters (R.T.), Allibone, Dring & Allen.

Driver, J.H., Sinclair, R. & Jack, K.H. Modulated substitutional-interstitial solute—atom clustering in nitrided austenitic Fe—34Ni—V-alloys. *Proc.* A **367**, 99–115 (1979).

Driver, J.S. Appendix to a paper by D.E. Cartwright. *Trans.* A **270**, 641–642 (1971).

Droogmans, G. *See* Casteels, Droogmans & Hendrickx.

Druce, O.J. *See* Carter (M.R.), Druce & Wake.

Drummond, I. & Ubbelohde, A.R. Electrical properties of graphite nitrates perpendicular to the layers. *Proc.* A **371**, 309–318 (1980).

Drummond, J.R. *See* Chaloner, Drummond, Houghton, Jarnot & Roscoe.

Drummond, J.R., Houghton, J.T., Peskett, G.D., Rodgers, C.D., Wale, M.J., Whitney, J. & Williamson, E.J. The stratospheric and mesospheric sounder on Nimbus 7 (Discussion). *Trans.* A **296**, 219–241 (1980).

Drummond, J.R. & Jarnot, R.F. Infrared measurements of stratospheric composition. II. Simultaneous NO and NO_2 measurements. *Proc.* A **364**, 237–254 (1978).

Drysdale, A.E. *See* Ditchburn & Drysdale.

Duck, P.W. Oscillatory flow through constricted or dilated channels and axisymmetric pipes. *Proc.* A **363**, 335–355 (1978).

Duckett, J.G. & Bell, P.R. An ultrastructural study of the mature spermatozoid of *Equisetum*. *Trans.* B **277**, 131–158 (1977).

Duckworth, R.F., Paul, G.R. & Cameron, A. The elastic properties of films in e.h.l. contacts. *Proc.* A **372**, 155–168 (1980).

Duckworth, W.E. *See* Brook (G.B.) & Duckworth.

Dudeney, J.R., Jones, T.B., Kressman, R.I. & Spracklen, C.T. Radio wave Doppler studies of the Antarctic ionosphere (Discussion). *Trans.* B **279**, 239–246 (1977).

Dudley, T.R. *See* Brooks (R.R.), Morrison, Reeves, Dudley & Akman.

Dudney, P.J. An approach to the growth analysis of perennial plants. *Proc.* B **184**, 217–220 (1973).

Duesbery, M.S., Vitek, V. & Bowen, D.K. The effect of shear stress on the screw dislocation core structure in body-centred cubic lattices. *Proc.* A **332**, 85–111 (1973).

Duffy, H. *See* Chalmers, Duffy & Tedford.

Dufton, P.L. *See* Bates (B.), Carson, Dufton, McKeith and others; Boksenberg, Kirkham, Michelson, Pettini and others.

Duinker, P. *See* Bartel, Duinker, Heintze, Heinzelmann and others.

Dulbecco, R. The Leeuwenhoek Lecture, 1974. The control of cell growth regulation by tumour-inducing viruses: a challenging problem. *Proc.* B **189**, 1–14 (1975).

Report on a discussion meeting on the biology of chemical carcinogenesis. *Proc.* B **196**, 117–130 (1977).

Dulbecco, R. & Okada, Sharon. Differentiation and morphogenesis of mammary cells *in vitro*. *Proc.* B **208**, 399–408 (1980).

Dulberger, Rivka. Intermorph structural differences between stigmatic papillae and pollen grains in relation to incompatibility in Plumbaginaceae (Discussion). *Proc.* B **188**, 257–274 (1975).

Duley, W.W. *See* Mirza & Duley; Sayer (R.J.), Prince & Duley.

Dullien, F.A.L. *See* Ertl & Dullien.

Dumoulin, P. *See* Lemblé, Pineau, Castagné, Dumoulin & Guttmann.
Dunbar, Noreen. *See* Woodruff (Sir Michael), Dunbar & Ghaffar.
Duncan, A.R. *See* Erlank, Allsopp, Duncan & Bristow.
Duncan, G.W. *See* Perkin (G.W.), Duncan, Mahoney & Smith.
Duncan, K.P. Occupational experience (Discussion). *Proc.* B **205**, 157—164 (1979).
Dundurs, J. *See* Comninou & Dundurs.
Dungey, J.C. & Hui, W.H. Nonlinear energy transfer in a narrow gravity-wave spectrum.
    Proc. A **368**, 239—265 (1979).
Dungey, J.W. & Southwood, D.J. Ultra-low frequency waves in the magnetosphere (Discussion).
    Trans. A **280**, 131—136 (1975).
Dunham, K.C. The regional setting (Discussion). *Trans.* A **272**, 81—86 (1972).
    See also Dunham (Sir Kingsley).
Dunham, Sir Kingsley. Geological setting of the useful minerals in Britain (Discussion).
    Proc. A **339**, 273—288 (1974).
    Progress in mineralogy [abstract] (Discussion). *Trans.* A **286**, 235—237 (1977).
    William Bullerwell. *Biogr. Mem.* **24**, 1—13 (1978).
    Introductory remarks to a Discussion. *Trans.* A **297**, 137—138 (1980).
    See also Dunham (K.C.).
Dunigan, J.M. *See* Hatfield, Fisher, Dunigan, Burchfield and others.
Dunitz, J.D. Chemical reaction paths (Discussion). *Trans.* B **272**, 99—108 (1975).
Dunn, I.G. *See* Burgis, Darlington, Dunn, Ganf, Gwahaba & McGowan; Moriarty (D.J.W.), Darlington,
    Dunn, Moriarty & Tevlin.
Dunn, P.J. *See* Kolenkiewicz, Smith, Rubincam, Dunn & Torrence; Smith (D.E.), Kolenkiewicz,
    Wyatt, Dunn & Torrence.
Dunnet, D. Some aspects of the Panantarctic cratonic margin in Australia (Discussion).
    Trans. A **280**, 641—654 (1976).
    Mt Isa — reconstruction of a faulted ore body (Discussion). *Trans.* A **283**, 333—344 (1976).
Dunnet, G.M. The place of the Antarctic in biological sciences (Discussion).
    Trans. B **279**, 105—112 (1977).
Dunnill, P. Immobilized cell and enzyme technology (Discussion). *Trans.* B **290**, 409—420 (1980).
    The provision of pharmaceuticals by appropriate technology (Discussion).
    Proc. B **209**, 153—157 (1980).
Dunstan, Diana R., Grant, Anne M.S., Marshall, R.D. & Neuberger, A. A protein, immunologically
    similar to Tamm—Horsfall glycoprotein, produced by cultured baby hamster kidney cells.
    Proc. B **186**, 297—316 (1974). Corrigenda. *Proc.* B **187**, 485 (1974).
DuPeuble, P.A. *See* Roberts (D.G.), Montadert, Thompson, Auffret and others.
Dupré, B. *See* Allègre, Brévart, Dupré & Minster.
Durand, H. Present status and future prospects of silicon solar cell arrays and systems (Discussion).
    Trans. A **295**, 435—443 (1980).
Duraud, J.P. *See* Dran, Duraud, Klossa, Langevin & Maurette.
Durão, D.F.G. & Whitelaw, J.H. Instantaneous velocity and temperature measurements in oscillating
    diffusion flames. *Proc.* A **338**, 479—501 (1974).
Durham, J.J. *See* Beckinsale, Bowie & Durham.
Durney, D.W. Pressure-solution and crystallization deformation (Discussion).
    Trans. A **283**, 229—240 (1976).
Durnin, J.V.G.A. Nutrition (Discussion). *Trans.* B **274**, 447—455 (1976).
    See also Norgan, Ferro-Luzzi & Durnin.
Durrani, S.A. Charged-particle track analysis, thermoluminescence and microcratering studies of lunar
    samples (Discussion). *Trans.* A **285**, 309—317 (1977).
Durrani, S.A., Bull, R.K. & McKeever, S.W.S. Solar-flare exposure and thermoluminescence of Luna 24
    core material. *Trans.* A **297**, 41—50 (1980).
Durrant, A.V. A derivation of optical field quantization from absorber theory.
    Proc. A **370**, 41—59 (1980).
Durst, F. & Whitelaw, J.H. Optimization of optical anemometers. *Proc.* A **324**, 157—181 (1971).

Duru, I.H. *See* Barut & Duru.
Duruz, J.J., Michels, H.J. & Ubbelohde, A.R. Molten fatty acid salts as model ionic liquids. I. Thermodynamic and transport parameters of some organic sodium salts.
  Proc. A **322**, 281—299 (1971).
Duruz, J.J. & Ubbelohde, A.R. Structure of organic ionic melt meosphases.
  Proc. A **330**, 1—13 (1972).
Magnetic susceptibility of alkali *n*-butyrates and isovalerates near their melting points.
  Proc. A **342**, 39—49 (1975).
Some novel electromagnetic effects in the conductivity of molten organic salts.
  Proc. A **347**, 301—310 (1976).
Dus, R. & Tompkins, F.C. Mechanism of the hydrogen—oxygen reaction on platinum films from surface potential measurements. *Proc.* A **343**, 477—488 (1975).
Duxbury, G., Horani, M. & Rostas, J. Rotational analysis of the electronic emission spectrum of the H_2S^+ ion radical. *Proc.* A **331**, 109—137 (1972).
Dwek, R.A., Jones, R., Marsh, D., McLaughlin, A.C., Press, Elizabeth M., Price, N.C. & White, A.I. Antibody—hapten interactions in solution (Discussion). *Trans.* B **272**, 53—74 (1975).
Dworetsky, M.M. *See* Carnochan, Dworetsky, Todd, Willis & Wilson.
Dyball, R.E.J. *See* Poulain, Wakerley & Dyball.
Dyer, P. *See* Bush & Dyer.
Dyer, R.G., MacLeod, N.K. & Ellendorff, F. Electrophysiological evidence for sexual dimorphism and synaptic convergence in the preoptic and anterior hypothalamic areas of the rat.
  Proc. B **193**, 421—440 (1976).
Dymoke, L.D. Future engineering systems for naval ships (Discussion).
  Trans. A **273**, 129—135 (1972).
Dyson, B.F., Loveday, M.S. & Rodgers, M.J. Grain boundary cavitation under various states of applied stress. *Proc.* A **349**, 245—259 (1976).
Dyson, J. & Schutz, B.F. Perturbations and stability of rotating stars. I. Completeness of normal modes. *Proc.* A **368**, 389—410 (1979).
Dysthe, K.B. Note on a modification to the nonlinear Schrödinger equation for application to deep water waves. *Proc.* A **369**, 105—114 (1979).
Dziewonski, A.M. *See* Gilbert & Dziewonski.

Eades, J.A. *See* Buxton (B.F.), Eades, Steeds & Rackham.
Eagar, R.M.C. Some new Namurian bivalve faunas and their significance in the origin of *Carbonicola* and in the colonization of Carboniferous deltaic environments. *Trans.* B **280**, 535—567 (1977).
Eagles, P.M. On the stability of slowly varying flow between concentric cylinders.
  Proc. A **355**, 209—224 (1977).
A Bénard convection problem with a perturbed lower wall. *Proc.* A **371**, 359—379 (1980).
Eakins, M.N. *See* Bullock (G.), Eakins, Sawyer & Slater.
Earnshaw, J.C. *See* Dixon (H.E.), Earnshaw, Hook, Hough, Smith, Stephenson & Turver.
Earnshaw, W. *See* King (J.), Botstein, Casjens, Earnshaw, Harrison & Lenk.
Eastburn, P. See Hartley (A.J.), Eastburn & Leece.
Easterday, B.C. Animals in the influenza world (Discussion). *Trans.* B **288**, 433—437 (1980).
Easterling, K.E. *See* Loberg, Johansson & Easterling.
Eastmond, G.C. *See* Bamford (C.H.), Eastmond & Fildes.
Easton, A.J., Hamilton, D., Kempe, D.R.C. & Sheppard, S.M.F. Low-temperature metasomatic garnets in marine sediments (Discussion). *Trans.* A **286**, 253—271 (1977).
Eastwood, Sir Eric. Clifford Paterson Lecture. Radar: new techniques and applications.
  Proc. A **354**, 137—155 (1977).
Eaton, M.A.W. *See* Stebbing (N.), Lindley & Eaton.
Ebashi, S. The Croonian Lecture, 1979. Regulation of muscle contraction.
  Proc. B **207**, 259—286 (1980). Erratum. *Proc.* B **208**, 483 (1980).
Eberhardt, P. *See* Geiss, Eberhardt, Grögler, Guggisberg, Maurer & Stettler.

Eccles, Sir John. *See* Allen (G.I.), Eccles, Nicoll, Oshima & Rubia.

Eccles, Sir John & Feindel, W. Wilder Graves Penfield. *Biogr. Mem.* **24**, 473—513 (1978).

Eccles, Sir John, Nicoll, R.A., Oshima, T. & Rubia, F.J. The anionic permeability of the inhibitory postsynaptic membrane of hippocampal pyramidal cells. *Proc.* B **198**, 345—361 (1977).

Echlin, P. The application of scanning electron microscopy to biological research (Discussion). *Trans.* B **261**, 51—59 (1971).

Eckert, H.-G. *See* Franck, Rowold, Wegner & Eckert.

Eckhart, W. Induced cellular DNA synthesis by 'early' and 'late' temperature-sensitive mutants of polyoma virus (Discussion). *Proc.* B **177**, 59—63 (1971).

Edden, Anne C. *See* Cartwright (D.E.), Edden, Spencer & Vassie.

Edelson, B.I. & Davis, R.C. Satellite communications in the 1980s and after (Discussion). *Trans.* A **289**, 159—174 (1978).

Edelstein, W.A. *See* Drever (R.W.P.), Hough, Pugh, Edelstein and others; Mallard, Hutchison, Edelstein, Ling, Foster & Johnson.

Edgar, J.A. Danainae (Lep.) and 1,2-dehydropyrrolizidine alkaloid-containing plants — with reference to observations made in the New Hebrides (Discussion). *Trans.* B **272**, 467—476 (1975).

Edge, A.V.J. *See* Ash, Barrer, Chio & Edge.

Edgley, P.D. *See* Franklin (R.N.), Edgley, Hamberger & Motley; Franklin (R.N.), Mackinley, Edgley, & Wall.

Edgley, P.D. & von Engel, A. Theory of positive columns in electronegative gases. *Proc.* A **370**, 375—387 (1980).

Edholm, O.G. *See* McCance, Abu Rabiyah, Beer, Edholm, Even-Paz, Luff & Samueloff.

Edholm, O.G., Humphrey, S., Lourie, J.A., Tredre, Barbara E. & Brotherhood, J. Biological studies of Yemenite and Kurdish Jews in Israel and other groups in southwest Asia. VI. Energy expenditure and climatic exposure of Yemenite and Kurdish Jews in Israel. *Trans.* B **266**, 127—140 (1973).

Edholm, O.G. & Samueloff, S. Biological studies of Yemenite and Kurdish Jews in Israel and other groups in southwest Asia. I. Introduction, background and methods. *Trans.* B **266**, 85—95 (1973).

Edholm, O.G., Samueloff, S., Mourant, A.E., Fox, R.H., Lourie, J.A., Lehmann, H., Lehmann, E.E., Bavly, Sarah, Beaven, G. & Even-Paz, Z. Biological studies of Yemenite and Kurdish Jews in Israel and other groups in southwest Asia. XIII. Conclusion and summary. *Trans.* B **266**, 221—224 (1973).

Edholm, P. Computed tomography — a new technique in diagnostic radiology (Discussion). *Proc.* B **195**, 277—279 (1977).

Edington, J.W. *See* Doig & Edington; Porter (D.A.) & Edington.

Edmonds, D.T. Membrane ion channels and ionic hydration energies. *Proc.* B **211**, 51—62 (1980). [Abstract] *Proc.* A **373**, 285 (1980).

Edmonds, D.V. *See* Muddle & Edmonds.

Edmunds, D.E. & Evans, W.D. Orlicz and Sobolev spaces on unbounded domains. *Proc.* A **342**, 373—400 (1975).

Edmunds, D.E., Potter, A.J.B. & Stuart, C.A. Non-compact positive operators. *Proc.* A **328**, 67—81 (1972).

Edmunds, D.E. & Webb, J.R.L. Quasilinear elliptic problems in unbounded domains. *Proc.* A **334**, 397—410 (1973).

Edson, E.F. Crop protection (Discussion). *Trans.* B **267**, 93—100 (1973).

Edwards, G.J. *See* Blaney, Bradley, Edwards, Jolliffe and others.

Edwards, J.S. *See* Palka & Edwards.

Edwards, J.S. & Palka, A.J. The cerci and abdominal giant fibres of the house cricket, *Acheta domesticus.* I. Anatomy and physiology of normal adults. *Proc.* B **185**, 83—103 (1974).

Edwards, O.E. *See* Lemieux & Edwards.

Edwards, R.H.T., Miller, G.J., Hearn, C.E.D. & Cotes, J.E. Pulmonary function and exercise responses in relation to body composition and ethnic origin in Trinidadian males. *Proc.* B **181**, 407—420 (1972).

Edwards, R.N., Law, L.K. & White, A. Geomagnetic variations in the British Isles and their relation to electrical currents in the ocean and shallow seas. *Trans.* A **270**, 289–323 (1971).

Edwards, R.W. Future research needs (Discussion). *Proc.* B **177**, 463–468 (1971).

Edwards, S.F. *See* Allen (G.), Burgess, Edwards & Walsh; Deam & Edwards; Dolan & Edwards; Edwards (Sir Sam).

Edwards, S.F. & McComb, W.D. A local energy transport equation for isotropic turbulence. *Proc.* A **325**, 313–321 (1971).

Local transport, equations for turbulent shear flow. *Proc.* A **330**, 495–516 (1972).

Edwards, S.F. & Stockmayer, W.H. The equation of state of materials intermediate to rubbers and glasses. *Proc.* A **332**, 439–442 (1973).

Edwards, S.F. & Taylor, J.B. Negative temperature states of two-dimensional plasmas and vortex fluids. *Proc.* A **336**, 257–271 (1974).

Edwards, Sir George. Review Lecture. The technical aspects of supersonic civil transport aircraft. *Trans.* A **275**, 529–565 (1974).

Edwards, Sir Sam. The theory of rubber elasticity (Discussion). *Proc.* A **351**, 297–406 (1976). *See also* Edwards (S.F.).

van Eek, W.H. The challenge of producing oil and gas in deep water (Discussion). *Trans.* A **290**, 113–124 (1978).

Eggitt, P.W. Russell. Choosing between crops: aspects that affect the user (Discussion). *Trans.* B **281**, 93–106 (1977).

Eglin, R.P. *See* Wilkie (N.M.), Eglin, Sanders & Clements.

Eglinton, G. *See* Pillinger & Eglinton; Pillinger, Eglinton, Gowar & Jull.

Eglinton, G., Hajlbrahim, S.K., Maxwell, J.R., Quirke, J.M.E., Shaw, G.J., Volkman, J.K. & Wardroper, A.M.K. Lipids of aquatic sediments, Recent and ancient (Discussion). *Trans.* A **293**, 69–91 (1979).

Eglinton, G., Simoneit, B.R.T. & Zoro, J.A. The recognition of organic pollutants in aquatic sediments (Discussion). *Proc.* B **189**, 415–442 (1975).

Ehhalt, D.H. *In situ* observations (Discussion). *Trans.* A **296**, 175–189 (1980).

Ehinger, B. *See* Dowling (J.E.) & Ehinger.

Ehlers, J. *See* Breuer & Ehlers.

Ehrburger, P. & Donnet, J.B. Interface in composite materials (Discussion). *Trans.* A **294**, 495–505 (1980).

Eide, S.A. *See* Caldwell (D.R.) & Eide.

Einsele, G. *See* von Rad & Einsele.

Eisenberg, H., Borochov, N., Kam, Z. & Voordouw, G. Conformation of plasmid DNA and of DNA–histone chromatin-like complexes by laser light scattering (Discussion). *Trans.* A **293**, 303–313 (1979).

Eisenberg, S. The role of gene A and *E. coli rep* protein in the replication of ϕX 174 replicative form DNA (Discussion). *Proc.* B **210**, 337–349 (1980).

Eisler, R. Behavioural responses of marine poikilotherms to pollutants (Discussion). *Trans.* B **286**, 507–521 (1979).

Eisner, D. *See* Attwell, Cohen & Eisner.

El-Baz, F. Lunar stratigraphy (Discussion). *Trans.* A **285**, 549–553 (1977).

Elder, G.H. *See* Jackson (A.H.), Sancovich, Ferramola, Evans and others.

Elder, K. *See* Skehel, Waterfield, McCauley, Elder & Wiley.

Elder, M., Hitchcock, P., Mason, R. & Shipley, G.G. A refinement analysis of the crystallography of the phospholipid, 1,2-dilauroyl-DL-phosphatidylethanolamine, and some remarks on lipid–lipid and lipid–protein interactions. *Proc.* A **354**, 157–170 (1977).

Elderfield, H. Authigenic silicate minerals and the magnesium budget in the oceans (Discussion). *Trans.* A **286**, 273–281 (1977).

Eldholm, O. & Sundvor, E. The continental margins of the Norwegian–Greenland Sea: recent results and outstanding problems (Discussion). *Trans.* A **294**, 77–86 (1980).

El Din, Nasr. *See* McCance, El Neil, El Din, Widdowson, Southgate and others.

Eley, D.D. Eric Keightley Rideal. *Biogr. Mem.* **22**, 381–413 (1976).

See also Ashwell (G.J.), Eley, Wallwork & Willis; Batt, Eley & Norton; Caggiano, Eley & Hey.

Eley, D.D., Hey, M.J. & Ward, A.J.I. Proton magnetic relaxation of water sorbed by methaemoglobin crystals. *Proc.* A **331**, 457—468 (1973).

Eley, D.D., Rochester, C.H. & Scurrell, M.S. The polymerization of ethylene on chromium oxide catalysts.

I. Kinetic studies. *Proc.* A **329**, 361—373 (1972).

II. Infrared studies. *Proc.* A **329**, 375—390 (1972).

Eley, D.D. & Russell, S.H. The nitrogen isotope mixing reaction catalysed by polycrystalline metal wires. *Proc.* A **341**, 31—44 (1974).

El Gomati, M.M. *See* Browning (R.), Bassett, El Gomati & Prutton.

Elias, D. The possible development of telecommunications and its effects on the telecommunications industry (Discussion). *Trans.* A **289**, 19—28 (1978).

Elias, P.S. The medical significance of marine pollution by organic chemicals (Discussion). *Proc.* B **189**, 443—458 (1975).

Ellendorff, F. *See* Dyer (R.G.), MacLeod & Ellendorff.

Elliott, A. Direct demonstration of the helical nature of paramyosin filaments (Discussion). *Trans.* B **261**, 197—199 (1971).

The arrangement of myosin on the surface of paramyosin filaments in the white adductor muscle of *Crassostrea angulata*. *Proc.* B **186**, 53—66 (1974).

Elliott, A., Offer, G. & Burridge, K. Electron microscopy of myosin molecules from muscle and non-muscle sources. *Proc.* B **193**, 45—53 (1976).

Elliott, C. Economic aspects of village health (Discussion). *Proc.* B **209**, 71—82 (1980).

Elliott, C.J. *See* Wass, Henderson & Elliott.

Elliott, D. The energy balance and deformation mechanisms of thrust sheets (Discussion). *Trans.* A **283**, 289—312 (1976).

Elliott, H. The search for cosmic-ray anisotropies (Discussion). *Trans.* A **277**, 381—393 (1974).

Elliott, J.P. Review Lecture. Calculations of nuclear structure. *Proc.* A **326**, 199—213 (1972).

Elliott, Katherine. *See* Nugroho & Elliott; Tyrrell & Elliott.

Elliott, L.A. Computing experience with hyperbolic partial differential equations (Discussion). *Proc.* A **323**, 263—270 (1971).

Elliott, R.J., Harley, R.T., Hayes, W. & Smith, S.R.P. Raman scattering and theoretical studies of Jahn—Teller induced phase transitions in some rare-earth compounds. *Proc.* A **328**, 217—266 (1972).

Elliott, R.J., Hayes, W., Kleppmann, W.G., Rushworth, A.J. & Ryan, J.F. Experimental and theoretical studies of effects of anharmonicity and high-temperature disorder on Raman scattering in fluorite crystals. *Proc.* A **360**, 317—345 (1978).

Ellis, P., Holah, G., Houghton, J.T., Jones, T.S., Peckham, G., Peskett, G.D., Pick, D.R., Rodgers, C.D., Roscoe, H., Sandwell, R., Smith, S.D. & Williamson, E.J. Remote sounding of atmospheric temperature from satellites. IV. The selective chopper radiometer for Nimbus 5. *Proc.* A **334**, 149—170 (1973).

Ellis, R.S. Searches for optical evidence of galaxy evolution (Discussion). *Trans.* A **296**, 355—366 (1980).

Ellis, Susan T. *See* Hunt (S.V.), Ellis & Gowans.

Ellis, Susan T. & Gowans, J.L. The role of lymphocytes in antibody formation. V. Transfer of immunological memory to tetanus toxoid: the origin of plasma cells from small lymphocytes, stimulation of memory cells *in vitro* and the persistence of memory after cell-transfer. *Proc.* B **183**, 125—139 (1973).

Ellis, T., McClintock, P.V.E., Bowley, R.M. & Allum, D.R. The breakdown of superfluidity in liquid ^{4}He. II. An investigation of excitation emission from negative ions travelling at extreme super-critical velocities. *Trans.* A **296**, 581—595 (1980).

Ellison, D.H., Salmon, G.A. & Wilkinson, F. Nanosecond pulse radiolysis of methanolic and aqueous solutions of readily oxidizable solutes. *Proc.* A **328**, 23—36 (1972).

Ellory, J.C. *See* Smith (M.W.) & Ellory; Tucker (E.M.), Ellory, Wooding, Morgan & Herbert; Young (J.D.), Jones & Ellory.

El Neil, Hamad. *See* McCance, El Neil, El Din, Widdowson, Southgate and others.

El-Ruwayih, A. *See* Bishop (A.W.), Kumapley & El-Ruwayih.

Elston, J. & Dennett, M.D. A weather watch for semi-arid lands within the tropics (Discussion).
Trans. B 278, 593–609 (1977).

Elston, J., Karamanos, A.J., Kassam, A.H. & Wadsworth, R.M. The water relations of the field bean crop (Discussion). *Trans.* B 273, 581–591 (1976).

Elsworth, Yvonne P., Taylor, A.R. & James, J.F. A field compensated interference spectrometer for spectroscopy of the night airglow in the visible region. *Proc.* A 369, 335–349 (1980).

Eltayeb, I.A. Hydromagnetic convection in a rapidly rotating fluid layer.
Proc. A 326, 229–254 (1972).

On linear wave motions in magnetic-velocity shears. *Trans.* A 285, 607–636 (1977).

Nonlinear thermal convection in an elasticoviscous layer heated from below.
Proc. A 356, 161–176 (1977).

See also Ahmed & Eltayeb.

Eltayeb, I.A. & Kumar, S. Hydromagnetic convective instability of a rotating, self-gravitating fluid sphere containing a uniform distribution of heat sources. *Proc.* A 353, 145–162 (1977).

Elton, G.A.H. Food technology and the law (Discussion). *Proc.* B 191, 99–110 (1975).

Elwell, D. Priorities in the initial use of *Spacelab* for crystal growth (Discussion).
Proc. A 361, 151–156 (1978).

Embleton, B.J.J. *See* McElhinny & Embleton.

Embury, Susan. *See* Seller, Embury, Polani & Adinolfi.

Embury, Susan, Seller, Mary J., Adinolfi, M. & Polani, P.E. Neural tube defects in curly-tail mice. I. Incidence, expression and similarity to the human condition. *Proc.* B 206, 85–94 (1979).

Emerson, D. *See* Coulson & Emerson; Rood, Emerson & Milledge.

Emson, P.C. *See* Iversen (L.L.), Lee, Gilbert, Hunt & Emson.

Endacott, J.D. Underground power cables (Discussion). *Trans.* A 275, 193–203 (1973).

Endean, R.P.J. *See* Beynon (W.J.G.), Barrett & Endean.

Enderby, J.E. Neutron and X-ray scattering from aqueous solutions (Discussion).
Proc. A 345, 107–117 (1975).

Neutron diffraction, isotopic substitution and the structure of aqueous solutions (Discussion).
Trans. B 290, 553–566 (1980).

Eneroth, P. *See* Gustafsson & Eneroth.

von Engel, A. *See* Dixon (A.J.), von Engel & Harrison; Edgley & von Engel; Tan (K.L.) & von Engel.

Engel, A.E.J. Characteristics, occurrence and origins of basalts of the oceans [abstract] (Discussion).
Trans. A 268, 493 (1971).

Engel, A.R. *See* Carpenter (G.F.), Coe, Engel & Quenby; Quenby, Coe & Engel.

von Engelhardt, W. & Stengelin, R. Chemical changes at impact-induced phase transitions on the lunar surface (Discussion). *Trans.* A 285, 285–291 (1977).

Engelland, A. *See* Trench, Pool, Logan & Engelland.

England, Marjorie A. *See* Wakely & England.

England, R.E. The underwater contractor: his rôle and development (Discussion).
Trans. A 290, 153–159 (1978).

Engler, G. *See* Schell, Van Montagu, De Beuckeleer, De Block and others; Van Montagu, Holsters, Zambryski, Hernalsteens, Depicker and others.

English, C.A. & Venables, J.A. The structure of the diatomic molecular solids.
Proc. A 340, 57–80 (1974).

English, C.A., Venables, J.A. & Salahub, D.R. The use of a spin dependent intermolecular potential to predict the crystal structure and antiferromagnetism of solid α-oxygen.
Proc. A 340, 81–90 (1974).

English, H. *See* Sozou & English.

Engstrom, S.F.T. *See* Boland (B.C.), Jones, Wilson, Engstrom & Noci.

Epler, D. *See* Castle & Epler.

Ercoli, B. *See* Garton, Reeves, Tomkins & Ercoli.

Erginsav, A. *See* Barlow (A.J.) & Erginsav.

Erickson, H.P. & Klug, A. Measurement and compensation of defocusing and aberrations by Fourier processing of electron micrographs (Discussion). *Trans.* B **261**, 105–118 (1971).

Erickson, W.D. & Linnett, J.W. An *'ab initio'* Gaussian orbital calculation of the (100) surface of crystalline lithium hydride (Discussion). *Proc.* A **331**, 347–359 (1972).

Erikson, E. *See* Erikson (R.L.), Collett, Erikson & Purchio.

Erikson, R.L., Collett, M.S., Erikson, E. & Purchio, A.F. Towards a molecular description of cell transformation by avian sarcoma virus (Discussion). *Proc.* B **210**, 387–396 (1980).

Erlank, A.J., Allsopp, H.L., Duncan, A.R. & Bristow, J.W. Mantle heterogeneity beneath southern Africa: evidence from the volcanic record (Discussion). *Trans.* A **297**, 295–307 (1980).

Erlij, D. Salt transport across isolated frog skin (Discussion). *Trans.* B **262**, 153–161 (1971).
See also Grinstein & Erlij.

Ertl, H. & Dullien, F.A.L. Cluster sizes in the pre-freezing region from momentum transfer considerations and the relation of viscosity anomalies to the reduced freezing temperature.
Proc. A **335**, 235–250 (1973).

Escher, A. *See* Bridgwater, Escher & Watterson.

Escher, A., Jack, S. & Watterson, J. Tectonics of the North Atlantic Proterozoic dyke swarm (Discussion). *Trans.* A **280**, 529–539 (1976).

Eshelby, J.D. Dislocation theory for geophysical application (Discussion).
Trans. A **274**, 331–338 (1973).

Esposito, M.S. *See* Moens, Mowat, Esposito & Esposito.

Esposito, R.E. *See* Moens, Mowat, Esposito & Esposito.

Esposito, V. *See* Metafora, Felsani, Cotrufo, Tajana, Del Rio, De Prisco and others; Metafora, Felsani, Cotrufo, Tajana, Di Iorio, Del Rio and others.

Essex, B. A new approach to decision making in primary health care (Discussion).
Proc. B **209**, 89–96 (1980).

Etique, Phillippe. *See* Signer, Baur, Etique, Frick & Funk.

Etkin, B. & Goering, P.L.E. Air-curtain walls and roofs – 'dynamic' structures (Discussion).
Trans. A **269**, 527–543 (1971).

Ettles, C.M.M. *See* Ranger, Ettles & Cameron; Stokes (M.J.) & Ettles.

Eusebi, F. *See* de Santis, Eusebi & Miledi.

Evans, A.G., Gulden, M.E. & Rosenblatt, M. Impact damage in brittle materials in the elastic–plastic response régime. *Proc.* A **361**, 343–365 (1978).

Evans, D.C.B., Nye, J.F. & Cheeseman, K.J. The kinetic friction of ice. *Proc.* A **347**, 493–512 (1976).

Evans, D.E. *See* Gibson (A.F.), Kimmitt, Koohian, Evans & Levy.

Evans, D.G. Cyril Leslie Oakley. *Biogr. Mem.* **22**, 295–305 (1976).

Evans, E. *See* Miller (D.S.), Baker, Bowden, Evans, Holt and others.

Evans, E.L. *See* Thomas (J.M.), Evans & Williams.

Evans, E.P. *See* Ford (C.E.), Evans, Burtenshaw, Clegg, Tuffrey & Barnes.

Evans, G.B. *See* Archer (A.S.), Cundall, Evans & Palmer.

Evans, L.T. Physiological adaptation to performance as crop plants (Discussion).
Trans. B **275**, 71–83 (1976).

Evans, M.J.B. *See* Boucher (E.A.) & Evans; Boucher (E.A.), Evans & Kent.

Evans, N. *See* Jackson (A.H.), Sancovich, Ferramola, Evans and others.

Evans, R. *See* Schilling, Bergeron & Evans.

Evans, R.G. *See* Burton (W.M.), Evans & Griffin.

Evans, R.G., Nandy, K. & Wilson, R. Preliminary observations of the Magellanic Clouds with the ultraviolet sky survey telescope (Discussion). *Trans.* A **279**, 473–477 (1975).

Evans, S. Depth profiles of ion-induced structural changes in diamond from X-ray photoelectron spectroscopy. *Proc.* A **360**, 427–443 (1978).
Some effects of heteroatom size and reactivity on low-energy ion implantation in diamond, revealed by X-ray induced electron spectroscopy and oxidative depth profiling.
Proc. A **370**, 107–129 (1980).

Evans, S., Adams, J.M. & Thomas, J.M. The surface structure and composition of layered silicate minerals: novel insights from X-ray photoelectron diffraction, K-emission spectroscopy and

cognate techniques. *Trans.* A **292**, 563–591 (1979).

Evans, S. & Thomas, J.M. The chemical nature of ion-bombarded carbon: a photoelectron spectroscopic study of 'cleaned' surfaces of diamond and graphite. *Proc.* A **353**, 103–120 (1977).

Evans, T. *See* Brozel, Evans & Stephenson; Davies (G.) & Evans.

Evans, T. & Rainey, P. Changes in the defect structure of diamond due to high temperature + high pressure treatment. *Proc.* A **344**, 111–130 (1975).

Evans, W.D. *See* Edmunds & Evans.

Eve, M. The use of path integrals in guided wave theory. *Proc.* A **347**, 405–417 (1976).
 See also Budden & Eve.

Even-Paz, Z. *See* Edholm (O.G.), Samueloff, Mourant, Fox and others; Fox (R.H.), Even-Paz, Woodward & Jack; McCance, Abu Rabiyah, Beer, Edholm, Even-Paz, Luff & Samueloff.

Evensen, N.M. *See* O'Nions, Evensen & Hamilton; O'Nions, Evensen, Hamilton & Carter.

Everitt, W.N. & Jones, D.S. On an integral inequality. *Proc.* A **357**, 271–288 (1977).

Everson, I. Antarctic marine secondary production and the phenomenon of cold adaptation (Discussion). *Trans.* B **279**, 55–66 (1977).

Ewing, M. *See* Miyashiro, Shido & Ewing.

Ewins, P.D. & Potter, R.T. Some observations on the nature of fibre reinforced plastics and the implications for structural design (Discussion). *Trans.* A **294**, 507–517 (1980).

Exton, H. Basic Fourier series. *Proc.* A **369**, 115–136 (1979).

Eyre, B.L. *See* Bullough (R.), Eyre & Krishan; Kumar (A.) & Eyre.

Ezra, Sir Derek. Operational research within the National Coal Board (Discussion).
 Trans. A **287**, 467–486 (1977).

Faber, D.S. *See* Korn, Triller & Faber.

Faber, T.E. A continuum theory of disorder in nematic liquid crystals.
 I. *Proc.* A **353**, 247–259 (1977).
 II. Intermolecular correlations. *Proc.* A **353**, 261–275 (1977).
 III. Nuclear magnetic relaxation. *Proc.* A **353**, 277–288 (1977).
 IV. Application to lattice models. *Proc.* A **370**, 509–521 (1980).
 See also Horn (R.G.) & Faber.

Fabian, A.C. Theories of the nuclei of active galaxies (Discussion). *Proc.* A **366**, 449–459 (1979).

Fabian, D.J. *See* Padalia, Lang, Norris, Watson & Fabian.

Fabian, D.M. *See* Pillinger & Fabian.

Fall, S.M. Dynamical aspects of galaxy clustering (Discussion). *Trans.* A **296**, 339–345 (1980).

Faller, A.M. *See* Briden, Rex, Faller & Tomblin.

Fan, L.T. & Hwang, W.S. Corrigenda to *Proc.* A **283**, 576–582 (1965). *Proc.* A **327**, 574 (1972).

Fang, M.T.C. *See* Allen (J.E.), Fang & Fraser.

Fargie, D. & Martin, B.W. Developing laminar flow in a pipe of circular cross-section.
 Proc. A **321**, 461–476 (1971).

Farinas, Blanca R. *See* van Breemen, Farinas, Casteels, Gerba, Wuytack & Deth.

Farman, J.C. Ozone measurements at British Antarctic Survey stations (Discussion).
 Trans. B **279**, 261–271 (1977).

Farrant, J. Water transport and cell survival in cryobiological procedures (Discussion).
 ⋅ *Trans.* B **278**, 191–205 (1977).

Farrell, R.A. *See* Wilson (S.), Silver & Farrell.

Farrell, W.E. Earth tides, ocean tides and tidal loading (Discussion). *Trans.* A **274**, 253–259 (1973).

Farrow, G.E. The climate of Aldabra Atoll (Discussion). *Trans.* B **260**, 67–91 (1971).
 See also Stoddart (D.R.), Taylor, Fosberg & Farrow.

Farrow, G.E. & Brander, K.M. Tidal studies on Aldabra (Discussion). *Trans.* B **260**, 93–121 (1971).

Fastie, W.G., Moos, H. Warren, Henry, R.C. & Feldman, P.D. Rocket and spacecraft studies of ultraviolet emissions from astrophysical targets (Discussion). *Trans.* A **279**, 391–400 (1975).

Faulk, W. Page, Yeager, Carol, McIntyre, J.A. & Ueda, Mirthes. Oncofoetal antigens of human tropho-blast. *Proc.* B **206**, 163–182 (1979).

Faulk, W.P. & Galbraith, Gillian M.P. Trophoblast transferrin and transferrin receptors in the host–parasite relationship of human pregnancy. *Proc.* B **204**, 83–97 (1979).

Faulkner, R.G. *See* Scarlett, Buxton & Faulkner.

Fawzy, I. *See* Bishop (R.E.D.) & Fawzy.

Fawzy, I. & Bishop, R.E.D. On the dynamics of linear non-conservative systems.
Proc. A **352**, 25–40 (1976).

Fay, P. *See* Kulasooriya, Lang & Fay; Lang & Fay; Rother & Fay.

Fay, P. & Lang, Norma J. The heterocysts of blue-green algae. I. Ultrastructural integrity after isolation. *Proc.* B **178**, 185–192 (1971).

Feachem, R.G. Community participation in appropriate water supply and sanitation technologies: the mythology for the Decade (Discussion). *Proc.* B **209**, 15–29 (1980).

Fearn, D.R. Thermally driven hydromagnetic convection in a rapidly rotating sphere.
Proc. A **369**, 227–242 (1979).

Fearon, E.O. *See* Bevis & Fearon.

Feather, N. Richard Whiddington. *Biogr. Mem.* **17**, 741–756 (1971).
Rutherford Memorial Lecture, 1977. Some episodes of the α-particle story, 1903–1977.
Proc. A **357**, 117–129 (1977).
See also Massey & Feather.

Fee, E.J. *See* Mortimer (C.H.) & Fee.

Feeney, J. The use of three-bond spin–spin coupling constants in the determination of conformations of molecules in solution (Discussion). *Proc.* A **345**, 61–72 (1975).
See also Birdsall (B.), Griffiths, Roberts, Feeney & Burgen.

Feeney, J., Roberts, G.C.K., Birdsall, B., Griffiths, D.V., King, R.W., Scudder, P. & Burgen, Sir Arnold. [1]H nuclear magnetic resonance studies of the tyrosine residues of selectively deuterated *Lactobacillus casei* dihydrofolate reductase. *Proc.* B **196**, 267–290 (1977).

Feilden, G.B.R. Introductory remarks to a Discussion. *Trans.* A **275**, 311 (1973).
Looking ahead (Discussion). *Trans.* A **276**, 611–615 (1974).

Feindel, W. *See* Eccles & Feindel.

Feinstein, Graciela. *See* Frydman (B.), Frydman, Valasinas, Levy & Feinstein.

Fejer, J.A. Alteration of the ionosphere by man-made waves (Discussion).
Trans. A **280**, 151–165 (1975).

Feldberg, W. The Ferrier Lecture, 1974. Body temperature and fever: changes in our views during the last decade. *Proc.* B **191**, 199–229 (1975).

Feldman, G. & Matthews, P.T. Colour symmetry and the ψ particles (Discussion).
Proc. A **355**, 621–627 (1977).

Feldman, Joan D. *See* Keating, Beazley, Feldman & Gaze; Keating & Feldman.

Feldman, P.D. *See* Fastie, Moos, Henry & Feldman.

Fell, Dame Honor. The role of mucopolysaccharides in the protection of cartilage cells against im-mune reactions (Discussion). *Trans.* B **271**, 325–341 (1975).

Felsani, A. *See* Metafora, Felsani, Cotrufo, Tajana, Del Rio, De Prisco and others; Metafora, Felsani, Cotrufo, Tajana, Di Iorio, Del Rio and others.

Fendall, N.R.E. Training and management for primary health care (Discussion).
Proc. B **209**, 97–109 (1980).

Fender, B.E.F., Hobbis, L.C.W. & Manning, G. The U.K. Spallation Neutron Source (Discussion).
Trans. B **290**, 657–672 (1980).

Fendt, A. *See* Kaiser (W.), Fendt, Kranitzky & Laubereau.

Fenner, R.T. *See* Tan (C.L.) & Fenner.

Fenner, T.I. & Loizou, G. Optimally scalable matrices. *Trans.* A **287**, 307–349 (1977).

Fennessy, Sir Edward. The global picture (Discussion). *Trans.* A **289**, 5–17 (1978).

Fenstermacher, C. High-energy short-pulse carbon dioxide lasers (Discussion).
Trans. A **298**, 377–391 (1980).

Fenton, J.D. *See* Longuet-Higgins (M.S.) & Fenton.

Fenton,. J.D. & Abbott, J.E. Initial movement of grains on a stream bed: the effect of relative protrusion. *Proc.* A **352**, 523—537 (1977).

Ferenci, T. & Kornberg, H.L. The role of phosphotransferase-mediated syntheses of fructose 1-phosphate and fructose 6-phosphate in the growth of *Escherichia coli* on fructose. *Proc.* B **187**, 105—119 (1974).

Fereres, E. *See* Hsiao, Acevedo, Fereres & Henderson.

Ferguson, J.M., Hicks, F.G., Hoaksey, A., Rowlands, P.C. & Lloyd, B. The effects of small additions of silicon on the oxidation resistance of Fe-based alloys and steels in high pressure CO_2 [abstract] (Discussion). *Trans.* A **295**, 333 (1980).

Ferramola, A.M. *See* Jackson (A.H.), Sancovich, Ferramola, Evans and others.

Ferrar, A.N. *See* Bamford (C.H.) & Ferrar.

Ferrara, G. *See* Bonatti, Honnorez & Ferrara.

Ferreira, H.G. *See* Lew, Ferreira & Moura.

Ferro-Luzzi, A. *See* Norgan, Ferro-Luzzi & Durnin.

Fersht, A.R. Catalysis, binding and enzyme—substrate complementarity. *Proc.* B **187**, 397—407 (1974).

Fettiplace, R. & Crawford, A.C. The coding of sound pressure and frequency in cochlear hair cells of the terrapin. *Proc.* B **203**, 209—218 (1978).

Fichtel, C.E. Solar cosmic rays (Discussion). *Trans.* A **270**, 167—174 (1971).
Primary γ-rays (Discussion). *Trans.* A **277**, 365—379 (1974).

Fiddy, M.A. *See* Burge, Fiddy, Greenaway & Ross; Ross (G.), Fiddy, Nieto-Vesperinas & Wheeler.

Fidler, R.S. *See* Croker, Fidler & Smith.

Fiecchi, A., Kienle, M. Galli, Scala, A., Galli, G., Paoletti, E. Grossi, Cattabeni, F. & Paoletti, R. Hydrogen-exchange and double bond formation in cholesterol biosynthesis (Discussion). *Proc.* B **180**, 147—165 (1972).

Field, D. *See* Dixon (R.N.) & Field.

Field, J.E. *See* Chaudhri & Field; Coley & Field; Fuller (K.N.G.), Fox & Field; Heavens & Field; Hutchings, Winter & Field; Winter (R.E.) & Field.

Field, J.E. & Lesser, M.B. On the mechanics of high speed liquid jets. *Proc.* A **357**, 143—162 (1977).

Fielder, G. *See* Hulme (G.) & Fielder; Telfer & Fielder.

Fielder, G., Fryer, R.J., Titulaer, C., Herring, A.K. & Wise, B. Lunar crater origin in the maria from analysis of Orbiter photographs. *Trans.* A **271**, 361—409 (1972).

Fielding, J. Keynote speech (Symposium). *Trans.* A **282**, 118—122 (1976).

Fife, P.C. & Nicholes, K.R.K. Dispersion in flow through small tubes. *Proc.* A **344**, 131—145 (1975).

Fifield, Susan M. & Finlayson, L.H. Peripheral neurons and peripheral neurosecretion in the stick insect, *Carausius morosus*. *Proc.* B **200**, 63—85 (1978).

de Figueiredo, Isabel M.B. & Raab, R.E. A pictorial approach to macroscopic space—time symmetry, with particular reference to light scattering. *Proc.* A **369**, 501—516 (1980).

Fildes, F.J.T. *See* Bamford (C.H.), Eastmond & Fildes.

Fillenz, Marianne. Fine structure of noradrenaline storage vesicles in nerve terminals of the rat vas deferens (Discussion). *Trans.* B **261**, 319—323 (1971).

Filshie, G.M. Medical and surgical methods of early termination of pregnancy (Discussion). *Proc.* B **195**, 115—127 (1976).

Finch, J.T. & Klug, A. Three-dimensional reconstruction of the stacked-disk aggregate of tobacco mosaic virus protein from electron micrographs (Discussion). *Trans.* B **261**, 211—219 (1971).

Finch, J.T., Lewit-Bentley, A., Bentley, G.A., Roth, M. & Timmins, P.A. Neutron diffraction from crystals of nucleosome core particles (Discussion). *Trans.* B **290**, 635—638 (1980).

Finch, R.A. *See* Bennett (M.D.), Finch, Smith & Rao.

Finean, J.B., Freeman, R. & Limbrick, A.R. Structural studies relating to the distribution of molecular components in erythrocyte membranes (Discussion). *Trans.* B **268**, 15—21 (1974).

Fineran, B.A. & Bullock, Suzanne. Ultrastructure of graniferous tracheary elements in the haustorium of *Exocarpus bidwillii*, a root hemi-parasite of the Santalaceae. *Proc.* B **204**, 329—343 (1979).
Erratum: *Proc.* B **206**, 489 (1980).

Fineran, B.A. & Nicol, J.A.C. Studies on the eyes of New Zealand parrot-fishes (Labridae). *Proc.* B **186**, 217—247 (1974).

Studies on the eyes of anchovies *Anchoa mitchilli* and *A. hepsetus* (Engraulidae) with particular reference to the pigment epithelium. *Trans.* B **276**, 321—350 (1977).

Studies on the photoreceptors of *Anchoa mitchilli* and *A. hepsetus* (Engraulidae) with particular reference to the cones. *Trans.* B **283**, 25—60 (1978).

Finger, H.B. Recent developments in building systems (Discussion). *Trans.* A **272**, 503—531 (1972).

Fink, P.T. & Soh, W.K. A new approach to roll-up calculations of vortex sheets.
Proc. A **362**, 195—209 (1978).

Finlayson, L.H. *See* Fifield & Finlayson.

Finn, R. *See* Concus & Finn.

Finney, J.L. The organization and function of water in protein crystals (Discussion).
Trans. B **278**, 3—32 (1977).

Finniston, H.M. Review Lecture. Steel: an industry with a future. *Proc.* A **326**, 1—22 (1971).

Metallurgical processes (Discussion). *Trans.* A **275**, 313—327 (1973).

The Sixth Royal Society Technology Lecture. Nuclear energy for the steel industry.
Proc. A **340**, 129—146 (1974).

Fischer, B. & Poggio, G.F. Depth sensitivity of binocular cortical neurons of behaving monkeys (Discussion). *Proc.* B **204**, 409—414 (1979).

Fisher, A.M. *See* Best (C.H.) & Fisher.

Fisher, J., Belasco, J.G., Charnas, R.L., Khosla, S. & Knowles, J.R. β-Lactamase inactivation by mechanism-based reagents (Discussion). *Trans.* B **289**, 309—319 (1980).

Fisher, J.W. *See* Hatfield, Fisher, Dunigan, Burchfield and others.

Fisher, M. *See* Rämme, Fisher, Claesson & Szwarc.

Fisher, M., Rämme, G., Claesson, S. & Szwarc, M. Capture of solvated electrons and the collapse of solvated electron—sodium cation pair into sodium atom, studied by flash photolysis.
Proc. A **327**, 481—490 (1972).

Fisher, M.E. *See* Longuet-Higgins (H.C.) & Fisher.

Fisher, R. *See* Stimson & Fisher.

Fisher, R.F. & Wakely, Jennifer. The elastic constants and ultrastructural organization of a basement membrane (lens capsule). *Proc.* B **193**, 335—358 (1976).

Fisher, R.M., Huffman, G.P., Nagata, T. & Schwerer, F.C. Electrical conductivity and the thermocline of the Moon (Discussion). *Trans.* A **285**, 517—521 (1977).

Fisher, S.K. *See* Boycott, Dowling, Fisher, Kolb & Laties; Goldman (Karen A.) & Fisher; Kinney & Fisher.

Fisher, S.K. & Boycott, B.B. Synaptic connexions made by horizontal cells within the outer plexiform layer of the retina of the cat and the rabbit. *Proc.* B **186**, 317—331 (1974).

Fiske, R.S. & Jackson, E.D. Orientation and growth of Hawaiian volcanic rifts: the effect of regional structure and gravitational stresses. *Proc.* A **329**, 299—326 (1972).

Fisken, R.A., Garey, L.J. & Powell, T.P.S. The intrinsic, association and commissural connections of area 17 of the visual cortex. *Trans.* B **272**, 487—536 (1975).

Fitz-Gerald, G.F. The reflexion of plane gravity waves travelling in water of variable depth.
Trans. A **284**, 49—89 (1976).

Fitz-Gerald, J.M. Appendix to a paper by Caro, Fitz-Gerald & Schroter.
Proc. B **177**, 138—159 (1971).
See also Caro, Fitz-Gerald & Schroter.

Fitzsimons, J.T. Angiotensin and other peptides in the control of water and sodium intake (Discussion). *Proc.* B **210**, 165—182 (1980).

Flavell, R.B. *See* Riley & Flavell.

Flavill, R.P. *See* McDonnell, Ashworth, Flavill, Carey, Bateman & Jennison.

Fleck, R.J. *See* Greenwood (W.R.), Hadley, Anderson, Fleck & Schmidt.

Fleig, A.J. *See* Krueger, Guenther, Fleig, Heath, Hilsenrath and others.

Fleming, C.A. Ernest Marsden. *Biogr. Mem.* **17**, 463—496 (1971).
See also Watters & Fleming.

Fleming, G.R. *See* Beddard, Fleming, Gijzeman & Porter; Beddard, Fleming, Porter & Robbins.

Flemming, N.C. Holocene eustatic changes and coastal tectonics in the northeast Mediterranean:

implications for models of crustal consumption. *Trans.* A **289**, 405–458 (1978).

Flenley, J.R. *See* Walker (D.) & Flenley.

Fletcher, A.W. Review Lecture: Metal recycling from scrap and waste materials. *Proc.* A **351**, 151–178 (1976).

Fletcher, B.N. *See* Andreieff, Bouysse, Curry, Fletcher and others.

Flewitt, P.E.J. *See* Doig & Flewitt.

Flinn, R.M. *See* Clegg (E.J.), Pawson, Ashton & Flinn.

Flood, P.G. *See* Orme, Flood & Sargent; Scoffin, Stoddart, McLean & Flood.

Flood, P.G., Orme, G.R. & Scoffin, T.P. An analysis of the textural variability displayed by inter-reef sediments of the Impure Carbonate Facies in the vicinity of the Howick Group (Discussion). *Trans.* A **291**, 73–83 (1978).

Flood, P.G. & Scoffin, T.P. Reefal sediments of the northern Great Barrier Reef (with an appendix by A.B. Cribb) (Discussion). *Trans.* A **291**, 55–68 (1978).

Florides, P.S. A new interior Schwarzschild solution. *Proc.* A **337**, 529–535 (1974).

Florides, P.S. & Synge, J.L. Coordinate conditions in a Riemannian space for coordinates based on a subspace. *Proc.* A **323**, 1–10 (1971).

Flory, P.J. Statistical thermodynamics of random networks (Discussion). *Proc.* A **351**, 351–380 (1976).

Flower, D.A. *See* Chapman (W.A.), Cross, Flower, Peckham & Smith.

Flowers, T.J., Ward, Margaret E. & Hall, J.L. Salt tolerance in the halophyte *Suaeda maritima*: some properties of malate dehydrogenase (Discussion). *Trans.* B **273**, 523–540 (1976).

Flygare, W.H. The structure of simple organic liquids (Discussion). *Trans.* A **293**, 277–285 (1979).

Fofonoff, N.P. & Webster, F. Current measurements in the western Atlantic (Discussion). *Trans.* A **270**, 423–436 (1971).

Fogg, G.E. Recycling through algae (Discussion). *Proc.* B **179**, 201–207 (1971).
Aquatic primary production in the Antarctic (Discussion). *Trans.* B **279**, 27–38 (1977).

Foglia, V.G. *See* Young (Sir Frank) & Foglia.

Foglio, M.E., Sekerka, R.F. & Van Vleck, J.H. Theory of the width of the ferromagnetic resonance line of europium iron garnet. *Proc.* A **344**, 21–50 (1975).

Foglio, M.E. & Van Vleck, J.H. Theory of the magnetic anisotropy and nuclear magnetic resonance of europium iron garnet. *Proc.* A **336**, 115–140 (1974).

Fokkema, J.T. Diffraction of elastic waves by the periodic rigid boundary of a semi-infinite solid. *Proc.* A **363**, 487–502 (1978).

Foley, T., Klinowski, J. & Meares, P. Differential conductance coefficients in a cation-exchange membrane. *Proc.* A **336**, 327–354 (1974).

Folland, R. *See* Bridges, Charlesby & Folland; Charlesby, Folland & Steven.

Follett, B.K. *See* Davies (D.T.) & Follett.

Foo, V.Y., Brion, C.E. & Hasted, J.B. Electron energy-loss spectra of some triatomic molecules. *Proc.* A **322**, 535–554 (1971).

Foot, A.S. Under-used products from crops and animals (Discussion). *Trans.* B **281**, 221–230 (1977).

Ford, C.E., Evans, E.P., Burtenshaw, M.D., Clegg, H.M., Tuffrey, M. & Barnes, R.D. A functional 'sex-reversed' oocyte in the mouse. *Proc.* B **190**, 187–197 (1975).

Ford, C.E., O'Hara, M.J. & Spencer, P.N. The origin of lunar felspathic liquids (Discussion). *Trans.* A **285**, 193–197 (1977).

Ford, E.B. Theodosius Grigorievich Dobzhansky. *Biogr. Mem.* **23**, 59–89 (1977).
See also Clarke (Sir Cyril) & Ford.

Ford, G.M. *See* Drever (R.W.P.), Hough, Pugh, Edelstein and others.

Ford, L.H. The interaction of an atom with electromagnetic vacuum fluctuations in the presence of a pair of perfectly conducting plates. *Proc.* A **362**, 559–571 (1978).
Quantum coherence effects and the second law of thermodynamics. *Proc.* A **364**, 227–236 (1978).
Casimir effect for a self-interacting scalar field. *Proc.* A **368**, 305–310 (1979).
Comment on the interaction of an atom with a pair of perfectly conducting plates. *Proc.* A **368**, 311–312 (1979).

Forey, P.L. *Latimeria*: a paradoxical fish. *Proc.* B **208**, 369–384 (1980).

Fork, R.L. *See* Shank, Ippen, Fork, Migus & Kobayashi.
Formosinho, S.J. *See* Ashpole, Formosinho & Porter.
Formosinho, S.J., Porter, Sir George & West, M.A. Vibrational relaxation in the triplet state.
  Proc. A **333**, 289–296 (1973).
Fornberg, B. & Whitham, G.B. A numerical and theoretical study of certain nonlinear wave phenomena. *Trans.* A **289**, 373–404 (1978).
Forrest, A.R. Computational geometry (Discussion). *Proc.* A **321**, 187–195 (1971).
Forrest, J.S. Introductory remarks to a Discussion. *Trans.* A **275**, 35–37 (1973).
Forster, G.R. The ecology of *Latimeria chalumnae* Smith: results of field studies from Grande Comore. *Proc.* B **186**, 291–296 (1974). Corrigendum. *Proc.* B **187**, 485 (1974).
Forsyth, J.B. *See* Brown (P. Jane), Forsyth & Mason.
Forsyth, P.J.E. Fatigue as a design limitation (Symposium). *Trans.* A **282**, 167–180 (1976).
Fosberg, F.R. Preliminary survey of Aldabra vegetation (Discussion). *Trans.* B **260**, 215–225 (1971). *See also* Stoddart (D.R.), Taylor, Fosberg & Farrow.
Fosdick, R.L. & Rajagopal, K.R. Thermodynamics and stability of fluids of third grade.
  Proc. A **369**, 351–377 (1980).
Fosdick, R.L. & Serrin, J. Rectilinear steady flow of simple fluids. *Proc.* A **332**, 311–333 (1973).
Foster, C.D. *See* Cullis & Foster.
Foster, M.A. *See* Mallard, Hutchison, Edelstein, Ling, Foster & Johnson.
Foster, R.E. *See* Waxman (S.G.) & Foster.
Foster, S.S.D. *See* Gray (D.A.) & Foster.
Foster, W.A. & Treherne, J.E. Feeding, predation and aggregation behaviour in a marine insect, *Halobates robustus* Barber (Hemiptera: Gerridae), in the Galapagos Islands.
  Proc. B **209**, 539–553 (1980).
Foucher, J.-P. & Sibuet, J.C. Thermal régime of the northern Bay of Biscay continental margin in the vicinity of the D.S.D.P. Sites 400–402 (Discussion). *Trans.* A **294**, 157–167 (1980).
Fountain, J.A. *See* Horai, Winkler, Keihm, Langseth, Fountain & West.
Fourcans, B. *See* Schroepfer, Lutsky, Martin, Huntoon and others.
Fowden, L. *See* Lea (P.J.) & Fowden.
Fowler, A.C. & Larson, D.A. On the flow of polythermal glaciers.
  I. Model and preliminary analysis. *Proc.* A **363**, 217–242 (1978).
  II. Surface wave analysis. *Proc.* A **370**, 155–171 (1980).
Fowler, P.H. The Rutherford Memorial Lecture, 1971. Evolution of the elements.
  Proc. A **329**, 1–16 (1972).
Fox, E.N. Limit analysis for plates: a simple loading problem involving a complex exact solution.
  Trans. A **272**, 463–492 (1972).
  Limit analysis for plates: the exact solution for a clamped square plate of isotropic homogeneous material obeying the square yield criterion and loaded by uniform pressure.
  Trans. A **277**, 121–155 (1974).
Fox, L. Some experiments with singularities in linear elliptic partial differential equations (Discussion).
  Proc. A **323**, 179–190 (1971).
Fox, M.J.H. On the nonlinear transfer of energy in the peak of a gravity-wave spectrum. II.
  Proc. A **348**, 467–483 (1976).
Fox, P.G. *See* Fuller (K.N.G.), Fox & Field.
Fox, P.J. *See* Ryan (W.B.F.) & Fox.
Fox, R.H. *See* Beaven (G.H.), Fox & Hornabrook; Budd, Fox, Hendrie & Hicks; Edholm (O.G.), Samueloff, Mourant, Fox and others.
Fox, R.H., Budd, G.M., Woodward, Patricia M., Hackett, A.J. & Hendrie, A.L. A study of temperature regulation in New Guinea people (Discussion). *Trans.* B **268**, 375–391 (1974).
Fox, R.H., Even-Paz, Z., Woodward, Patricia M. & Jack, J.W. Biological studies of Yemenite and Kurdish Jews in Israel and other groups in southwest Asia. VIII. A study of temperature regulation in Yemenite and Kurdish Jews in Israel. *Trans.* B **266**, 149–168 (1973).
Francis, J.R.D. Experiments on the motion of solitary grains along the bed of a water-stream.
  Proc. A **332**, 443–471 (1973).
  See also Abbott & Francis.

Franck, B., Rowold, A., Wegner, Ch. & Eckert, H.-G. Synthesis of probable and improbable precursors for porphyrin biosynthesis (Discussion). *Trans.* B 273, 181–189 (1976).

Frank, F.C. Dislocation models for fault creep processes (Discussion). *Trans.* A 274, 351–354 (1973). *See also* Frank (Sir Charles).

Frank, F.C. & Perkins, D.H. Cecil Frank Powell. (With an appendix by A.M. Tyndall.) *Biogr. Mem.* 17, 541–563 (1971).

Frank, L.A. *See* Craven & Frank.

Frank, Sir Charles. The Frank–Read source (Symposium). *Proc.* A 371, 136–138 (1980). *See also* Frank (F.C.).

Franke, W.W. Structure and biochemistry of the nuclear envelope (Discussion). *Trans.* B 268, 67–93 (1974).

Franke, W.W. & Scheer, U. Morphology of transcriptional units at different states of activity (Discussion). *Trans.* B 283, 333–342 (1978).

Frankel, Sir Otto. Base-sterile speltoids: the location of the *Bs* gene of *Triticum aestivum*. *Proc.* B 188, 163–166 (1975).
Floral initiation in wheat. *Proc.* B 192, 273–298 (1976).

Frankel, Sir Otto & Roskams, Mary. Stability of floral differentiation in *Triticum*. *Proc.* B 188, 139–162 (1975).

Frankenhaeuser, B. & Århem, P. Steady state current rectification in potential clamped nodes of Ranvier (*Xenopus laevis*). *Trans.* B 270, 515–525 (1975).

Frankevich, E.L. Morrow, T. & Salmon, G.A. The radiation-induced formation of excited states of aromatic hydrocarbons in benzene and cyclohexane. I. Radiation induced fluorescence from solutions of aromatic hydrocarbons in cyclohexane. *Proc.* A 328, 445–456 (1972).

Franklin, R.M., Hinnen, Rosmarie, Schäefer, R. & Tsukagoshi, N. Structure and assembly of lipid-containing viruses, with special reference to bacteriophage PM2 as one type of model system (Discussion). *Trans.* B 276, 63–80 (1976).

Franklin, R.N., Edgley, P.D , Hamberger, S.M. & Motley, R.W. Nonlinear behaviour of a finite amplitude electron plasma wave. IV. Decay to ion cyclotron waves. *Proc.* A 363, 547–557 (1978).

Franklin, R.N., Hamberger, S.M., Lampis, G. & Smith, G.J. Nonlinear behaviour of a finite amplitude electron plasma wave.
I. Electron trapping effects. *Proc.* A 347, 1–24 (1975).
II. Wave–wave interactions. *Proc.* A 347, 25–46 (1975).

Franklin, R.N., MacKinlay, R.R., Edgley, P.D. & Wall, D.N. Nonlinear behaviour of a finite amplitude electron plasma wave. III. The sideband instability. *Proc.* A 360, 229–242 (1978).

Franks, A., Lindsey, K., Bennett, J.M., Speer, R.J., Turner, D. & Hunt, D.J. The theory, manufacture, structure and performance of N.P.L. X-ray gratings. *Trans.* A 277, 503–543 (1975).

Franks, F. Solvation and conformational effects in aqueous solutions of biopolymer analogues (Discussion). *Trans.* B 278, 33–57 (1977).
Solvation interactions of proteins in solution (Discussion). *Trans.* B 278, 89–96 (1977).

Fraser, D.A. *See* Allen (J.E.), Fang & Fraser.

Frazier, E.N. The photosphere – magnetic and dynamic state (Discussion). *Trans.* A 281, 295–303 (1976).

Frazier, J. Observations on sea turtles at Aldabra Atoll (Discussion). *Trans.* B 260, 373–410 (1971).

Fredga, K. Chromosomal changes in vertebrate evolution (Discussion). *Proc.* B 199, 377–397 (1977).

Fredholm, B. *See* Hökfelt, Lundberg, Schultzberg, Johansson, Skirboll, Anggård and others.

Freedman, P.A. *See* Till (S.M.), Freedman, Tuckett & Jones.

Freeman, F.F., Gabriel, A.H., Jones, B.B. & Jordan, Carole. Helium-like ion forbidden line emission, and solar active regions (Discussion). *Trans.* A 270, 127–133 (1971).

Freeman, M.A.R. Unlinked surface replacement (Discussion). *Proc.* B 192, 199–205 (1976).

Freeman, N.C. A two dimensional distributed soliton solution of the Korteweg–de Vries equation. *Proc.* A 366, 185–204 (1979). *See also* Anker & Freeman.

Freeman, N.C. & Davey, A. On the evolution of packets of long surface waves. *Proc.* A 344, 427–433 (1975).

Freeman, N.C. & Johnson, R.S. A note on stellar winds and breezes. *Proc.* A 329, 241–249 (1972).

Freeman, R. Review Lecture. Nuclear magnetic resonance spectroscopy in two frequency dimensions. *Proc.* A 373, 149–178 (1980).
  See also Finean, Freeman & Limbrick.

French, J.E. *See* Sheppard (B.L.) & French.

Frère, J.-M. *See* Ghuysen, Frère, Leyh-Bouille, Perkins & Nieto.

Frick, U. *See* Signer, Baur, Etique, Frick & Funk.

Fricker, H.S. The effects on rubber elasticity of the addition and scission of cross-links under strain. *Proc.* A 335, 267–287 (1973).
  On the theory of stress relaxation by cross-link reorganization. *Proc.* A 335, 289–300 (1973).

Friday, A.E. *See* Romero-Herrera, Lehmann, Joysey & Friday.

Friedlander, E. High voltage a.c. power transmission development (Discussion). *Trans.* A 275, 189–192 (1973).

Friedman, H. A survey of the X-ray sky with HEAO A-1 (Discussion). *Proc.* A 366, 423–434 (1979).

Friedman, J.L. On the Born approximation for perturbations of a spherical star and the Newman–Penrose constants. *Proc.* A 335, 163–190 (1973).

Friedman, M., Teufel, L.W. & Morse, J.D. Strain and stress analyses from calcite twin lamellae in experimental buckles and faulted drape-folds (Discussion). *Trans.* A 283, 87–107 (1976).

Fries, W. *See* Zeki (S.) & Fries.

Fripp, R.E.P. *See* Gay (N.C.) & Fripp.

Frisby, J.P. & Mayhew, J.E.W. Spatial frequency tuned channels: implications for structure and function from psychophysical and computational studies of stereopsis (Discussion). *Trans.* B 290, 95–116 (1980).

Frischer, Ruth. *See* Lubell & Frischer.

Frith, C.B. Feeding ecology of land birds on West Island, Aldabra Atoll, Indian Ocean: a preliminary survey (Discussion). *Trans.* B 286, 195–210 (1979).

Frith, Dawn W. A twelve month study of insect abundance and composition at various localities on Aldabra Atoll (Discussion). *Trans.* B 286, 119–126 (1979).

Froesch, D. Antigen-induced secretion in the optic gland of *Octopus vulgaris*. *Proc.* B 205, 379–384 (1979).

Froesch, D. & Marthy, H.-J. The structure and function of the oviducal gland in octopods (Cephalopoda). *Proc.* B 188, 95–101 (1975).

Fröhlich, A. *See* Cassels (J.W.S.) & Fröhlich.

Fröhlich, A. & Taylor, M.J. The arithmetic theory of local Galois Gauss sums for tame characters. *Trans.* A 298, 141–181 (1980).

Fröhlich, H. Recollections of the development of solid state physics (Symposium). *Proc.* A 371, 102–103 (1980).

Froidevaux, C. & Souriau, M. Thermo-mechanical models of lithosphere and asthenosphere: can a change in plate velocity induce magmatic activity? (Discussion). *Trans.* A 288, 387–392 (1978).

Froines, J.R. *See* Costa (S.M. de B.), Froines, Harris, Leblanc, Orger & Porter.

Frost, A.J., Hill, A.W. & Brooker, B.E. The early pathogenesis of bovine mastitis due to *Escherichia coli*. *Proc.* B 209, 431–439 (1980).

Frost, B.W. *See* Steele (J.H.) & Frost.

Frost, J.C., Leadbetter, A.J. & Richardson, R.M. Molecular crystals and liquid crystals: new results for *t*-butyl chloride (Discussion). *Trans.* B 290, 567–582 (1980).

Fry, T.M. *See* Bacon (F.T.) & Fry.

Frydman, B., Frydman, Rosalia B., Valasinas, Aldonia, Levy, Estrella S. & Feinstein, Graciela. Biosynthesis of uroporphyrinogens from porphobilinogen: mechanism and the nature of the process (Discussion). *Trans.* B 273, 137–160 (1976).

Frydman, Rosalia B. *See* Frydman (B.), Frydman, Valasinas, Levy & Feinstein.

Fryer, G. Evolution and adaptive radiation in the Macrothricidae (Crustacea: Cladocera): a study in comparative functional morphology and ecology. *Trans.* B 269, 137–274 (1974).
  Studies on the functional morphology and ecology of the atyid prawns of Dominica. *Trans.* B 277, 57–128 (1977).

Sidnie Milana Manton. *Biogr. Mem.* **26**, 327–356 (1980).

Fryer, G.M. Theory of stress-induced diffusion in ionic crystals. *Proc.* A **327**, 81–96 (1972).

Fryer, R.J. *See* Fielder, Fryer, Titulaer, Herring & Wise.

Fuchs, Sir Vivian. Introductory remarks to a Discussion. *Trans.* B **279**, 3 (1977).

Fuhrhop, J.-H. & Subramaian, J. Chemical reactivities of tetrapyrrole pigments: a comparison of experimental behaviour with the results of s.c.f.–π–m.o. calculations (Discussion). *Trans.* B **273**, 335–352 (1976).

Fuller, K.N.G., Fox, P.G. & Field, J.E. The temperature rise at the tip of fast-moving cracks in glassy polymers. *Proc.* A **341**, 537–557 (1975).

Fuller, K.N.G. & Tabor, D. The effect of surface roughness on the adhesion of elastic solids. *Proc.* A **345**, 327–342 (1975).

Fuller, M.D. Review of effects of shock (< 60 kbar; $< 6 \times 10^9$ Pa) on magnetism of lunar samples (Discussion). *Trans.* A **285**, 409–416 (1977).

Fulling, S.A. *See* Davies (P.C.W.) & Fulling.

Fulling, S.A. & Davies, P.C.W. Radiation from a moving mirror in two dimensional space-time: conformal anomaly. *Proc.* A **348**, 393–414 (1976).

Fulton, Barbara P., Miledi, R. & Szczepaniak, C. Effects of a *Dendroaspis* neurotoxin on synaptic transmission in the spinal cord and the neuromuscular junction of the frog. *Proc.* B **207**, 491–497 (1980).

Fulton, Barbara P., Miledi, R. & Takahashi, T. Electrical synapses between motoneurons in the spinal cord of the newborn rat. *Proc.* B **208**, 115–120 (1980).

Funk, H. *See* Signer, Baur, Etique, Frick & Funk.

Funnell, B.M., Norton, P.E.P. & West, R.G. The crag at Bramerton, near Norwich, Norfolk (with an appendix by D.F. Mayhew). *Trans.* B **287**, 489–534 (1979).

Furness, J.B. & Costa, M. The nervous release and the action of substances which affect intestinal muscle through neither adrenoreceptors nor cholinoreceptors (Discussion). *Trans.* B **265**, 123–133 (1973).

Fyfe, W.S. The granulite facies, partial melting and the Archaean crust (Discussion). *Trans.* A **273**, 457–461 (1973).
Heat flow and magmatic activity in the Proterozoic (Discussion). *Trans.* A **280**, 655–660 (1976).
Chemical aspects of rock deformation (Discussion). *Trans.* A **283**, 221–228 (1976).
The geochemical cycle of uranium (Discussion). *Trans.* A **291**, 433–445 (1979).

Fynn, M., Onomakpome, N. & Peart, W.S. The effect of ionophores (A23187 and RO2–2985) on renin secretion and renal vasoconstriction. *Proc.* B **199**, 199–212 (1977).

Gabbay, K.H. *See* Steiner (D.F.), Patzelt, Chan, Quinn, Tager, Nielsen and others.

Gabella, G. Fine structure of smooth muscle (Discussion). *Trans.* B **265**, 7–16 (1973).
The sphincter pupillae of the guinea-pig: structure of muscle cells, intercellular relations and density of innervation. *Proc.* B **186**, 369–386 (1974).

Gabor, D. & Brown, J. Willis Jackson – Baron Jackson of Burnley. *Biogr. Mem.* **17**, 379–398 (1971).

Gabriel, A.H. A magnetic model of the solar transition region (Discussion). *Trans.* A **281**, 339–352 (1976).
See also Freeman (F.F.), Gabriel, Jones & Jordan.

Gadian, D.G. *See* Ackerman, Bore, Gadian, Grove & Radda; Dawson (M. Joan), Gadian & Wilkie.

Gadoth, N. *See* Lehmann (E.E.), Gadoth & Samueloff.

Gadsby, D.C., Niedergerke, R. & Ogden, D.C. The dual nature of the membrane potential increase associated with the activity of the sodium/potassium exchange pump in skeletal muscle fibres. *Proc.* B **198**, 463–472 (1977).

Gagliardi, D. Digital coaxial cable systems (Discussion). *Trans.* A **289**, 113–121 (1978).

Gait, P.D. *See* Simpson (C.J.S.M.), Gait & Simmie.

Gakis, N. *See* Hartland & Gakis.

Galbraith, Gillian M.P. *See* Faulk (W.P.) & Galbraith.

Gale, B. *See* McCartney & Gale.

Gallagher, J.T. *See* Phipps, Richardson, Corfield, Gallagher and others.

Gallagher, J.T., Kent, P.W., Passatore, M., Phipps, R.J. & Richardson, P.S. The composition of tracheal mucus and the nervous control of its secretion in the cat (with an appendix by D. Lamb). *Proc.* B **192**, 49–76 (1975).

Galli, G. *See* Fiecchi, Kienle, Scala, Galli and others.

Gallop, Angela, Bartrop, Julie & Smith, D.C. The biology of chloroplast acquisition by *Elysia viridis*. *Proc.* B **207**, 335–349 (1980).

Gamble, J.C. *See* Davies (J.M.) & Gamble; Hughes (R.N.) & Gamble.

Gambling, W.A. & Payne, D.N. Optical fibre systems (Discussion). *Trans.* A **289**, 135–150 (1978).

Games, D.E. *See* Jackson (A.H.), Sancovich, Ferramola, Evans and others.

Gane, N., Pfaelzer, P.F. & Tabor, D. Adhesion between clean surfaces at light loads. *Proc.* A **340**, 495–517 (1974).

Ganellin, C.R. *See* Richards (W.G.), Clarkson & Ganellin.

Ganf, G.G. *See* Burgis, Darlington, Dunn, Ganf, Gwahaba & McGowan.

Ganf, G.G. & Viner, A.B. Ecological stability in a shallow equatorial lake (Lake George, Uganda) (Discussion). *Proc.* B **184**, 321–346 (1973).

Gaposchkin, E.M. Gravity-field determination from laser observations (Discussion). *Trans.* A **284**, 515–527 (1977).

Garamvölgyi, N., Vizi, E.S. & Knoll, J. The site and state of myosin in intestinal smooth muscle (Discussion). *Trans.* B **265**, 219–222 (1973).

Garcia, A.G. *See* Dixon (W.R.), Garcia & Kirpekar.

García-Moliner, F. & Rubio, J. The quantum theory of one-electron states at surfaces and interfaces. *Proc.* A **324**, 257–273 (1971).

Garcia-Romeu, F. Anionic and cationic exchange mechanisms in the skin of anurans, with special reference to Leptodactylidae *in vivo* (Discussion). *Trans.* B **262**, 163–174 (1971).

Gardner, A.W. & Glueckauf, E. The activity coefficients of electrolytes with particular reference to aqueous mixtures of 2:2 with 1:1 electrolytes. *Proc.* A **321**, 515–543 (1971).

Gardner, W.E. Image processing (Discussion). *Trans.* A **292**, 251–256 (1979).

Gardner-Medwin, A.R. The recall of events through the learning of associations between their parts. *Proc.* B **194**, 375–402 (1976).

Garey, L.J. A light and electron microscopic study of the visual cortex of the cat and monkey. *Proc.* B **179**, 21–40 (1971).
See also Fisken, Garey & Powell.

Garey, L.J. & Powell, T.P.S. An experimental study of the termination of the lateral geniculo-cortical pathway in the cat and monkey. *Proc.* B **179**, 41–63 (1971).

Garland, J.A. The dry deposition of sulphur dioxide to land and water surfaces. *Proc.* A **354**, 245–268 (1977).

Garland, J.H.N. Forecasting the effects of polluting discharges on rivers with special reference to the River Trent (Discussion) [abstract]. *Proc.* B **180**, 437–438 (1972).

Garlick, G.F.J. Lunar surface movements — the evidence and the causes (Discussion). *Trans.* A **285**, 325–329 (1977).

Garms, R. *See* Le Berre, Garms, Davies, Walsh & Philippon.

Garner, R.C. Carcinogen prediction in the laboratory: a personal view (Discussion). *Proc.* B **205**, 121–134 (1979).

Garnham, P.C.C. Edward Hindle. *Biogr. Mem.* **20**, 217–234 (1974).
Alexander John Haddow. *Biogr. Mem.* **26**, 225–254 (1980).
See also Peters (W.), Garnham, Killick-Kendrick, Rajapaksa and others; Shortt & Garnham.

Garnham, P.C.C. & Kuttler, K.L. A malaria parasite of the white-tailed deer (*Odocoileus virginianus*) and its relation with known species of *Plasmodium* in other ungulates. *Proc.* B **206**, 395–402 (1980). Erratum. *Proc.* B **208**, 483 (1980).

Garrett, S.D. William Brown. *Biogr. Mem.* **21**, 155–174 (1975).

Gartner, E.M. & Thrush, B.A. Infrared emission by active nitrogen.
I. The kinetic behaviour of $N_2(B'\ ^3\Sigma_u^-)$. *Proc.* A **346**, 103–119 (1975).

II. The kinetic behaviour of $N_2(B\,^3\Pi_g)$. *Proc.* A **346**, 121—137 (1975).

Garton, W.R.S. *See* Connerade, Baig, Garton & Newsom; Connerade, Garton, Mansfield & Martin; Lu, Tomkins & Garton.

Garton, W.R.S. & Parkinson, W.H. Series of autoionization resonances in Ba I converging on Ba II 6 ^{2}P. *Proc.* A **341**, 45—48 (1974).

Garton, W.R.S., Reeves, E.M. & Tomkins, F.S. Hyperfine structure and isotope shift of the $6s6p^2$ $^4P_{1/2}$ level of Tl I. *Proc.* A **341**, 163—166 (1974).

Garton, W.R.S., Reeves, E.M., Tomkins, F.S. & Ercoli, B. Rydberg series and autoionization resonances in the Sc I absorption spectrum. *Proc.* A **333**, 1—16 (1973).
   Rydberg series and autoionization resonances in the Y I absorption spectrum.
      Proc. A **333**, 17—24 (1973).

Garton, W.R.S., Tomkins, F.S. & Crosswhite, H.M. Magnetic effects in Ba I and Sr I absorption spectra. *Proc.* A **373**, 189—197 (1980).

Garvey, J.P., Gold, V., McAdam, M.E. & Cooper, A. Kinetics of hydrogen isotope exchange reactions. XXVIII. Steric course of radiation-induced tritium exchange between water and cyclohexane-1,2-diols. *Proc.* A **346**, 427—441 (1975).

Gaskell, T.F. Environmental pollution in offshore operations (Discussion). *Trans.* A **290**, 179—185 (1978).

Gass, I.G. Proposals concerning the variation of volcanic products and processes within the oceanic environment (Discussion). *Trans.* A **271**, 131—140 (1972).

Gass, I.G., Chapman, D.S., Pollack, H.N. & Thorpe, R.S. Geological and geophysical parameters of mid-plate volcanism (Discussion). *Trans.* A **288**, 581—597 (1978).

Gast, P.W. Dispersed element chemistry of oceanic ridge basalts [abstract] (Discussion). *Trans.* A **268**, 467 (1971).

van Gastel, A.J.G. *See* de Nettancourt, Devreux, Carluccio and others.

Gaster, M. A theoretical model of a wave packet in the boundary layer on a flat plane. *Proc.* A **347**, 271—289 (1975).
   See also Roberts (J.B.) & Gaster.

Gaster, M. & Grant, I. An experimental investigation of the formation and development of a wave packet in a laminar boundary layer. *Proc.* A **347**, 253—269 (1975).

Gaster, M. & Roberts, J.B. The spectral analysis of randomly sampled records by a direct transform. *Proc.* A **354**, 27—58 (1977).

Gates, D.J., O'Connor, A.J. & Westcott, M. Partitioning the union of disks in plant competition models. *Proc.* A **367**, 59—79 (1979). [Abstract]. *Proc.* B **205**, 423 (1979).

Gatski, T.B. & Liu, J.T.C. On the interactions between large-scale structure and fine-grained turbulence in a free shear flow. III. A numerical solution. *Trans.* A **293**, 473—509 (1980).

Gaughwin, M. *See* Brooks (D.E.), Gaughwin & Mann.

Gauthier, F. Field and laboratory studies of the rheology of Mount Etna lava (Symposium). *Trans.* A **274**, 83—98 (1973).

Gawler, J. Clinical experience with the E.M.I. scanner (Discussion). *Proc.* B **195**, 291—297 (1977).

Gay, I.D. *See* Mason (R.), Textor, Iwasawa & Gay; Textor, Gay & Mason.

Gay, I.D., Textor, M., Mason, R. & Iwasawa, Y. Photoelectron spectroscopy, and low energy electron diffraction studies of the adsorption of dinitrogen, hydrogen and ammonia on a Fe(111) single crystal surface. *Proc.* A **356**, 25—36 (1977).

Gay, N.C. & Fripp, R.E.P. The control of ductility on the deformation of pebbles and conglomerates (Discussion). *Trans.* A **283**, 109—128 (1976).

Gaze, R.M. *See* Barlow (H.B.) & Gaze; Hope (R.A.), Hammond & Gaze; Keating, Beazley, Feldman & Gaze.

Gaze, R.M., Keating, M.J. & Chung, S.H. The evolution of the retinotectal map during development in *Xenopus. Proc.* B **185**, 301—330 (1974).

Geake, J.E. *See* Dollfus & Geake.

Geake, J.E., Walker, G., Telfer, D.J. & Mills, A.A. The cause and significance of luminescence in lunar plagioclase (Discussion). *Trans.* A **285**, 403—408 (1977).

Gebbie, H.A. Light scattering studies of the atmosphere (Discussion). *Trans.* A **293**, 413—417 (1979).

Gee, G. Introductory remarks to a Discussion. *Proc.* A **351**, 297–299 (1976).

Gee, J.V., Hayes, W. & O'Brien, M.C.M. Magneto-optical properties of interstitial hydrogen atoms in alkali halides. *Proc.* A **322**, 27–44 (1971).

Geffen, L.B. *See* Livett, Geffen & Rush.

Geiss, J., Eberhardt, P., Grögler, N., Guggisberg, S., Maurer, P. & Stettler, A. Absolute time scale of lunar mare formation and filling (Discussion). *Trans.* A **285**, 151–158 (1977).

Gelbart, W.M. Collision-induced and multiple light scattering by simple fluids (Discussion). *Trans.* A **293**, 359–375 (1979).

van Gelder, Monique. *See* Ito (Y.), Miledi, Molenaar, Vincent, Polak and others.

Gemne, G. Ontogenesis of corneal surface ultrastructure in nocturnal Lepidoptera. *Trans.* B **262**, 343–363 (1971).

Genat, B.R. & Mark, R.F. Electrophysiological experiments on the mechanism and accuracy of neuromuscular specificity in the axolotl (Discussion). *Trans.* B **278**, 335–347 (1977).

Genetello, C. *See* Schell, Van Montagu, De Beuckeleer, De Block and others.

de Gennes, P.G. Light scattering and dynamics of flexible polymers [abstract] (Discussion). *Trans.* A **293**, 391 (1979).

Gentner, W. *See* Krätschmer & Gentner.

George, T. Neville. Mid-Dinantian (Chadian) limestones in Gower. *Trans.* B **282**, 411–463 (1978).

Georgeson, M.A. Spatial frequency analysis in early visual processing (Discussion). *Trans.* B **290**, 11–22 (1980).

Georgiev, G.P. *See* Varshavsky, Bakayev, Bakayeva, Chumackov *et al.*

Georgiou, A.S., Legon, A.C. & Millen, D.J. Spectroscopic investigations of hydrogen bonding interactions in the gas phase. III. The identification of the hydrogen-bonded heterodimer $(CH_3)_3 CCN\cdots HF$ and the determination of its geometry by microwave and infrared spectroscopy. *Proc.* A **370**, 257–268 (1980).

Gérard, A. La tectonique du socle sous la Manche occidentale d'après les données du magnétisme aéroporté (Discussion). *Trans.* A **279**, 55–68 (1975).

Gerard, V.B. New Zealand Earth strain measurements (Discussion). *Trans.* A **274**, 311–321 (1973).

Gerba, Peggy. *See* van Breemen, Farinas, Gasteels, Gerba, Wuytack & Deth.

Gerhard, W. *See* Laver, Air, Webster, Gerhard, Ward & Dopheide.

Gerisch, G., Hülser, D., Malchow, D. & Wick, U. Cell communication by periodic cyclic-AMP pulses (Discussion). *Trans.* B **272**, 181–192 (1975).

Gerjuoy, E. Configuration space theory of 'truly three-body' scattering rates. *Trans.* A **270**, 197–287 (1971).

Gerloch, M. *See* Boyd (P.D.W.), Davies & Gerloch; Boyd (P.D.W.), Gerloch, Harding & Woolley.

Gerloch, M. & Harding, J.H. Superexchange in copper acetates. *Proc.* A **360**, 211–227 (1978).

Germond, J.E. *See* Oudet, Germond, Bellard, Spadafora & Chambon.

Geroch, R., Kronheimer, E.H. & Penrose, R. Ideal points in space–time. *Proc.* A **327**, 545–567 (1972).

Gerrard, J.H. The wakes of cylindrical bluff bodies at low Reynolds number. *Trans.* A **288**, 351–382 (1978).

Gerratt, J. The calculation of intermolecular potential energy surfaces. I. Basic theory. *Proc.* A **350**, 363–380 (1976).
See also Pyper & Gerratt.

Gerratt, J. & Raimondi, M. The spin-coupled valence bond theory of molecular electronic structure. I. Basic theory and application to the $^2\Sigma^+$ states of BeH. *Proc.* A **371**, 525–552 (1980).

Gerratt, J. & Wilson, I.D.L. L^2 R-matrix studies of molecular collision processes — energy dependence of $\sigma_{vj} \to v'j'$ for ^{4}He + H$_2$. *Proc.* A **372**, 219–241 (1980).

Gerschenfeld, H.M. *See* Piccolino & Gerschenfeld.

Gerschenfeld, H.M. & Piccolino, M. Sustained feedback effects of L-horizontal cells on turtle cones. *Proc.* B **206**, 465–480 (1980).

Getreuer, K.W. *See* Bovée, Creyghton, Getreuer, Korbee, Lobregt and others.

Ghaffar, A. *See* Woodruff (Sir Michael), Dunbar & Ghaffar.

Ghendon, Y.Z. & Markushin, S.G. Studies on mutation lesions and physiology of fowl plague virus *ts* mutants (Discussion). *Trans.* B **288**, 383–392 (1980).

Ghosh, A., Basu, A.N. & Sengupta, S. Lattice statics and dynamics of the NaF crystal.
Proc. A **340**, 199—211 (1974).

Ghuysen, J.-M., Frère, J.-M., Leyh-Bouille, M., Perkins, H.R. & Nieto, M. The active centres in penicillin-sensitive enzymes (Discussion). *Trans.* B **289**, 285—301 (1980).

Giachardi, D.J. & Wayne, R.P. The photolysis of ozone by ultraviolet radiation. VI. Reactions of $O(^1D)$. *Proc.* A **330**, 131—146 (1972).

Gibb, O. & Richards, H.J. Planning for development of groundwater and surface water resources (Discussion). *Proc.* A **363**, 109—130 (1978).

Gibb, T.C., Greatrex, R. & Greenwood, N.N. Mössbauer studies of Luna 16 and 20 lunar soils.
Trans. A **284**, 157—165 (1977).
 An assessment of results obtained from Mössbauer spectra of lunar samples (Discussion).
Trans. A **285**, 235—240 (1977).

Gibbard, P.L. Pleistocene history of the Vale of St Albans. *Trans.* B **280**, 445—483 (1977).

Gibbon, J.D. *See* Caudrey, Dodd & Gibbon; Dodd (R.K.) & Gibbon.

Gibbon, J.D., James, I.N. & Moroz, I.M. An example of soliton behaviour in a rotating baroclinic fluid. *Proc.* A **367**, 219—237 (1979).

Gibbons, G.W. Non-existence of equilibrium configurations of charged black holes.
Proc. A **372**, 535—538 (1980).

Gibbons, G.W. & Perry, M.J. Black holes and thermal Green functions.
Proc. A **358**, 467—494 (1978).

Gibbons, M.P. *See* Acheson (D.J.) & Gibbons.

Gibbons, T.B. *See* Thomas (G.B.) & Gibbons.

Gibbons, W. *See* Bullough (K.), Denby, Gibbons, Hughes and others.

Gibbs, C.F. Quantitative studies on marine biodegradation of oil. I. Nutrient limitation at 14°C.
Proc. B **188**, 61—82 (1975).

Gibbs, C.F., Pugh, K.B. & Andrews, A.R. Quantitative studies on marine biodegradation of oil. II. Effect of temperature. *Proc.* B **188**, 83—94 (1975).

Gibbs, P.E. Macrofauna of the intertidal sand flats on low wooded islands, northern Great Barrier Reef (Discussion). *Trans.* B **284**, 81—97 (1978).
 See also Stoddart (D.R.), McLean, Scoffin & Gibbs.

Gibson, A.F., Kimmitt, M.F., Koohian, A.O., Evans, D.E. & Levy, G.F.D. A study of radiation pressure in a refractive medium by the photon drag effect. *Proc.* A **370**, 303—311 (1980).

Gibson, E.J. The uses of wood: long term prospects (Discussion). *Trans.* B **271**, 91—100 (1975).

Gibson, I.L. *See* Morrison (M. Ann), Thompson, Gibson & Marriner.

Gibson, I.L. & Piper, J.D.A. Structure of the Icelandic basalt plateau and the process of drift (Discussion). *Trans.* A **271**, 141—150 (1972).

Gibson, Q.H. Francis John Worsley Roughton. *Biogr. Mem.* **19**, 563—582 (1973).

Gibson, T.S.H. Green turtle (*Chelonia mydas* (L.)) nesting activity at Aldabra Atoll (Discussion).
Trans. B **286**, 255—263 (1979).

Giddings, C. *See* Horridge, Giddings & Stange; Horridge, Giddings & Wilson.

Giddings, Caroline. *See* Horridge & Giddings.

Giebisch, G., Boulpaep, E.L. & Whittembury, G. Electrolyte transport in kidney tubule cells (Discussion). *Trans.* B **262**, 175—196 (1971).

Gieskes, W.W.C. *See* van Bennekom, Gieskes & Tijssen.

Gifford, D.R. *See* Ratter, Askew, Montegomery & Gifford; Ratter, Richards, Argent & Gifford.

Gijzeman, O.L.J. *See* Beddard, Fleming, Gijzeman & Porter.

Gilbert, F. Derivation of source parameters from low-frequency spectra (Discussion).
Trans. A **274**, 369—371 (1973).

Gilbert, F. & Dziewonski, A.M. An application of normal mode theory to the retrieval of structural parameters and source mechanisms from seismic spectra. *Trans.* A **278**, 187—269 (1975).

Gilbert, P.F.C. The reconstruction of a three-dimensional structure from projections and its application to electron microscopy. II. Direct methods. *Proc.* B **182**, 89—102 (1972).
 [Abstract]. *Proc.* A **330**, 147 (1972).

Gilbert, R.F. *See* Iversen (L.L.), Lee, Gilbert, Hunt & Emson.

Gilbertson, D.D. The palaeoecology of Middle Pleistocene mollusca from Sugworth, Oxfordshire. *Trans.* B **289**, 107—118 (1980).

Gill, A.E. Ocean models (Discussion). *Trans.* A **270**, 391—413 (1971).

A simple model for showing effects of geometry on the ocean tides. *Proc.* A **367**, 549—571 (1979).

Gill, W.N. *See* Sankarasubramanian & Gill.

Gill, W.N. & Sankarasubramanian, R. Dispersion of a non-uniform slug in time-dependent flow. *Proc.* A **322**, 101—117 (1971).

Dispersion of non-uniformly distributed time-variable continuous sources in time-dependent flow. *Proc.* A **327**, 191—208 (1972).

Gille, J.C., Bailey, P.L. & Russell, J.M., III. Temperature and composition measurements from the l.r.i.r. and l.i.m.s. experiments on Nimbus 6 and 7 (Discussion). *Trans.* A **296**, 205—218 (1980).

Gillespie, J.I. The effect of repetitive stimulation on the passive electrical properties of the presynaptic terminal of the squid giant synapse. *Proc.* B **206**, 293—306 (1979).

Gillespie, J.S., Creed, Kate E. & Muir, T.C. Electrical changes underlying excitation and inhibition in intestinal and related smooth muscle (Discussion). *Trans.* B **265**, 95—106 (1973).

Gillett, J.D. *See* Cole (S.J.) & Gillett.

Gillett, J.D., Cole, S.J. & Reeves, D. The influence of the brain hormone on retention of blood in the mid-gut of the mosquito *Aedes aegypti* (L.). *Proc.* B **190**, 359—367 (1975).

Gillett, J.D., Roman, E.A. & Phillips, V. Erratic hatching in *Aedes* eggs: a new interpretation. *Proc.* B **196**, 223—232 (1977).

Gillison, A.N. Phytogeographical relationships of the northern islands of the New Hebrides (Discussion). *Trans.* B **272**, 385—390 (1975).

Gilmour, R.S. Chromatin transcription with mercurated nucleotides (Discussion). *Trans.* B **283**, 379—380 (1978).

Gilpin-Brown, J.B. *See* Denton, Gilpin-Brown & Wright.

Gilson, J.C. Medicine and mineralogy (Discussion). *Trans.* A **286**, 585—592 (1977).

Gimblett, C.G. & Peckover, R.S. On the mutual interaction between rotation and magnetic fields for axisymmetric bodies. *Proc.* A **368**, 75—97 (1979).

Ginzburg, V.L. On the origin of cosmic rays (Discussion). *Trans.* A **277**, 463—479 (1975).

Girdler, R.W., Brown, C., Noy, D.J.M. & Styles, P. A geophysical survey of the westernmost Gulf of Aden. *Trans.* A **298**, 1—43 (1980).

Giresse, P. *See* Huault, Lefebvre, Guyader, Giresse and others.

Gittus, J.H. Theoretical equation for steady-state dislocation creep in a material having a threshold stress. *Proc.* A **342**, 279—287 (1975).

Dirac's Large Numbers theory and the structure of rocks. *Proc.* A **343**, 155—158 (1975).

Dirac's Large Numbers theory and a theoretical upper limit to the viscosity of crystalline materials. *Proc.* A **348**, 95—99 (1976).

High-temperature deformation of two phase structures (Discussion). *Trans.* A **288**, 121—146 (1978).

Gladman, T. *See* May (M.J.), Gladman & Walker.

Gladstone, G.P., Knight, B.C.J.G. & Wilson, Sir Graham. Paul Gordon Fildes. *Biogr. Mem.* **19**, 317—347 (1973).

Glass, D.V. Review Lecture. Recent and prospective trends in fertility in developed countries. *Trans.* B **274**, 1—52 (1976).

Glass, M. See Spizzichino & Glass.

Glauert, Audrey M. *See* Sanderson (C.J.) & Glauert; Thornley, Glauert & Sleytr.

Glencross, W.M. Extreme ultraviolet emission during flares (Discussion). *Trans.* A **270**, 117—125 (1971).

See also Brabban & Glencross; Herring (J.R.H.), Glencross, Parkinson & Pounds.

Glenister, P.H. *See* Lyon & Glenister.

Glenny, R.J.E. & Hopkins, B.E. Gas turbine requirements (Symposium). *Trans.* A **282**, 105—118 (1976).

Glimcher, M.J. *See* Herzfeld, Roufosse, Haberkorn, Griffin & Glimcher.

Glitsch, H.G. *See* Baker (P.F.) & Glitsch.

Glover, J., Pennock, J.F., Pitt, G.A.J. & Goodwin, T.W. Richard Alan Morton. *Biogr. Mem.* **24**, 409—442 (1978).

Gluckman, M.J. *See* Leichtberg, Weinbaum, Pfeffer & Gluckman.

Glucksman, J. *See* Ball (E.) & Glucksman.

Glueckauf, E. Further studies of 2:2 electrolytes: osmotic and activity coefficients.
Proc. A 351, 471—479 (1976).
Robert Spence. *Biogr. Mem.* 23, 501—528 (1977).
See also Gardner (A.W.) & Glueckauf.

Glynn, I.M., Hoffman, J.F. & Lew, V.L. Some 'partial reactions' of the sodium pump (Discussion).
Trans. B 262, 91—102 (1971).

Goad, L.J., Rubinstein, I. & Smith, A.G. The sterols of echinoderms (Discussion).
Proc. B 180, 223—246 (1972).

Goda, M.A.A. *See* Rao (C.R.A.) & Goda.

Godber, Marilyn J. *See* Mourant, Godber, Kopeć, Tills & Woodhead.

Godber, M.J., Kopeć, A.C., Mourant, A.E., Tills, D. & Lehmann, E.E. Biological studies of Yemenite and Kurdish Jews in Israel and other groups in southwest Asia. IX. The hereditary blood factors of the Yemenite and Kurdish Jews. *Trans.* B 266, 169—184 (1973).

Godwin, Sir Harry. History of the natural forests of Britain: establishment, dominance and destruction (Discussion). *Trans.* B 271, 47—67 (1975).
Concluding remarks to a Discussion. *Trans.* B 280, 373—374 (1977).

Godwin, Sir Harry & Vishnu-Mittre. Studies of the post-glacial history of British vegetation. XVI. Flandrian deposits of the fenland margin at Holme Fen and Whittlesey Mere, Hunts.
Trans. B 270, 561—604 (1975).

Goede, A. *See* Colhoun & Goede.

de Goër, A.M. *See* Challis (L.J.), de Goër, Guckelsberger & Slack.

Goering, P.L. *See* Etkin & Goering.

Goetze, C. The mechanisms of creep in olivine (Discussion). *Trans.* A 288, 99—119 (1978).

Gold, Elizabeth, Hammersley, R.E. & Richards, W.G. Possible new interstellar masers.
Proc. A 373, 269—284 (1980).

Gold, T. Pulsars and the origin of cosmic rays (Discussion). *Trans.* A 277, 453—461 (1974).
Origin and evolution of the lunar surface: the major questions remaining (Discussion).
Trans. A 285, 555—559 (1977).

Gold, T., Bilson, E. & Baron, R.L. The relationship of surface chemistry and albedo of lunar soil samples (Discussion). *Trans.* A 285, 427—431 (1977).

Gold, V. *See* Garvey, Gold, McAdam & Cooper.

Gold, V. & McAdam, M.E. Kinetics of hydrogen isotope exchange reactions. XXIX. Radiation-induced tritation of 1, 4-dioxan in aqueous solutions. *Proc.* A 346, 443—467 (1975).

Goldberg, E.D. Synthetic organohalides in the sea (Discussion). *Proc.* B 189, 277—289 (1975).

Goldblith, S.A. Processing, catering and cooking of foods by means of electromagnetic radiation (Discussion). *Proc.* B 191, 49—69 (1975).

Golde, M.F. & Thrush, B.A. Vacuum ultraviolet emission by active nitrogen.
I. The formation and removal of N_2 ($a^1\Pi_g$). *Proc.* A 330, 79—95 (1972).
II. The excitation of singlet and triplet states of carbon monoxide by active nitrogen.
Proc. A 330, 97—108 (1972).
III. The absolute rates of population of $N_2(a^1\Pi_g)$ and $CO(A^1\Pi)$. *Proc.* A 330, 109—120 (1972).
IV. The kinetic behaviour of $N_2(B'^3\Sigma_u^-)$. *Proc.* A 330, 121—130 (1972).

Goldflam, P., Hinz, K., Weigel, W. & Wissmann, G. Some features of the northwest African margin and magnetic quiet zone (Discussion). *Trans.* A 294, 87—96 (1980).

Golding, R.M. & Stubbs, L.C. The evaluation of the hyperfine interaction tensor components in molecular systems. *Proc.* A 354, 223—244 (1977).
Higher order hyperfine terms in the spin Hamiltonian. *Proc.* A 362, 525—536 (1978).

Goldman, E.C. Offshore subsea engineering (Discussion). *Trans.* A 290, 99—111 (1978).

Goldman, Karen A. & Fisher, S.K. Synaptic organization of the inner plexiform layer of the retina of *Xenopus laevis*. *Proc.* B 201, 57—72 (1978).

Goldring, D.C. British iron ores: their future use (Discussion). *Proc.* A 339, 313—328 (1974).

Goldschmidt-Clermont, M. *See* Moran, Mirault, Arrigo, Goldschmidt-Clermont & Tissières.

Goldsmith, H.L. *See* Karing & Goldsmith.

Goldsmith, H.L. & Marlow, Jean. Flow behaviour of erythrocytes. I. Rotation and deformation in dilute suspensions. *Proc.* B **182**, 351–384 (1972).

Goldstein, J.L. *See* Buchsbaum & Goldstein.

Goldstein, M. *See* Hökfelt, Lundberg, Schultzberg, Johansson, Skirboll, Anggård and others.

Goldstein, M.E. *See* Dowling (A.P.), Williams & Goldstein.

Goldstein, M.E. & Reid, R.L. Effect of fluid flow on freezing and thawing of saturated porous media. *Proc.* A **364**, 45–73 (1978).

Goldsworthy, F.A. *See* Dixon-Lewis, Goldsworthy & Greenberg.

Golladay, F.L. & Koch-Waser, C.K. The new policies for rural health: institutional, social and financial challenges to large-scale implementation (Discussion). *Proc.* B **199**, 169–178 (1977).

Golladay, F.L. & Liese, B.H. Issues in the institutionalization and management of rural health care: making technology more appropriate (Discussion). *Proc.* B **209**, 173–180 (1980).

Golub, L. X-ray bright points and the solar cycle (Discussion). *Trans.* A **297**, 595–604 (1980).

Gombrich, Sir Ernst. Review Lecture. Mirror and map: theories of pictorial representation. *Trans.* B **270**, 119–149 (1975).

Gomez-Crespo, G. Rural radiology: training (Discussion). *Proc.* B **209**, 131–138 (1980).

Gondhalekar, P.M. & Wilson, R. The interstellar ionization balance due to ultraviolet radiation (Discussion). *Trans.* A **279**, 331–336 (1975).

Gooch, D.J. Creep crack growth in 2¼CrMo weld metals: the suppression of trace element embrittlement by creep strength effects [abstract] (Discussion). *Trans.* A **295**, 295 (1980).

Good, I.J. Explicativity: a mathematical theory of explanation with statistical applications. *Proc.* A **354**, 303–330 (1977).

Good, I.J. & Tideman, T.N. From individual to collective ordering through multidimensional attribute space. *Proc.* A **347**, 371–385 (1976).

Goodall, C.V., Hopkins, H.D., Tulunay, Y. Kabasakal & D'Arcy, R.J. Topside ionosphere electron density measurements on Ariel 4 (Discussion). *Proc.* A **343**, 189–206 (1975).

Gooday, G.W. Functions of trisporic acid (Discussion). *Trans.* B **284**, 509–520 (1978).

Goodeve, Sir Charles. Frank Edward Smith. *Biogr. Mem.* **18**, 525–548 (1972).
  See also Tompkins & Goodeve.

Goodfellow, P.N. *See* Crumpton, Snary, Walsh, Barnstable and others.

Gooding, R.H. The orbits of Ariel 4 and Prospero (Discussion). *Proc.* A **343**, 257–264 (1975).

Goodman, G.M. *See* Crangle & Goodman.

Goodman, G.T. How do chemical substances affect the environment? (Discussion). *Proc.* B **185**, 127–148 (1974).
  Ecology and the problems of rehabilitating wates from mineral extraction (Discussion). *Proc.* A **339**, 373–387 (1974).

Goodman, M. *See* Dene, Goodman & Romero-Herrera.

Goodman, R.H. Systems health building (Discussion). *Trans.* A **272**, 611–619 (1972).

Goodman, R.M., McDonald, J.G., Horne, R.W. & Bancroft, J.B. Assembly of flexuous plant viruses and their proteins (Discussion). *Trans.* B **276**, 173–179 (1976).

Goodrich, F.C., Allen, L.H. & Chatterjee, A.K. The theory of absolute surface shear viscosity. III. The rotating ring problem. *Proc.* A **320**, 537–547 (1971).

Goodstein, D.L. & Saffman, P.G. The two fluid model of the helium film. *Proc.* A **325**, 447–468 (1971).

Goodwin, B.C. Mechanics, fields and statistical mechanics in developmental biology (Discussion). *Proc.* B **199**, 407–414 (1977).

Goodwin, G.K. *See* Bryant, Goodwin & Hagston.

Goodwin, T.W. *See* Glover, Pennock, Pitt & Goodwin.

Goody, R.M. Mars and Venus (Symposium). *Proc.* A **336**, 35–61 (1974).

Goos, H.J.T. *See* Wilson (J.F.), Goos & Dodd.

Gopal, E.S.R., Chandra Sekhar, P., Ananthakrishna, G., Ramachandra, R. & Subramanyam, S.V. Two-phase asymmetry in the phase diagram of critical binary liquid systems: carbon disulphide+nitromethane and cyclohexane+acetic anhydride. *Proc.* A **350**, 91–106 (1976).

Gordon, C. Gigantic land tortoises of Seychelles (Appendix to a paper by D.R. Stoddart & J.F. Peake) (Discussion). *Trans.* B **286**, 159—160 (1979).

Gordon, J. *See* Shapley & Gordon.

Gordon, J.E., Hall, R.O.A., Lee, J.A. & Mortimer, M.J. Heat capacities of plutonium and neptunium. *Proc.* A **351**, 179—196 (1976).

Gordon, J.E. & Jeronimidis, G. Composites with high work of fracture (Discussion). *Trans.* A **294**, 545—550 (1980).

Gordon, L.G.M. & Haydon, D.A. Potential-dependent conductances in lipid membranes containing alamethicin (Discussion). *Trans.* B **270**, 433—447 (1975).

Gordon, M., Leonis, C.G. & Suzuki, H. Ultracentrifuge study of critically branched polycondensates. III. Sedimentation equilibrium. *Proc.* A **345**, 207—230 (1975).

Gordon-Smith, G.W. *See* Jones (O.C.) & Gordon-Smith.

Goreau, N.I. *See* Goreau (T.F.), Goreau, Goreau & Carter.

Goreau, T.F., Goreau, N.I., Goreau, T.J. & Carter, J.G. *Fungiacava eilatensis* burrows in fossil *Fungia* (Pleistocene) from the Sinai Peninsula. *Proc.* B **193**, 245—252 (1976).

Goreau, T.J. Coral skeletal chemistry: physiological and environmental regulation of stable isotopes and trace metals in *Montastrea annularis. Proc.* B **196**, 291—315 (1977).

See also Goreau (T.F.), Goreau, Goreau & Carter.

Gorenstein, C. & Snyder, S.H. Enkephalinases (Discussion). *Proc.* B **210**, 123—132 (1980).

Gosling, J.T. *See* MacQueen, Gosling, Hildner, Munro, Poland & Ross.

Gossauer, A. *See* Inhoffen, Gossauer, Heise & Laas.

Goss-Custard, Susan, Jones, Jane, Kitching, J.A. & Norton, T.A. Tide pools of Carrigathorna and Barloge Creek. *Trans.* B **287**, 1—44 (1979).

Gostelow, T.P. *See* Hutchinson (J.N.) & Gostelow.

Gottesfeld, J.M. Organization of transcribed regions of chromatin (Discussion). *Trans.* B **283**, 343—357 (1978).

Gottlieb, H.P.W. Towards a unified nonlinear theory of massless bosons. *Proc.* A **368**, 429—440 (1979).

Goudie, A.S. *See* Shotton (F.W.), Goudie, Briggs & Osmaston.

Gough, R.J. *See* Ashkenazi, Sykes, Gough & Williams.

Gould, G.W. & Measures, J.C. Water relations in single cells (Discussion). *Trans.* B **278**, 151—166 (1977).

Gould, R.P. *See* Tait (J.F.), Tait, Gould & Mee.

Gould, S.J. *See* Hallam (A.) & Gould.

Gould, S.J. & Lewontin, R.C. The spandrels of San Marco and the Panglossian paradigm: a critique of the adaptationist programme (Discussion). *Proc.* B **205**, 581—598 (1979).

Gould, W.J. Spectral characteristics of some deep current records from the eastern North Atlantic (Discussion). *Trans.* A **270**, 437—450 (1971).

Currents on continental margins and beyond (Discussion). *Trans.* A **290**, 87—98 (1978).

Gourlay, A.R. Some recent methods for the numerical solution of time-dependent partial differential equations (Discussion). *Proc.* A **323**, 219—235 (1971).

Gow, C.E. The dentitions of the Tritheledontidae (Therapsida : Cynodontia). *Proc.* B **208**, 461—481 (1980).

Gow, R.S. *See* Broom & Gow.

Gowans, J.L. *See* Ellis & Gowans; Howard (J.C.) & Gowans; Hunt (S.V.), Ellis & Gowans.

Gowar, A.P. *See* Pillinger, Eglinton, Gowar & Jull; Pillinger & Gowar.

Goymour, C.G. *See* King (D.A.), Goymour & Yates.

Graas, H. *See* Leroy (V.), Richelmi & Graas.

Grabke, H.J., Petersen, E.M. & Paulitschke, W. Adsorption and segregation of sulphur, and its influence on the carburization and nitrogenation of iron and steel [abstract] (Discussion). *Trans.* A **295**, 128 (1980).

Graf, J. & Petersen, O.H. Electrogenic sodium pump in mouse liver parenchymal cells. *Proc.* B **187**, 363—367 (1974).

Graf, T. *See* Hayman (M.J.), Ramsay, Kitchener, Graf, Beug and others.

Graham, A.L. & Hutchison, R. Mineralogy and petrology of fragments from the Luna 24 core. *Trans.* A **297**, 15—22 (1980).

Graham, C. Symmetry indications of the polarization state of light scattered by fluids in electric and magnetic fields. *Proc.* A **369**, 517—535 (1980).

See also Buckingham (A.D.) & Graham.

Graham, R.H. *See* Coward, Graham, James & Wakefield.

Graham, S.C., Homer, J.B. & Rosenfeld, J.L.J. The formation and coagulation of soot aerosols generated by the pyrolysis of aromatic hydrocarbons. *Proc.* A **344**, 259—285 (1975).

Graham-Bryce, I.J. Crop protection: a consideration of the effectiveness and disadvantages of current methods and of the scope for improvement (Discussion). *Trans.* B **281**, 163—179 (1977).

Graham-Smith, W. On some variations in the latero-sensory lines of the placoderm fish *Bothriolepis*. *Trans.* B **282**, 1—39 (1978).

On the lateral lines and dermal bones in the parietal region of some crossopterygian and dipnoan fishes. *Trans.* B **282**, 41—105 (1978).

Grainger, L. Future trends in utilization of coal energy conversion (Discussion). *Trans.* A **276**, 527—539 (1974).

Grampp, W. *See* Bevan (S.), Crampp & Miledi.

Grant, Anne M.S. *See* Dunstan, Grant, Marshall & Neuberger.

Grant, E.H. *See* South & Grant.

Grant, I. *See* Gaster & Grant.

Grant, I.P. *See* Pyper & Grant; Rose (S.J.), Grant & Connerade.

Grant, I.S. *See* Chapman (J.A.), Grant, Taylor, Mahmud, Sardar-ul-Mulk & Shahid.

Grant, Janine. *See* Bolton (H.C.), Grant, McWilliam, Nicholson & Swingler.

Grant, K.R. *See* Borrell (P.), Borrell, Pedley & Grant.

Graves-Morris, P.R. *See* Chisholm & Graves-Morris.

Gray, C.H. *See* Rimington & Gray.

Gray, D.A. & Foster, S.S.D. Urban influences upon groundwater conditions in the Thames Flood Plain deposits of Central London (Discussion). *Trans.* A **272**, 242—257 (1972).

Gray, E.G. Presynaptic microtubules and their association with synaptic vesicles. *Proc.* B **190**, 369—372 (1975).

Synaptic vesicles and microtubules in frog motor endplates. *Proc.* B **203**, 219—227 (1978).

Gray, J.S. Pollution-induced changes in populations (Discussion). *Trans.* B **286**, 545—561 (1979).

Gray, P. *See* Boddington, Gray & Harvey; Boddington, Gray & Robinson; Boddington, Gray & Wake; Boddington, Gray & Walker; Humphreys & Gray.

Gray, P., Jones, D.T. & MacKinven, R. Thermal effects accompanying spontaneous ignition in gases. IV. The decomposition of diethyl peroxide in a cylindrical vessel and the effect of diluents on self-heating. *Proc.* A **325**, 175—196 (1971).

Gray, R.W., Harrison, G. & Lamb, J. Dynamic viscoelastic behaviour of low-molecular-mass polystyrene melts. *Proc.* A **356**, 77—102 (1977).

Grdenić, D. *See* De Sanctis, Grdenić, Taylor & Hodgkin.

Greatrex, R. *See* Gibb, Greatrex & Greenwood.

Greeley, R., Iversen, J.D., Pollack, J.B., Udovich, Nancy & White, B. Wind tunnel studies of Martian aeolian processes. *Proc.* A **341**, 331—360 (1974).

Green, A.E. *See* Buckley (C.P.) & Green.

Green, A.E., Laws, N. & Naghdi, P.M. On the theory of water waves. *Proc.* A **338**, 43—55 (1974).

Green, A.E. & Naghdi, P.M. Directed fluid sheets. *Proc.* A **347**, 447—473 (1976).

On thermodynamics and the nature of the second law. *Proc.* A **357**, 253—270 (1977).

On thermal effects in the theory of shells. *Proc.* A **365**, 161—190 (1979).

Corrigenda. *Proc.* A **367**, 572 (1979).

Green, A.E., Naghdi, P.M. & Wenner, M.L. On the theory of rods.

I. Derivations from the three-dimensional equations. *Proc.* A **337**, 451—483 (1974).

II. Developments by direct approach. *Proc.* A **337**, 485—507 (1974).

Green, B.N. *See* Craig (R.D.), Bateman, Green & Millington.

Green, D.H. Composition of basaltic magmas as indicators of conditions of origin (Discussion). *Trans.* A **268**, 707—725 (1971).

Green, D.P.L. *See* Tsien, Green, Levinson, Rudy & Sanders.

Green, D.P.L., Ito, Y., Miledi, R. & Vincent, Angela. A note on the structure of immunized end-plates. *Proc.* B 195, 323–326 (1977).

Green, D.P.L., Miledi, R., de la Mora, M. Perez & Vincent, Angela. Acetylcholine receptors (Discussion). *Trans.* B 270, 551–559 (1975).

Green, D.P.L., Miledi, R. & Vincent, Angela. Neuromuscular transmission after immunization against acetylcholine receptors. *Proc.* B 189, 57–68 (1975).

Green, J.S.A. Large-scale motion in the upper stratosphere and mesosphere: an evaluation of data and theories (Discussion). *Trans.* A 271, 577–583 (1972).

Green, Kerie F. *See* Boyle (L.L.) & Green.

Green, N.M. *See* Crumpton, Allan, Auger, Green & Maino.

Green, P. *See* Brookes (C.A.) & Green.

Greenaway, A.H. *See* Burge, Fiddy, Greenaway & Ross.

Greenberg, J.B. *See* Dixon-Lewis, Goldsworthy & Greenberg.

Greene, J.M. *See* Hatfield, Fisher, Dunigan, Burchfield and others.

Greenland, D.J. Soil damage by intensive arable cultivation: temporary or permanent? (Discussion). *Trans.* B 281, 193–208 (1977).

Greenwood, G.W. Fracture during creep (Discussion). *Trans.* A 288, 213–227 (1978).

Greenwood, N.N. *See* Gibb, Greatrex & Greenwood.

Greenwood, P.H. Lake George, Uganda (Discussion). *Trans.* B 274, 375–391 (1976).

Greenwood, P.H. & Lund, J.W.G. Introductory remarks to a Discussion. *Proc.* B 184, 229–233 (1973).

Greenwood, W.R., Hadley, D.G., Anderson, R.E., Fleck, R.J. & Schmidt, D.L. Late Proterozoic cratonization in southwestern Saudi Arabia (Discussion). *Trans.* A 280, 517–527 (1976).

Greger, G. Why and how is the Federal Government of Germany promoting the utilization of *Spacelab?* (Discussion). *Proc.* A 361, 143–150 (1978).

Gregory, G.E. Neuroanatomy of the mesothoracic ganglion of the cockroach *Periplaneta americana* (L.). I. The roots of the peripheral nerves. *Trans.* B 267, 421–465 (1974).

Gregory, P.H. The Leeuwenhoek Lecture, 1970. Airborne microbes: their significance and distribution. *Proc.* B 177, 469–483 (1971).

The recognition of microscopic objects (Discussion). *Proc.* B 189, 161–165 (1975).

Gregory, R.A. *See* Dockray & Gregory.

Gregory, R.L. Stereo vision and isoluminance (Discussion). *Proc.* B 204, 467–476 (1979).

Perception as hypotheses (Discussion). *Trans.* B 290, 181–197 (1980).

Gregson, A.K., Martin, R.L. & Mitra, S. The magnetic anisotropy and electronic structure of binuclear copper (II) acetate monohydrate. *Proc.* A 320, 473–486 (1971).

Grew, W.J.S. & Cameron, A. Thermodynamics of boundary lubrication and scuffing. *Proc.* A 327, 47–59 (1972).

Griffin, Beverly E., Dilworth, S.M., Ito, Y. & Novak, Ulrike. Polyoma virus: some considerations on its transforming genes (Discussion). *Proc.* B 210, 465–476 (1980).

Griffin, R.F. & Woolley, Sir Richard. Roderick Oliver Redman. *Biogr. Mem.* 22, 335–357 (1976).

Griffin, R.G. *See* Herzfeld, Roufosse, Haberkorn, Griffin & Glimcher.

Griffin, R.L. *See* Raymont, Krishnaswamy, Woodhouse & Griffin.

Griffin, W.G. *See* Burton (W.M.), Evans & Griffin.

Griffith, H.B. Endoneurosurgery: endoscopic intracranial surgery (Discussion). *Proc.* B 195, 261–268 (1977).

Griffith, R.W. Chemistry of the body fluids of the coelacanth, *Latimeria chalumnae*. *Proc.* B 208, 329–347 (1980).

Griffith, T.C. *See* Coleman (P.G.), Griffith & Heyland.

Griffiths, D.H. *See* Barker (P.F.) & Griffiths.

Griffiths, D.V. *See* Birdsall (B.), Griffiths, Roberts, Feeney & Burgen; Feeney, Roberts, Birdsall, Griffiths and others.

Griffiths, H.B., Miller, K.W., Paton, W.D.M. & Smith, E.B. On the role of separated gas in decompression procedures. *Proc.* B 178, 389–406 (1971).

Griffiths, R.K. *See* Bacon (G.E.), Bacon & Griffiths.
Grimes, N.W. On the specific heat of compounds with spinel structure.
 I. The ferrites. *Proc.* A **338**, 209—221 (1974).
 II. Zinc ferrite, a paramagnetic compound with magnetic ion occupying the octahedral site.
   Proc. A **338**, 223—233 (1974).
Grimshaw, R. Mean flows induced by internal gravity wave packets propagating in a shear flow.
   Trans. A **292**, 391—417 (1979).
 Slowly varying solitary waves.
 I. Korteweg—de Vries equation. *Proc.* A **368**, 359—375 (1979).
 II. Nonlinear Schrödinger equation. *Proc.* A **368**, 377—388 (1979).
Grindley, G.W. Structural control of volcanism at Mount Etna (Discussion).
   Trans. A **274**, 165—175 (1973).
Grinsted, J. *See* Richmond (M.H.), Bennett, Choi, Brown, Brunton and others.
Grinstein, S. & Erlij, D. Intracellular calcium and the regulation of sodium transport in the frog skin.
   Proc. B **202**, 353—360 (1978).
Grögler, N. *See* Geiss, Eberhardt, Grögler, Guggisberg, Maurer & Stettler.
Groh, G. Tomosynthesis and coded aperture imaging: new approaches to three-dimensional imaging
   in diagnostic radiography (Discussion). *Proc.* B **195**, 299—306 (1977).
Groome, I.J. *See* Carabine, Cullis & Groome.
Gross, G.F. The land invertebrates of the New Hebrides and their relationships (Discussion).
   Trans. B **272**, 391—421 (1975).
Grove, A.T. The geography of semi-arid lands (Discussion). *Trans.* B **278**, 457—475 (1977).
Grove, T.H. *See* Ackerman, Bore, Gadian, Grove & Radda.
Groves, G.V. Review Lecture. Rocket studies of atmospheric tides. *Proc.* A **351**, 437—469 (1976).
 Seasonal and diurnal variations of middle atmosphere winds (Discussion).
   Trans. A **296**, 19—40 (1980).
Grubb, P. The growth, ecology and population structure of giant tortoises on Aldabra (Discussion).
   Trans. B **260**, 327—372 (1971).
 Ecology of terrestrial decapod crustaceans on Aldabra (Discussion). *Trans.* B **260**, 411—416 (1971).
Grüneberg, H. The tabby syndrome in the mouse (with an Appendix by A.J. Lee).
   Proc. B **179**, 139—156 (1971).
 Population studies on a polymorphic prosobranch snail (*Clithon (Pictoneritina) oualaniensis* Lesson)
   (with an appendix by L. Nugaliyadde). *Trans.* B **275**, 385—437 (1976).
 Micro-evolution in a polymorphic prosobranch snail (*Clithon oualaniensis* (Lesson)).
   Proc. B **200**, 419—440 (1978).
 A search for causes of polymorphism in *Clithon oualaniensis* (Lesson) (Gastropoda; Prosobranchia).
   Proc. B **203**, 379—386 (1979).
 On pseudo-polymorphism. *Proc.* B **210**, 533—548 (1980).
Grüneberg, H., Cattanach, B.M., McLaren, Anne, Wolfe, H.G. & Bowman, Patricia. The molars of
   tabby chimaeras in the mouse. *Proc.* B **182**, 183—192 (1972).
Grüneberg, H. & McLaren, Anne. The skeletal phenotype of some mouse chimaeras.
   Proc. B **182**, 9—23 (1972).
Grunwald, C. Function of sterols (Discussion). *Trans.* B **284**, 541—558 (1978).
Grutzner, J.B. *See* Neuss, Nash, Lemke & Grutzner.
Grynszpan-Winograd, Odile. Morphological aspects of exocytosis in the adrenal medulla (Discussion).
   Trans. B **261**, 291—292 (1971).
Gualtierotti, T. The vestibular function research programme as a part of the *Spacelab* project: an
   investigation of the effect of free fall on unitary and integrated vestibular activity (Discussion).
   Proc. B **199**, 493—503 (1977).
Gubbins, D. Numerical solutions of the kinematic dynamo problem. *Trans.* A **274**, 493—521 (1973).
Guckelsberger, K. *See* Challis (L.J.), de Goër, Guckelsberger & Slack.
Guenther, B. *See* Krueger, Guenther, Fleig, Heath, Hilsenrath and others.
Guest, J.E. The summit of Mount Etna prior to the 1971 eruptions (symposium).
   Trans. A **274**, 63—78 (1973).

Guest, M.F. *See* Baybutt, Guest & Hillier.

Guest, M.F., Hillier, I.H., Saunders, V.R. & Wood, M.H. The theoretical description of ionic states observed by high energy photoelectron spectroscopy. *Proc.* A 333, 201–215 (1973).

Gugan, D. The electrical resistivity of potassium below 4.2 K. *Proc.* A 325, 223–249 (1971).

Guggisberg, S. *See* Geiss, Eberhardt, Grögler, Guggisberg, Maurer & Stettler.

Guil, J.M., Hayward, D.O. & Taylor, N. Absorption and diffusion of hydrogen and deuterium in tantalum at low temperatures. *Proc.* A 335, 141–161 (1973).

Guilley, H., Jonard, G., Richards, K.E. & Hirth, L. Specific encapsidation of fragments of TMV RNA (Discussion). *Trans.* B 276, 181–188 (1976).

Guinot, B. Determination of the motion of the pole, and comparison with astrometry (Discussion). *Trans.* A 294, 329–334 (1980).

Guiu, F. & Shadrake, L.G. Elastic displacements produced by cross-links in polyethylene crystals. *Proc.* A 346, 305–327 (1975).

Gulden, M.E. *See* Evans (A.G.), Gulden & Rosenblatt.

Gull, T.R., York, D.G., Snow, P., Jr, & Henize, K.G. On the distance to the candidate star coincident with A0620–00 (Discussion). *Proc.* A 350, 487–490 (1976).

Gunasekera, J.S. *See* Alexander (J.M.) & Gunasekera.

Gunn, D.L. The development of aircraft attack on locust swarms in Africa since 1945 and the start of operational research on control systems (Discussion). *Trans.* B 287, 251–261 (1979).

Strategies, systems, value judgements and dieldrin in control of locust hoppers (Discussion). *Trans.* B 287, 429–445 (1979).

See also Rainey (R.C.) & Gunn.

Gunn, J.E. The dynamics of galaxies and the 'missing mass' problem (Discussion). *Trans.* A 296, 313–318 (1980).

Gunter, D.L. A study of the coupled gravitational and electromagnetic perturbations to the Reissner–Nordström black hole: the scattering matrix, energy conversion, and quasi-normal modes. *Proc.* A 296, 497–526 (1980).

Gunton, D.J. & Saunders, G.A. Stability limits on the Poisson ratio: application to a martensitic transformation. *Proc.* A 343, 63–83 (1975).

Gupta, A.S. *See* Annapurna & Gupta; Gupta (P.S.) & Gupta.

Gupta, H.N. *See* Singh (R.K.) & Gupta.

Gupta, J.C. *See* Chapman (S.), Gupta & Malin.

Gupta, P.S. & Gupta, A.S. Effect of homogeneous and heterogeneous reactions on the dispersion of a solute in the laminar flow between two plates. *Proc.* A 330, 59–63 (1972).

Gupta, R.P., Tse, J.S. & Bancroft, G.M. Core level ligand field splittings in photoelectron spectra. *Trans.* A 293, 535–569 (1980).

Gurdon, J.B. The Croonian Lecture, 1976. Egg cytoplasm and gene control in development. *Proc.* B 198, 211–247 (1977).

Genes and the structure of organisms (Discussion). *Proc.* B 199, 399–406 (1977).

See also De Robertis, Partington & Gurdon.

Gurdon, J.B., Wyllie, A.H. & De Robertis, E.M. The transcription and translation of DNA injected into oocytes (Discussion). *Trans.* B 283, 367–372 (1978).

Gurney, C., Mai, Y.W. & Owen, R.C. Quasistatic cracking of materials with high fracture toughness and low yield stress. *Proc.* A 340, 213–231 (1974).

Gurney, C. & Ngan, K.M. Quasistatic crack propagation in nonlinear structures. *Proc.* A 325, 207–222 (1971).

Gurney, J.J. & Harte, B. Chemical variations in upper mantle nodules from southern African kimberlites (Discussion). *Trans.* A 297, 273–293 (1980).

Gustafsson, J.-Å. & Eneroth, P. Steroids in meconium and faeces from newborn infants (Discussion). *Proc.* B 180, 179–186 (1972).

Guttmann, M. The role of residuals and alloying elements in temper embrittlement (Discussion). *Trans.* A 295, 169–196 (1980).

See also Lemblé, Pineau, Castagné, Dumoulin & Guttmann.

Güven, R. The solution of Dirac's equation in a class of type D vacuum space–times.

Proc. A **356**, 465—470 (1977).
Guyader, J. *See* Huault, Lefebvre, Guyader, Giresse and others.
Gwahaba, J.J. The distribution, population density and biomass of fish in an equatorial lake, Lake George, Uganda. *Proc.* B **190**, 393—414 (1975).
See also Burgis, Darlington, Dunn, Ganf, Gwahaba & McGowan.

Habell, K.J. Thomas Smith. *Biogr. Mem.* **17**, 681—687 (1971).
Haber, S. *See* Baldwin (J.E.), Jung, Singh, Wan, Haber and others.
Haberich, F.J. *See* Lucas (M.L.), Schneider, Haberich & Blair.
Haberkorn, R.A. *See* Herzfeld, Roufosse, Haberkorn, Griffin & Glimcher.
Habib, D. *See* Sibuet, Ryan, Arthur, Barnes, Blechsmidt and others.
Hackett, A.J. *See* Fox (R.H.), Budd, Woodward, Hackett & Hendrie.
Haddock, B.A. Microbial energetics (Discussion). *Trans.* B **290**, 329—339 (1980).
Hadley, D.G. *See* Greenwood (W.R.), Hadley, Anderson, Fleck & Schmidt.
Hadley, R.F. Evaluation of land-use and land-treatment practices in semi-arid western United States (Discussion). *Trans.* B **278**, 543—554 (1977).
Haeusler, G. *See* Reuter, Blaustein & Haeusler.
Hagan, P.J. *See* Andersen, Hagan, Phillips & Powell.
Hagman, M. Incompatibility in forest trees (Discussion). *Proc.* B **188**, 313—326 (1975).
Hagston, W.E. *See* Bryant, Goodwin & Hagston; Bryant, Hagston & Radford; Jefferson (J.H.) & Hagston.
Hagstrum, H.D. & **Becker, G.E.** Energy spectra of electrons in surface orbitals (Discussion). *Proc.* A **331**, 395—402 (1972).
Hahn, J. The cycle of atmospheric nitrous oxide (Discussion). *Trans.* A **290**, 495—504 (1979).
Haidar, M. Abou. Tobacco rattle virus RNA—protein interactions (Discussion). *Trans.* B **276**, 165—172 (1976).
Hails, J.R. Offshore morphology and sediment distribution, Start Bay, Devon (Discussion). *Trans.* A **279**, 221—228 (1975).
Hailwood, E.A. *See* Roberts (D.G.), Montadert, Thompson, Auffret and others.
Hailwood, E.A., Hamilton, N. & **Morgan, G.E.** Magnetic polarity dating of tectonic events at passive continental margins (Discussion). *Trans.* A **294**, 189—208 (1980).
HajIbrahim, S.K. *See* Eglinton, HajIbrahim, Maxwell, Quirke, Shaw and others.
Halberstam, H. *See* Rogers (C.A.), Burgess, Halberstam & Birch.
Hales, B.J. *See* Creed (D.), Hales & Porter.
Hall, A.M. *See* Cotes, Dabbs, Hall, Lakhera, Saunders & Malhotra.
Hall, A.R. Late Pleistocene deposits at Wing, Rutland. *Trans.* B **289**, 135—164 (1980).
Hall, A. Rupert. The Wilkins Lecture, 1973. Newton and his editors. *Proc.* A **338**, 397—417 (1974).
Hall, C., Richards, R.E. & **Sharp, R.R.** Further studies of chemical shifts in the nuclear resonances of caesium ions in solution. *Proc.* A **337**, 297—315 (1974).
Hall, D.G. & **Pethica, B.A.** Thermodynamics of the Volta effect for surface films. *Proc.* A **354**, 425—439 (1977).
The thermodynamics of parallel flat plate condensers. *Proc.* A **364**, 457—472 (1978).
Hall, D.O., Adams, M.W.W., Morris, P. & **Rao, K.K.** Photolysis of water for H_2 production with the use of biological and artificial catalysts. *Trans.* A **295**, 473—476 (1980).
Hall, J.L. *See* Flowers, Ward & Hall.
Hall, M.A. *See* Jerie & Hall.
Hall, P. The stability of Poiseuille flow modulated at high frequencies. *Proc.* A **344**, 453—464 (1975).
The linear stability of flat Stokes layers. *Proc.* A **359**, 151—166 (1978).
The effect of external forcing on the stability of plane Poiseuille flow. *Proc.* A **359**, 453—478 (1978).
Centrifugal instabilities of circumferential flows in finite cylinders: nonlinear theory. *Proc.* A **372**, 317—356 (1980).
See also Blennerhassett & Hall; Seminara & Hall.

Hall, P. & Walton, I.C. The smooth transition to a convective régime in a two-dimensional box.
Proc. A **358**, 199—221 (1977).

Hall, P.J. *See* Sanderson, Hall & Thomas.

Hall, R.O.A. *See* Gordon (J.E.), Hall, Lee & Mortimer.

Hall, Sir Arnold. *See* Thomson (Sir George) & Hall.

Hall, Sir Arnold & Morgan, Sir Morien. Bennett Melvill Jones. *Biogr. Mem.* **23**, 253—282 (1977).

Hallam, A. & Gould, S.J. The evolution of British and American Middle and Upper Jurassic *Gryphaea*:
a biometric study. *Proc.* B **189**, 511—542 (1975).

Hallam, C. *See* Whittam, Hallam & Wattam.

Hallam, C. & Whittam, R. The role of sodium ions in ATP formation by the sodium pump.
Proc. B **198**, 109—128 (1977).

Hallam, H.E. *See* Barnes (A.J.), Hallam & Jones.

Halmshaw, R. The present role of radiological methods in engineering (Discussion).
Trans. A **292**, 157—162 (1979).

Halsey, M.J. *See* Brown (F.F.), Halsey & Richards.

Halstead, M.P., Kirsch, L.J., Prothero, A. & Quinn, C.P. A mathematical model for hydrocarbon
autoignition at high pressures. *Proc.* A **346**, 515—538 (1975).

Halstead, M.P., Prothero, A. & Quinn, C.P. A mathematical model of the cool-flame oxidation of
acetaldehyde. *Proc.* A **322**, 377—403 (1971).

Hamberger, B. *See* Hökfelt, Lundberg, Schultzberg, Johansson, Skirboll, Anggård and others.

Hamberger, S.M. *See* Franklin (R.N.), Edgley, Hamberger & Motley; Franklin (R.N.), Hamberger,
Lampis & Smith.

Hamblin, P.F. A theory of short period tides in a rotating basin. *Trans.* A **281**, 97—111 (1976).

Hamer, M.F. *See* Davies (G.) & Hamer; Davies (G.), Nazaré & Hamer.

Hamilton, D. *See* Andreieff, Bouysse, Curry, Fletcher and others; Channon & Hamilton; Easton,
Hamilton, Kempe & Sheppard.

Hamilton, D., Hommeril, P., Larsonneur, C. & Smith, A.J. Geological bibliography for the English
Channel (Part 2). Bibliographie géologique de la Manche (Partie deuxieme).
Trans. A **279**, 289—295 (1975).

Hamilton, G.M. & Moore, S.L. Deformation and pressure in an elastohydrodynamic contact.
Proc. A **322**, 313—330 (1971).

Hamilton, N. *See* Hailwood, Hamilton & Morgan.

Hamilton, P.J. *See* O'Nions, Evensen & Hamilton; O'Nions, Evensen, Hamilton & Carter.

Hamilton, W.R. Fossil giraffes from the Miocene of Africa and a revision of the phylogency of the
Giraffoidea. *Trans.* B **283**, 165—229 (1978).

Hamlin, M.J. & Wright, C.E. The effects of drought on the river systems (Discussion).
Proc. A **363**, 69—96 (1978).

Hammersley, R.E. *See* Gold (Elizabeth), Hammersley & Richards.

Hammond, B.J. *See* Hope (R.A.), Hammond & Gaze.

Hammond, Norman. The early history of American agriculture: recent research and current contro-
versy (Discussion). *Trans.* B **275**, 120—128 (1976).

Hamoir, G. Extractability and properties of the contractile proteins of vertebrate smooth muscle
(Discussion). *Trans.* B **265**, 169—181 (1973).

Hanby, J.A. *See* Lemon, Hanby & Porter.

Hancock, P. *See* Chubb (J.P.), Billingham, Hancock, Dimbylow & Newcombe.

Handford, P.T. Patterns of variation in a number of genetic systems in *Maniola jurtina:* the boundary
region. *Proc.* B **183**, 265—284 (1973).

Patterns of variation in a number of genetic systems in *Maniola jurtina:* the Isles of Scilly.
Proc. B **183**, 285—300 (1973).

Hands, B.A. *See* Bentley (P.D.) & Hands.

Hanks, M.J. & Mather, K. Genetics of coxal chaetae in *Drosophila melanogaster*. II. Responses to
selection. *Proc.* B **202**, 211—230 (1978).

Hanley, P.L., Kiflawi, I. & Lang, A.R. On topographically identifiable sources of cathodoluminescence
in natural diamonds. *Trans.* A **284**, 329—368 (1977).

Hannant, D.J. & Zonsveld, J.J. Polyolefin fibrous networks in cement matrices for low cost sheeting (Discussion). *Trans.* A **294**, 591—597 (1980).

Hannink, R.H.J., Kohlstedt, D.L. & Murray, M.J. Slip system determination in cubic carbides by hardness anisotropy. *Proc.* A **326**, 409—420 (1972).

Hansen, N. *See* Jones (A.R.), Ralph & Hansen.

Hansen, R.O., Newman, E.T., Penrose, R. & Tod, K.P. The metric and curvature properties of $\mathcal{H}$-space. *Proc.* A **363**, 445—468 (1978).

Hanson, G.N. *See* Langmuir & Hanson.

Hanson, Jean. Evidence from electron microscope studies on actin paracrystals concerning the origin of the cross-striation in the thin filaments of vertebrate skeletal muscle.
Proc. B **183**, 39—58 (1973).
See also O'Brien, Bennett & Hanson.

Hardie, R.C. *See* Horridge, Mimura & Hardie.

Harding, B.N. & Powell, T.P.S. An electron microscopic study of the centre-median and ventrolateral nuclei of the thalamus in the monkey. *Trans.* B **279**, 357—412 (1977).

Harding, G.L., Pippard, A.B. & Tomlinson, J.R. Resistance of superconducting—normal interfaces.
Proc. A **340**, 1—31 (1974).

Harding, J.H. *See* Boyd (P.D.W.), Gerloch, Harding & Woolley; Gerloch & Harding.

Hardman, J.S. & Lilley, B.A. Mechanisms of compaction of powdered materials.
Proc. A **333**, 183—199 (1973).

Hardy, G.W. *See* Beddell, Clark, Hardy, Lowe, Ubatuba, Vane & Wilkinson.

Hardy, R.N., Hockaday, A.R. & Tapp, R.L. Observations on the structure of the small intestine in foetal, neo-natal and suckling pigs. *Trans.* B **259**, 517—531 (1971).

Harkavy, O., Jaffe, F.S., Koblinsky, Marjorie A. & Segal, S.J. Funding of contraceptive research (Discussion). *Proc.* B **195**, 37—55 (1976).

Harlan, J.R. Plant and animal distribution in relation to domestication (Discussion).
Trans. B **275**, 13—25 (1976).

Harley, J.L. The objectives of conservation (Discussion). *Proc.* B **197**, 3—10 (1977).
Review Lecture. Ectomycorrhizes as nutrient absorbing organs. *Proc.* B **203**, 1—21 (1978).
See also Clapham & Harley.

Harley, R.T. *See* Elliott (R.J.), Harley, Hayes & Smith.

Harper, D.R. Research and development needs (Discussion). *Trans.* A **272**, 651—657 (1972).

Harper, W.J. The use of steel in bridge construction (Symposium). *Trans.* A **282**, 37—40 (1976).

Harries, D.R. & Marwick, A.D. Non-equilibrium segregation in metals and alloys (Discussion).
Trans. A **295**, 197—207 (1980).

Harries, J.E. Spectroscopic observations of middle atmosphere composition (Discussion).
Trans. A **296**, 161—173 (1980).

Harrington, J.M. & Rowlinson, J.S. The gas—liquid surface of the penetrable sphere model. III.
Proc. A **367**, 15—28 (1979).

Harris, A.J. *See* Dennis (M.J.), Harris & Kuffler; Kuffler, Dennis & Harris.

Harris, A.J., Kuffler, S.W. & Dennis, M.J. Differential chemosensitivity of synaptic and extrasynaptic areas on the neuronal surface membrane in parasympathetic neurons of the frog, tested by micro-application of acetylcholine. *Proc.* B **177**, 541—553 (1971).

Harris, B. & Ankara, A.O. Cracking in composites of glass fibres and resin.
Proc. A **359**, 229—250 (1978).

Harris, G.W. *See* Burrows (J.P.), Cliff, Harris, Thrush & Wilkinson.

Harris, Harry. Lionel Sharples Penrose. *Biogr. Mem.* **19**, 521—561 (1973).

Harris, Henry. The Croonian Lecture, 1971. Cell fusion and the analysis of malignancy.
Proc. B **179**, 1—20 (1971).
See also Bramwell (M.E.) & Harris.

Harris, J.E. The role of intergranular precipitates in controlling creep cavitation [abstract] (Discussion). *Trans.* A **295**, 307 (1980).

Harris, J.M. *See* Costa (S.M. de B.), Froines, Harris, Leblanc, Orger & Porter.

Harris, J.R. Some electron microscopic studies on intact nuclear 'ghosts' and nuclear membrane

fragments (Discussion). *Trans.* B **268**, 109–117 (1974).

Harris, L.R.F. Strategic systems planning (Discussion). *Trans.* A **289**, 213–225 (1978).

Harris, P. & Kornberg, H.L. The uptake of glucose by a thermophilic *Bacillus* sp.
Proc. B **182**, 159–170 (1972).

Harris, P.G., Hutchison, R. & Paul, D.K. Plutonic xenoliths and their relation to the upper mantle (Discussion). *Trans.* A **271**, 313–323 (1972).

Harris, P.M., Thurrell, R.C., Healing, R.A. & Archer, A.A. Aggregates in Britain (Discussion).
Proc. A **339**, 329–353 (1974).

Harris, S.C. Microwave studies of superconducting two-phase IN-SN. *Proc.* A **350**, 267–279 (1976).

Harrison, A.J. & Weinberg, F.J. Flame stabilization by plasma jets. *Proc.* A **321**, 95–103 (1971).

Harrison, F.A. Ion transport across rumen and omasum epithelium (Discussion).
Trans. B **262**, 301–305 (1971).

Harrison, G. *See* Barlow (A.J.), Harrison, Irving, Kim, Lamb & Pursley; Gray (R.W.), Harrison & Lamb.

Harrison, G.A. Genetic and anthropological studies in the Human Adaptability section of the International Biological Programme (Discussion). *Trans.* B **274**, 437–445 (1976).
See also Boyce (A.J.), Harrison, Platt, Hornabrook and others; Clegg (E.J.), Jeffries & Harrison.

Harrison, G.A., Hiorns, R.W. & Boyce, A.J. Movement, relatedness and the genetic structure of the population of Karkar Island (Discussion). *Trans.* B **268**, 241–249 (1974).

Harrison, L.G. & Koga, Y. Chloride ion substitution in alkali bromide surfaces: cooperative interactions, including a surface transition. *Proc.* A **327**, 97–122 (1972).

Harrison, L.G. & Lacalli, T.C. Hyperchirality: a mathematically convenient and biochemically possible model for the kinetics of morphogenesis. *Proc.* B **202**, 361–397 (1978).

Harrison, M.F.A. *See* Dixon (A.J.), von Engel & Harrison.

Harrison, R.G., Key, P.Y. & Little, V.I. Stimulated scattering and induced Bragg reflexion of light in liquid media.
I. Theoretical. *Proc.* A **334**, 193–214 (1973).
II. Experimental. *Proc.* A **334**, 215–229 (1973).

Harrison, S. *See* King (J.), Botstein, Casjens, Earnshaw, Harrison & Lenk.

Harrison, V.A.W. *See* Walsh (D.), Hayes & Harrison.

Harrison, W. *See* Roberts (D.G.), Montadert, Thompson, Auffret and others.

Harrowell, R.V. Superconducting reciprocating machines (Discussion). *Trans.* A **275**, 85–94 (1973).

Hart, M. Review Lecture. Ten years of X-ray interferometry. *Proc.* A **346**, 1–22 (1975).
See also Aldred & Hart; Cusatis & Hart.

Hart, S.R. K, Rb, Sr and Ba contents and Sr isotope ratios of ocean floor basalts (Discussion).
Trans. A **268**, 573–587 (1971).

Harte, B. Kimberlite nodules, upper mantle petrology, and geotherms (Discussion).
Trans. A **288**, 487–500 (1978).
See also Gurney (J.J.) & Harte.

Hartland, S. *See* Leidi & Hartland.

Hartland, S. & Gakis, N. A model for coalescence in a two-dimensional close-packed dispersion.
Proc. A **369**, 137–155 (1979).

Hartley, A.J., Eastburn, P. & Leece, N. Steelworks control of residuals (Discussion).
Trans. A **295**, 45–55 (1980).

Hartley, B.S. Evolution of enzyme structure (Discussion). *Proc.* B **205**, 443–452 (1979).
Introductory remarks to a Discussion. *Trans.* B **290**, 279–280 (1980).

Hartman, P. *See* Jenkins (H.D.B.) & Hartman.

Hartwieg, E.A. *See* Waxman (S.G.), Bradley & Hartwieg.

Harvey, D.I. *See* Boddington, Gray & Harvey.

Harvey, P.H. *Cepaea nemoralis* on clifftops in south-west England. *Proc.* B **181**, 375–393 (1972).
See also Clutton-Brock & Harvey.

Harvey, R.G. An anthropometric survey of growth and physique of the populations of Karkar Island and Lufa subdistrict, New Guinea (Discussion). *Trans.* B **268**, 279–292 (1974).

Harwood, R.S. Dynamical models of the middle atmosphere for tracer studies (Discussion).
Trans. A **296**, 103–127 (1980).

Hasegawa, M. *See* Nakanishi (H.), Jones, Thomas, Hasegawa & Rees.

Haselgrove, J.C. *See* Lowy, Vibert, Haselgrove & Poulsen.

Haskell, T.G. & Wybourne, B.G. A dynamical group for the harmonic oscillator.
Proc. A 334, 541—551 (1973).

Hasoon, M.A. & Martin, B.W. The stability of viscous axial flow in an annulus with a rotating inner cylinder. *Proc.* A 352, 351—380 (1977).

Hassard, B.D., Chang, T.S. & Ludford, G.S.S. An exact solution in the stability of m.h.d. Couette flow. *Proc.* A 327, 269—278 (1972).

Hast, N. Global measurements of absolute stress (Discussion). *Trans.* A 274, 409—419 (1973).

Hasted, J.B. *See* Foo, Brion & Hasted.

Hastie, N.D. *See* Bishop (J.O.), Beckmann, Campo, Hastie and others.

Hastie, R.J. *See* Connor, Hastie & Taylor.

Hastings, W.F. *See* Oxley & Hastings.

Hastings, W.F., Mathew, P. & Oxley, P.L.B. A machining theory for predicting chip geometry, cutting forces etc. from work material properties and cutting conditions. *Proc.* A 371, 569—587 (1980).

Hatfield, L.D., Fisher, J.W., Dunigan, J.M., Burchfield, R.W., Greene, J.M., Webber, J.A., Vasileff, R.T. & Kinnick, M.D. Cephalosporanic acids: a new look at reactions at the C-3′ position (Discussion). *Trans.* B 289, 173—179 (1980).

Hattangadi, A., Wagner, L.E. & Seth, B.B. Role of residual elements on through-thickness properties of carbon steel plates [abstract] (Discussion). *Trans.* A 295, 302 (1980).

Haus, H.A. Mode-locked semiconductor diode lasers (Discussion). *Trans.* A 298, 257—266 (1980).

Hausen, K. *See* Strausfeld & Hausen.

Hausen, K. & Strausfeld, N.J. Sexually dimorphic interneuron arrangements in the fly visual system.
Proc. B 208, 57—71 (1980).

Havner, K.S. & Shalaby, A.H. A simple mathematical theory of finite distortional latent hardening in single crystals. *Proc.* A 358, 47—70 (1977).

Haward, R.N. & Owen, D.R.J. The detergent stress-cracking of polyethylene.
Proc. A 352, 505—521 (1977).

Hawker, Lilian E. & Beckett, A. Fine structure and development of the zygospore of *Rhizopus sexualis* (Smith) Callen. *Trans.* B 263, 71—100 (1971).

Hawkes, A.G. *See* Colquhoun & Hawkes.

Hawkesworth, C.J. *See* Norry, Truckle, Lippard, Hawkesworth, Weaver & Marriner.

Hawkins, F.J. Introduction to the Copernicus satellite (Discussion). *Proc.* A 340, 397—402 (1974).

Hawkins, G.S. Astronomical alinements in Britain, Egypt and Peru (Discussion).
Trans. A 276, 157—167 (1974).

Hawkridge, D.G. Space for the Open University (Discussion). *Proc.* A 345, 567—573 (1975).

Haworth, E.Y. *See* Pennington, Haworth, Bonny & Lishman.

Haworth, R.D. & Whalley, W.B. Alexander Robertson. *Biogr. Mem.* 17, 617—642 (1971).

Hawthorne, Sir William. Introductory remarks to a Discussion. *Proc.* A 321, 145—146 (1971).
Harry Ralph Ricardo. *Biogr. Mem.* 22, 359—380 (1976).
Vote of thanks to C.L. Wilson for his Review Lecture. *Proc.* A 358, 136—139 (1977).
Introduction to a Discussion. *Trans.* A 295, 345—347 (1980).

Hawthorne, Sir William, Cohen, H. & Howell, A.R. Hayne Constant. *Biogr. Mem.* 19, 269—279 (1973).

Hay, A.J., Skehel, J.J. & McCauley, J. Structure and synthesis of influenza virus complementary RNAs (Discussion). *Trans.* B 288, 341—348 (1980).

Hayami, S. *See* Yamaguchi, Kobayashi, Matsumiya & Hayami.

Haydon, D.A. *See* Brooks (D.E.), Levine, Requena & Haydon; Gordon (L.G.M.) & Haydon; Requena, Billett & Haydon; Requena & Haydon.

Hayes, A.P. *See* Walsh (D.), Hayes & Harrison.

Hayes, B.P. *See* Roberts (Alan) & Hayes.

Hayes, M. A note on group velocity. *Proc.* A 354, 533—535 (1977).
Energy flux for trains of inhomogeneous plane waves. *Proc.* A 370, 417—429 (1980).

Hayes, M.G.W. Theory of the limiting polarization of radio waves emerging obliquely from the ionosphere. *Proc.* A 324, 369—390 (1971).

Hayes, W. *See* Elliott (R.J.), Harley, Hayes & Smith; Elliott (R.J.), Hayes, Kleppmann, Rushworth & Ryan; Gee (J.V.), Hayes & O'Brien.

Hayes, W.D. Group velocity and nonlinear dispersive wave propagation.
   Proc. A **332**, 199—221 (1973).

Hayhurst, A.N. *See* Burdett & Hayhurst.

Hayhurst, A.N. & Kittelson, D.B. Ionization of alkaline earth additives in hydrogen flames.
   I. Hydrogen atom concentrations and ion stabilities. *Proc.* A **338**, 155—173 (1974).
   II. Kinetics of production and recombination of ions. *Proc.* A **338**, 175—195 (1974).

Hayhurst, A.N. & Telford, N.R. The occurrence of chemical reactions in supersonic expansions of a gas into a vacuum and its relation to mass spectrometric sampling. *Proc.* A **322**, 483—507 (1971).

Hayhurst, D.R. *See* Leckie & Hayhurst; Leckie, Hayhurst & Morrison.

Hayhurst, D.R., Leckie, F.A. & Morrison, C.J. Creep rupture of notched bars.
   Proc. A **360**, 243—264 (1978).

Hayhurst, D.R. & Storåkers, B. Creep rupture of the Andrade shear disk.
   Proc. A **349**, 369—382 (1976).

Hayman, B. Aspects of creep buckling.
   I. The influence of post-buckling characteristics. *Proc.* A **364**, 393—414 (1978).
   II. The effects of small deflexion approximations on predicted behaviour.
   Proc. A **364**, 415—433 (1978).

Hayman, M.J., Ramsay, G., Kitchener, Gay, Graf, T., Beug, H., Roussel, Martine, Saule, S. & Stehelin, Dominique. Cell transformation by avian defective leukaemia viruses (Discussion).
   Proc. B **210**, 397—409 (1980).

Hayman, W.K. *See* Cartwright (Dame Mary) & Hayman.

Haynes, C.M. *See* Owen, Haynes & Bayley.

Hayns, M.R. *See* Willis (J.R.), Hayns & Bullough.

Hayns, M.R. & Wood, M.H. A model for the simultaneous heterogeneous and homogeneous nucleation of gas bubbles. *Proc.* A **368**, 331—343 (1979).

Hayter, J.B. *See* Leslie (M.), Jenkin, Hayter, White, Cox & Warner.

Hayward, D.O. *See* Guil, Hayward & Taylor.

Hayward, J.A. *See* Parish & Hayward.

Head, A.K. *See* Wood (W.W.) & Head.

Heading, J. General theorems relating to wave propagation governed by self-adjoint and Hermitian self-adjoint differential operators of order $2n$. *Proc.* A **360**, 279—300 (1978).

Heal, O.W. & Perkins, D.F. I.B.P. studies on montane grassland and moorlands (Discussion).
   Trans. B **274**, 295—314 (1976).

Healing, R.A. *See* Harris (P.M.), Thurrell, Healing & Archer.

Healy, M.J.R. Handling and interpreting multiple results (Discussion). *Proc.* B **184**, 369—374 (1973).
   What computers can and cannot do (Discussion). *Proc.* B **184**, 375—378 (1973).

Healy, W.P. The representation of microscopic charge and current densities in terms of polarization and magnetization fields. *Proc.* A **358**, 367—383 (1978).

Heap, Sara R. The ultraviolet spectrum of ζ Tauri (Discussion). *Trans.* A **279**, 371—377 (1975).

Heard, H.C. Comparison of the flow properties of rocks at crustal conditions (Discussion).
   Trans. A **283**, 173—186 (1976).

Heard, M.J. *See* Chamberlain, Clough, Heard, Newton, Stott & Wells; Chamberlain, Heard, Little & Wiffen.

Hearn, C.E.D. *See* Edwards (R.H.T.), Miller, Hearn & Cotes.

Hearn, J.P. Immunization against pregnancy (Discussion). *Proc.* B **195**, 149—160 (1976).

Heath, D.F. *See* Krueger, Guenther, Fleig, Heath, Hilsenrath and others.

Heatherly, L. *See* White (C.L.), Clausing & Heatherly.

Heavens, S.N. & Field, J.E. The ignition of a thin layer of explosive by impact.
   Proc. A **338**, 77—93 (1974).

Hebden, D. *See* Hutchison (Sir Kenneth) & Hebden.

Hedden, W.L., Jr, & Dowling, J.E. The interplexiform cell system. II. Effects of dopamine on goldfish retinal neurones. *Proc.* B **201**, 27—55 (1978).

Heddle, D.W.O. High resolution studies of electron excitation. VI. Resonance series in helium. *Proc.* A 352, 441–449 (1977).

Heddle, D.W.O., Keesing, R.G.W. & Kurepa, Jelena M. High resolution studies of electron excitation.
   I. The 4S states of helium and the energy scale. *Proc.* A 334, 135–147 (1973).
   II. The 3^3D, 4^3D and 4^1D states of helium: features in the $n = 4$ excitation functions.
   Proc. A 337, 435–441 (1974).

Heddle, D.W.O., Keesing, R.G.W. & Parkin, A. High resolution studies of electron excitation. IV. The $n = 3$ states of helium. *Proc.* A 352, 419–428 (1977). Corrigendum. *Proc.* A 353, 589 (1977).

Heddle, D.W.O., Keesing, R.G.W. & Watkins, R.D. High resolution studies of electron excitation. III. Polarization near threshold of light from the 4D states of helium. *Proc.* A 337, 443–450 (1974).

Hedeyat, S. *See* Lehmann (H.), Ala, Hedeyat, Montazemi and others.

Hegde, M.S. *See* Rao (C.N.R.), Sarma & Hegde; Rao (C.N.R.), Sarma, Vasudevan & Hegde.

Hegh, V. *See* Morton (H.), Hegh & Clunie.

Heier, K.S. Geochemistry of granulite facies rocks and problems of their origin (Discussion). *Trans.* A 273, 429–442 (1973).
   The distribution and redistribution of heat-producing elements in the continents (Discussion). *Trans.* A 288, 393–400 (1978).
   The movement of uranium during higher grade metamorphic processes (Discussion). *Trans.* A 291, 413–421 (1979).

Heilbronn, H. *See* Davenport (H.) & Heilbronn.

Heine, V. Electrons at surfaces of solids (Discussion). *Proc.* A 331, 307–320 (1972).

Heintze, J. *See* Bartel, Duinker, Heintze, Heinzelmann and others.

Heinzelmann, G. *See* Bartel, Duinker, Heintze, Heinzelmann and others.

Heise, K.P. *See* Inhoffen, Gossauer, Heise & Laas.

Heiser, C.B. *See* Pickersgill & Heiser.

Helander, E. Provision of rehabilitation of the disabled on the community level (Discussion). *Proc.* B 209, 139–140 (1980).

Helbrough, K. *See* Huston & Helbrough.

Held, A. A coordinate system based on a twisting null geodesic congruence. *Proc.* A 332, 415–417 (1973).

Hellawell, A. *See* Double, Hellawell & Perry.

Hellawell, J.M. Change in natural and managed ecosystems: detection, measurement and assessment (Discussion). *Proc.* B 197, 31–57 (1977).

Helle, K.B. *See* Banks (P.) & Helle.

Helliwell, R.A. Coherent v.l.f. waves in the magnetosphere (Discussion). *Trans.* A 280, 137–149 (1975).
   Active very low frequency experiments on the magnetosphere from Siple Station, Antarctica (Discussion). *Trans.* B 279, 213–224 (1977).

Hellon, R.F. *See* Mitchell (D.) & Hellon.

Hemming, C.F., Popov, G.B., Roffey, J. & Waloff, Zena. Characteristics of Desert Locust plague upsurges (Discussion). *Trans.* B 287, 375–386 (1979).

Hemming, F.W. Polyprenyl phosphates as coenzymes in protein and oligosaccharide glycosylation (Discussion). *Trans.* B 284, 559–568 (1978).

Hemmings, C., Hemmings, W.A., Patey, A.L. & Wood, C. The ingestion of dietary protein as large molecular mass degradation products in adult rats. *Proc.* B 198, 439–453 (1977).

Hemmings, W.A. The entry into the brain of large molecules derived from dietary protein. *Proc.* B 200, 175–192 (1978).
   See also Hemmings (C), Hemmings, Patey & Wood; Williams (E.W.) & Hemmings.

Hemmings, W.A. & Williams, E.W. The attachment of IgG to cell components: a reconsideration of Brambell's receptor hypothesis of protein transmission. *Proc.* B 187, 209–219 (1974).
   Quantitative and visualization studies of the transport of rat and bovine IgG and ferritin across the segments of the small intestine of the suckling rat. *Proc.* B 197, 425–440 (1977).

Hemp, J. *See* Al-Rabeh, Baker & Hemp.

Hempleman, H.V. *See* Hennessy (T.R.) & Hempleman.

Hempstead, C.A. Statement from R. Pohl, 25 July 1974, in Krefeld (Symposium).
  Proc. A **371**, 112—115 (1980).
Henderson, B. *See* McGeehin, Henderson & Benson.
Henderson, D.A. Smallpox eradication (Discussion). *Proc.* B **199**, 83—97 (1977).
Henderson, D.W. *See* Hsiao, Acevedo, Fereres & Henderson.
Henderson, I. *See* Horridge & Henderson.
Henderson, J.C. de C. *See* Cassell, Henderson & Ramachandran.
Henderson, L.F. & Siegenthaler, A. Experiments on the diffraction of weak blast waves: the von
  Neumann paradox. *Proc.* A **369**, 537—555 (1980).
Henderson, P. *See* Wass, Henderson & Elliott.
Henderson, Sir William. Opening remarks to a Discussion. *Proc.* B **209**, 3—4 (1980).
Hendrickse, R.G. Paediatrics (Discussion). *Proc.* B **199**, 73—82 (1977).
Hendrickx, H. *See* Casteels, Droogmans & Hendrickx.
Hendrie, A.L. *See* Budd, Fox, Hendrie & Hicks; Fox (R.H.), Budd, Woodward, Hackett & Hendrie.
Hendry, A. *See* Coates & Hendry.
Hendry, R.M. Observations of the semidiurnal internal tide in the western North Atlantic Ocean.
  Trans. A **286**, 1—24 (1977).
Henize, K.G. *See* Gull, York, Snow & Henize.
Hennessy, J. *See* Brown (G.C.) & Hennessy.
Hennessy, J. & Turner, G. ^{40}Ar—^{39}Ar ages and irradiation history of Luna 24 basalts.
  Trans. A **297**, 27—39 (1980).
Hennessy, T.R. & Hempleman, H.V. An examination of the critical released gas volume concept in
  decompression sickness. *Proc.* B **197**, 299—313 (1977).
Henry, R.C. *See* Fastie, Moos, Henry & Feldman.
Hensens, O.D., Hill, H.A.O., Thornton, J., Turner, A.M. & Williams, R.J.P. The structures of some
  cobalamins in solution (Discussion). *Trans.* B **273**, 353—357 (1976).
Herbert, J. Hormones and behaviour (Discussion). *Proc.* B **199**, 425—443 (1977).
  See also Tucker (E.M.), Ellory, Wooding, Morgan & Herbert.
Herbstein, F.H. *See* Bernstein (J.), Regev, Herbstein, Main and others.
Herbstein, F.H. & Kapon, M. The crystal structures of the polyiodide salts (phenacetin)$_2$·HI_5 and
  (theobromine)$_2$·H_2I_8. *Trans.* A **291**, 199—218 (1979).
Herchen, S. *See* Baldwin (J.E.), Jung, Singh, Wan, Haber and others.
Hermans, P.W. *See* Knaap, van den Hout & Hermans.
Hernalsteens, J.P. *See* Schell, Van Montagu, De Beuckeleer, De Block and others; Van Montagu,
  Holsters, Zambryski, Hernalsteens, Depicker and others.
Herndon, J.M. Re-evaporation of condensed matter during the formation of the solar system.
  Proc. A **363**, 283—288 (1978).
  The nickel silicide inner core of the Earth. *Proc.* A **368**, 495—500 (1979).
  The chemical composition of the interior shells of the Earth. *Proc.* A **372**, 149—154 (1980).
Heroux, L. & Cohen, M. Measurements of electron temperature in the solar chromosphere and corona
  (Discussion). *Trans.* A **270**, 99—107 (1971).
Herrick, C.S. Electroconvection cells in dielectric liquids interfaced with conducting fluids.
  Proc. A **336**, 487—494 (1974).
Herring, A.K. *See* Fielder, Fryer, Titulaer, Herring & Wise.
Herring, C. Recollections (Symposium). *Proc.* A **371**, 67—76 (1980).
Herring, J.R.H., Glencross, W.M., Parkinson, J.H. & Pounds, K.A. A satellite-borne X-ray spectro-
  heliograph. *Proc.* A **321**, 493—502 (1971).
Herring, P.J. *See* Zagalsky & Herring.
Hersey, S.J. The energetic coupling of acid secretion in gastric mucosa (Discussion).
  Trans. B **262**, 261—275 (1971).
Hersom, A. Thermal processing (Discussion). *Proc.* B **191**, 87—98 (1975).
Hervig, R.L. *See* Dawson (J.B.), Smith & Hervig.
Herzberg, Agnes M. On a statistical problem of E.A. Milne. II. *Proc.* A **336**, 223—227 (1974).
  See also Cox (D.R.) & Herzberg.

Herzfeld, J., Roufosse, A., Haberkorn, R.A., Griffin, R.G. & Glimcher, M.J. Magic angle sample spinning in inhomogeneously broadened biological systems (Discussion).
Trans. B **289**, 459–469 (1980).

Herzog, M. *See* Pfeiffer, Herzog & Hirth.

Heslop, Barbara F. & Lyttle, Vivienne A. Hyperplasia of donor cells in rejecting weakly antigenic skin grafts. *Proc.* B **193**, 209–215 (1976).

Heslop-Harrison, J. The Croonian Lecture, 1974. The physiology of the pollen grain surface.
Proc. B **190**, 275–299 (1975).

Genetics and physiology of angiosperm incompatibility systems (Discussion).
Proc. B **202**, 73–92 (1978).

See also Dickinson (H.G.) & Heslop-Harrison; Howlett, Knox, Paxton & Heslop-Harrison.

Heslop-Harrison, J., Heslop-Harrison, Y. & Barber, J. The stigma surface in incompatibility responses (Discussion). *Proc.* B **188**, 287–297 (1975).

Heslop-Harrison, Y. *See* Heslop-Harrison(J.), Heslop-Harrison & Barber.

Hessberg, H., Niekerke, J. & Stephan, K.-H. Preliminary results of absolute u.v. rocket photometry of γ Ori (Discussion). *Trans.* A **279**, 457 (1975).

Heuckroth, L.E. & Karim, R.A. Afghan seismotectonics (Discussion). *Trans.* A **274**, 389–395 (1973).

Heuer, R.D. *See* Bartel, Duinker, Heintze, Heinzelmann and others.

Heumann, H.-G. Smooth muscle: contraction hypothesis based on the arrangement of actin and myosin filaments in different states of contraction (Discussion). *Trans.* B **265**, 213–217 (1973).

Heuser, J., Katz, Sir Bernard & Miledi, R. Structural and functional changes of frog neuromuscular junctions in high calcium solutions. *Proc.* B **178**, 407–415 (1971).

Heuser, J. & Miledi, R. Effect of lanthanum ions on function and structure of frog neuromuscular junctions. *Proc.* B **179**, 247–260 (1971).

Hewetson, Valerie P. *See* Turner (Judith), Hewetson, Hibbert, Lowry & Chambers.

Hewitt, R.A. *See* Williams (A.) & Hewitt.

Hewson-Browne, R.C. *See* Sozou & Hewson-Browne.

Hey, M.J. *See* Caggiano, Eley & Hey; Eley, Hey & Ward.

Heyland, G.R. *See* Coleman (P.G.), Griffith & Heyland.

Heyman, J. The development of new analytical techniques (Discussion).
Trans. A **272**, 565–572 (1972).

Heymès, R. *See* Bucourt, Heymès, Lutz, Penasse & Perronnet.

Heywood, R.B. A limnological survey of the Ablation Point area, Alexander Island, Antarctica (Discussion). *Trans.* B **279**, 39–54 (1977).

Hibberd, F.H. The origin of the Earth's magnetic field. *Proc.* A **369**, 31–45 (1979).

Hibbert, D.B. & Robertson, A.J.B. The emission of electrons from glass induced by a strong electric field and the mechanism of the silent electric discharge. *Proc.* A **349**, 63–79 (1976).

Hibbert, F.A. *See* Turner (Judith), Hewetson, Hibbert, Lowry & Chambers.

Hibbert, F.A., Switsur, V.R. & West, R.G. Radiocarbon dating of Flandrian pollen zones at Red Moss, Lancashire. *Proc.* B **177**, 161–176 (1971).

Hickman, M.E. *See* Cusack, Hickman & Born.

Hicks, F.G. *See* Ferguson, Hicks, Hoaksey, Rowlands & Lloyd.

Hicks, K.E. *See* Budd, Fox, Hendrie & Hicks.

Hicks, R. Marian. *See* Markham, Horne & Hicks.

Hicks, R. Marian, Ketterer, B. & Warren, R.C. The ultrastructure and chemistry of the luminal plasma membrane of the mammalian urinary bladder: a structure with low permeability to water and ions (Discussion). *Trans.* B **268**, 23–38 (1974).

Hide, R. Jupiter and Saturn (Symposium). *Proc.* A **336**, 63–84 (1974).

Towards a theory of irregular variations in the length of the day and core–mantle coupling (Discussion). *Trans.* A **284**, 547–554 (1977).

Higgs, E.S. The history of European agriculture – the uplands (Discussion).
Trans. B **275**, 159–173 (1976).

Hildner, E. *See* MacQueen, Gosling, Hildner, Munro, Poland & Ross.

Hildreth, E. *See* Marr (D.) & Hildreth.

Hill, A.E. Solute—solvent coupling in epithelia: a critical examination of the standing-gradient osmotic flow theory. *Proc.* B **190**, 99—114 (1975).

Solute—solvent coupling in epithelia: an electro-osmotic theory of fluid transfer.
Proc. B **190**, 115—134 (1975).

Solute—solvent coupling in epithelia: contribution of the junctional pathway to fluid production.
Proc. B **191**, 537—547 (1975).

See also Hill (Bruria S.) & Hill.

Hill, A.E. & Hill, Bruria S. Sucrose fluxes and junctional water flow across *Necturus* gall bladder epithelium. *Proc.* B **200**, 163—174 (1978).

Hill, A.W. *See* Frost (A.J.), Hill & Brooker.

Hill, Bruria S. *See* Hill (A.E.) & Hill.

Hill, Bruria S. & Hill, A.E. Fluid transfer by *Necturus* gall bladder epithelium as a function of osmolarity. *Proc.* B **200**, 151—162 (1978).

Hill, C.R. The future of ultrasonics in diagnostic medicine (Discussion).
Trans. A **292**, 299—305 (1979).

Hill, D. A formalism of the indirect Auger effect. II. *Proc.* A **347**, 565—573 (1976).

Hill, D. & Landsberg, P.T. A formalism for the indirect Auger effect. I. *Proc.* A **347**, 547—564 (1976).

Hill, H.A.O. *See* Barry (C.D.), Hill, Sadler & Williams; Hensens, Hill, Thornton, Turner & Williams.

Hill, H.A.O., Sammes, P.G. & Waley, S.G. Active sites of β-lactamases from *Bacillus cereus* (Discussion). *Trans.* B **289**, 333—344 (1980).

Hill, J.E. The bats of Aldabra Atoll, western Indian Ocean (Discussion).
Trans. B **260**, 573—576 (1971).

Hill, J.G. Experience with the linear Boltzmann equation (Discussion). *Proc.* A **323**, 293—304 (1971).

Hill, P., Campbell, J. Allan & Petrie, Isobel A. *Rhodnius prolixus* and its symbiotic actinomycete: a microbiological, physiological and behavioural study. *Proc.* B **194**, 501—525 (1976).

Hill, R. On constitutive macro-variables for heterogeneous solids at finite strain.
Proc. A **326**, 131—147 (1972).

Hill, Robert. Leslie William Mapson. *Biogr. Mem.* **18**, 427—444 (1972).

Hill, Sir John. Future trends in nuclear power generation (Discussion).
Trans. A **276**, 587—601 (1974).

Hillas, A.M. Survey of data on primary cosmic-ray nuclei above 10^{14} eV (Discussion).
Trans. A **277**, 413—428 (1974).

Hille, B., Woodhull, Ann M. & Shapiro, B.I. Negative surface charge near sodium channels of nerve: divalent ions, monovalent ions, and pH (Discussion). *Trans.* B **270**, 301—318 (1975).

Hillier, I.H. *See* Baybutt, Guest & Hillier; DeKock, Lloyd, Hillier & Saunders; Guest (M.F.), Hillier, Saunders & Wood.

Hills, G.J. *See* Carrington (A.), Hills & Webb.

Hilsenrath, E. *See* Krueger, Guenther, Fleig, Heath, Hilsenrath and others.

Hiltbrand, E. *See* Béné, Borcard, Hiltbrand & Magnin.

Himsworth, Sir Harold & Pitt-Rivers, Rosalind. Charles Robert Harington.
Biogr. Mem. **18**, 267—308 (1972).

Hinde, R.A. Mother—infant separation and the nature of inter-individual relationships: experiments with rhesus monkeys. *Proc.* B **196**, 29—50 (1977).

Hines, C.O. Motions in the ionospheric D and E regions (Discussion). *Trans.* A **271**, 457—471 (1972).

Hinkley, R.K., Walker, T.E.H. & Richards, W.G. On the e.p.r. spectrum of vibrationally excited hydroxyl radicals. *Proc.* A **331**, 553—560 (1973).

Hinnen, Rosmarie. *See* Franklin (R.M.), Hinnen, Schäefer & Tsukagoshi.

Hinotani, K. *See* Krishnaswamy (M.R.), Menon, Narasimham, Hinotani, Ito and others.

Hinshaw, V.S. *See* Webster (R.G.), Hinshaw, Bean & Sriram.

Hinshaw, W.S. *See* Andrew, Hinshaw, Hutchins & Jasinski.

Hinz, K. *See* Goldflam, Hinz, Weigel & Wissmann.

Hiorns, R.W. *See* Harrison (G.A.), Hiorns & Boyce; Sloper, Hiorns & Powell; Winfield, Hiorns & Powell.

Hirsch, Sir Peter. Direct observations of dislocations by transmission electron microscopy:

recollections of the period 1946–56 (Symposium). *Proc.* A **371**, 160–164 (1980).
See also Cherns, Hirsch & Saka.
Hirschler, M.M. *See* Cullis & Hirschler.
Hirst, W. *See* Adams (D.R.) & Hirst; Crompton (D.), Hirst & Howse.
Hirst, W. & Hollander, A.E. Surface finish and damage in sliding. *Proc.* A **337**, 379–394 (1974).
Hirst, W. & Lewis, M.G. The rheology of oils during impact. III. Elastic behaviour.
    Proc. A **334**, 1–18 (1973).
Hirst, W. & Moore, A.J. Non-Newtonian behaviour in elastohydrodynamic lubrication.
    Proc. A **337**, 101–121 (1974).
  The elastohydrodynamic behaviour of polyphenyl ether. *Proc.* A **344**, 403–426 (1975).
  Elastohydrodynamic lubrication at high pressures. *Proc.* A **360**, 403–425 (1978).
  Corrigendum. *Proc.* A **365**, 565 (1979).
  Elastohydrodynamic lubrication at high pressures. II. Non-Newtonian behaviour.
    Proc. A **365**, 537–565 (1979).
  The effect of temperature on traction in elastohydrodynamic lubrication.
    Trans. A **298**, 183–208 (1980).
Hirth, J.P. Adsorption at grain boundaries and its effect on decohesion (Discussion).
    Trans. A **295**, 139–149 (1980).
Hirth, L. *See* Guilley, Jonard, Richards & Hirth; Jonard, Briand, Bouley, Witz & Hirth; Pfeiffer,
    Herzog & Hirth.
Hitchcock, P. *See* Elder (M.), Hitchcock, Mason & Shipley.
Hitchin, N.J. Linear field equations on self-dual spaces. *Proc.* A **370**, 173–191 (1980).
See also Atiyah, Hitchin & Singer.
Hnatiuk, R.J. Temporal and spatial variations in precipitation on Aldabra (Discussion).
    Trans. B **286**, 25–33 (1979).
Hnatiuk, R.J. & Merton, L.F.H. A perspective of the vegetation of Aldabra (Discussion).
    Trans. B **286**, 79–84 (1979).
Hnatiuk, Sarah H. Numbers of plant species on the islands of Aldabra Atoll (Discussion).
    Trans. B **286**, 247–254 (1979).
Hoaksey, A. *See* Ferguson, Hicks, Hoaksey, Rowlands & Lloyd.
Hoar, P.T. Review Lecture. Corrosion of metals: its cost and control. *Proc.* A **348**, 1–18 (1976).
Hobbis, L.C.W. *See* Fender, Hobbis & Manning.
Hobbs, L.W., Hughes, A.E. & Pooley, D. A study of interstitial clusters in irradiated alkali halides
    using direct electron microscopy. *Proc.* A **332**, 167–185 (1973).
Hobbs, P.V. *See* Levin (Z.) & Hobbs.
Hockaday, A.R. *See* Hardy (R.N.), Hockaday & Tapp.
Hocking, L.M. *See* Bennetts & Hocking.
Hocking, L.M. & Stewartson, K. On the nonlinear response of a marginally unstable plane parallel
    flow to a two-dimensional disturbance. *Proc.* A **326**, 289–313 (1972).
Hoddeson, Lillian H. & Baym, G. The development of the quantum mechanical electron theory of
    metals: 1900–28 (Symposium). *Proc.* A **371**, 8–23 (1980).
Hodge, A.M. *See* Bassett (D.C.) & Hodge.
Hodge, Sir William. Solomon Lefschetz. *Biogr. Mem.* **19**, 433–453 (1973).
Hodgkin, A.L. Anniversary Address. 1971. *Proc.* A **326**, v–xx (1971); *also Proc.* B **180**, v–xx (1972).
See also Hodgkin (Sir Alan).
Hodgkin, Dorothy M. Crowfoot. The Bakerian Lecture, 1972. Insulin, its chemistry and biochemistry.
    Proc. B **186**, 191–215 (1974); *Proc.* A **338**, 251–275 (1974).
  Kathleen Lonsdale. *Biogr. Mem.* **21**, 447–484 (1975).
  John Desmond Bernal. *Biogr. Mem.* **26**, 17 (1980).
  See also De Sanctis, Gdenis, Taylor & Hodgkin; De Sanctis & Hodgkin; Venkatesan, Dale, Hodgkin,
    Nockolds, Moore & O'Connor.
Hodgkin, R.A. *See* Wigner & Hodgkin.
Hodgkin, Sir Alan. Anniversary Addresses:
    1972 *Proc.* A **331**, 285–304 (1972); *also Proc.* B **183**, 1–19 (1973).

1973 *Proc.* A **336**, v–xx (1974); *also Proc.* B **185**, v–xx (1974).

1974 *Proc.* A **342**, 1–17 (1975); *also Proc.* B **188**, 103–119 (1975).

1975 *Proc.* A **348**, 153–173 (1976); *also Proc.* B **192**, 371–391 (1976).

The optimum density of sodium channels in an unmyelinated nerve (Discussion). *Trans.* B **270**, 297–300 (1975).

Edgar Douglas Adrian (Baron Adrian of Cambridge). *Biogr. Mem.* **25**, 1–73 (1979). *See also* Hodgkin (A.L.).

Hodgson, A.A. Nature and paragenesis of asbestos minerals (Discussion). *Trans.* A **286**, 611–624 (1977).

Hoekstra, R. *See* de Jager, Hoekstra, van der Hucht, Kamperman & Lamers.

Hoffman, J.F. *See* Glynn, Hoffman & Lew.

Hoffmann, K.-P. & Cynader, M. Functional aspects of plasticity in the visual system of adult cats after early monocular deprivation (Discussion). *Trans.* B **278**, 411–424 (1977).

Hoffman, P. Evolution of an early Proterozoic continental margin: the Coronation geosyncline and associated aulacogens of the northwestern Canadian shield (Discussion). *Trans.* A **273**, 547–581 (1973).

Hoffmann, Roald. *See* Whangbo, Hoffmann & Woodward.

Hogan, M.J. *See* Steinberg, Wood & Hogan.

Hogarth, W.L. & McElwain, D.L.S. Internal relaxation, ionization and recombination in a dense hydrogen plasma.

I. Application of singular perturbation theory. *Proc.* A **345**, 251–263 (1975).

II. Population distributions and rate coefficients. *Proc.* A **345**, 265–276 (1975).

Hogben, Lancelot. Francis Albert Eley Crew. *Biogr. Mem.* **20**, 135–153 (1974).

Hogenboom, N.G. Incompatibility and incongruity: two different mechanisms for the non-functioning of intimate partner relationships (Discussion). *Proc.* B **188**, 361–375 (1975).

Hohn, Barbara. *See* Hohn (T.), Wurtz & Hohn.

Hohn, T. Packaging of genomes in bacteriophages: a comparison of ssRNA bacteriophages and dsDNA bacteriophages (Discussion). *Trans.* B **276**, 143–150 (1976).

Hohn, T., Wurtz, M. & Hohn, Barbara. Capsid transformation during packaging of bacteriophage λ DNA (Discussion). *Trans.* B **276**, 51–61 (1976).

Hökfelt, T., Lundberg, J.M., Schultzberg, M., Johansson, O., Skirboll, L., Änggård, A., Fredholm, B., Hamberger, B., Pernow, B., Rehfeld, J. & Goldstein, M. Cellular localization of peptides in neural structures (Discussion). *Proc.* B **210**, 63–77 (1980).

Holah, G. *See* Ellis (P.), Holah, Houghton, Jones, Peckham and others.

Holden, A.V. The effects of pesticides on life in fresh waters (Discussion). *Proc.* B **180**, 383–394 (1972).

Holdgate, M.W. Terrestrial ecosystems in the Antarctic (Discussion). *Trans.* B **279**, 5–25 (1977).

Targets of pollutants in the atmosphere (Discussion). *Trans.* A **290**, 591–607 (1979).

Holdsworth, S.R. *See* Batte, Brear, Holdsworth, Myers & Reynolds.

Holford, Lord. Problems for the architect and town planner caused by air in motion (Discussion). *Trans.* A **269**, 335–341 (1971).

Holland, G.N. *See* Moore (W.S.) & Holland.

Hollander, A.E. *See* Hirst & Hollander.

Holliday, R. Recombination and meiosis (Discussion). *Trans.* B **277**, 359–370 (1977). *See also* Kirkwood & Holliday.

Hollingworth, T. & Berry, M. Network analysis of dendritic fields of pyramidal cells in neocortex and Purkinje cells in the cerebellum of the rat. *Trans.* B **270**, 227–264 (1975).

Hollis, D.P. & Nunnally, R.L. Recent ^{31}P n.m.r. studies of myocardium [abstract] (Discussion). *Trans.* B **289**, 437–439 (1980).

Holm, A.V. A study of variable stars in the ultraviolet (Discussion). *Trans.* A **279**, 459–471 (1975).

Holm, P. & Lothe, J. The topological nature of the polarization field for body waves in anisotropic elastic media. *Proc.* A **370**, 331–350 (1980).

Holman, Mollie E. A survey of the new findings presented and the discussions arising during sessions I, II and III (Discussion). *Trans.* B **265**, 157–165 (1973).

Holmes, A.W. Substitute foods – a practical alternative? (Discussion). *Trans.* B **267**, 157–166 (1973).

Holmes, G.D. History of forestry and forest management (Discussion). *Trans.* B **271**, 69–80 (1975).

Holmes, K.C., Tregear, R.T. & Leigh, J. Barrington. Interpretation of the low angle X-ray diffraction from insect flight muscle in rigor. *Proc.* B **207**, 13–33 (1980).

Holmes, P. A nonlinear oscillator with a strange attractor. *Trans.* A **292**, 419–448 (1979).

Holmes, W. Choosing between animals (Discussion). *Trans.* B **281**, 121–137 (1977).

Holroyd, Sir Ronald. Alexander Fleck (Baron Fleck of Saltcoats). *Biogr. Mem.* **17**, 243–254 (1971).

Holsters, M. *See* Schell, Van Montagu, De Beuckeleer, De Block and others; Van Montagu, Holsters, Zambryski, Hernalsteens, Depicker and others.

Holt, J. *See* Miller (D.S.), Baker, Bowden, Evans, Holt and others.

Holt, M. & Modarress, D. Application of the method of integral relations to laminar boundary layers in three dimensions. *Proc.* A **353**, 319–347 (1977).

Holt, S.S. Results from the Ariel 5 All-Sky Monitor (Discussion). *Proc.* A **350**, 505–519 (1976).

Holten, D. *See* Windsor & Holten.

Holton, J.R. Wave propagation and transport in the middle atmosphere (Discussion). *Trans.* A **296**, 73–85 (1980).

Holzer, T.E. *See* Hundhausen & Holzer.

Holtzman, E. Cytochemical studies of protein transport in the nervous system (Discussion). *Trans.* B **261**, 407–421 (1971).

Homer, J.B. *See* Graham (S.C.), Homer & Rosenfeld.

Homer, J.B. & Hurle, I.R. Shock-tube studies on the decomposition of tetramethyl-lead and the formation of lead oxide particles. *Proc.* A **327**, 61–79 (1972).

Hommel, M. *See* Killick-Kendrick, Molyneux, Hommel, Leaney & Robertson.

Hommeril, P. *See* Hamilton (D.), Hommeril, Larsonneur & Smith; Huault, Lefebvre, Guyader, Giresse and others.

Hondros, E.D. Residuals and properties (Discussion). *Trans.* A **295**, 9–23 (1980).
See also Lea (C.), Seah & Hondros; Seah & Hondros; Seah, Spencer & Hondros.

Honeycombe, R.W.K. Keynote speech (Symposium). *Trans.* A **282**, 426–437 (1976).
See also Davenport (A.T.) & Honeycombe; Dippenaar & Honeycombe.

Hong, N.-S., Jones, A.R. & Weinberg, F.J. Doppler velocimetry within turbulent phase boundaries. *Proc.* A **353**, 77–85 (1977).

Honnorez, J. *See* Bonatti, Honnorez & Ferrara.

Hook, J.R. *See* Dixon (H.E.), Earnshaw, Hook, Hough, Smith, Stephenson & Turver.

Hook, J.R. & Waldram, J.R. A Ginzburg–Landau equation with non-local correction for superconductors in zero magnetic field. *Proc.* A **334**, 171–192 (1973).

Hook, O. & Johnels, A.G. The breeding and distribution of the grey seal (*Halichoerus grypus* Fab.) in the Baltic Sea, with observations on other seals of the area. *Proc.* B **182**, 37–58 (1972).

Hope, D.B. *See* Buffoni, Della Corte & Hope; Livett, Uttenthal & Hope; Uttenthal & Hope; Uttenthal, Livett & Hope.

Hope, R.A., Hammond, B.J. & Gaze, R.M. The arrow model: retinotectal specificity and map formation in the goldfish visual system. *Proc.* B **194**, 447–466 (1976).

Hopfield, Helen S. Improvements in the tropospheric refraction correction for range measurement (Discussion). *Trans.* A **294**, 341–352 (1980).

Hopkins, B.E. *See* Glenny & Hopkins.

Hopkins, H.D. *See* Goodall, Hopkins, Tulunay & D'Arcy.

Hopley, D. Sea level change on the Great Barrier Reef: an introduction (Discussion). *Trans.* A **291**, 159–166 (1978).
See also McLean (R.F.), Stoddart, Hopley & Polach; Stoddart (D.R.), McLean & Hopley; Stoddart (D.R.), McLean, Scoffin, Thom & Hopley.

Hoppe, W. Use of zone correction plates and other techniques for structure determination of aperiodic objects at atomic resolution using a conventional electron microscope (Discussion). *Trans.* B **261**, 71–94 (1971).

Hopwood, D.A. *See* Beringer, Brewin, Johnston, Schulman & Hopwood.

Hopwood, D.A. & Chater, K.F. Fresh approaches to antibiotic production (Discussion). *Trans.* B **290**, 313–328 (1980).

Horai, K., Winkler, J.L., Jr, Keihm, S.J., Langseth, M.G., Fountain, J.A. & West, E.A. Thermal conduction in a composite circular cylinder: a new technique for thermal conductivity measurements of lunar core samples. *Trans.* A **293**, 571–598 (1980).

Horani, M. *See* Duxbury, Horani & Rostas.

Horder, T.J. *See* Cook (J.E.) & Horder.

Horgan, R. Nature and distribution of cytokinins (Discussion). *Trans.* B **284**, 439–447 (1978).

Horn, F. On a connexion between stability and graphs in chemical kinetics.
  I. Stability and the reaction diagram. *Proc.* A **334**, 299–312 (1973).
  II. Stability and the complex graph. *Proc.* A **334**, 313–330 (1973).
  Stability and complex balancing in mass-action systems with three short complexes.
      Proc. A **334**, 331–342 (1973).

Horn, P. & Kirsten, T. Lunar highland stratigraphy and radiometric dating (Discussion).
    Trans. A **285**, 145–150 (1977).

Horn, R. *See* Bouysse, Horn, Lefort & Le Lann; Larsonneur, Horn & Auffret.

Horn, R.G. & Faber, T.E. Molecular alignment in nematic liquid crystals: a comparison between the results of experiments at high pressure and predictions based on mean field theories.
    Proc. A **368**, 199–223 (1979).

Hornabrook, R.W. The demography of the population of Karkar Island (Discussion).
    Trans. B **268**, 229–239 (1974).
    See also Beavan (G.H.), Fox & Hornabrook; Boyce (A.J.), Harrison, Platt, Hornabrook and others.

Hornabrook, R.W., Crane, G.G. & Stanhope, J.M. Karkar and Lufa: an epidemiological and health background to the human adaptability studies of the International Biological Progamme (Discussion). *Trans.* B **268**, 293–308 (1974).

Horne, J.G. *See* Doyle (E.D.), Horne & Tabor.

Horne, R.W. *See* Goodman (R.M.), McDonald, Horne & Bancroft; Markham, Horne & Hicks.

Horner, R.W. Current proposals for the Thames barrier and the organization of the investigations (Discussion). *Trans.* A **272**, 179–185 (1972).

Horridge, G.A. Alternatives to superposition images in clear-zone compound eyes.
    Proc. B **179**, 97–124 (1971).
  Further observations on the clear zone eye of *Ephestia*. *Proc.* B **181**, 157–173 (1972).
  The ommatidium of the dorsal eye of *Cloeon* as a specialization for photoreisomerization.
      Proc. B **193**, 17–29 (1976).
  The separation of visual axes in apposition compound eyes. *Trans.* B **285**, 1–59 (1978).
  Review Lecture. Apposition eyes of large diurnal insects as organs adapted to seeing.
      Proc. B **207**, 287–309 (1980).
    See also Burrows (M.) & Horridge; Diesendorf & Horridge; Ioannides & Horridge; Meyer-Rochow & Horridge; Walcott & Horridge.

Horridge, G.A. & Burrows, M. Synapses upon motoneurons of locusts during retrograde degeneration.
    Trans. B **269**, 95–108 (1974).

Horridge, G.A. & Giddings, Caroline. Movement on dark–light adaptation in beetle eyes of the neuropteran type. *Proc.* B **179**, 73–85 (1971).
  The retina of *Ephestia* (Lepidoptera). *Proc.* B **179**, 87–95 (1971).

Horridge, G.A., Giddings, C. & Stange, G. The superposition eye of skipper butterflies.
    Proc. B **182**, 457–495 (1972).

Horridge, G.A., Giddings, C. & Wilson, M. The eye of the soldier beetle *Chauliognathus pulchellus* (Cantharidae). *Proc.* B **203**, 361–378 (1979).

Horridge, G.A. & Henderson, I. The ommatidium of the lacewing *Chrysopa* (Neuroptera).
    Proc. B **192**, 259–271 (1976).

Horridge, G.A. & McLean, Miriam. The dorsal eye of the mayfly *Atalophlebia* (Ephemeroptera).
    Proc. B **200**, 137–150 (1978).

Horridge, G.A., McLean, M., Stange, G. & Lillywhite, P.G. A diurnal moth superposition eye with high resolution *Phalaenoides tristifica* (Agaristidae). *Proc.* B **196**, 233–250 (1977).

Horridge, G.A. & Mimura, K. Fly photoreceptors. I. Physical separation of two visual pigments in *Calliphora* retinula cells 1–6. *Proc.* B **190**, 211–224 (1975).

Horridge, G.A., Mimura, K. & Hardie, R.C. Fly photoreceptors. III. Angular sensitivity as a function of wavelength and the limits of resolution. *Proc.* B **194**, 151–177 (1976).

Horridge, G.A., Mimura, K. & Tsukahara, Y. Fly photoreceptors. II. Spectral and polarized light sensitivity in the drone fly *Eristalis*. *Proc.* B **190**, 225–237 (1975).

Horridge, G.A., Ninham, B.W. & Diesendorf, M.O. Theory of the summation of scattered light in clear zone compound eyes. *Proc.* B **181**, 137–156 (1972).

Horswill, P. & Horton, A. Cambering and valley bulging in the Gwash valley at Empingham, Rutland (with an appendix by P.R. Vaughan) (Discussion). *Trans.* A **283**, 427–462 (1976).

Horton, A. *See* Horswill & Horton.

Horton, E.W. Review Lecture. The prostaglandins. *Proc.* B **182**, 411–426 (1972).

Hothem, L.D. *See* Strange & Hothem.

Hotta, Y. *See* Stern & Hotta.

Hough, J. *See* Drever (R.W.P.), Hough, Pugh, Edelstein and others.

Hough, J.H. *See* Dixon (H.E.), Earnshaw, Hook, Hough, Smith, Stephenson & Turver.

Houghton, J.T. Greenhouse effects of some atmospheric constituents (Discussion).
   Trans. A **290**, 515–521 (1979).
Introduction to a Discussion. *Trans.* A **296**, 3–5 (1980).
   See also Chaloner, Drummond, Houghton, Jarnot & Roscoe; Curtis (P.D.), Houghton, Peskett & Rodgers; Drummond (J.R.), Houghton, Peskett, Rodgers, Wale and others; Ellis (P.), Holah, Houghton, Jones, Peckham and others.

Houghton, J.T. & Walshaw, C.D. Gordon Miller Bourne Dobson. *Biogr. Mem.* **23**, 41–57 (1977).

Hoult, D.I. Rotating frame zeugmatography (Discussion). *Trans.* B **289**, 543–547 (1980).

Hoult, D.I. & Richards, R.E. Critical factors in the design of sensitive high resolution nuclear magnetic resonance spectrometers. *Proc.* A **344**, 311–340 (1975). Corrigenda. *Proc.* A **346**, 577 (1975).

Hounsfield, G.N. The E.M.I. scanner (Discussion). *Proc.* B **195**, 281–289 (1977).
Computer reconstructed X-ray imaging (Discussion). *Trans.* A **292**, 223–232 (1979).

House, W.A. & Jaycock, M.J. The application of the gas–solid virial expansion to argon absorbed on the (100) face of sodium chloride. *Proc.* A **348**, 317–337 (1976).

Housley, R.M. Solar wind and micrometeorite effects in the lunar regolith (Discussion).
   Trans. A **285**, 363–367 (1977).

van den Hout, K.D. *See* Knaap, van den Hout & Hermans.

Howard, A.J. *See* Birchall, Howard & Bailey.

Howard, J. & Lloyd, B. The Oxfam sanitation unit (Discussion). *Proc.* B **199**, 179–182 (1977).

Howard, J.C. & Gowans, J.L. The role of lymphocytes in antibody formation. III. The origin from small lymphocytes of cells forming direct and indirect haemolytic plaques to sheep erythrocytes in the rat. *Proc.* B **182**, 193–209 (1972).

Howard, J.G., Christie, G.H. & Courtenay, Barbara M. Studies on immunological paralysis.
   IV. The relative contributions of continuous antibody neutralization and central inhibition to paralysis with type III pneumococcal polysaccharide. *Proc.* B **178**, 417–438 (1971).
   VII. Rapid reversal of Felton's paralysis as evidence for 'tolerant' cells.
      Proc. B **180**, 347–361 (1972).

Howard, P. Gravity and the circulation (Discussion). *Proc.* B **199**, 485–491 (1978).

Howard, W.E. *See* Metz, Howard, Wunsch, Neusser & Schlag.

Howarth, J.V. Heat production in non-myelinated nerves (Discussion).
   Trans. B **270**, 425–432 (1975).

Howarth, J.V., Ritchie, J.M. & Stagg, D. The initial heat production in garfish olfactory nerve fibres.
   Proc. B **205**, 347–367 (1979).

Howarth, R.D. & Whalley, W.B. Alexander Robertson. *Biogr. Mem.* **17**, 617–642 (1971).

Howarth, R.J. *See* Webb (J.S.) & Howarth.

Howe, M.S. Conservation of energy in random media, with application to the theory of sound absorption by an inhomogeneous flexible plate. *Proc.* A **331**, 479–496 (1973).
On the kinetic theory of wave propagation in random media. *Trans.* A **274**, 523–549 (1973).
Contributions to the theory of scattering by randomly irregular surfaces.
   Proc. A **337**, 413–433 (1974).

The attenuation of sound by a randomly irregular impedance layer. *Proc.* A **347**, 513—535 (1976).

On the added mass of a perforated shell, with application to the generation of aerodynamic sound by a perforated trailing edge. *Proc.* A **365**, 209—233 (1979).

On the theory of unsteady high Reynolds number flow through a circular aperture.
Proc. A **366**, 205—223 (1979).

On the diffraction of sound by a screen with circular apertures in the presence of a low Mach number grazing flow. *Proc.* A **370**, 523—544 (1980).

Aerodynamic sound generated by a slotted trailing edge. *Proc.* A **373**, 235—252 (1980).

Howe, M.S. & Williams, J.E. Ffowcs. On the noise generated by an imperfectly expanded supersonic jet. *Trans.* A **289**, 271—314 (1978).

Howell, A.R. *See* Hawthorne, Cohen & Howell.

Howells, Gwyneth P. The estuary of the Hudson River, U.S.A. (Discussion).
Proc. B **180**, 521—534 (1972).

Howells, Susan & O'Hara, M.J. Low solubility of alumina in enstatite and uncertainties in estimated palaeogeotherms (Discussion). *Trans.* A **288**, 471—486 (1978).

Howlett, B.J., Knox, R.B., Paxton, J.D. & Heslop-Harrison, J. Pollen-wall proteins: physicochemical characterization and role in self-incompatibility in *Cosmos bipinnatus*.
Proc. B **188**, 167—182 (1975).

Howse, M.G.W. *See* Crompton (D.), Hirst & Howse.

Hoyle, G. *See* Tosney & Hoyle; Woollacott & Hoyle.

Hoyle, G. & Williams, Melissa. The musculature of *Peripatus* and its innervation.
Trans. B **288**, 481—510 (1980).

Hoyle, Sir Fred. The work of Nicolaus Copernicus (Symposium). *Proc.* A **336**, 105—114 (1974).

Hsiao, T.C., Acevedo, E., Fereres, E. & Henderson, D.W. Water stress, growth and osmotic adjustment (Discussion). *Trans.* B **273**, 479—500 (1976).

Hu, S.-L. *See* Sambrook, Botchan, Hu, Mitchison & Stringer.

Huault, Marie-Francoise, Lefebvre, Dominique, Guyader, J., Giresse, P., Hommeril, P. & Larsonneur, C. Evolution of the estuary of the Seine since the last glaciation (Discussion).
Trans. A **279**, 229—231 (1975).

Hubbard, C.E. William Bertram Turrill. *Biogr. Mem.* **17**, 689—712 (1971).
John Hutchinson. *Biogr. Mem.* **21**, 345—365 (1975).

Hubel, D.H. & Wiesel, T.N. Ferrier Lecture. Functional architecture of macaque monkey visual cortex. *Proc.* B **198**, 1—59 (1977).

Hubel, D.H., Wiesel, T.N. & LeVay, S. Plasticity of ocular dominance columns in monkey striate cortex (Discussion). *Trans.* B **278**, 377—409 (1977).

Huber, M.C.E. & Sandeman, R.J. Oscillator strengths of Cr I lines lying between 200 and 541 nm from hook-method and absorption measurements in a furnace. *Proc.* A **357**, 355—379 (1977).

Huber, M.C.E., Sandeman, R.J. & Tubbs, E.F. The spectrum of Cr I between 179.8 and 200 nm wavelengths, absorption cross sections, and oscillator strengths. *Proc.* A **342**, 431—438 (1975).

Huberman, A. *See* Aréchiga, Huberman & Naylor.

van der Hucht, K.A. The continuum energy distribution of the Wolf-Rayet star in γ^2 Velorum (Discussion). *Trans.* A **279**, 451—455 (1975).
See also de Jager, Hoekstra, van der Hucht, Kamperman & Lamers.

Hucknall, D.J. *See* Cullis, Hucknall & Shepherd.

Huddleston, J.A. *See* O'Sullivan (J.), Huddleston & Abraham.

Hudson, R.C.L. *See* Veron & Hudson.

Hudson, Sir William. Review Lecture. The Snowy Mountains hydroelectric and irrigation scheme (Australia). *Proc.* A **326**, 23—37 (1971).

Huerre, P. The nonlinear stability of a free shear layer in the viscous critical layer régime.
Trans. A **293**, 643—672 (1980).

Huerre, P. & Scott, J.F. Effects of critical layer structure on the nonlinear evolution of waves in free shear layers. *Proc.* A **371**, 509—524 (1980).

Huffman, G.P. *See* Fisher (R.M.), Huffman, Nagata & Schwerer.

Hughes, A.E. *See* Hobbs (L.W.), Hughes & Pooley; Jain (S.C.) & Hughes.

Hughes, A.R.W. *See* Bullough (K.), Denby, Gibbons, Hughes and others.

Hughes, C.P. *See* Whittington & Hughes.

Hughes, C.P., Ingham, J.K. & Addison, R. The morphology, classification and evolution of the Trinucleidae (Trilobita). *Trans.* B **272**, 537–604 (1975).

Hughes, D.E. & McKenzie, P. The microbial degradation of oil in the sea (Discussion). *Proc.* B **189**, 375–390 (1975).

Hughes, E.O. *See* Bamford (C.H.) & Hughes.

Hughes, G.M. Ultrastructure and morphometry of the gills of *Latimeria chalumnae*, and a comparison with the gills of associated fishes. *Proc.* B **208**, 309–328 (1980).

Hughes, J. Biologically active peptides: prospects for drug development (Discussion). *Trans.* B **290**, 387–394 (1980).

See also McKnight, Hughes & Kosterlitz.

Hughes, J.G. *See* Crothers & Hughes.

Hughes, R.N. & Gamble, J.C. A quantitative survey of the biota of intertidal soft substrata on Aldabra Atoll, Indian Ocean. *Trans.* B **279**, 327–355 (1977).

Hughes, S. Earth satellite orbits with resonant lunisolar perturbations. I. Resonances dependent only on inclination. *Proc.* A **372**, 243–264 (1980).

Hughes, S. & Meadows, A.J. A study of near-circular satellite orbits: with an application to lunisolar perturbations. *Proc.* A **355**, 131–140 (1977).

Hughes, T.H. Effect of the environment on processing and handling materials at sea (Discussion). *Trans.* A **290**, 161–177 (1978).

Hühnermann, H. *See* Brimicombe, Stacey, Stacey, Hühnermann & Menzel.

Hui, W.H. Supersonic and hypersonic flow with attached shock waves over delta wings. *Proc.* A **325**, 251–268 (1971).

See also Dungey (J.C.) & Hui.

Hukins, D.W.L. *See* Aspden & Hukins.

Hulet, W.H. Structure and functional development of the eel leptocephalus *Ariosoma balearicum* (De La Roche, 1809). *Trans.* B **282**, 107–138 (1978).

Hull, D. *See* Beahan, Bevis & Hull.

Hüller, A. *See* Bomchil, Hüller, Rayment, Roser, Smalley, Thomas & White.

Hulme, B. *See* Pessina, Hulme & Peart.

Hulme, E.C. *See* Birdsall (N.J.M.), Hulme & Burgen.

Hulme, G. & Fielder, G. Effusion rates and rheology of lunar lavas (Discussion). *Trans.* A **285**, 227–234 (1977).

Hulmes, D.J.S. *See* Doyle (B.B.), Hulmes, Miller, Parry, Piez & Woodhead-Galloway.

Hülser, D. *See* Gerisch, Hülser, Malchow & Wick.

Humphrey, S. *See* Edholm (O.G.), Humphrey, Lourie, Tredre & Brotherhood.

Humphreys, A.E. & Gray, P. Thermal diffusion as a probe of binary diffusion coefficients at elevated temperatures. II. Methane + nitrogen and methane + carbon dioxide. *Proc.* A **322**, 89–100 (1971).

Humphries, D.J. *See* O'Hara & Humphries.

Humphries, R.L. & Luckhurst, G.R. A statistical theory of liquid crystalline mixtures: phase separation. *Proc.* A **352**, 41–56 (1976).

Hundhausen, A.J. & Holzer, T.E. Large-scale solar magnetic fields, coronal holes and high-speed solar wind streams (Discussion). *Trans.* A **297**, 521–529 (1980).

Huneke, J.C. *See* Wasserburg, Papanastassiou, Tera & Huneke.

Hunsicker, H.Y. Development of Al–Zn–Mg–Cu alloys for aircraft (Symposium). *Trans.* A **282**, 359–376 (1976).

Hunt, C.P. *See* Stoddart (C.T.H.) & Hunt.

Hunt, D.J. *See* Franks (A.), Lindsey, Bennett, Speer, Turner & Hunt.

Hunt, G.E. A new look to the Martian atmosphere. *Proc.* A **341**, 317–330 (1974).

Hunt, G.W. Imperfection-sensitivity of semi-symmetric branching. *Proc.* A **357**, 193–211 (1977).

Hunt, G.W., Reay, N.A. & Yoshimura, T. Local diffeomorphisms in the bifurcational manifestations of the umbilic catastrophes. *Proc.* A **369**, 47–65 (1979).

Hunt, J.C.R. The effect of single buildings and structures (Discussion). *Trans.* A **269**, 457–467 (1971).
Hunt, N.A. *See* Baldwin (J.E.), Jung, Singh, Wan, Haber and others.
Hunt, R.D. *See* Smith (F.B.) & Hunt.
Hunt, R.T. *See* Williamson (J.B.P.) & Hunt.
Hunt, Ruth. *See* Medawar, Hunt & Mertin.
Hunt, S. *See* Iversen (L.L.), Lee, Gilbert, Hunt & Emson.
Hunt, S. & Oates, K. Fine structure and molecular organization of the periostracum in a gastropod mollusc *Buccinum undatum* L. and its relation to similar structural protein systems in other invertebrates. *Trans.* B **283**, 417–459 (1978).
Hunt, S.V., Ellis, Susan T. & Gowans, J.L. The role of lymphocytes in antibody formation. IV. Carriage of immunological memory by lymphocyte fractions separated by velocity sedimentation and on glass bead columns. *Proc.* B **182**, 211–231 (1972).
Hunter, G. & Kuriyan, Mary. Two-centre wavefunctions in the theory of electron scattering by hydrogen atoms. *Proc.* A **341**, 491–515 (1975).
Proton collisions with hydrogen atoms at low energies: quantum theory and integrated cross-sections. *Proc.* A **353**, 575–588 (1977).
The scattering states of HD^+. *Proc.* A **358**, 321–333 (1978).
Huntingdon, A.T. The collection and analysis of volcanic gases from Mount Etna (Symposium). *Trans.* A **274**, 119–128 (1973).
Huntoon, S. *See* Schroepfer, Lutsky, Martin, Huntoon and others.
Hurle, I.R. *See* Homer & Hurle.
Hurst, R. Shipbuilding in the future (Discussion). *Trans.* A **273**, 13–24 (1972).
Hursthouse, M.B. *See* Bonnett (R.), Davies, Hursthouse & Sheldrick; Neidle, Rogers & Hursthouse.
Huston, A.E. & Helbrough, K. The Synchroscan picosecond streak camera system (Discussion). *Trans.* A **298**, 287–293 (1980).
Hutchings, I.M., Winter, R.E. & Field, J.E. Solid particle erosion of metals: the removal of surface material by spherical projectiles. *Proc.* A **348**, 379–392 (1976).
Hutchins, M.G. *See* Andrew, Hinshaw, Hutchins & Jasinski.
Hutchinson, J.N. & Gostelow, T.P. The development of an abandoned cliff in London Clay at Hadleigh, Essex (Discussion). *Trans.* A **283**, 557–604 (1976).
Hutchinson, J.W. Bounds and self-consistent estimates for creep of polycrystalline materials. *Proc.* A **348**, 101–127 (1976).
Hutchinson, M.H.R. *See* Bradley (D.J.), Hutchinson & Koetser; Bradley (D.J.), Hutchinson, Koetser, Morrow, New & Petty.
Hutchinson, Sir Joseph. Closing remarks to a Discussion. *Trans.* B **267**, 167–172 (1973).
India: local and introduced crops (Discussion). *Trans.* B **275**, 129–141 (1976).
Hutchinson, W.B. *See* Dillamore, Morris, Smith & Hutchinson.
Hutchison, D.L. *See* Raudkivi & Hutchison.
Hutchison, J.L. *See* Anderson (J.S.), Bevan, Cheetham, Von Dreele and others; Anderson (J.S.), Hutchison & Lincoln; McConnell (J.D.M.), Hutchison & Anderson; Mallinson (L.G.), Jefferson, Thomas & Hutchison.
Hutchison, J.L., Anderson, J.S. & Rao, C.N.R. Electron microscopy of ferroelectric bismuth oxides containing perovskite layers. *Proc.* A **355**, 301–312 (1977).
Hutchison, J.M.S. *See* Mallard, Hutchison, Edelstein, Ling, Foster & Johnson.
Hutchison, R. *See* Graham (A.L.) & Hutchison; Harris (P.G.), Hutchison & Paul.
Hutchison, Sir Kenneth & Hebden, D. Frederick James Dent. *Biogr. Mem.* **20**, 155–180 (1974).
Hutson, A.M. *See* Cogan, Hutson & Shaffer.
Hutter, K. Floating sea ice plates and the significance of the dependence of the Poisson ratio on brine content. *Proc.* A **343**, 85–108 (1975).
Hutter, K. & Olunloyo, V.O.S. On the distribution of stress and velocity in an ice strip, which is partly sliding over and partly adhering to its bed, by using a Newtonian viscous approximation. *Proc.* A **373**, 385–403 (1980).
Huxley, A.F. The Croonian Lecture, 1967. The activation of striated muscle and its mechanical response. *Proc.* B **178**, 1–27 (1971).

A note suggesting that the cross-bridge attachment during muscle contraction may take place in two stages. *Proc.* B **183**, 83–86 (1973).

Huxley, C.R. The tortoise and the rail (Discussion). *Trans.* B **286**, 225–230 (1979).

Huxley, C.R. & Wilkinson, R. Vocalizations of the Aldabra white-throated rail *Dryolimnas cuvieri aldabranus. Proc.* B **197**, 315–331 (1977).

Huxley, H.E. Introductory remarks to a Discussion. *Trans.* B **261**, 3–4; 119 (1971).

Croonian Lecture, 1970. The structural basis of muscular contraction.
Proc. B **178**, 131–149 (1971).

Concluding remarks to a Discussion. *Trans.* B **265**, 231 (1973).

Huxley, John. The coloration of *Papilio zalmoxis* and *P. antimachus,* and the discovery of Tyndall blue in butterflies. *Proc.* B **193**, 441–453 (1976).

Hyatt, P.J. *See* McDougall, Williams, Hyatt, Bell, Tait & Tait.

Hyde, B.G. *See* Merritt, Hyde, Bursill & Philp; O'Keefe (M.) & Hyde.

Hyde, P.J.W. *See* Rickards, Hyde & Krinsley.

Iaccarino, S. *See* Sibuet, Ryan, Arthur, Barnes, Blechsmidt and others.

Iguchi, E. & Tilley, R.J.D. The elastic strain energy of crystallographic shear planes in reduced tungsten trioxide. *Trans.* A **286**, 55–85 (1977).

Ihrig, E. *See* Rosensteel, Ihrig & Trainor.

Ihrig, E., Rosensteel, G., Chow, H. & Trainor, L.E.H. Group theory and many body diagrams. II. Enumeration methods and number approximations. *Proc.* A **348**, 339–357 (1976).

Ikehara, N. *See* Bannwarth, Ikehara & Schweiger.

Iles, J.F. Organization of motoneurones in the prothoracic ganglion of the cockroach *Periplaneta americana* (L.). *Trans.* B **276**, 205–219 (1976).

The speed of passive dendritic conduction of synaptic potentials in a model motoneurone.
Proc. B **197**, 225–229 (1977).

See also Attwell & Iles.

Ilyin, N.P. *See* Vinogradov, Yaroshevsky & Ilyin.

Imrie, W.M. Undercarriage material requirements (Symposium). *Trans.* A **282**, 91–104 (1976).

Infeld, E. & Rowlands, G. Stability of nonlinear ion sound waves and solitons in plasmas.
Proc. A **366**, 537–554 (1979).

Ingerslev, F. Noise and the sound insulation of buildings (Discussion). *Trans.* A **272**, 595–602 (1972).

Ingham, J.K. *See* Hughes (C.P.), Ingham & Addison.

Inglis, S.C. & Almond, J.W. An influenza virus gene encoding two different proteins (Discussion).
Trans. B **288**, 375–381 (1980).

Ingram, D.S. *See* Sargent (J.A.), Ingram & Tommerup.

Ingram, D.S. & Tommerup, Inez C. The life history of *Plasmodiophora brassicae* Woron.
Proc. B **180**, 103–112 (1972).

Ingram, V.M. *See* Bruns & Ingram.

Inhoffen, H.H. Gossauer, A., Heise, K.P. & Laas, H. Chemical behaviour of dicyanocob(III)yrinic acid heptamethyl ester and cob(I)yrinic acid heptamethyl ester in some preparative experiments (Discussion). *Trans.* B **273**, 327–333 (1976).

Inuzuka, E. *See* Suzuki (Y.), Tsuchiya, Kinoshita, Sugiyama & Inuzuka.

Ioannides, A.C. & Horridge, G.A. The organization of visual fields in the hemipteran acone eye.
Proc. B **190**, 373–391 (1975).

Ion, D.C. General problems of collecting and understanding world energy data (Discussion).
Trans. A **276**, 431–438 (1974).

Ions, W.D. *See* Block (H.), Ions, Powell, Singh & Walker.

Ipohorski, M. *See* Woolhouse & Ipohorski.

Ippen, E.P. *See* Shank, Ippen, Fork, Migus & Kobayashi.

Ippen, E.P., Shank, C.V., Wiesenfeld, J.M. & Migus, A. Subpicosecond pulse techniques (Discussion).
Trans. A **298**, 225–232 (1980).

Iredale, R. Putting things together in the 1980s (Discussion). *Trans.* A **275**, 401–416 (1973).

Ireland, F.E. & Bryce, D.J. The philosophy of control of air pollution in the United Kingdom (Discussion). *Trans.* A **290**, 625–637 (1979).

Irvine, H.M. The linear theory of free vibrations of suspended membranes.
   Proc. A **350**, 317–334 (1976).
   A note on luffing in sails. *Proc.* A **365**, 345–347 (1979).

Irvine, H.M. & Caughey, T.K. The linear theory of free vibrations of a suspended cable.
   Proc. A **341**, 299–315 (1974).

Irvine, K.J. Keynote speech (Symposium). *Trans.* A **282**, 339–346 (1976).

Irving, E. & McGlynn, J.C. Proterozoic magnetostratigraphy and the tectonic evolution of Laurentia (Discussion). *Trans.* A **280**, 433–467 (1976).

Irving, J.B. *See* Barlow (A.J.), Harrison, Irving, Kim, Lamb & Pursley.

Irving, P.E. & Kurzfeld, A. Interaction effects between trace element impurities and environment in fatigue of high strength steels [abstract] (Discussion). *Trans.* A **295**, 210–211 (1980).

Isaacs, N.W. *See* Kennard, Isaacs, Motherwell, Coppola and others.

Ise, N. *See* Okubo & Ise; Okubo, Kitano, Ishiwatari & Ise; Okubo, Maruno & Ise; Shikata, Kim, Mita, Ise & Kunugi.

Ise, N., Maruno, T. & Okubo, T. Role of solvation and desolvation in polymer 'catalysis'. I. The influence of high pressures on the spontaneous and Ag^+-induced aquations of $Co(NH_3)_5 Br^{2+}$ calalysed by macro-ions. *Proc.* A **370**, 485–500 (1980).

Isham, C.J. Some quantum field theory aspects of the superspace quantization of general relativity.
   Proc. A **351**, 209–232 (1976).
   Twisted quantum fields in a curved space–time. *Proc.* A **362**, 383–404 (1978).
   Spinor fields in four dimensional space–time. *Proc.* A **364**, 591–599 (1978).
   Quantum gravity [abstract] (Discussion). *Proc.* A **368**, 33–36 (1979).
   See also Avis & Isham.

Isham, Valerie. *See* Cox (D.R.) & Isham.

Ishiwatari, T. *See* Okubo, Kitano, Ishiwatari & Ise.

Islam, J.N. A class of exact interior solutions of the Einstein–Maxwell equations.
   Proc. A **353**, 523–531 (1977).
   On rotating charged dust in general relativity. *Proc.* A **362**, 329–340 (1978).
   On rotating charged dust in general relativity. II. *Proc.* A **367**, 271–280 (1979).
   On rotating charged dust in general relativity. III. *Proc.* A **372**, 111–115 (1980).

Islam, M.N. *See* Palmer (S.B.), Lee & Islam.

Islam, M.R. *See* Nicholls (G.D.) & Islam.

Isler, O. Paul Karrer. *Biogr. Mem.* **24**, 245–321 (1978).

Isles, G.L. *See* Dixon-Lewis, Isles & Walmsley.

Ison, H.C.K. *See* Butler (G.) & Ison.

Israel, H.W. *See* Steward, Israel, Mott, Wilson & Krikorian.

Israel, M., Lesbats, B., Meunier, F.M. & Stinnakre, J. Postsynaptic release of adenosine triphosphate induced by single impulse transmitter action. *Proc.* B **193**, 461–468 (1976).

Israel, W. & Stewart, J.M. On transient relativistic thermodynamics and kinetic theory. II.
   Proc. A **365**, 43–52 (1979).

Israelachvili, J.N. The calculation of van der Waals dispersion forces between macroscopic bodies.
   Proc. A **331**, 39–55 (1972). Erratum. *Proc.* A **332**, 565 (1973).

Israelachvili, J.N. & Tabor, D. The measurement of van der Waals dispersion forces in the range 1.5 to 130 nm. *Proc.* A **331**, 19–38 (1972).

Ito, N. *See* Krishnaswamy (M.R.), Menon, Narasimham, Hinotani, Ito and others.

Ito, S. Form and function of the glycocalyx on free cell surfaces (Discussion).
   Trans. B **268**, 55–66 (1974).
   See also Arnott, Best, Ito & Nicol.

Ito, S. & Nicol, J.A.C. Identification of decarboxylated *S*-adenosylmethionine in the tapetum lucidum of the catfish. *Proc.* B **190**, 33–43 (1975).

Ito, S., Thurston, E.L. & Nicol, J.A.C. Melanoid tapeta lucida in teleost fishes.
   Proc. B **191**, 369–385 (1975).

Jaffre, T. *See* Kelly (P.C.), Brooks, Dilli & Jaffre.

Jaffré, T., Kersten, W., Brooks, R.R. & Reeves, R.D. Nickel uptake by Flacourtiaceae of New Caledonia. *Proc.* B **205**, 385–394 (1979).

de Jager, C. The production of solar and stellar chromospheres and coronae (Discussion). *Trans.* A **270**, 175–182 (1971).

Some remarks on supergiant photospheres (Discussion). *Trans.* A **279**, 421–427 (1975).

de Jager, C., Hoekstra, R., van der Hucht, K.A., Kamperman, T.M. & Lamers, H.J.G.L.M. Two years of operation of the ultraviolet stellar spectrophotometer S59 in E.S.R.O.'s TD1A satellite (Discussion). *Trans.* A **279**, 413–420 (1975).

de Jager, C., Kuperus, M. & Rosenberg, H. Physics of the solar atmosphere (Discussion). *Trans.* A **281**, 415–426 (1976).

Solar flares (Discussion). *Trans.* A **281**, 507–513 (1976).

Jahn, B.-M., Vidal, P. & Tilton, G. Archaean mantle heterogeneity: evidence from chemical and isotopic abundances in Archaean igneous rocks (Discussion). *Trans.* A **297**, 353–364 (1980).

Jahr, C.E. *See* Nicoll, Alger & Jahr.

Jain, J.K. India: underground water resources (Discussion). *Trans.* B **278**, 507–524 (1977).

Jain, S.C. *See* Sobell, Tsai, Jain & Sakore; Tsai (C.-C.), Jain & Sobell.

Jain, S.C. & Hughes, A.E. The effect of particle solubility and inhomogeneities on the kinetics of precipitation in crystals. *Proc.* A **360**, 47–70 (1978).

Jain, S.L., Kutty, T.S., Roy-Chowdhury, T. & Chatterjee, S. The sauropod dinosaur from the Lower Jurassic Kota formation of India. *Proc.* B **188**, 221–228 (1975).

Jamar, C. *See* Nandy, Thompson, Jamar, Monfils & Wilson.

James, D. Airframe material requirements (Symposium). *Trans.* A **282**, 83–89 (1976).

James, I.N. *See* Gibbon, James & Moroz.

James, J.F. *See* Elsworth, Taylor & James.

James, Margaret O., Smith, R.L., Williams, R.T. & Reidenberg, M. The conjugation of phenylacetic acid in man, sub-human primates and some non-primate species. *Proc.* B **182**, 25–35 (1972).

James, P.R. *See* Coward, Graham, James & Wakefield.

James, R.W. The Adams and Elsasser dynamo integrals. *Proc.* A **331**, 469–478 (1973).

The spectral form of the magnetic induction equation. *Proc.* A **340**, 287–299 (1974).

New tensor spherical harmonics, for application to the partial differential equations of mathematical physics. *Trans.* A **281**, 195–221 (1976).

Jan, Lily Y. *See* Jan (Y.N.), Jan & Dennis.

Jan, Y.N., Jan, Lily Y. & Dennis, M.J. Two mutations of synaptic transmission in *Drosophila*. *Proc.* B **198**, 87–108 (1977).

Janovský, I. *See* Dainton, Janovský & Salmon.

Jansen, J.K.S. *See* Kuffler (D.P.), Thompson & Jansen.

Jardine, N. Patterns of differentiation between human local populations. *Trans.* B **263**, 1–33 (1971).

Jarman, Heather N. & Bay-Petersen, J.L. Agriculture in prehistoric Europe – the lowlands (Discussion). *Trans.* B **275**, 175–186 (1976).

Jarman, M.R. Early animal husbandry (Discussion). *Trans.* B **275**, 85–97 (1976).

Jarnot, R.F. *See* Chaloner, Drummond, Houghton, Jarnot & Roscoe; Drummond (J.R.) & Jarnot.

Jarvis, D.A. *See* Chadwick (P.) & Jarvis.

Jarvis, L.G. *See* Smith (M.W.) & Jarvis.

Jarvis, P.G. The interpretation of the variations in leaf water potential and stomatal conductance found in canopies in the field (Discussion). *Trans.* B **273**, 593–610 (1976).

Jasinski, A. *See* Andrew, Hinshaw, Hutchins & Jasinski.

Jaycock, M.J. *See* House & Jaycock.

Jean, M. & Pritchard, W.G. The flow of fluids from nozzles at small Reynolds numbers. *Proc.* A **370**, 61–72 (1980).

Jeans, C.V. The origin of the Triassic clay assemblages of Europe with special reference to the Keuper Marl and Rhaetic of parts of England. *Trans.* A **289**, 549–636 (1978).

Jedziniak, Judith A. *See* Cohen (R.J.), Jedziniak & Benedek.

Jefferies, R.P.S. The Ordovician fossil *Lagynocystis pyramidalis* (Barrande) and the ancestry of amphioxus. *Trans.* B 265, 409—469 (1973).

Jefferies, R.P.S. & Lewis, D.N. The English Silurian fossil *Placocystites forbesianus* and the ancestry of the vertebrates. *Trans.* B 282, 205—323 (1978).

Jefferson, D.A. *See* Mallinson (L.G.), Jefferson, Thomas & Hutchison.

Jefferson, D.A. & Thomas, J.M. High resolution electron microscopic and X-ray studies of non-random disorder in an unusual layered silicate (chloritoid). *Proc.* A 361, 399—411 (1978).

Jefferson, J.H. & Hagston, W.E. Renormalization theory of charged particles in an insulator and the effective-mass approximation. *Proc.* A 355, 355—376 (1977).

Jeffery, P.K. *See* Phipps, Richardson, Corfield, Gallagher and others.

Jeffrey, D.J. Conduction through a random suspension of spheres. *Proc.* A 335, 355—367 (1973).
  Group expansions for the bulk properties of a statistically homogeneous, random suspension. *Proc.* A 338, 503—516 (1974).

Jeffreys, Sir Harold. Robert Stoneley. *Biogr. Mem.* 22, 555—564 (1976).
  Keith Edward Bullen. *Biogr. Mem.* 23, 19—39 (1977).

Jeffries, D.J. *See* Clegg (E.J.), Jeffries & Harrison.

Jeger, O. *See* Prelog & Jeger.

Jellum, E. Application of mass spectrometry and metabolite profiling to the study of human diseases (Discussion). *Trans.* A 293, 13—19 (1979).

Jenkin, G.T. *See* Leslie (M.), Jenkin, Hayter, White, Cox & Warner.

Jenkins, D.B. *See* Dickinson (P.H.G.), Bain, Thomas, Williams, Jenkins & Twiddy.

Jenkins, F.A. & Parrington, F.R. The postcranial skeletons of the Triassic mammals *Eozostrodon, Magazostrodon* and *Erythrotherium. Trans.* B 273, 387—431 (1976).

Jenkins, G.M., Kawamura, K. & Ban, L.L. Formation and structure of polymeric carbons. *Proc.* A 327, 501—517 (1972).

Jenkins, H.D.B. & Hartman, P. A new approach to the calculation of electrostatic energy relations in minerals: the dioctahedral and trioctahedral phyllosilicates. *Trans.* A 293, 169—208 (1979).

Jenkins, H.D.B. & Pratt, K.F. On 'basic' radii of simple and complex ions and the repulsion energy of ionic crystals. *Proc.* A 356, 115—134 (1977).

Jenkins, J.G. Machinery — capitalization, organization and management (Discussion). *Trans.* B 267, 71—79 (1973).

Jennings, B.R. *See* Morris (V.J.) & Jennings.

Jennings, B.R. & Coles, H.J. Laser induced birefringence in liquids and solutions. *Proc.* A 348, 525—538 (1976).

Jennings, K.R. Negative ions (Discussion). *Trans.* A 293, 125—133 (1979).

Jennings, R.E. *See* Clark (T.A.), Courts & Jennings.

Jennings, S.G. *See* Brazier-Smith, Jennings & Latham.

Jennison, R.C. *See* McDonnell, Ashworth, Flavill, Carey, Bateman & Jennison.

Jensen, D.E. Prediction of soot formation rates: a new approach. *Proc.* A 338, 375—396 (1974).

Jensen, D.E. & Jones, G.A. Alkaline earth flame chemistry. *Proc.* A 364, 509—535 (1978).

Jensen, S., Lange, R., Berge, G., Palmork, K.H. & Renberg, L. On the chemistry of EDC-tar and its biological significance in the sea (Discussion). *Proc.* B 189, 333—346 (1975).

Jéquier, N. Appropriate technology: the challenge of the second generation (Discussion). *Proc.* B 209, 7—14 (1980).

Jerie, P.H. & Hall, M.A. The identification of ethylene oxide as a major metabolite of ethylene in *Vicia faba* L. *Proc.* B 200, 87—94 (1978).

Jeronimidis, G. The fracture behaviour of wood and the relations between toughness and morphology. *Proc.* B 208, 447—460 (1980).
  See also Gordon (J.E.) & Jeronimidis.

Jewell, W.S. The analytic methods of operations research (Discussion). *Trans.* A 287, 373—404 (1977).

Jiménez, F. *See* Viñuela, Camacho, Jiménez, Carrascosa, Ramírez & Salas.

Jimenez, J. & Whitham, G.B. An averaged Lagrangian method for dissipative wavetrains. *Proc.* A 349, 277—287 (1976).

Jobe, C.E. & Burggraf, O.R. The numerical solution of the asymptotic equations of trailing edge flow. *Proc.* A **340**, 91–111 (1974).

Johansen, J. Contribution to a paper by Pennington, Haworth, Bonny & Lishman. *Trans.* B **264**, 244–245 (1972).

Johansson, E.D.B. Advantages and disadvantages of the intrauterine device and the hormone implant (Discussion). *Proc.* B **195**, 81–91 (1976).

Johansson, O. *See* Hökfelt, Lundberg, Schultzberg, Johansson, Skirboll, Ånggård and others.

Johansson, T. *See* Loberg, Johansson & Easterling.

Johari, G.P. & Jones, S.J. Dielectric properties of polycrystalline D_2O ice Ih (hexagonal). *Proc.* A **349**, 467–495 (1976).

John, P. *See* Whatley (Jean M.), John & Whatley.

Johnels, A.G. *See* Hook (O.) & Johnels.

Johnson, A.W. Synthesis of corrins and related macrocycles based on pyrrolic intermediates (Discussion). *Trans.* B **273**, 319–326 (1976).

John Donald Rose. *Biogr. Mem.* **23**, 449–463 (1977).

Johnson, B.F. *See* Coles (E.C.), Beilin, Bulpitt, Dollery, Johnson and others.

Johnson, D. *See* Sibuet, Ryan, Arthur, Barnes, Blechsmidt and others.

Johnson, D.J. Recent advances in studies of carbon fibre structure (Discussion). *Trans.* A **294**, 443–449 (1980).

See also Dobb, Johnson & Saville.

Johnson, F.A. A bond charge model of lattice dynamics. I. *Proc.* A **339**, 73–83 (1974).

Elastic constants.

I. *Proc.* A **357**, 289–295 (1977).

II. *Proc.* A **357**, 297–307 (1977).

III. *Proc.* A **357**, 309–321 (1977).

Johnson, F.A. & Moore, Katharine. A bond charge model of lattice dynamics. II. *Proc.* A **339**, 85–96 (1974).

Johnson, G. *See* Mallard, Hutchison, Edelstein, Ling, Foster & Johnson.

Johnson, J.W. Resin matrices and their contribution to composite properties (Discussion). *Trans.* A **294**, 487–494 (1980).

Johnson, K.L., Kendall, K. & Roberts, A.D. Surface energy and the contact of elastic solids. *Proc.* A **324**, 301–313 (1971).

Johnson, K.L., O'Connor, J.J. & Woodward, A.C. The effect of the indenter elasticity on the Hertzian fracture of brittle materials. *Proc.* A **334**, 95–117 (1973).

Johnson, K.L. & Roberts, A.D. Observations of viscoelastic behaviour of an elastohydrodynamic lubricant film. *Proc.* A **337**, 217–242 (1974).

Johnson, K.L. & Tevaarwerk, J.L. Shear behaviour of elastohydrodynamic oil films. *Proc.* A **356**, 215–236 (1977).

Johnson, P. & McKenzie, G.H. A laser light-scattering study of haemoglobin systems. *Proc.* B **199**, 263–278 (1977).

Johnson, P. & Perrella, M. On the dissociation of the sheep haemoglobin molecule at neutral pH.

I. Osmotic pressure measurements. *Proc.* B **176**, 445–460 (1971).

[Abstract] *Proc.* A **321**, 141 (1971).

II. Sedimentation equilibrium measurements. *Proc.* B **176**, 461–480 (1971).

[Abstract] *Proc.* A **321**, 141 (1971).

Johnson, P.C. *See* Parker & Johnson.

Johnson, P.C. & Parker, A.B. The dielectric breakdown of low density gases. II. Experiments on mercury vapour. *Proc.* A **325**, 529–541 (1971).

Johnson, R.S. On the modulation of water waves on shear flows. *Proc.* A **347**, 537–546 (1976).

On the continuum theory of the one-fluid solar wind for small Prandtl number. *Proc.* A **348**, 129–142 (1976).

On the continuum theory of the two-fluid solar wind for small mass ratio. *Proc.* A **348**, 511–523 (1976).

On the modulation of water waves in the neighbourhood of $kh \approx 1.363$. *Proc.* A **357**, 131–141 (1977).

On the propagation of long waves on the surface of helium II: linearized theory.
Proc. A **362**, 97—111 (1978).
A note on the effects of healing and relaxation in helium II due to heat transfer at a wall.
Proc. A **362**, 375—382 (1978).
See also Freeman (N.C.) & Johnson.

Johnson, R.T., Mullinger, A.M. & Skaer, R.J. Perturbation of mammalian cell division: human mini segregants derived from mitotic cells. *Proc.* B **189**, 591—602 (1975).

Johnsson, A. Circumnutations under free-fall conditions in space? (Discussion).
Proc. B **199**, 505—512 (1977).

Johnston, A.W.B. *See* Beringer, Brewin, Johnston, Schulman & Hopwood.

Johnston, Bernadette T., Schrameck, Joan E. & Mark, R.F. Re-innervation of axolotl limbs. II. Sensory nerves. *Proc.* B **190**, 59—75 (1975).

Johnston, D.R. Tree growth and wood production in Britain (Discussion).
Trans. A **271**, 101—114 (1975).

Johnston, R. *See* Donaldson (C.H.), Drever & Johnston.

Jolley, Elizabeth & Smith, D.C. The green hydra symbiosis. II. The biology of the establishment of the association. *Proc.* B **207**, 311—333 (1980).

Jolliffe, B.W. *See* Blaney, Bradley, Edwards, Jolliffe and others.

Jonah, D.A. & King, M.B. A method for predicting the variation with temperature of the solubilities of gases in non-polar liquids. *Proc.* A **323**, 361—375 (1971). Errata. *Proc.* A **325**, 561 (1971).

Jonard, G. *See* Guilley, Jonard, Richards & Hirth.

Jonard, G., Briand, J.P., Bouley, J.P., Witz, J. & Hirth, L. Nature and specificity of the RNA—protein interaction in the case of the tymoviruses (Discussion). *Trans.* B **276**, 123—129 (1976).

Jones, A. & Rosenfeld, J.L.J. Monte-Carlo simulation of hydrogen-atom recombination.
Proc. A **333**, 419—434 (1973).

Jones, A.R. Electromagnetic wave scattering by assemblies of particles in the Rayleigh approximation.
Proc. A **366**, 111—127 (1979).
See also Cox (J.B.), Jones & Weinberg; Hong, Jones & Weinberg.

Jones, A.R., Lloyd, S.A. & Weinberg, F.J. Combustion in heat exchangers.
Proc. A **360**, 97—115 (1978).

Jones, A.R., Ralph, B. & Hansen, N. Subgrain coalescence and the nucleation of recrystallization at grain boundaries in aluminium. *Proc.* A **368**, 345—357 (1979).

Jones, A.R., Schwar, M.J.R. & Weinberg, F.J. Generalizing variable shear interferometry for the study of stationary and moving refractive index fields with the use of laser light.
Proc. A **322**, 119—135 (1971). Corrigendum, p. 555.

Jones, B. Switching surges and air insulation (Discussion). *Trans.* A **275**, 165—180 (1973).

Jones, B.B. *See* Boland (B.C.), Jones, Wilson, Engstrom & Noci; Freeman (F.F.), Gabriel, Jones & Jordan.

Jones, B.J.T. Galaxy formation after $z = 1000$ (Discussion). *Trans.* A **296**, 289—298 (1980).

Jones, D. *See* Barnes (A.J.), Hallam & Jones.

Jones, D.A. Nonlinear autoregressive processes. *Proc.* A **360**, 71—95 (1978).

Jones, D.K.C. *See* Brunsden & Jones.

Jones, D.S. The scattering of sound by a simple shear layer. *Trans.* A **284**, 287—328 (1977).
Arthur Erdélyi. *Biogr. Mem.* **25**, 267—286 (1979).
Infinite integrals and convolution. *Proc.* A **371**, 479—508 (1980).
See also Everitt & Jones.

Jones, D.S. & Morgan, J.D. A linear model of a finite amplitude Helmholtz instability.
Proc. A **338**, 17—41 (1974).

Jones, D.T. *See* Gray (P.), Jones & MacKinven.

Jones, E.A. *See* Bodmer (W.F.), Jones, Barnstable & Bodmer; Crumpton, Snary, Walsh, Barnstable and others.

Jones, G.A. *See* Jensen (D.E.) & Jones.

Jones, G. Melvill. Plasticity in the adult vestibulo-ocular reflex arc (Discussion).
Trans. B **278**, 319—334 (1977).

Jones, H. The dynamics of spinning detonation waves. *Proc.* A **348**, 299–316 (1976).
The mechanics of vibrating flames in tubes. *Proc.* A **353**, 459–473 (1977).
The generation of sound by flames. *Proc.* A **367**, 291–309 (1979).
Notes on work at the University of Bristol, 1930–7 (Symposium). *Proc.* A **371**, 52–55 (1980).

Jones, I.T.N. & Bayes, K.D. The kinetics and mechanism of the reaction of atomic oxygen with acetylene. *Proc.* A **335**, 547–562 (1973).

Jones, I.T.N. & Wayne, R.P. The photolysis of ozone by ultraviolet radiation. V. Photochemical formation of $O_2(^1\Delta_g)$. *Proc.* A **321**, 409–424 (1971).

Jones, Jane. *See* Goss-Custard, Jones, Kitching & Norton.

Jones, Janet E. The cosmological effects of generalized field equations in relativity.
Proc. A **340**, 263–286 (1974).

Jones, J.K.N. Arthur Charles Neish. *Biogr. Mem.* **20**, 295–315 (1974).

Jones, J.S. The genetic structure of a southern peripheral population of the snail *Cepaea nemoralis*.
Proc. B **183**, 371–384 (1973).

Jones, K.W. *See* Bradley (D.J.), Jones & Sibbett.

Jones, M. On the use of the Breit–Pauli approximation in the study of relativistic effects in electron–atom scattering. *Trans.* A **277**, 587–622 (1975).

Jones, M.N. Group-theoretical methods in geophysics. *Proc.* A **348**, 81–93 (1976).
A group-theoretical formulation of geophysical elastodynamics. *Proc.* A **356**, 549–568 (1977).

Jones, M.N. & Zala, C. Thermochemical studies of human red blood cells: enthalpy changes on dilution in buffers. *Proc.* B **185**, 73–82 (1974).

Jones, O.C. & Gordon-Smith, G.W. Absolute radiometry by means of a black body source.
Proc. A **335**, 369–386 (1973).

Jones, O.T.G. Chlorophyll *a* biosynthesis (Discussion). *Trans.* B **273**, 207–225 (1976).

Jones, P.B. Astrophysical significance of the dissipation of turbulence in a dense baryon fluid.
Proc. A **323**, 111–125 (1971).

Jones, P.M., Rackham, G.M. & Steeds, J.W. Higher order Laue zone effects in electron diffraction and their use in lattice parameter determination. *Proc.* A **354**, 197–222 (1977).

Jones, R. *See* Dwek, Jones, Marsh, McLaughlin, Press and others; Martin (A.W.), Jones & Mann.

Jones, R. & Mann, T. Lipid peroxidation in spermatozoa. *Proc.* B **184**, 103–107 (1973).
Lipid peroxides in spermatozoa: formation, role of plasmalogen, and physiological significance.
Proc. B **193**, 317–333 (1976).

Jones, R., Mann, T. & Sherins, R.J. Adverse effects of peroxidized lipid on human spermatozoa.
Proc. B **201**, 413–417 (1978).

Jones, R.E. De-iodination of labelled protein during intestinal transmission in the suckling rat.
Proc. B **199**, 279–290 (1977).

Jones, R. Hughes. *See* Chisholm & Jones.

Jones, R.V. 'Fresnel aether drag' in a transversely moving medium. *Proc.* A **328**, 337–352 (1972).
Introduction to a Discussion. *Proc.* A **342**, 441–445 (1975).
Research establishments (Discussion). *Proc.* A **342**, 481–490 (1975).
Epilogue to a Discussion. *Proc.* A **342**, 549–554 (1975).
Appendix B: A scientific staff college (Discussion). *Proc.* A **342**, 575–579 (1975).
Appendix C: Bibliography (Discussion). *Proc.* A **342**, 581–586 (1975).
'Aether drag' in a transversely moving medium. *Proc.* A **345**, 351–364 (1975).
Rotary 'aether drag'. *Proc.* A **349**, 423–439 (1976).
Radiation pressure of light in a dispersive medium. *Proc.* A **360**, 365–371 (1978).

Jones, R.V. & Leslie, B. The measurement of optical radiation pressure in dispersive media.
Proc. A **360**, 347–363 (1978).

Jones, Siân E.M. *See* Young (J.D.), Jones & Ellory.

Jones, S.J. See Johari & Jones.

Jones, T.B. *See* Dudeney, Jones, Kressman & Spracklen.

Jones, T.S. *See* Ellis (P.), Holah, Houghton, Jones, Peckham and others.

Jones, W. *See* Nakanishi (H.), Jones, Thomas, Hasegawa & Rees.

Jones, W.C. The ultrastructure of *Gymnosphaera albida* Sassaki, a marine axopodiate protozoon.
Trans. B **275**, 349–384 (1976).

Jones, W.J. Molecular structure and properties from high resolution Raman spectra of gases (Discussion). *Trans.* A **293**, 249–256 (1979).

See also Butcher, Willetts & Jones; Till (S.M.), Freedman, Tuckett & Jones; Till (Susan M.), Jones & Shotton.

Jones-Mortimer, M.C. & Kornberg, H.L. Genetical analysis of fructose utilization by *Escherichia coli. Proc.* B **187**, 121–131 (1974).

Order of genes adjacent to *ptsX* on the *E. coli* genome. *Proc.* B **193**, 313–315 (1976).

de Jong, J.J. Setting objectives in a business enterprise — a cybernetic approach (Discussion). *Trans.* A **287**, 493–507 (1977).

Jope, E.M. The evolution of plants and animals under domestication: the contribution of studies at the molecular level (Discussion). *Trans.* B **275**, 99–116 (1976).

Jopson, W.P. *See* Crapper, Dombrowski & Jopson.

Jordan, Carole. The structure and energy balance of solar active regions (Discussion). *Trans.* A **281**, 391–404 (1976).

Helium line emission: its relation to atmospheric structure (Discussion). *Trans.* A **297**, 541–554 (1980).

See also Burton (W.M.), Jordan, Ridgeley & Wilson; Freeman (F.F.), Gabriel, Jones & Jordan.

Jordan, P. *See* Akhtar, Abboud, Barnard, Jordan & Zaman.

Jorna, S. & Springer, C. Derivation of Green-type, transitional and uniform asymptotic expansions from differential equations. V. Angular oblate spheroidal wavefunctions $\overline{ps}_n^r(\eta,h)$ and $\overline{qs}_n^r(\eta,h)$ for large h. *Proc.* A **321**, 545–555 (1971).

Joron, J.L. *See* Bougault, Joron & Treuil.

Jortner, J. *See* Nitzan, Jortner & Rentzepis.

Joshi, R.D. *See* Singh (G.), Joshi, Chopra & Singh.

Jowett, J.K.S. Technology, economic constraints and possible solutions for providing wide-coverage educational systems by means of satellites (Discussion). *Proc.* A **345**, 511–529 (1975).

Joyce, G.S. On the simple cubic lattice Green function. *Trans.* A **273**, 583–610 (1973).

Analytic properties of the Ising model with triplet interactions on the triangular lattice. *Proc.* A **343**, 45–62 (1975).

On the magnetization of the triangular lattice Ising model with triplet interactions. *Proc.* A **345**, 277–293 (1975).

Joyce, R.J.V. The evolution of an aerial application system for the control of Desert Locusts (Discussion). *Trans.* B **287**, 305–314 (1979).

Joyner, R.W. *See* Somorjai, Joyner & Lang.

Joyner, R.W., Kishi, K. & Roberts, M.W. Low energy electron diffraction and electron spectroscopic studies of the oxidation and sulphidation of Pb(100) and Pb(110) surfaces. *Proc.* A **358**, 223–241 (1977).

Joyner, R.W. & Roberts, M.W. Photoelectron spectroscopic investigation of the adsorption and catalytic decomposition of formic acid by copper, nickel and gold. *Proc.* A **350**, 107–126 (1976).

Joysey, K.A. *See* Romero-Herrera, Lehmann, Joysey & Friday.

Judd, B.R. & Runciman, W.A. Transverse Zeeman effect for ions in uniaxial crystals. *Proc.* A **352**, 91–108 (1976).

Julesz, B. Spatial nonlinearities in the instantaneous perception of textures with identical power spectra (Discussion). *Trans.* B **290**, 83–94 (1980).

Julian, F.J., Sollins, M.R. & Moss, R.L. Sarcomere length non-uniformity in relation to tetanic responses of stretched skeletal muscle fibres. *Proc.* B **200**, 109–116 (1978).

Jull, A.J.T. *See* Pillinger, Eglinton, Gowar & Jull.

Jung, M. *See* Baldwin (J.E.), Jung, Singh, Wan, Haber and others.

Juniper, B.E. The perception of gravity by a plant (Discussion). *Proc.* B **199**, 537–550 (1977).

Kabir, P.K. *See* Dass & Kabir; Dass, Kabir & Kenny.

Kagami, H. *See* Roberts (D.G.), Montadert, Thompson, Auffret and others.

Kaiser, T.R. *See* Bullough (K.), Denby, Gibbons, Hughes and others.

Kaiser, T.R., Orr, D. & Smith, A.J. Very low frequency electromagnetic phenomena: 'whistlers' and micropulsations (Discussion). *Trans.* B **279**, 225–238 (1977).

Kaiser, W., Fendt, A., Kranitzky, W. & Laubereau, A. Infrared picosecond pulses and applications (Discussion). *Trans.* A **298**, 267–271 (1980).

Kaiser, W.A. The excitation functions of Ba (p, X) ^{M}Xe (M = 124–136) in the energy range 38–600 MeV; the use of 'cosmogenic' xenon for estimating 'burial' depths and 'real' exposure ages (Discussion). *Trans.* A **285**, 337–362 (1977).

Kajzar, F. *See* Ray (D.K.) & Kajzar.

Kalyanasundaram, K. *See* Darwent, Kalyanasundaram & Porter.

Kalyanasundaram, K. & Porter, Sir George. Model systems for photosynthesis. VI. Chlorophyll *a* sensitized reduction of methyl viologen in non-ionic micelles. *Proc.* A **364**, 29–44 (1978).

Kam, Z. *See* Eisenberg (H.), Borochov, Kam & Voordouw.

Kamperman, T.M. *See* de Jager, Hoekstra, van der Hucht, Kamperman & Lamers.

Kanazawa, T. *See* Kihara, Kanazawa & Tamura.

Kanwal, S.S. *See* Banerjee, Bhatnagar, Choudhry & Kanwal.

Kapitza, P.L. The Bernal Lecture, 1976. Scientific and social approaches for the solution of global problems. *Proc.* A **357**, 1–14 (1977). [Abstract]. *Proc.* B **199**, 327–328 (1977).

Kaplan, I.R. Stable isotopes as a guide to biogeochemical processes (Discussion). *Proc.* B **189**, 183–211 (1975).

Kaplan, M.M. The role of the World Health Organization in the study of influenza (Discussion). *Trans.* B **288**, 417–421 (1980).

Kaplanis, J.N. *See* Thompson (M.J.), Svoboda, Kaplanis & Robbins.

Kapon, M. *See* Herbstein & Kapon.

Karamanos, A.J. *See* Elston, Karamanos, Kassam & Wadsworth.

Karasawa, Y., Levin, G. & Szwarc, M. Electron-transfer and proton-transfer equilibria in systems involving different ion-pairs. *Proc.* A **326**, 53–71 (1971).

Karel, M. Membrane separation processes and freeze concentration in the 1980s (Discussion). *Proc.* B **191**, 21–48 (1975).

Karihaloo, B.L. Dislocation mobility and the concept of flow stress. *Proc.* A **344**, 375–385 (1975).
Fracture characteristics of solids containing doubly-periodic arrays of cracks. *Proc.* A **360**, 373–387 (1978).
A note on complexities of compression failure. *Proc.* A **368**, 483–493 (1979).

Karim, R.A. *See* Heuckroth & Karim.

Karing, T. & Goldsmith, H.L. Flow behaviour of blood cells and rigid spheres in an annular vortex. *Trans.* B **279**, 413–445 (1977).

Kasahara, K. Earthquake fault studies in Japan (Discussion). *Trans.* A **274**, 287–296 (1973).

Kassam, A.H. *See* Elston, Karamanos, Kassam & Wadsworth.

Katasyev, L.A. *See* Andreeva, Katasyev & Uvarov; Andreeva, Katasyev, Uvarov, Nesterov & Chasovitin.

Katayama, Y., North, R.A. & Williams, J.T. The action of substance P on neurons of the myenteric plexus of the guinea-pig small intestine. *Proc.* B **206**, 191–208 (1979).

Katz, J.J., Oettmeier, W. & Norris, J.R. Organization of antenna and photo-reaction centre chlorophylls on the molecular level (Discussion). *Trans.* B **273**, 227–253 (1976).

Katz, Sir Bernard. Physiological evidence for quantal transmitter release [abstract] (Discussion). *Trans.* B **261**, 381 (1971).
Archibald Vivian Hill. *Biogr. Mem.* **24**, 71–149 (1978).
See also Bevan (S.J.), Katz & Miledi; Heuser, Katz & Miledi.

Katz, Sir Bernard & Miledi, R. The effect of atropine on acetylcholine action at the neuromuscular junction. *Proc.* B **184**, 221–226 (1973).
The nature of the prolonged endplate depolarization in anti-esterase treated muscle. *Proc.* B **192**, 27–38 (1975).
Transmitter leakage from motor nerve endings. *Proc.* B **196**, 59–72 (1977).

121

Suppression of transmitter release at the neuromuscular junction. *Proc.* B **196**, 465–469 (1977).

The reversal potential at the desensitized endplate. *Proc.* B **199**, 329–334 (1977).

A re-examination of curare action at the motor endplate. *Proc.* B **203**, 119–133 (1978).

Estimates of quantal content during 'chemical potentiation' of transmitter release.
Proc. B **205**, 369–378 (1979).

Kauffman, E.G. Evolutionary rates and patterns among Cretaceous Bivalvia (Discussion).
Trans. B **284**, 277–304 (1978).

Kaula, W.M. Potentialities of lunar laser ranging for measuring tectonic motions (Discussion).
Trans. A **274**, 185–193 (1973).

Kaup, D.J. & Newell, A.C. Solitons as particles, oscillators, and in slowly changing media: a singular perturbation theory. *Proc.* A **361**, 413–446 (1978).

Kawamura, K. *See* Jenkins (G.M.), Kawamura & Ban.

Kay, H.D. William Kershaw Slater. *Biogr. Mem.* **17**, 663–680 (1971).

John Boyd Orr (Baron Boyd Orr of Brechin Mearns). *Biogr. Mem.* **18**, 43–81 (1972).

Kazakia, J.Y. *See* Varley, Kazakia & Blythe.

Kazakia, J.Y. & Varley, E. Large amplitude waves in bounded media.

 II. The deformation of an impulsively loaded slab: the first reflexion.
Trans. A **277**, 191–237 (1974).

 III. The deformation of an impulsively loaded slab: the second reflexion.
Trans. A **277**, 239–250 (1974).

Kazimirovsky, E.S. & Kokourov, V.D. The measurement of ionospheric drifts over east Siberia, U.S.S.R. (Discussion). *Trans.* A **271**, 499–508 (1972).

Keady, G. & Norbury, J. The jet from a horizontal slot under gravity. *Proc.* A **344**, 471–487 (1975).

Keating, M.J. The time course of experience-dependent synaptic switching of visual connections in *Xenopus laevis. Proc.* B **189**, 603–610 (1975).

Evidence for plasticity of intertectal neuronal connections in adult *Xenopus* (Discussion).
Trans. B **278**, 277–294 (1977).

See also Chung, Bliss & Keating; Chung, Keating & Bliss; Gaze, Keating & Chung.

Keating, M.J., Beazley, Lynda, Feldman, Joan D. & Gaze, R.M. Binocular interaction and intertectal neuronal connexions: dependence upon developmental stage. *Proc.* B **191**, 445–466 (1975).

Keating, M.J. & Feldman, Joan D. Visual deprivation and intertectal neuronal connexions in *Xenopus laevis. Proc.* B **191**, 467–474 (1975).

Kebede, A. *See* Miller (D.S.), Baker, Bowden, Evans, Holt and others.

Keene, D.E. *See* Cullis, Keene & Trimm.

Keesing, R.G.W. High resolution studies of electron excitation. V. The metastable states of helium.
Proc. A **352**, 429–439 (1977).

See also Heddle, Keesing & Kurepa; Heddle, Keesing & Parkin; Heddle, Keesing & Watkins.

Keihm, S.J. *See* Horai, Winkler, Keihm, Langseth, Fountain & West.

Kekwick, R.A. & Pedersen, Kai O. Arne Tiselius. *Biogr. Mem.* **20**, 401 (1974).

Kell, G.S., McLaurin, G.E. & Whalley, E. The *PVT* properties of water. IV. Liquid water in the range 150–350°C, from saturation to 1 kbar. *Proc.* A **360**, 389–402 (1978).

Kelland, N.C. *See* Orme, Webb, Kelland & Sargent.

Kellaway, G.A. *See* Chandler (R.J.), Kellaway, Skempton & Wyatt.

Kellaway, G.A., Redding, J.H., Shephard-Thorn, E.R. & Destombes, J.-P. The Quaternary history of the English Channel (Discussion). *Trans.* A **279**, 189–218 (1975).

Kellenberger, E. Cooperativity and regulation through conformational changes as features of phage assembly (Discussion). *Trans.* B **276**, 3–13 (1976).

Keller, A. *See* Diamant, Keller, Baer, Litt & Arridge; Mackley & Keller.

Keller, W. *See* Zentgraf, Keller & Müller.

Kelly, A. Walter Rosenhain and materials research at Teddington (Symposium).
Trans. A **282**, 5–36 (1976).

Summary and closing remarks (Symposium). *Trans.* A **282**, 479–482 (1976).

Introduction to a Discussion. *Trans.* A **288**, 3–8 (1978).

Welcoming remarks (Discussion). *Trans.* A **295**, 7–8 (1980).

See also Aveston & Kelly; Macmillan & Kelly.

Kelly, A. & Street, K.N. Creep of discontinuous fibre composites.
  I. Experimental behaviour of lead–phosphor bronze. *Proc.* A **328**, 267–282 (1972).
  II. Theory for the steady-state. *Proc.* A **328**, 283–293 (1972).

Kelly, Angela R. & Patterson, L.K. Model systems for photosynthesis. II. Concentration quenching of chlorophyll *b* fluorescence in solid solutions. *Proc.* A **324**, 117–126 (1971).

Kelly, J.A. *See* DeLucia, Kelly, Mangion, Moews & Knox.

Kelly, P.C., Brooks, R.R., Dilli, S. & Jaffré, T. Preliminary observations on the ecology and plant chemistry of some nickel-accumulating plants from New Caledonia. *Proc.* B **189**, 69–80 (1975).

Kelly, R. & Padley, P.J. Measurement of collisional ionization cross-sections for metal atoms in flames. *Proc.* A **327**, 345–366 (1972).

Kelly, S.S. *See* Robbins (N.), Olek, Kelly, Takach & Christopher.

Kelman, A. & Sequeira, L. Resistance in plants to bacteria (Discussion).
  Proc. B **181**, 247–266 (1972).

Kelsey, J. Geodetic aspects concerning possible subsidence in southeastern England (Discussion).
  Trans. A **272**, 141–149 (1972).

Kelsey, T. & Petty, H.C. The air-blast circuit breaker (Discussion). *Trans.* A **275**, 131–138 (1973).

Kemball, C. Hugh Stott Taylor. *Biogr. Mem.* **21**, 517–547 (1975).

See also Campbell (N.) & Kemball; Dowie, Kemball, Kempling & Whan.

Kemball, C. & Kempling, J.C. Reactions of neopentane and 1,1-dimethylcyclopropane with deuterium over metal films. *Proc.* A **329**, 391–403 (1972).

Kemball, C. & McCosh, R. Hydrogen–deuterium exchange reactions on ion-exchanged X-type zeolites.
  Proc. A **321**, 249–257 (1971).
  Exchange reactions of olefins with deuterium oxide on ion-exchanged X-type zeolites.
  Proc. A **321**, 259–273 (1971).

Kemball, C., Nisbet, J.D., Robertson, P.J. & Scurrell, M.S. Development of a gas chromatograph–mass spectrometer with 'on-line' computer for studies in catalysis. Reactions of ethylene and deuterium on oxides. *Proc.* A **338**, 299–310 (1974).

Kemmer, N. & Schlapp, R. Max Born. *Biogr. Mem.* **17**, 17–52 (1971).

Kemp, D.R. & Porter, G. Photochemistry of methylated *p*-benzoquinones.
  Proc. A **326**, 117–130 (1971).

Kemp, Janet M. & Powell, T.P.S. The structure of the caudate nucleus of the cat: light and electron microscopy. *Trans.* B **262**, 383–401 (1971).
  The synaptic organization of the caudate nucleus. *Trans.* B **262**, 403–412 (1971).
  The site of termination of afferent fibres in the caudate nucleus. *Trans.* B **262**, 413–427 (1971).
  The termination of fibres from the cerebral cortex and thalamus upon dendritic spines in the caudate nucleus: a study with the Golgi method. *Trans.* B **262**, 429–439 (1971).
  The connexions of the striatum and globus pallidus: synthesis and speculation.
  Trans. B **262**, 441–457 (1971).

Kemp, T.S. Whaitsiid Therocephalia and the origin of cynodonts. *Trans.* B **264**, 1–54 (1972).
  The primitive cynodont *Procynosuchus*: functional anatomy of the skull and relationships.
  Trans. B **285**, 73–122 (1979).
  The primitive cynodont *Procynosuchus*: structure, function and evolution of the postcranial skeleton. *Trans.* B **288**, 217–258 (1980).

Kempe, D.R.C. *See* Easton, Hamilton, Kempe & Sheppard.

Kempling, J.C. *See* Dowie, Kemball, Kempling & Whan; Kemball & Kempling.

Kendal, A.P. *See* Murphy (B.R.), Markoff, Chanock, Spring and others.

Kendall, D.G. Hunting quanta (Discussion). *Trans.* A **276**, 231–266 (1974).
  Review Lecture. The recovery of structure from fragmentary information.
  Trans. A **279**, 547–582 (1975).
  Appendix to a paper by B.H.P. Rivett. *Proc.* A **354**, 422–423 (1977).

Kendall, D.G. & Piggot, S. Preface to a Discussion. *Trans.* A **276**, 3 (1974).

Kendall, K. Control of cracks by interfaces in composites. *Proc.* A **341**, 409–428 (1975).
  Transition between cohesive and interfacial failure in a laminate. *Proc.* A **344**, 287–302 (1975).

Complexities of compression failure. *Proc.* A **361**, 245—263 (1978).

See also Johnson (K.L.), Kendall & Roberts.

Kendall, K. & Tabor, D. An ultrasonic study of the area of contact between stationary and sliding surfaces. *Proc.* A **323**, 321—340 (1971).

Kendall, Marion D. EMMA—4 analysis of iron in cells of the thymic cortex of a weaver-bird (*Quelea quelea*). *Trans.* B **273**, 79—82 (1975).

See also Bacchus & Kendall; Ward (P.) & Kendall.

Kendrick, Mary P. Siltation problems in relation to the Thames barrier (Discussion).
Trans, A **272**, 223—243 (1972).

Kendrick-Jones, J. The subunit structure of gizzard myosin (Discussion).
Trans. B **265**, 183—189 (1973).

Kennard, Olga, Isaacs, N.W., Motherwell, W.D.S., Coppola, J.C., Wampler, D.L., Larson, A.C. & Watson, D.G. The crystal and molecular structure of adenosine triphosphate.
Proc. A **325**, 401—436 (1971).

Kennedy, W.J. *See* Braithwaite (C.J.R.), Taylor & Kennedy.

Kenner, G.W. The Bakerian Lecture. Towards synthesis of proteins. *Proc.* A **353**, 441—457 (1977);
Proc. B **197**, 237—253 (1977).

Kenner, G.W., Rimmer, J., Smith, K.M. & Unsworth, J.F. Studies on the biosynthesis of the *Chlorobium* chlorophylls (Discussion). *Trans.* B **273**, 255—276 (1976).

Kennett, H.M., Lee, A.E. & Wilson, J.M. Faceting in dilute oxygen atmospheres of group VIA metals (Discussion). *Proc.* A **331**, 429—443 (1972).

Kenney-Wallace, Geraldine A. Picosecond relaxation processes in liquids (Discussion).
Trans. A **298**, 309—319 (1980).

Kenny, B.G. *See* Dass, Kabir & Kenny.

Kent, B. *See* Lloyd (G.O.), Kent, Saunders & Lea.

Kent, H.J. *See* Boucher (E.A.), Evans & Kent; Boucher (E.A.) & Kent.

Kent, P.W. *See* Gallagher, Kent, Passatore, Phipps & Richardson; Phipps, Richardson, Corfield, Gallagher and others.

Kent, Sir Peter. Vertical tectonics associated with rifting and spreading (Discussion).
Trans. A **294**, 125—135 (1980).

Kenyon, N.H. *See* Stride, Belderson & Kenyon.

Keppler, E. The chemical composition of energetic charged particles in interplanetary space (Discussion). *Trans.* A **297**, 621—627 (1980).

Kerensky, O.A. Bridges and other large structures (Discussion). *Trans.* A **269**, 343—351 (1971).
Gilbert Roberts. *Biogr. Mem.* **25**, 477—503 (1979).

Kernahan, J.A. *See* Butts, Tinker & Kernahan.

Kerney, M.P., Preece, R.C. & Turner, C. Molluscan and plant biostratigraphy of some Late Devensian and Flandrian deposits in Kent. *Trans.* B **291**, 1—43 (1980).

Kerr, K. Ann., Ashmore, J.P. & Speakman, J.C. The crystal and molecular structure of quaterrylene: a redetermination. *Proc.* A **344**, 199—215 (1975).

Kerr, R.P. *See* Weir (G.J.) & Kerr.

Kersten, W. *See* Jaffré, Kersten, Brooks & Reeves.

Kessler, A. & Standley, C.C. The W.H.O. Expanded Programme of Research, Development and Research Training in Human Reproduction (Discussion). *Proc.* B **195**, 129—136 (1976).

Kessler, C. SF_6 switchgear for outdoor and indoor switching stations (Discussion).
Trans. A **275**, 109—120 (1973).

Kesson, Susan E. Mare basalt petrogenesis (Discussion). *Trans.* A **285**, 159—167 (1977).

Ketchledge, R.W. Electronic switching for trunk systems (Discussion). *Trans.* A **289**, 79—91 (1978).

Ketterer, B. *See* Hicks (R. Marian), Ketterer & Warren.

Key, M.H. Some topical issues in research on short-pulse laser-produced plasmas (Discussion).
Trans. A **298**, 351—364 (1980).

Key, P.Y. *See* Cutter, Key & Little; Harrison (R.G.), Key & Little.

Keynes, R.D. *See* Rojas & Keynes.

Keynes, R.D., Bezanilla, F., Rojas, E. & Taylor, R.E. The rate of action of tetrodotoxin on sodium

conductance in the squid giant axon (Discussion). *Trans.* B **270**, 365–375 (1975).

Khare, P.K. *See* Pant & Khare.

Khosla, S. *See* Fisher (J.), Belasco, Charnas, Khosla & Knowles.

Kien, J. *See* Altman (J.S.) & Kien.

Kienle, M. Galli. *See* Fiecchi, Kienle, Scala, Galli and others.

Kiepenheuer, K.O. The role and necessity of optical space observations in solar physics (Discussion). *Trans.* A **270**, 109–116 (1971).

Kiflawi, I. *See* Hanley, Kiflawi & Lang.

Kihara, H., Kanazawa, T. & Tamura, H. Weldability and toughness specifications for structural steels in Japan — with special reference to WES-135 and -136 (Symposium). *Trans.* A **282**, 247–258 (1976).

Kilbourne, E.D. Influenza: viral determinants of the pathogenicity and epidemicity of an invariant disease of variable occurrence (Discussion). *Trans.* B **288**, 291–297 (1980).

Kilburn, T. & Piggott, L.S. Frederic Calland Williams. *Biogr. Mem.* **24**, 583–604 (1978).

Killick-Kendrick, R. *See* Molyneux, Killick-Kendrick & Ashford; Peters (W.), Garnham, Killick-Kendrick, Rajapaksa and others.

Killick-Kendrick, R., Leaney, A.J., Ready, P.D. & Molyneux, D.H. *Leishmania* in phlebotomid sandflies. IV. The transmission of *Leishmania mexicana amazonensis* to hamsters by the bite of experimentally infected *Lutzomyia longipalpis*. *Proc.* B **196**, 105–115 (1977).

Killick-Kendrick, R., Molyneux, D.H. & Ashford, R.W. *Leishmania* in phlebotomid sandflies. I. Modification of the flagellum associated with attachment to the mid-gut oesophageal valve of the sandfly. *Proc.* B **187**, 409–419 (1974).

Killick-Kendrick, R., Molyneux, D.H., Hommel, M., Leaney, A.J. & Robertson, E.S. *Leishmania* in phlebotomid sandflies. V. The nature and significance of infections of the pylorus and ileum of the sandfly by leishmaniae of the *braziliensis* complex. *Proc.* B **198**, 191–199 (1977).

Killick-Kendrick, R. & Ward, R.D. *Leishmania* in phlebotomid sandflies. II. The insusceptibility of sandfly larvae to *Leishmania*. *Proc.* B **188**, 229–231 (1975).

Kim, M.G. *See* Barlow (A.J.), Harrison, Irving, Kim, Lamb & Pursley.

Kim, S. *See* Shikata, Kim, Mita, Ise & Kunugi.

Kimmitt, M.F. *See* Gibson (A.F.), Kimmitt, Koohian, Evans & Levy.

King, B.C. & Chapman, G.R. Volcanism of the Kenya rift valley (Discussion). *Trans.* A **271**, 185–208 (1972).

King, B.L. Intergranular embrittlement in CrMoV steels: an assessment of the effects of residual impurity elements on high temperature ductility and crack growth (Discussion). *Trans.* A **295**, 235–251 (1980).
See also Brear & King.

King, C.-Y., Nason, R.D. & Tocher, D. Kinematics of fault creep (Discussion). *Trans.* A **274**, 355–360 (1973).

King, D.A., Goymour, C.G. & Yates, Jr, J.T. Chemisorption of carbon monoxide on tungsten (Discussion). *Proc.* A **331**, 361–376 (1972).

King, D.A. & Wells, M.G. Reaction mechanism in chemisorption kinetics: nitrogen on the $\{100\}$ plane of tungsten. *Proc.* A **339**, 245–269 (1974).

King, E.A. The lunar regolith: physical characteristics and dynamics (Discussion). *Trans.* A **285**, 273–278 (1977).

King, Felicity M.A. *See* King (M.H.), King & Martodipoero.

King, G.C.P. Geological faults: fracture, creep and strain (Discussion). *Trans.* A **288**, 197–212 (1978).

King, G.C.P. & Bilham, R.G. Strain measurement instrumentation and technique (Discussion). *Trans.* A **274**, 209–217 (1973).

King, J., Botstein, D., Casjens, S., Earnshaw, W., Harrison, S. & Lenk, Elaine. Structure and assembly of the capsid of bacteriophage P22 (Discussion). *Trans.* B **276**, 37–49 (1976).

King, M.B. *See* Jonah & King.

King, M.H., King, Felicity M.A. & Martodipoero, Soebagio. Health microplanning: a systems approach to appropriate technology (Discussion). *Proc.* B **199**, 61–68 (1977).

King, P.J. & Wyatt, A.F.G. Surface waves on liquid helium-4. *Proc.* A **322**, 355–359 (1971).

King, P.W. & Learner, R.C.M. The application of the Zeeman effect to the precise measurement of pulsed magnetic fields. *Proc.* A **323**, 431–442 (1971).

King, R.W. *See* Feeney, Roberts, Birdsall, Griffiths and others.

King, R.W. & Burgen, A.S.V. Kinetic aspects of structure—activity relations: the binding of sulphonamides by carbonic anhydrase. *Proc.* B **193**, 107–125 (1976).

King-Hele, D.G. Analysis of the orbit of Cosmos 316 (1969–108A). *Proc.* A **330**, 467–494 (1972); Errata. *Proc.* A **332**, 565 (1973).

Concluding remarks to a Discussion. *Trans.* A **276**, 273–275 (1974).

The Bakerian Lecture, 1974. A view of Earth and air. *Trans.* A **278**, 67–109 (1975).

Prologue (Discussion). *Trans.* A **284**, 421–430 (1977).

Upper-atmosphere studies by ranging to satellites (Discussion). *Trans.* A **284**, 555–563 (1977).

The gravity field of the Earth (Discussion). *Trans.* A **294**, 317–328 (1980).

Skylab 1 rocket (1973–27B): orbit determination and analysis. *Trans.* A **296**, 597–637 (1980).

King-Hele, D.G. & Walker, Doreen M.C. The effect of atmospheric winds on satellite orbits of high eccentricity. *Proc.* A **350**, 281–298 (1976).

Kingman, J.F.C. On the Chapman—Kolmogorov equation. *Trans.* A **276**, 341–369 (1974).

Coherent random walks arising in some genetical models. *Proc.* A **351**, 19–31 (1976).

Random partitions in population genetics. *Proc.* A **361**, 1–20 (1978);
[Abstract] *Proc.* B **201**, 217 (1978).

The dynamics of neutral mutation. *Proc.* A **363**, 135–146 (1978);
[Abstract] *Proc.* B **203**, 91 (1978).

Kinloch, A.J. *See* Andrews (E.H.) & Kinloch.

Kinney, Marion S. & Fisher, S.K. The photoreceptors and pigment epithelium of the adult *Xenopus* retina: morphology and outer segment renewal. *Proc.* B **201**, 131–147 (1978).

The photoreceptors and pigment epithelium of the larval *Xenopus* retina: morphogenesis and outer segment renewal. *Proc.* B **201**, 149–167 (1978).

Changes in length and disk shedding rate of *Xenopus* rod outer segments associated with metamorphosis. *Proc.* B **201**, 169–177 (1978).

Kinnick, M.D. *See* Hatfield, Fisher, Dunigan, Burchfield and others.

Kinoshita, K. *See* Suzuki (Y.), Tsuchiya, Kinoshita, Sugiyama & Inuzuka.

Kirby, R. & Oele, E. The geological history of the Sandettie—Fairy Bank area, southern North Sea (Discussion). *Trans.* A **279**, 257–267 (1975).

Kirk, K.L. *See* Costa (J.L.), Dobson, Kirk, Poulsen, Valeri & Vecchione.

Kirk, R.L. *See* Boyce (A.J.), Harrison, Platt, Hornabrook and others.

Kirkham, B. *See* Bates (B.), Carson, Dufton, McKeith and others; Boksenberg, Kirkham, Michelson, Pettini and others.

Kirkwood, T.B.L. & Holliday, R. The evolution of ageing and longevity (Discussion).
Proc. B **205**, 531–546 (1979).

Kirpekar, S.M. *See* Dixon (W.R.), Garcia & Kirpekar.

Kirsch, L.J. *See* Halstead, Kirsch, Prothero & Quinn.

Kirshner, A.G. *See* Kirshner (N.) & Kirshner.

Kirshner, N. & Kirshner, A.G. Chromogranin A, dopamine β-hydroxylase and secretion from the adrenal medulla (Discussion). *Trans.* B **261**, 279–289 (1971).

Kirsner, N. Summary and closing remarks (Symposium). *Trans.* A **282**, 479 (1976).

Kirsten, T. Rare gases implanted in lunar fines (Discussion). *Trans.* A **285**, 391–395 (1977).
See also Horn (P.) & Kirsten.

Kishi, K. *See* Joyner, Kishi & Roberts.

Kishi, K. & Roberts, M.W. The adsorption of nitric oxide by iron surfaces studied by photoelectron spectroscopy. *Proc.* A **352**, 289–302 (1976).

Kisiel, Z. *See* Bevan (J.W.), Kisiel, Legon, Millen & Rogers.

Kistiakowsky, G.B. & Westheimer, F.H. James Bryant Conant. *Biogr. Mem.* **25**, 209–232 (1979).

Kitano, H. *See* Okubo, Kitano, Ishiwatari & Ise.

Kitchener, Gay. *See* Hayman (M.J.), Ramsay, Kitchener, Graf, Beug and others.

Kitchin, J. *See* Baldwin (J.E.), Jung, Singh, Wan, Haber and others.

Kitching, J.A. *See* Goss-Custard, Jones, Kitching & Norton.

Kittelson, D.B. *See* Hayhurst (A.N.) & Kittelson.

Klein, G. & Mulholland, H.P. Repeated elastic reflexions of a particle in an expanding sphere.
Proc. A 361, 447–461 (1978).

Klein, J. *See* Streilein & Klein.

Klein, J. & Briscoe, B.J. The diffusion of long-chain molecules through bulk polyethelene.
Proc. A 365, 53–73 (1979).

Kleinschmidt, A.K. Electron microscopic studies of macromolecules without appositional contrast
(Discussion). *Trans.* B 261, 143–149 (1971).

Kléman, M. Remarks on a possible elasticity of membranes and lamellar media: disordered layers.
Proc. A 347, 387–404 (1976).

Kleppmann, W.G. *See* Elliott (R.J.), Hayes, Kleppmann, Rushworth & Ryan.

Klimpke, C. *See* Landsberg & Klimpke.

Klinowski, J. *See* Barrer & Klinowski; Foley, Klinowski & Meares.

Klossa, J. *See* Dran, Kuraud, Klossa, Langevin & Maurette.

Klug, A. Optical diffraction and filtering and three-dimensional reconstructions from electron micro-
graphs (Discussion). *Trans.* B 261, 173–179 (1971).
Introductory remarks (Discussion). *Trans.* B 283, 233–239 (1978).
See also Erickson (H.P.) & Klug; Finch (J.T.) & Klug.

Knaap, H.F.P., van den Hout, K.D. & Hermans, P.W. Dynamic aspects of light scattering by gases
(Discussion). *Trans.* A 293, 407–412 (1979).

Knapp, M.F. *See* Lowe (D.A.), Mill & Knapp.

Knight, B.C.J.G. *See* Gladstone, Knight & Wilson.

Knight, D.E. *See* Baker (P.F.), Knight & Whitaker; Woodgate (B.E.), Knight, Uribe, Sheather, Bowles
& Nettleship.

Knight, D.J.E. *See* Blaney, Bradley, Edwards, Jolliffe and others.

Knight, D.P. *See* Woodhead-Galloway & Knight.

Knoll, J. *See* Garamvölgyi, Vizi & Knoll.

Knott, G.F. & Mackley, M.R. On eddy motions near plates and ducts, induced by water waves and
periodic flows. *Trans.* A 294, 599–623 (1980).

Knowles, F. *See* Axon, Nasir & Knowles.

Knowles, J.R. *See* Fisher (J.), Belasco, Charnas, Khosla & Knowles.

Knowles, Sir Francis. Introductory remarks to a Discussion (second day): neurosecretion.
Trans. B 261, 389–390 (1971).

Knowles, Sir Francis, Vollrath, L. & Meurling, P. Cytology and neuroendocrine relations of the
pituitary of the dogfish, *Scyliorhinus canicula. Proc.* B 191, 507–525 (1975).

Knox, J.R. *See* DeLucia, Kelly, Mangion, Moews & Knox.

Knox, R.B. *See* Howlett, Knox, Paxton & Heslop-Harrison.

Kobayashi, H. *See* Yamaguchi, Kobayashi, Matsumiya & Hayami.

Kobayashi, T. *See* Shank, Ippen, Fork, Migus & Kobayashi.

Koblinsky, Marjorie A. *See* Harkavy, Jaffe, Koblinsky & Segal.

Koch-Weser, C.K. *See* Golladay & Koch-Weser.

Koetser, H. *See* Bradley (D.J.), Hutchinson & Koetser; Bradley (D.J.), Hutchinson, Koetser, Morrow,
New & Petty.

Koga, Y. *See* Harrison (L.G.) & Koga.

Kohlstedt, D.L. *See* Hannink, Kohlstedt & Murray.

Kohsaka, M. *See* Baldwin (J.E.), Jung, Singh, Wan, Haber and others.

Kokourov, V.D. *See* Kazimirovsky & Kokourov.

Kolb, H. *See* Boycott, Dowling, Fisher, Kolb & Laties.

Kolbuszewski, M.L. *See* Bilby & Kolbuszewski.

Kolenkiewicz, R. *See* Smith (D.E.), Kolenkiewicz, Wyatt, Dunn & Torrence.

Kolenkiewicz, R., Smith, D.E., Rubincam, D.P., Dunn, P.J. & Torrence, M.H. Polar motion and Earth
tides from laser tracking (Discussion). *Trans.* A 284, 485–494 (1977).

Komori, S. *See* Mizushina, Ogino, Ueda & Komori.

Kondow, T. *See* Watanabe (K.), Kondow, Soma, Onishi & Tamaru.

Konomi, T. *See* Baldwin (J.E.), Jung, Singh, Wan, Haber and others.

Konopasek, M. *See* Buckley (C.P.), Lloyd & Konopasek.

Konstantinova, T.V. *See* Chailakhyan, Aksenova, Konstantinova & Bavrina.

van Konynenburg, P.H. & Scott, R.L. Critical lines and phase equilibria in binary van der Waals mixtures. *Trans.* A **298**, 495—540 (1980).

Koohian, A.O. *See* Gibson (A.F.), Kimmitt, Koohian, Evans & Levy.

Kopal, Z. Dynamical arguments which concern melting of the Moon (Discussion).
Trans. A **285**, 561—568 (1977).

Kopeć, A.C. *See* Godber, Kopeć, Mourant, Tills & Lehmann; Lehmann (H.), Ala, Hedeyat, Montazemi and others.

Kopeć, Ada C. *See* Mourant, Godber, Kopeć, Tills & Woodhead.

Kopp, I. *See* Bolman, Brown, Carrington, Kopp & Ramsay.

Korbee, D. *See* Bovée, Creyghton, Getreuer, Korbee, Lobregt and others.

Korda, A. *See* Daniels (A.), Korda, Tanswell, Williams & Williams.

Korn, H., Triller, A. & Faber, D.S. Structural correlates of recurrent collateral interneurons producing both electrical and chemical inhibitions of the Mauthner cell. *Proc.* B **202**, 533—538 (1978).

Kornberg, H.L. The Leeuwenhoek Lecture, 1972. Carbohydrate transport by micro-organisms.
Proc. B **183**, 105—123 (1973).
See also Ferenci & Kornberg; Harris (P.) & Kornberg; Jones-Mortimer & Kornberg; Lord, McFadden & Kornberg; Riordan & Kornberg.

Kornberg, H.L. & Miller, Elaine K. Role of phosphoenolpyruvate-phosphotransferase in glucose utilization by bacilli. *Proc.* B **182**, 171—181 (1972).

Kornberg, R.D. *See* Prunell & Kornberg.

Kosterlitz, H.W. *See* McKnight, Hughes & Kosterlitz; Robson (Linda E.) & Kosterlitz.

Kosterlitz, H.W. & Paterson, S.J. Characterization of opioid receptors in nervous tissue (Discussion).
Proc. B **210**, 113—122 (1980).

Kotas, T.J. An experimental study of the three dimensional boundary layer on the end wall of a vortex chamber. *Proc.* A **352**, 169—187 (1976).

Kouba, J. Geodetic satellite Doppler positioning and application to Canadian test adustments (Discussion). *Trans.* A **294**, 271—276 (1980).

Kovalevsky, J. Lunar orbital theory (Discussion). *Trans.* A **284**, 565—571 (1977).

Kovalsky, V.V. Geochemical ecology and problems of health (Discussion).
Trans. B **288**, 185—191 (1979).

Krakiwsky, E.J. Statistical techniques and Doppler satellite positioning (Discussion).
Trans. A **294**, 365—375 (1980).

Kranitzky, W. *See* Kaiser (W.), Fendt, Kranitzky & Laubereau.

Krätschmer, W. & Gentner, W. A long term change in the cosmic ray composition?: studies on fossil cosmic ray tracks in lunar samples (Discussion). *Trans.* A **285**, 593—599 (1977).

Krause, J.T. *See* Simpkins & Krause.

Krebs, J.R. *See* Dawkins & Krebs.

Krebs, Sir Hans. Otto Heinrich Warburg. *Biogr. Mem.* **18**, 629—699 (1972).

Kreisel, G. Bertrand Arthur William Russell, Earl Russell. *Biogr. Mem.* **19**, 583—620 (1973).
Kurt Gödel. *Biogr. Mem.* **26**, 149—224 (1980).

Kreiss, H.-O. Difference approximations for initial boundary-value problems (Discussion).
Proc. A **323**, 255—261 (1971).

Kressman, R.I. *See* Dudeney, Jones, Kressman & Spracklen.

Kressmann, A. *See* Birnstiel, Kressmann, Schaffner, Portmann & Busslinger.

de Kretser, D.M. Towards a pill for men (Discussion). *Proc.* B **195**, 161—174 (1976).

Kretzschmar, K.M. & Wilkie, D.R. The use of the Peltier effect for simple and accurate calibration of thermoelectric devices. *Proc.* B **190**, 315—321 (1975).

Krijgsman, P.C.J. *See* Driedonks, Krijgsman & Mellema.

Krikorian, A.D. *See* Steward, Israel, Mott, Wilson & Krikorian.

Krinsky, N.I. Non-photosynthetic functions of carotenoids (Discussion).
Trans. B **284**, 581—590 (1978).
Krinsley, D.H. *See* Rickards, Hyde & Krinsley.
Krishan, K. *See* Bullough (R.), Eyre & Krishan.
Krishna, P. *See* Pandey, Lele & Krishna.
Krishnaswamy, M.R., Menon, M.G.K., Narasimham, V.S., Hinotani, K., Ito, N., Miyake, S., Osborne, J.L., Parsons, A.J. & Wolfendale, A.W. The Kolar Gold Fields neutrino experiment.
I. The interactions of cosmic ray neutrinos. *Proc.* A **323**, 489—509 (1971).
II. Atmospheric muons at a depth of 7000 hg cm^{-2} (Kolar). *Proc.* A **323**, 511—522 (1971).
Krishnaswamy, S. *See* Raymont, Krishnaswamy, Woodhouse & Griffin.
Kroeger, E.A. *See* Marshall (Jean M.) & Kroeger.
Kröner, A. Proterozoic crustal evolution in parts of southern Africa and evidence for extensive sialic crust since the end of the Archaean (Discussion). *Trans.* A **280**, 541—553 (1976).
Kronheimer, E.H. *See* Geroch, Kronheimer & Penrose.
Krueger, A.J., Guenther, B., Fleig, A.J., Heath, D.F., Hilsenrath, E., McPeters, R., & Prabhakara, C. Satellite ozone measurements (Discussion). *Trans.* A **296**, 191—204 (1980).
Krug, R.M., Bouloy, Michele & Plotch, S.J. RNA primers and the role of host nuclear RNA polymerase II in influenza viral RNA transcription (Discussion). *Trans.* B **288**, 359—370 (1980).
Kruse, H. *See* Wänke, Palme, Baddenhausen, Dreibus, Kruse & Spettel.
Küchemann, D. *See* Damms & Küchemann; Thomas (H.H.B.M.) & Küchemann.
Kuffler, D.P., Thompson, W. & Jansen, J.K.S. The fate of foreign endplates in cross-innervated rat soleus muscle. *Proc.* B **208**, 189—222 (1980).
Kuffler, S.W. *See* Dennis (M.J.), Harris & Kuffler; Harris (A.J.), Kuffler & Dennis; McMahan (U.J.) & Kuffler.
Kuffler, S.W., Dennis, M.J. & Harris, A.J. The development of chemosensitivity in extrasynaptic areas of the neuronal surface after denervation of parasympathetic ganglion cells in the heart of the frog. *Proc.* B **177**, 555—563 (1971).
Kuhn, H.G. *See* Stacey (Virginia) & Kuhn.
Kuhn, H.G., Baird, P.E.G., Brimicombe, M.W.S.M., Stacey, D.N. & Stacey, V. Evidence for α-particle structure in medium-heavy nuclei from optical isotope shifts. *Proc.* A **342**, 51—54 (1975).
Kühne, M. *See* Smith (P.L.) & Kühne.
Kuhns, W.J. *See* Burger, Turner, Kuhns & Weinbaum.
Kuiken, H.K. The cooling of a low-heat-resistance sheet moving through a fluid.
Proc. A **341**, 233—252 (1974).
The cooling of a low-heat resistance cylinder moving through a fluid. *Proc.* A **346**, 23—35 (1975).
Kulasooriya, S.A., Lang, Norma J. & Fay, P. The heterocysts of blue-green algae. III. Differentiation and nitrogenase activity. *Proc.* B **181**, 199—209 (1972).
Kumapley, N.K. *See* Bishop (A.W.), Kumapley & El-Ruwayih.
Kumar, A. & Eyre, B.L. Grain boundary segregation and intergranular fracture in molybdenum.
Proc. A **370**, 431—458 (1980).
Kumar, Dharma. The edge of the desert: the problems of poor and semi-arid lands (Discussion).
Trans. B **278**, 477—491 (1977).
Kumar, I.J. On the asymptotic solution of a nonlinear Volterra integral equation.
Proc. A **324**, 45—61 (1971).
See also Trivedi & Kumar.
Kumar, Ram & Stephens, R.W.B. Dispersion of flexural waves in circular cylindrical shells.
Proc. A **329**, 283—297 (1972).
Kumar, S. *See* Eltayeb & Kumar.
Kumar, S. & Roberts, P.H. A three-dimensional kinematic dynamo. *Proc.* A **344**, 235—258 (1975).
Kunugi, S. *See* Shikata, Kim, Mita, Ise & Kunugi.
Kuperus, M. *See* de Jager, Kuperus & Rosenberg.
Kurepa, Jelena M. *See* Heddle, Keesing & Kurepa.
Kuriyama, H. *See* Bülbring & Kuriyama.
Kuriyan, Mary. *See* Hunter & Kuriyan.

Kurzfeld, A. *See* Irving (P.E.) & Kurzfeld.
Kusano, K., **Miledi, R. & Stinnakre, J.** Microinjection of calcium into droplets of aequorin.
   Proc. B **189**, 39–47 (1975).
   Postsynaptic entry of calcium induced by transmitter action. *Proc.* B **189**, 49–56 (1975).
Kutsch, W. *See* Altman (Jennifer S.), Anselment & Kutsch.
Kuttler, K.L. *See* Garnham & Kuttler.
Kutty, T.S. *See* Jain (S.L.), Kutty, Roy-Chowdhury & Chatterjee.

Laas, H. *See* Inhoffen, Gossauer, Heise & Laas.
Labitzke, Karin. Climatology of the stratosphere and mesosphere (Discussion).
   Trans. A **296**, 7–18 (1980).
Lacalli, T.C. *See* Harrison (L.G.) & Lacalli.
Lachmann, P.J. Lymphocyte cooperation (Discussion). *Proc.* B **176**, 425–426 (1971).
La Cour, L.F. The constitutive heterochromatin in chromosomes of *Fritillaria* sp., as revealed by
   Giemsa banding. *Trans.* B **285**, 61–71 (1978).
La Cour, L.F. & Wells, B. Nuclear pores at prophase of meiosis in plants (Discussion).
   Trans. B **268**, 95–100 (1974).
   The nucleolus at prophase of meiosis in three plants: an ultrastructural study.
   Proc. B **191**, 231–243 (1975).
   Some morphological aspects of the synaptonemal complex in higher plants (Discussion).
   Trans. B **277**, 259–266 (1977).
Lacy, D. *See* Bell (Janet B.G.) & Lacy; Bell (Janet B.G.), Vinson & Lacy; Collins & Lacy.
Laemmli, U.K. *See* Wagner (J.) & Laemmli.
Laget, M. *See* Courtès, Laget, Sivan, Viton and others.
Laing, Sir Maurice. The inherent problems of effecting change (Discussion).
   Trans. A **272**, 497–502 (1972).
Lainson, R., Ready, P.D. & Shaw, J.J. *Leishmania* in phlebotomid sandflies. VII. On the taxonomic
   status of *Leishmania peruviana*, causative agent of Peruvian 'uta', as indicated by its development
   in the sandfly, *Lutzomyia longipalpis*. *Proc.* B **206**, 307–318 (1979).
Lainson, R., Ward, R.D. & Shaw, J.J. *Leishmania* in phlebotomid sandflies. VI. Importance of hind-
   gut development in distinguishing between parasites of the *Leishmania mexicana* and *L. brazil-
   iensis* complexes. *Proc.* B **199**, 309–320 (1977).
Lake, J.V. The behaviour of plants in various gas mixtures (Discussion).
   Proc. B **179**, 177–188 (1971).
   Comparison of natural and artificial sources of light (Discussion). *Proc.* B **179**, 189–192 (1971).
Lake, R.D. *See* Shephard-Thorn, Lake & Atitullah.
Lakhera, S.C. *See* Cotes, Dabbs, Hall, Lakhera, Saunders & Malhotra.
Lakin, W.D., Ng, B.S. & Reid, W.H. Approximations to the eigenvalue relation for the Orr–Sommer-
   feld problem. *Trans.* A **289**, 347–371 (1978).
Lal, D. Long-term variations in the cosmic-ray flux (Discussion). *Trans.* A **277**, 395–411 (1974).
   Irradiation and accretion of solids in space based on observations of lunar rocks and grains (Dis-
   cussion). *Trans.* A **285**, 69–95 (1977).
Laloue, M. Functions of cytokinins (Discussion). *Trans.* B **284**, 449–457 (1978).
Lamb, D. Appendix to a paper by Gallagher, Kent, Passatore, Phipps & Richardson.
   Proc. B **192**, 49–76 (1975).
Lamb, H.H. Climate, vegetation and forest limits in early civilized times (Discussion).
   Trans. A **276**, 195–230 (1974).
   Climatic analysis (Discussion). *Trans.* B **280**, 341–350 (1977).
Lamb, J. *See* Barlow (A.J.), Harrison, Irving, Kim, Lamb & Pursley; Gray (R.W.), Harrison & Lamb;
   Phillips (M.C.), Barlow & Lamb.
Lamb, M.M. *See* Daneholt, Case, Lamb, Nelson & Wieslander.
Lamb, R.A. & Choppin, P.W. A ninth unique influenza virus-coded polypeptide (Discussion).
   Trans. B **288**, 327–333 (1980).

Lambeck, K. Tidal dissipation in the oceans: astronomical, geophysical and oceanographic consequences. *Trans.* A **287**, 545—594 (1977).

Lambeck, Kurt. *See* Cazenave, Daillet & Lambeck.

Lambeck, Kurt & Cazenave, Anny. The Earth's variable rate of rotation: a discussion of some meteorological and oceanic causes and consequences (Discussion). *Trans.* A **284**, 495—506 (1977).

Lambert, D.L. A review of models of the solar photosphere and low chromosphere: the temperature—height profile (Discussion). *Trans.* A **270**, 3—21 (1971).

Lambert, G., Le Roulley, J.C. & Bristeau, P. Accumulation and circulation of gaseous radon between lunar fines (Discussion). *Trans.* A **285**, 331—336 (1977).

Lambert, R.M. *See* Bridge & Lambert.

Lambeth, P.J. Insulators for 1000 to 1500 kV systems (Discussion). *Trans.* A **275**, 153—163 (1973).

Lamers, H.J.G.L.M. Mass loss from the A-type supergiant α Cygni (Discussion).
Trans. A **279**, 445—450 (1975).
See also de Jager, Hoekstra, van der Hucht, Kamperman & Lamers.

Lammel, E., Niedergerke, R. & Page, Sally. Analysis of a rapid twitch facilitation in the frog heart.
Proc. B **189**, 577—590 (1975).

Lammlein, D.R. Lunar seismicity, structure, and tectonics (Discussion).
Trans. A **285**, 451—461 (1977).

Lampen, J.O., Nielsen, J.B.K., Izui, K. & Caulfield, M.P. *Bacillus licheniformis* β-lactamases: multiple forms and their roles (Discussion). *Trans.* B **289**, 345—348 (1980).

Lampis, G. *See* Franklin (R.N.), Hamberger, Lampis & Smith.

Lancelot, Y. Geological setting and results of Legs 41, 47a and 50 off northwest Africa [abstract] (Discussion). *Trans.* A **294**, 35 (1980).
Mesozoic palaeo-oceanography of rifted margins of the Atlantic [abstract] (Discussion).
Trans. A **294**, 123 (1980).

Land, M.F. *See* Blest & Land; Denton & Land.

Land, R.B. *See* Baird (D.T.), Land, Scaramuzzi & Wheeler.

Landsberg, P.T. On detailed balance between Auger recombination and impact ionization in semiconductors. *Proc.* A **331**, 103—108 (1972).
See also Hill (D.) & Landsberg; Mallinson (W.R.) & Landsberg; Schöll & Landsberg.

Landsberg, P.T. & Adams, M.J. Theory of donor-acceptor radiative and Auger recombination in simple semiconductors. *Proc.* A **334**, 523—539 (1973).

Landsberg, P.T. & Klimpke, C. Theory of the Schottky barrier solar cell.
Proc. A **354**, 101—118 (1977).

Landsberg, P.T. & Park, D. Entropy in an oscillating universe. *Proc.* A **346**, 485—495 (1975).

Lane, D.P. & Crawford, L.V. The complex between simian virus 40 T antigen and a specific host protein (Discussion). *Proc.* B **210**, 451—463 (1980).

Laneri, U. *See* de Nettancourt, Devreux, Carluccio and others.

Lang, A.R. On the growth-sectorial dependence of defects in natural diamonds.
Proc. A **340**, 233—248 (1974).
See also Hanley, Kiflawi & Lang.

Lang, A.R. & Zhen-Hong, Mai. Pendellösung interference in the Bragg reflexion of X-rays from a crystal surface. *Proc.* A **368**, 313—329 (1979).

Lang, B. *See* Somorjai, Joyner & Lang.

Lang, C.A. Communication between an engineer and a computer (Discussion).
Proc. A **321**, 243—248 (1971).

Lang, D. Individual macromolecules: preparation and recent results with DNA (Discussion).
Trans. B **261**, 151—158 (1971).

Lang, Norma J. *See* Fay & Lang; Kulasooriya, Lang & Fay.

Lang, Norma J. & Fay, P. The heterocysts of blue-green algae. II. Details of ultrastructure.
Proc. B **178**, 193—203 (1971).

Lang, W.C. *See* Padalia, Lang, Norris, Watson & Fabian.

Lange, R. *See* Jensen (S.), Lange, Berge, Palmork & Renberg.

Langer, A.M. *See* Bowes, Langer & Rohl.

Langevin, Y. *See* Dran, Duraud, Klossa, Langevin & Maurette.

Langford, T.E. & Aston, R.J. The ecology of some British rivers in relation to warm water discharges from power stations (Discussion). *Proc.* B **180**, 407—419 (1972).

Langmuir, C.H. & Hanson, G.N. An evaluation of major element heterogeneity in the mantle sources of basalts (Discussion). *Trans.* A **297**, 383—407 (1980).

Langseth, M.G. *See* Horai, Winkler, Keihm, Langseth, Fountain & West.

Lanham, N.W. *See* Pearlman, Lanham, Lehr & Wohn.

Lansdown, R. Moderately raised blood lead levels in children (Discussion).
Proc. B **205**, 145—151 (1979).

Lapierre, F. Contribution à l'étude géologique et sédimentologique de la Manche orientale (Discussion). *Trans.* A **279**, 177—187 (1975).

Lardge, M.G.C. *See* Schnurmann & Lardge.

Lardner, R.W. The development of plane shock waves in nonlinear viscoelastic media.
Proc. A **347**, 329—344 (1976).

Larson, A.C. *See* Kennard, Isaacs, Motherwell, Coppola and others.

Larson, D.A. *See* Fowler (A.C.) & Larson.

Larson, R.B. Star formation in young galaxies (Discussion). *Trans.* A **296**, 299—302 (1980).

Larsonneur, C. *See* Hamilton (D.), Hommeril, Larsonneur & Smith; Huault, Lefebvre, Guyader, Giresse and others.

Larsonneur, C., Horn, R. & Auffret, J.P. Géologie de la partie méridionale de la Manche centrale (Discussion). *Trans.* A **279**, 145—153 (1975).

Lasansky, A. Synaptic organization of cone cells in the turtle retina. *Trans.* B **262**, 365—381 (1971).
Organization of the outer synaptic layer in the retina of the larval tiger salamander.
Trans. B **265**, 471—489 (1973).

Lassen, H.H. *See* Levinton & Lassen.

Latham, J. *See* Brazier-Smith, Brook, Latham, Saunders & Smith; Brazier-Smith, Jennings & Latham.

Laties, A.M. *See* Boycott, Dowling, Fisher, Kolb & Laties.

Lau, J.C. The vortex-street structure of turbulent jets. Part 2. *Proc.* A **368**, 547—571 (1979).

Laubereau, A. Picosecond Raman scattering (Discussion). *Trans.* A **293**, 441—453 (1979).
See also Kaiser (W.), Fendt, Kranitzky & Laubereau.

Laughton, A.S. & Roberts, D.G. Morphology of the continental margin (Discussion).
Trans. A **290**, 75—85 (1978).

Launois, M. An ecological model for the study of the grasshopper *Oedaleus senegalensis* in west Africa (Discussion). *Trans.* B **287**, 345—355 (1979).

Laurila, S. *See* Carter (W.E.), Berg & Laurila.

Lauterbur, P.C. Progress in n.m.r. zeugmatographic imaging (Discussion).
Trans. B **289**, 483—486 (1980).

Laver, W.G., Air, G.M., Webster, R.G., Gerhard, W., Ward, C.W. & Dopheide, T.A.A. The mechanism of antigenic drift in influenza virus : sequence changes in the haemagglutinin of variants selected with monoclonal hybridoma antibodies (Discussion). *Trans.* B **288**, 313—326 (1980).

Laverack, M.S. *See* Macmillan, Neil & Laverack; Neil, Macmillan, Robertson & Laverack; Robertson (R.M.) & Laverack.

Laverack, M.S., Macmillan, D.L. & Neil, D.M. A comparison of beating parameters in larval and post-larval locomotor systems of the lobster *Homarus gammarus* (L.). *Trans.* B **274**, 87—99 (1976).

Laverack, M.S., Neil, D.M. & Robertson, R.M. Metachronal exopodite beating in the mysid *Praunus flexuosus:* a quantitative analysis. *Proc.* B **198**, 139—154 (1977).

Law, L.K. *See* Edwards (R.N.), Law & White.

Lawn, B.R. *See* Sinclair (J.E.) & Lawn.

Lawrence, M.J. The genetics of self-incompatibility in *Papaver rhoeas* (Discussion).
Proc. B **188**, 275—285 (1975).

Lawrie, T.D.V. *See* Macfarlane & Lawrie.

Laws, N. *See* Green (A.E.), Laws & Naghdi.

Laws, N. & McLaughlin, R. Self-consistent estimates for the viscoelastic creep compliances of composite materials. *Proc.* A **359**, 251—273 (1978).

Laws, R.M. Seals and whales of the Southern Ocean (Discussion). *Trans.* B 279, 81—96 (1977).

Lawson, Janice. *See* Dickinson (H.G.) & Lawson.

Lawson, T.V. Landscape effects with particular reference to urban situations (Discussion).
Trans. A 269, 493—501 (1971).

Lawther, P.J. Epidemics of non-infectious disease (Discussion). *Proc.* B 205, 63—75 (1979).

Lea, C. *See* Bullock (E.), Lea & McLean; Lloyd (G.O.), Kent, Saunders & Lea.

Lea, C., Sawle, R. & Sellars, C.M. Hot ductility and sulphur segregation in 1%C—1%Cr steels [abstract]
(Discussion). *Trans.* A 295, 121 (1980).

Lea, C., Seah, M.P. & Hondros, E.D. Categorizing the embrittling residuals in engineering alloys by
Auger electron spectroscopy [abstract] (Discussion). *Trans.* A 295, 136 (1980).

Lea, M.J. *See* Almond, Lea & Dobbs; Dobbs, Lea & Peck.

Lea, M.J., Llewellyn, J.D., Peck, D.R. & Dobbs, E.R. Ultrasonic studies of the electronic structure of
hexagonal metal crystals. I. Normal state in magnesium, zinc and cadmium.
Proc. A 334, 357—377 (1973).

Lea, P.J. & Fowden, L. The purification and properties of glutamine-dependent asparagine synthetase
isolated from *Lupinus albus. Proc.* B 192, 13—26 (1975).

Leach, H.F. *See* Leith & Leach.

Leach, S.J. Research at the Building Research Establishment into the applications of solar collectors
for space and water heating in buildings. *Trans.* A 295, 403—414 (1980).

Leadbeater, B.S.C. Developmental and ultrastructural observations on two stalked marine choanoflagellates, *Acanthoecopsis spiculifera* Norris and *Acanthoeca spectabilis* Ellis.
Proc. B 204, 57—66 (1979).

See also Manton (Irene) & Leadbeater; Manton (Irene), Sutherland & Leadbeater.

Leadbetter, A.J. *See* Frost (J.C.), Leadbetter & Richardson.

Leaf, A. & Sharp, G.W.G. The stimulation of sodium transport by aldosterone (Discussion).
Trans. B 262, 323—332 (1971).

Leakey, D.M. Current and medium term developments in switching (Discussion).
Trans. A 289, 43—63 (1978).

Leaney, A.J. *See* Killick-Kendrick, Leaney, Ready & Molyneux; Killick-Kendrick, Molyneux,
Hommel, Leaney & Robertson.

Leardini, T. Geothermal power (Discussion). *Trans.* A 276, 507—526 (1974).

Learner, R.C.M. *See* King (P.W.) & Learner.

Le Berre, R., Garms, R., Davies, J.B., Walsh, J.F. & Philippon, B. Displacements of *Simulium dam-
nosum* and strategy of control against onchocerciasis (Discussion). *Trans.* B 287, 277—288 (1979).

Leblanc, R.M. *See* Costa (S.M. de B.), Froines, Harris, Leblanc, Orger & Porter.

Leckie, F.A. The constitutive equations of continuum creep damage mechanics (Discussion).
Trans. A 288, 27—47 (1978).

See also Hayhurst (D.R.), Leckie & Morrison.

Leckie, F.A. & Hayhurst, D.R. Creep rupture of structures. *Proc.* A 340, 323—347 (1974).

Leckie, F.A., Hayhurst, D.R. & Morrison, C.J. The creep behaviour of sphere-cylinder shell inter-
sections subjected to internal pressure. *Proc.* A 349, 9—34 (1976).

Le Cren, E.D. The productivity of freshwater communities (Discussion).
Trans. B 274, 359—374 (1976).

Ledwith, A., Russell, P.J. & Sutcliffe, L.H. Alkoxy radical intermediates in the thermal and photochemical oxidation of alcohols. *Proc.* A 332, 151—166 (1973).

Lee, A.E. *See* Kennett, Lee & Wilson.

Lee, A.E. & Singer, K.E. Studies of the tungsten—oxygen surface reaction by means of reflexion
high energy electron diffraction. *Proc.* A 323, 523—539 (1971).

Lee, A.G., Birdsall, N.J.M., Metcalfe, J.C., Warren, G.B. & Roberts, G.C.K. A determination of the
mobility gradient in lipid bilayers by ^{13}C nuclear magnetic resonance.
Proc. B 193, 253—274 (1976).

Lee, A.J. Appendix to a paper by H. Grüneberg. *Proc.* B 179, 153—156 (1971).

Lee, C.M. *See* Iversen (L.L.), Lee, Gilbert, Hunt & Emson.

Lee, C.W. Topology change in general relativity. *Proc.* A 364, 295—308 (1978).

Lee, D.N. The optic flow field: the foundation of vision (Discussion). *Trans.* B **290**, 169–179 (1980).

Lee, D.L., Nixon, P.E. & North, A.C.T. Structural studies on crystals found in the intestine of *Nematodirus battus*. *Proc.* B **208**, 409–414 (1980).

Lee, E.P.F. & Potts, A.W. An investigation of the valence shell electronic structure of alkaline earth halides by using *ab initio* s.c.f. calculations and photoelectron spectroscopy. *Proc.* A **365**, 395–411 (1979).

Lee, E.W. *See* Palmer (S.B.) & Lee; Palmer (S.B.), Lee & Islam.

Lee, E.W. & Asgar, M.A. The magnetostriction of nickel. *Proc.* A **326**, 73–85 (1971).

Lee, J.A. *See* Gordon (J.E.), Hall, Lee & Mortimer.

Lee, K.E. Introductory remarks to a Discussion. *Trans.* B **272**, 269–276 (1975).
Conclusions to a Discussion.~*Trans.* B **272**, 477–486 (1975).

Lee, W.-H. *See* Schroepfer, Lutsky, Martin, Huntoon and others.

Leece, N. *See* Hartley (A.J.), Eastburn & Leece.

Lees, D.G. *See* Skeldon, Calvert & Lees.

de Leeuw, S.W., Perram, J.W. & Smith, E.R. Simulation of electrostatic systems in periodic boundary conditions.
  I. Lattice sums and dielectric constants. *Proc.* A **373**, 27–56 (1980).
  II. Equivalence of boundary conditions. *Proc.* A **373**, 57–66 (1980).

Lefebvre, A.H. *See* Ballal & Lefebvre.

Lefebvre, Dominique. *See* Huault, Lefebvre, Guyader, Giresse and others.

Lefkowitz, I. Integrated control of industrial systems (Discussion). *Trans.* A **287**, 443–465 (1977).

Lefort, J.P. Étude géologique du socle ante-Mésozoïque au nord du Massif Armoricain: limites et structures de la Domnonée (Discussion). *Trans.* A **279**, 123–135 (1975).
Le controle du socle dans l'évolution de la sedimentation en Manche occidentale apres le Paléozoïque (Discussion). *Trans.* A **279**, 137–143 (1975).
See also Bouysse, Horn, Lefort & Le Lann.

Legon, A.C. *See* Bevan (J.W.), Kisiel, Legon, Millen & Rogers; Bevan (J.W.), Legon, Millen & Rogers; Georgiou, Legon & Millen.

Legon, A.C., Millen, D.J. & Rogers, S.C. Spectroscopic investigations of hydrogen bonding interactions in the gas phase. I. The determination of the geometry, dissociation energy, potential constants and electric dipole moment of the hydrogen-bonded heterodimer HCN·· ·HF from its microwave rotational spectrum. *Proc.* A **370**, 213–237 (1980).

Le Guern, F. The collection and analysis of volcanic gases (Symposium).
Trans. A **274**, 129–135 (1973).

Lehmann, E.E. *See* Edholm (O.G.), Samueloff, Mourant, Fox and others; Godber, Kopeć, Mourant, Tills & Lehmann.

Lehmann, E.E., Gadoth, N. & Samueloff, S. Biological studies of Yemenite and Kurdish Jews in Israel and other groups in southwest Asia. II. A clinical survey of Yemenite and Kurdish Jews in Israel. *Trans.* B **266**, 97–100 (1973).

Lehmann, H. *See* Edholm (O.G.), Samueloff, Mourant, Fox and others; Romero-Herrera & Lehmann; Romero-Herrera, Lehmann, Joysey & Friday.

Lehmann, H., Ala, F., Hedeyat, S., Montazemi, K., Nejad, H. Karini, Lightman, S., Kopeć, A.C., Mourant, A.E., Teesdale, P. & Tills, D. Biological studies of Yemenite and Kurdish Jews in Israel and other groups in southwest Asia. XI. The hereditary blood factors of the Kurds of Iran. *Trans.* B **266**, 195–205 (1973).

Lehner, P. & Lutz, M.B.K. Geology of the Iberian Shelf [abstract] (Discussion).
Trans. A **294**, 63 (1980).

Lehouelleur, J. & Schmidt, H. Extracellular recording of localized electrical activity in denervated frog slow muscle fibres. *Proc.* B **209**, 403–413 (1980).

Lehr, C.G. *See* Pearlman, Lanham, Lehr & Wohn.

Leichtberg, S., Weinbaum, S., Pfeffer, R. & Gluckman, M.J. A study of unsteady forces at low Reynolds number: a strong interaction theory for the coaxial settling of three or more spheres. *Trans.* A **282**, 585–610 (1976).

Leidi, M. & Hartland, S. Rows of two dimensional drops at fluid/liquid interfaces.
Proc. A **347**, 75–84 (1975).

The effect of vertical forces on the coalescence of two dimensional drops.
Proc. A **349**, 343—354 (1976).

Leigh, J. Barrington. *See* Holmes (K.C.), Tregear & Leigh.

Leigh, J.S., Jr. *See* Chance, D'Ambrosia, Leigh & McDonald.

Leigh, R.S., Szigeti, B. & Tewary, V.K. Force constants and lattice frequencies.
Proc. A **320**, 505—526 (1971).

Leisegang, E.C. *See* Thomas (R.K.), Leisegang & Thompson.

Leith, I.R. & Leach, H.F. Adsorbate interactions on copper-exchanged X-type zeolite catalysts studied by e.p.r. spectroscopy. *Proc.* A **330**, 247—263 (1972).

Leithead, C.S. Medicine in Ethiopia (Discussion). *Proc.* B **194**, 49—56 (1976).

Le Lann, F. *See* Bouysse, Horn, Lefort & Le Lann.

Lele, Shrikant. *See* Pandey, Lele & Krishna.

Lemblé, P., Pineau, A., Castagné, J.L., Dumoulin, P. & Guttmann, M. Temper embrittlement at high alloy contents: a 12% Cr martensitic steel [abstract] (Discussion). *Trans.* A **295**, 209 (1980).

Lemeunier, Francoise. *See* Ashburner & Lemeunier.

Lemeunier, Francoise & Ashburner, M. Relationships within the *melanogaster* species subgroup of the genus *Drosophila (Sophophora)*. II. Phylogenetic relationships between six species based upon polytene chromosome banding sequences. *Proc.* B **193**, 275—294 (1976).

Lemieux, R.U. & Edwards, O.E. Léo Edmond Marion. *Biogr. Mem.* **26**, 357—370 (1980).

Lemke, P.A. *See* Neuss, Nash, Lemke & Grutzner.

Lemmers, M. *See* Van Montagu, Holsters, Zambryski, Hernalsteens, Depicker and others.

Lemon, R.N., Hanby, J.A. & Porter, R. Relationship between the activity of precentral neurones during active and passive movements in conscious monkeys. *Proc.* B **194**, 341—373 (1976).

Lemon, R.N. & Porter, R. Afferent input to movement-related precentral neurones in conscious monkeys. *Proc.* B **194**, 313—339 (1976).

Leng, C.A., Rowlinson, J.S. & Thompson, S.M. The gas-liquid surface of the penetrable sphere model. *Proc.* A **352**, 1—23 (1976).

Leng, Christine A., Rowlinson, J.S. & Thompson, S.M. The gas—liquid surface of the penetrable sphere model. II. *Proc.* A **358**, 267—280 (1978).

Lenherr, A.D. & Omerod, M.G. Electron reactions with thymidine in γ-irradiated frozen aqueous glasses. *Proc.* A **325**, 81—99 (1971).

Lenk, Elaine. *See* King (J.), Botstein, Casjens, Earnshaw, Harrison & Lenk.

Lennie, P. Perceptual signs of parallel pathways (Discussion). *Trans.* B **290**, 23—37 (1980).

Lennon, G.W. & Baker, T.F. Earth tides and their place in geophysics [abstract] (Discussion).
Trans. A **274**, 199—202 (1973).

Leonis, C.G. *See* Gordon (M.), Leonis & Suzuki.

Leppard, N.A.G. Satellite Doppler fixation and international boundaries (Discussion).
Trans. A **294**, 289—298 (1980).

Leppington, F.G. *See* Crighton & Leppington; Davis (A.M.J.) & Leppington.

Le Ribault, L. Application de l'exoscopie des quartz à quelques échantillons prélevés en Manche orientale (Discussion). *Trans.* A **279**, 279—288 (1975).

Lernmark, Å. *See* Steiner (D.F.), Patzelt, Chan, Quinn, Tager, Nielson and others.

Le Roulley, J.C. *See* Lambert (G.), Le Roulley & Bristeau.

Leroy, J. *See* McLenaghan & Leroy.

Leroy, V., Richelmi, J. & Graas, H. Surface composition of heat-treated steel sheets [abstract] (Discussion). *Trans.* A **295**, 126—127 (1980).

Lesbats, B. *See* Israel (M.), Lesbats, Meunier & Stinnakre.

Leslie, B. *See* Jones (R.V.) & Leslie.

Leslie, F.M. *See* Clark (M.G.) & Leslie.

Leslie, M., Jenkin, G.T., Hayter, J.B., White, J.W., Cox, S. & Warner, G. Precise location of hydrogen atoms in complicated structures by diffraction of polarized neutrons from dynamically polarized nuclei (Discussion). *Trans.* B **290**, 497—503 (1980).

Leslie, R.B. *See* Atkinson (D.), Davis & Leslie.

Lesser, M.B. *See* Field (J.E.) & Lesser.

Lett, J.T. *See* Nagasawa, Cox & Lett.

Leuchars, E. *See* Davies (A.J.S.), Leuchars, Wallis & Doenhoff.

LeVay, S. *See* Hubel, Wiesel & LeVay.

Levene, H. *See* Dobzhansky, Levene & Spassky.

Levin, G. *See* Karasawa, Levin & Szwarc.

Levin, R.J. *See* Syme & Levin.

Levin, Z. & Hobbs, R.V. Splashing of water drops on solid and wetted surfaces: hydrodynamics and charge separation. *Trans.* A **269**, 555—585 (1971).

Levine, B.A. & Williams, R.J.P. The determination of the conformations of small molecules in solution by means of paramagnetic shift and relaxation perturbations of n.m.r. spectra (Discussion). *Proc.* A **345**, 5—22 (1975).

Levine, J. & Stebbins, R.T. Ultra sensitive laser interferometers and their application to problems of geophysical interest (Discussion). *Trans.* A **274**, 279—284 (1973).

Levine, M.M. *See* Murphy (B.R.), Markoff, Chanock, Spring and others.

Levine, Y.K. *See* Brooks (D.E.), Levine, Requena & Haydon.

Levinson, S.R. The purity of tritiated tetrodotoxin as determined by bioassay (Discussion). *Trans.* B **270**, 337—348 (1975).

    See also Tsien, Green, Levinson, Rudy & Sanders.

Levinson, S.R. & Meves, H. The binding of tritiated tetrodotoxin to squid giant axons (Discussion). *Trans.* B **270**, 349—352 (1975).

Levinton, J.S. & Lassen, H.H. Selection, ecology and evolutionary adjustment within bivalve mollusc populations (Discussion). *Trans.* B **284**, 403—415 (1978).

Levy, Estrella S. *See* Frydman (B.), Frydman, Valasinas, Levy & Feinstein.

Levy, G.F.D. *See* Gibson (A.F.), Kimmitt, Koohian, Evans & Levy.

Lew, V.L. *See* Glynn, Hoffman & Lew.

Lew, V.L., Ferreira, H.G. & Moura, Teresa. The behaviour of transporting epithelial cells. I. Computer analysis of a basic model. *Proc.* B **206**, 53—83 (1979).

Lewins, J. Linear stochastic neutron transport theory. *Proc.* A **362**, 537—558 (1978).

Lewis, A. Resonance Raman evidence for secondary protein—Schiff base interactions in bacterio-rhodopsin: correlation of the primary excitation mechanism with a model for proton pumping and visual transduction (Discussion). *Trans.* A **293**, 315—327 (1979).

Lewis, D. Voyaging stars: aspects of Polynesian and Micronesian astronomy (Discussion). *Trans.* A **276**, 133—148 (1974).

    Heteromorphic incompatibility system under disruptive selection (Discussion). *Proc.* B **188**, 247—256 (1975).

    Sporophytic incompatibility with 2 and 3 genes. *Proc.* B **196**, 161—170 (1977).

    See also Dickinson (H.G.) & Lewis.

Lewis, D. & Rao, A.N. Evolution of dimorphism and population polymorphism in *Pemphis acidula* Forst. *Proc.* B **178**, 79—94 (1971).

Lewis, D.H. & Smith, D.C. The autotrophic nutrition of symbiotic marine coelenterates with special reference to hermatypic corals. I. Movement of photosynthetic products between the symbionts. *Proc.* B **178**, 111—129 (1971).

Lewis, D.L. *See* Losty & Lewis.

Lewis, D.N. *See* Jefferies & Lewis.

Lewis, G. Polymorphism and selection in *Cochlicella acuta*. *Trans.* B **276**, 399—451 (1977).

Lewis, I.M. The peoples and cultures of Ethiopia (Discussion). *Proc.* B **194**, 7—16 (1976).

Lewis, M.G. *See* Hirst & Lewis.

Lewis, Z.V. *See* Berry (M.V.) & Lewis.

Lewit-Bentley, A. *See* Finch (J.T.), Lewit-Bentley, Bentley, Roth & Timmins.

Lewontin, R.C. *See* Gould (S.J.) & Lewontin.

Lex, M., Silvester, W.B. & Stewart, W.D.P. Photorespiration and nitrogenase activity in the blue-green alga, *Anabaena cylindrica*. *Proc.* B **180**, 87—102 (1972).

Lexton, M.J., Marshall, R.M. & Purnell, J.H. The reaction of hydrogen atoms with propylene. *Proc.* A **324**, 433—446 (1971).

The reaction of hydrogen atoms with isobutene. *Proc.* A **324**, 447–458 (1971).
Leyh-Bouille, M. *See* Ghuysen, Frére, Leyh-Bouille, Perkins & Nieto.
Lieb, E.H. *See* Temperley & Lieb.
Liese, B.H. *See* Golladay & Liese.
Lighthill, M.J. Introductory remarks (Discussion). *Trans.* A **269**, 323–326 (1971).
Time-varying currents (Discussion). *Trans.* A **270**, 371–390 (1971).
Large-amplitude elongated-body theory of fish locomotion. *Proc.* B **179**, 125–138 (1971).
See also Lighthill (Sir James).
Lighthill, Sir James. Introductory remarks to a Discussion. *Trans.* A **272**, 495–496 (1972).
Concluding remarks to a Discussion. *Trans.* A **272**, 659–661 (1972).
Introductory remarks to a Discussion. *Trans.* A **273**, 3–4 (1972).
Concluding remarks to a Discussion. *Trans.* A **273**, 183–184 (1972).
Opening remarks to a Discussion. *Proc.* A **345**, 433 (1975).
Closing remarks to a Discussion. *Proc.* A **345**, 601–605 (1975).
Introductory remarks to a Discussion. *Trans.* A **289**, 3–4 (1978).
See also Lighthill (M.J.).
Lightman, S. *See* Lehmann (H.), Ala, Hedeyat, Montazemi and others.
Lilley, B.A. *See* Hardman & Lilley.
Lilley, D.M.J. *See* Richards (B.M.), Pardon, Lilley, Cotter *et al.*
Lilley, T.H. & Briggs, C.C. Activity coefficients of calcium sulphate in water at 25 °C.
Proc. A **349**, 355–368 (1976).
Lillywhite, P.G. *See* Horridge, McLean, Stange & Lillywhite.
Limbrick, A.R. *See* Abe, Limbrick & Miledi; Finean, Freeman & Limbrick.
Lin, S.H. On the theory of non-radiative transfer of electronic excitation.
Proc. A **335**, 51–66 (1973).
Study of vibronic, spin-orbit and vibronic-spin-orbit couplings of formaldehyde with applications to radiative and non-radiative processes. *Proc.* A **352**, 57–71 (1976).
Linares, H. *See* Whittembury, de Martínez, Linares & Paz-Aliaga.
Lincoln, F.J. *See* Anderson (J.S.), Hutchison & Lincoln.
Lindeman, J. *See* Bovée, Creyghton, Getreuer, Korbee, Lobregt and others.
Lindenmann, J. Cross-priming and cross-inhibition by antibody in the influenza virus—host antigen system (Discussion). *Proc.* B **176**, 419–423 (1971).
Lindley, I.J.D. *See* Stebbing (N.), Lindley & Eaton.
Lindsay, B.G. Nuisance parameters, mixture models, and the efficiency of partial likelihood estimators. *Trans.* A **296**, 639–662 (1980).
Lindsey, K. *See* Franks, Lindsey, Bennett, Speer, Turner & Hunt.
Lindsley, D.L. & Sandler, L. The genetic analysis of meiosis in female *Drosophila melanogaster* (Discussion). *Trans.* B **277**, 295–312 (1977).
Lindvall, F.C. Machines and manpower (Discussion). *Trans.* B **267**, 37–49 (1973).
Ling, C.R. *See* Mallard, Hutchison, Edelstein, Ling, Foster & Johnson.
Ling, E.A. The proboscis apparatus of the nemertine *Lineus ruber. Trans.* B **262**, 1–22 (1971).
Linnett, J.W. *See* Erickson (W.D.) & Linnett.
Linskens, H.F. Incompatibility in *Petunia* (Discussion). *Proc.* B **188**, 299–311 (1975).
Lipman, P.W. *See* Christiansen & Lipman.
Lipman, P.W., Prostka, H.J. & Christiansen, R.L. Cenozoic volcanism and plate-tectonic evolution of the Western United States. I. Early and Middle Cenozoic (Discussion).
Trans. A **271**, 217–248 (1972).
Lippard, S.J. *See* Norry, Truckle, Lippard, Hawkesworth, Weaver & Marriner.
Lipson, H. Albert James Bradley. *Biogr. Mem.* **19**, 117–128 (1973).
Lisher, E.J. Comments on the use of the Korteweg–de Vries equation in the study of anharmonic lattices. *Proc.* A **339**, 119–126 (1974).
Lishman, J.P. *See* Pennington, Haworth, Bonny & Lishman.
Lissmann, H.W. James Gray. *Biogr. Mem.* **24**, 55–70 (1978).
Litt, M. *See* Diamant, Keller, Baer, Litt & Arridge.

Little, E.A., Bullough, R. & Wood, M.H. On the swelling resistance of ferritic steel.
Proc. A **372**, 565–579 (1980).

Little, P. *See* Chamberlain, Heard, Little & Wiffen.

Little, V.I. *See* Cutter, Key & Little; Harrison (R.G.), Key & Little.

Liu, J.T.C. *See* Alper & Liu; Gatski & Liu.

Liu, J.T.C. & Merkine, L. On the interactions between large-scale structure and fine-grained turbulence in a free shear flow. I. The development of temporal interactions in the mean.
Proc. A **352**, 213–247 (1976).

Livett, B.G. *See* Uttenthal, Livett & Hope.

Livett, B.G., Geffen, L.B. & Rush, R.A. Immunochemical methods for demonstrating macromolecules in sympathetic neurons (Discussion). *Trans.* B **261**, 359–361 (1971).

Livett, B.G., Uttenthal, L.O. & Hope, D.B. Localization of neurophysin-II in the hypothalamo-neurohypophysial system of the pig by immunofluorescence histochemistry (Discussion).
Trans. B **261**, 371–378 (1971).

Livingstone, D.R. *See* Bayne, Moore, Widdows, Livingstone & Salkeld.

Llewellyn, D.T., Marriott, J.B., Naylor, D.J. & Thewlis, G. The effects of residual elements on the properties of engineering steels (Discussion). *Trans.* A **295**, 69–85 (1980).

Llewellyn, J.D. *See* Lea (M.J.), Llewellyn, Peck & Dobbs.

Llinás, R. *See* Blight & Llinás.

Lloyd, A.J., Savage, R.J.G., Stride, A.H. & Donovan, D.T. The geology of the Bristol Channel floor.
Trans. A **274**, 595–626 (1973).

Lloyd, B. *See* Ferguson, Hicks, Hoaksey, Rowlands & Lloyd; Howard (J.) & Lloyd.

Lloyd, D.R. *See* DeKock, Lloyd, Hillier & Saunders.

Lloyd, D.W. *See* Buckley (C.P.), Lloyd & Konopasek.

Lloyd, G.O., Kent, B., Saunders, S.R.J. & Lea, C. The action of borates as inhibitors for the high temperature oxidation of alloys [abstract] (Discussion). *Trans.* A **295**, 334–335 (1980).

Lloyd, K.H. *See* Rees (D.), Roper, Lloyd & Low.

Lloyd, R. Problems in determining water quality criteria for freshwater fisheries (Discussion).
Proc. B **180**, 439–449 (1972).

Lloyd, S.A. *See* Jones (A.R.), Lloyd & Weinberg.

Loberg, B., Johansson, T. & Easterling, K.E. Improvement in toughness of welded constructional steels through titanium additions [abstract] (Discussion). *Trans.* A **295**, 306 (1980).

Lobregt, S. *See* Bovée, Creyghton, Getreuer, Korbee, Lobregt and others.

Locher, P.R. Computer simulation of selective excitation in n.m.r. imaging (Discussion).
Trans. B **289**, 537–542 (1980).

Lock, Alison & Collett, T. The three-dimensional world of a toad. *Proc.* B **206**, 481–487 (1980).

Locket, N.A. Retinal anatomy in some scopelarchid deep-sea fishes. *Proc.* B **178**, 161–184 (1971).
  The reflecting structure in the iridescent cornea of the serranid teleost *Nemanthias carberryi*.
  Proc. B **182**, 249–254 (1972).
  Retinal structure in *Latimeria chalumnae*. *Trans.* B **266**, 493–518 (1973).
  The choroidal tapetum lucidum of *Latimeria chalumnae*. *Proc.* B **186**, 281–290 (1974).
  Variation of architecture with size in the multiple-bank retina of a deep-sea teleost, *Chauliodus sloani*. *Proc.* B **208**, 223–242 (1980).
  Review Lecture. Some advances in coelacanth biology. *Proc.* B **208**, 265–307 (1980).

Loew, E.R. Light, and photoreceptor degeneration in the Norway lobster, *Nephrops norvegicus* (L.).
Proc. B **193**, 31–44 (1976).

Löf, G.O.G. Solar space heating with air and liquid systems (Discussion).
Trans. A **295**, 349–359 (1980).

Loftus, K.V. *See* Bleaney, Loftus & Rosenberg.

Logan, A.G., Tenyi, I., Peart, W.S., Breathnach, A.S. & Martin, B.G.H. The effect of lanthanum on renin secretion and renal vasoconstriction. *Proc.* B **195**, 327–342 (1977).

Logan, Jennifer A., Prather, M.J., Wofsy, S.C. & McElroy, M.B. Atmospheric chemistry: response to human influence. *Trans.* A **290**, 187–234 (1978).

Logan, M. *See* Trench, Pool, Logan & Engelland.

Loizou, G. *See* Fenner (T.I.) & Loizou.

Lømo, T., Westgaard, R.H. & Dahl, H.A. Contractile properties of muscle: control by pattern of muscle activity in the rat. *Proc.* B **187**, 99—103 (1974).

Loncarevic, B.D. *See* Aumento, Longcarevic & Ross.

Longuet-Higgins, H.C. Review Lecture. The algorithmic description of natural language.
Proc. B **182**, 255—276 (1972).

The intersection of potential energy surfaces in polyatomic molecules.
Proc. A **344**, 147—156 (1975).

Closing remarks to a Discussion. *Trans.* B **272**, 195—198 (1975).

Review Lecture. The perception of music. *Proc.* B **205**, 307—322 (1979).

See also Power (R.J.D.) & Longuet-Higgins; Sutherland (N.S.) & Longuet-Higgins.

Longuet-Higgins, H.C. & Fisher, M.E. Lars Onsager. *Biogr. Mem.* **24**, 443—471 (1978).

Longuet-Higgins, H.C. & Prazdny, K. The interpretation of a moving retinal image.
Proc. B **208**, 385—397 (1980).

Longuet-Higgins, M.S. A numerical disproof of a conjecture in projective geometry.
Proc. A **323**, 443—454 (1971).

Clifford's chain and its analogues in relation to the higher polytopes. *Proc.* A **330**, 443—466 (1972).

On the form of the highest progressive and standing waves in deep water.
Proc. A **331**, 445—456 (1973).

On the mass, momentum, energy and circulation of a solitary wave. *Proc.* A **337**, 1—13 (1974).

Integral properties of periodic gravity waves of finite amplitude. *Proc.* A **342**, 157—174 (1975).

On the nonlinear transfer of energy in the peak of a gravity-wave spectrum: a simplified model.
Proc. A **347**, 311—328 (1976).

The mean forces exerted by waves on floating or submerged bodies with applications to sand bars and wave power machines. *Proc.* A **352**, 463—480 (1977).

The instabilities of gravity waves of finite amplitude in deep water.

I. Superharmonics. *Proc.* A **360**, 471—488 (1978).

II. Subharmonics. *Proc.* A **360**, 489—505 (1978).

A technique for time-dependent free-surface flows. *Proc.* A **371**, 441—451 (1980).

On the forming of sharp corners at a free surface. *Proc.* A **371**, 453—478 (1980).

See also Byatt-Smith & Longuet-Higgins; Caldwell (D.R.) & Longuet-Higgins.

Longuet-Higgins, M.S. & Cokelet, E.D. The deformation of steep surface waves on water.

I. A numerical method of computation. *Proc.* A **350**, 1—26 (1976).

II. Growth of normal-mode instabilities. *Proc.* A **364**, 1—28 (1978).

Longuet-Higgins, M.S. & Fenton, J.D. On the mass, momentum, energy and circulation of a solitary wave. II. *Proc.* A **340**, 471—493 (1974).

Lopatin, B.G. *See* Sibuet, Ryan, Arthur, Barnes, Blechsmidt and others.

Lord, J.M., McFadden, B.A. & Kornberg, H.L. Changes in microbody-marker enzymes during growth of *Tetrahymena pyriformis* E. *Proc.* B **185**, 19—31 (1974).

Lorentzen, E.F. A world view of ship operations in the 1980s (Discussion).
Trans. A **273**, 23—34 (1972).

Lorient, M.F. *See* Boisseau, Lorient, Born & Michal.

Lorimer, J.P. & Pepper, D.C. The non-stationary polymerization of styrene by $HClO_4$ in CH_2Cl_2.
Proc. A **351**, 551—568 (1976).

Losty, H.H.W. & Lewis, D.L. Homopolar machines (Discussion). *Trans.* A **275**, 69—75 (1973).

Lothe, J. *See* Holm (P.) & Lothe.

Loudon, R. Theory of thermally induced surface fluctuations on simple fluids.
Proc. A **372**, 275—295 (1980).

Lourie, J.A. Biological studies of Yemenite and Kurdish Jews in Israel and other groups in southwest Asia.

III. Physical characteristics of Yemenite and Kurdish Jews in Israel. *Trans.* B **266**, 101—112 (1973).

IV. Spirometric studies of Yemenite and Kurdish Jews in Israel. *Trans.* B **266**, 113—119 (1973).

See also Edholm (O.G.), Humphrey, Lourie, Tredre & Brotherhood; Edholm (O.G.), Samueloff, Mourant, Fox and others.

Loutit, J.F. *See* Barnes (D.W.H.), Loutit & Sansom.

Love, E.R. *See* Bachelard, Daniel, Love & Pratt; Daniel, Love & Pratt.

Loveday, M.S. *See* Dyson (B.F.), Loveday & Rodgers.

Lovell, Sir Bernard. On the stellar origin of low energy cosmic rays (Discussion).
Trans. A **277**, 489—501 (1974).
Patrick Maynard Stuart Blackett (Baron Blackett). *Biogr. Mem.* **21**, 1—115 (1975).
See also Anderson (B.) & Lovell.

Lovelock, D. Vector—tensor field theories and the Einstein—Maxwell field equations.
Proc. A **341**, 285—297 (1974).

Lovelock, J.E. Thermodynamics and the recognition of alien biospheres (Discussion).
Proc. B **189**, 167—181 (1975).

Lovett, M. *See* Rigby, Chia, Clayton & Lovett.

Low, C.H. *See* Rees (D.), Roper, Lloyd & Low.

Lowe, D.A. *See* Mill & Lowe.

Lowe, D.A., Mill, P.J. & Knapp, M.F. The fine structure of the PD proprioceptor of *Cancer pagurus*.
II. The position sensitive cells. *Proc.* B **184**, 199—205 (1973).

Lowe, L.A. *See* Beddell, Clark, Hardy, Lowe, Ubatuba, Vane & Wilkinson.

Lowenstein, O. & Compton, G.J. A comparative study of the responses of isolated first-order semi-circular canal afferents to angular and linear acceleration, analysed in the time and frequency domains. *Proc.* B **202**, 313—338 (1978).

Lowenstein, O. & Saunders, R.D. Otolith-controlled responses from the first-order neurons of the labyrinth of the bullfrog (*Rana catesbeiana*) to changes in linear acceleration.
Proc. B **191**, 475—505 (1975).

Lowery, R.S. Blood parasites of vertebrates on Aldabra (Discussion). *Trans.* B **260**, 577—580 (1971).

Lowes, F.J. The torque on a magnet. *Proc.* A **337**, 555—567 (1974).

Lowndes, D.H., Miller, K.M., Poulsen, R.G. & Springford, M. Studies of the anisotropy of electron-impurity scattering in metals by the use of the de Haas — van Alphen effect: application to gold. *Proc.* A **331**, 497—523 (1973).

Lowry, Katharine H. *See* Turner (Judith), Hewetson, Hibbert, Lowry & Chambers.

Lowy, J., Vibert, P.J., Haselgrove, J.C. & Poulsen, F.R. The structure of the myosin elements in vertebrate smooth muscles (Discussion). *Trans.* B **265**, 191—196 (1973).

Lu, K.T. On the interaction between the 5s-p and 4d-p channels in Sr I.
Proc. A **353**, 431—440 (1977).

Lu, K.T., Tomkins, F.S. & Garton, W.R.S. Configuration interaction effect on diamagnetic phenomena in atoms: strong mixing and Landau regions. *Proc.* A **362**, 421—424 (1978).

Lubbock, R. Why are clownfishes not stung by sea anemones? *Proc.* B **207**, 35—61 (1980).

Lubell, I. & Frischer, Ruth. The current status of male and female sterilization procedures (Discussion). *Proc.* B **195**, 93—114 (1976).

Lucas, C.E. Closing remarks to a Discussion. *Trans.* B **274**, 509—511 (1976).

Lucas, M.L. & Blair, J.A. The magnitude and distribution of the acid microclimate in proximal jejunum and its relation to luminal acidification. *Proc.* B **200**, 27—41 (1978).

Lucas, M.L., Schneider, W., Haberich, F.J. & Blair, J.A. Direct measurement by pH-microelectrode of the pH microclimate in rat proximal jejunum. *Proc.* B **192**, 39—48 (1975).

Luckhurst, G.R. *See* Humphries (R.L.) & Luckhurst.

Luckhurst, G.R. & Romano, S. Computer simulation studies of anisotropic systems. IV. The effect of translational freedom. *Proc.* A **373**, 111—130 (1980).

Luckhurst, G.R. & Zannoni, C. A theory of dielectric relaxation in anisotropic systems.
Proc. A **343**, 389—398 (1975).
Line broadening in the electron resonance spectra of spin probes dissolved in anisotropic media: the effect of nuclear spin quantization. *Proc.* A **353**, 87—102 (1977).

Ludford, G.S.S. *See* Hassard, Chang & Ludford.

Ludmer, Z. *See* Cohen (M.D.), Ludmer, Thomas & Williams.

Luff, R. *See* McCance, Abu Rabiyah, Beer, Edholm, Even-Paz, Luff & Samueloff.

Lukes, T. & Tripathi, R.S. A new approach to the density of eigenvalues and localization in a general

disordered system. *Proc.* A **362**, 79–95 (1978).

Lumley, Ann. *See* Rainey (R.C.), Betts & Lumley.

Lumsden, D.N. *See* Roberts (D.G.), Montadert, Thompson, Auffret and others.

Lund, J.W.G. Eutrophication (Discussion). *Proc.* B **180**, 371–382 (1972).
See also Greenwood (P.H.) & Lund.

Lundberg, J.M. *See* Hökfelt, Lundberg, Schultzberg, Johannson, Skirboll, Anggård and others.

Lundqvist, A. Complex self-incompatibility systems in angiosperms (Discussion).
Proc. B **188**, 235–245 (1975).

Lutschak, W. *See* Berg (E.) & Lutschak.

Lutsky, B.N. *See* Schroepfer, Lutsky, Martin, Huntoon and others.

Lutwak-Mann, Cecilia. *See* Brooks (D.E.), Lutwak-Mann, Mann & Martin.

Lutz, A. *See* Bucourt, Heymès, Lutz, Penasse & Perronet.

Lutz, M.B.K. *See* Lehner & Lutz.

Lwoff, A. Jacques Lucien Monod. *Biogr. Mem.* **23**, 385–412 (1977).

Lycett, G.J. *See* Bolman, Brown, Carrington & Lycett.

Lyle, T.K., Miller, Sir Stephen & Ashton, N.H. William Stewart Duke-Elder.
Biogr. Mem. **26**, 85–105 (1980).

Lyon, Mary F. Review Lecture. Mechanisms and evolutionary origins of variable X-chromosome activity in mammals. *Proc.* B **187**, 243–268 (1974).

Lyon, Mary F. & Glenister, P.H. Reduced reproductive performance in androgen-resistant *Tfm/Tfm* female mice. *Proc.* B **208**, 1–12 (1980).

Lyons, C.G.R. *See* Scaife (W.G.S.) & Lyons.

Lysenko, I.A. *See* Sprenger & Lysenko.

Lysenko, I.A., Orlyansky, A.D. & Portnyagin, Yu.I. A study of the wind régime at an altitude of about 100 km by the meteor-radar method (Discussion). *Trans.* A **271**, 601–610 (1972).

Lyth, R.E. An experiment in community welfare (Discussion). *Proc.* B **199**, 151–160 (1977).

Lythgoe, J.N. The structure and function of iridescent corneas in teleost fishes.
Proc. B **188**, 437–457 (1975).

Lyttle, Vivienne A. *See* Heslop & Lyttle.

Lyzenga, G.A. *See* Bell (G.D.) & Lyzenga.

Maassab, H.F. *See* Murphy (B.R.), Markoff, Chanock, Spring and others.

McAdam, M.E. *See* Garvey, Gold, McAdam & Cooper; Gold & McAdam.

McAllister, G.T. *See* Rohde & McAllister.

McArthur, A.J. Air movement and heat loss from sheep. III. Components of insulation in a controlled environment. *Proc.* B **209**, 219–237 (1980).

McArthur, A.J. & Monteith, J.L. Air movement and heat loss from sheep.
I. Boundary layer insulation of a model sheep, with and without fleece.
Proc. B **209**, 187–208 (1980).
II. Thermal insulation of fleece in wind. *Proc.* B **209**, 209–217 (1980).

McBride, D.A. *See* Bates (B.), Bradley, McBride and others.

McBurney, R.N. *See* Barker (J.L.) & McBurney; Crawford (A.C.) & McBurney.

McCabe, A.M. *See* Colhoun, Dickson, McCabe & Shotton.

McCallum, J. Design procedures and their implementation (Discussion).
Trans. A **273**, 119–128 (1972).

McCance, R.A., El Neil, Hamad, El Din, Nasr, Widdowson, Elsie M., Southgate, D.A.T., Passmore, R., Shirling, D. & Wilkinson, R.T. The response of normal men and women to changes in their environmental temperatures and ways of life. *Trans.* B **259**, 533–561 (1971).

McCance, R.A., Rabiyah, Y.Abu, Beer, G., Edholm, O.G., Even-Paz, Z., Luff, R. & Samueloff, S. Have the Bedouin a special 'desert' physiology? *Proc.* B **185**, 263–271 (1974).

McCance, R.A. & Widdowson, Elsie M. Review Lecture. The determinants of growth and form.
Proc. B **185**, 1–17 (1974).

McCants, M. *See* Wang (R.T.), Nicol, Thurston & McCants.

McCarthy, P.J. Representations of the Bondi—Metzner—Sachs group.
 I. Determination of the representations. *Proc.* A **330**, 517—535 (1972).
 II. Properties and classification of the representations. *Proc.* A **333**, 317—336 (1973).
 The Bondi—Metzner—Sachs group in the nuclear topology. *Proc.* A **343**, 489—523 (1975).
 Lifting of projective representations of the Bondi—Metzner—Sachs group.
   Proc. A **358**, 141—171 (1977).
 Hyperfunctions and asymptotic symmetries. *Proc.* A **358**, 495—498 (1978).
 See also Crampin (M.) & McCarthy.
McCarthy, P.J. & Crampin, M. Representations of the Bondi—Metzner—Sachs group. III. Poincaré spin multiplicities and irreducibility. *Proc.* A **335**, 301—311 (1973).
McCartney, L.N. & Gale, B. A generalized theory of fatigue crack propagation.
   Proc. A **322**, 223—241 (1971).
 Two theoretical models of fatigue crack propagation. *Proc.* A **333**, 337—345 (1973).
Macau-Hercot, D. *See* Monfils & Macau-Hercot.
McCauley, J. *See* Hay, Skehel & McCauley.
McCauley, J.W. *See* Skehel, Waterfield, McCauley, Elder & Wiley.
McCavert, P. & Rudge, M.R.H. On the use of regional wavefunctions in bound-state calculations.
   Proc. A **328**, 429—444 (1972).
McCleave, J.A. *See* Brooks (R.R.), McCleave & Malaisse.
McClintock, P.V.E. *See* Allum, McClintock, Phillips & Bowley; Ellis (T.), McClintock, Bowley & Allum; Phillips (A.) & McClintock.
Maccoll, A. Concluding remarks to a Discussion. *Trans.* A **293**, 167—168 (1979).
McColl, I. & Rendall, M. A critical assessment of the place of endoscopy in gastroenterology (Discussion). *Proc.* B **195**, 251—259 (1977).
McColl, I.R. & Morley, J.G. Damage tolerant fibre reinforced sheet metal composites.
   Trans. A **287**, 17—43 (1977).
McComb, W.D. *See* Edwards (S.F.) & McComb.
McConnell, G. *See* Pearson (C.R.) & McConnell.
McConnell, J.D.M., Hutchison, J.L. & Anderson, J.S. Electron microscopy of the barium ferrite layer structures. *Proc.* A **339**, 1—12 (1974).
McCormick, R.A. Air pollution in the locality of buildings (Discussion).
   Trans. A **269**, 515—526 (1971).
McCosh, R. *See* Kemball & McCosh.
McCrae, J.M. *See* Owen (G.) & McCrae.
McCrea, W.H. Introductory remarks to a Discussion. *Trans.* A **296**, 271—272 (1980).
McCrea, W.H. Introductory remarks to a Discussion. *Trans.* A **296**, 271—272 (1980).
 Appendix to the Biographical Memoir of William Ogilvy Kermack. *Biogr. Mem.* **17**, 420—423 (1971).
 See also Massey & McCrea.
MacCuaig, R.D. Pesticides for locust control (Discussion). *Trans.* B **287**, 447—455 (1979).
McCutcheon, D.B. *See* Clay & McCutcheon.
McDonald, E. *See* Battersby & McDonald.
McDonald, G. *See* Chance, D'Ambrosia, Leigh & McDonald.
McDonald, I.R. *See* Adams (Eveline M.), McDonald & Singer.
McDonald, J.G. *See* Goodman (R.M.), McDonald, Horne & Bancroft.
Macdonald, P.D.M. *See* Barlow (P.W.) & Macdonald.
Macdonald, R. Trace element evidence for mantle heterogeneity beneath the Scottish Midland Valley in the Carboniferous and Permian (Discussion). *Trans.* A **297**, 245—257 (1980).
Macdonald, R.G. *See* Curran, Macdonald, Stone & Thrush.
McDonnell, J.A.M., Ashworth, D.G., Flavill, R.P., Carey, W.C., Bateman, D.C. & Jennison, R.C. The characterization of lunar surface impact erosion and solar wind sputter processes on the lunar surface (Discussion). *Trans.* A **285**, 303—308 (1977).
McDougall, J.G., Williams, B.C., Hyatt, P.J., Bell, J.B.G., Tait, J.F. & Tait, Sylvia A.S. Purification of dispersed rat adrenal cells by column filtration. *Proc.* B **206**, 15—31 (1979).
McDowell, E.M. *See* Peters, Shorthouse, Ward & McDowell.

McEachran, R.P. *See* Rosenthal, McEachran & Cohen.

McElhinny, M.W. *See* Barbetti & McElhinny.

McElhinny, M.W. & Embleton, B.J.J. Precambrian and Early Palaeozoic palaeomagnetism in Australia (Discussion). *Trans.* A 280, 417–431 (1976).

McElroy, M.B. *See* Logan (Jennifer A.), Prather, Wofsy & McElroy.

McElroy, M.B., Wofsy, S.C. & Yung, Y.L. The nitrogen cycle: perturbations due to man and their impact on atmospheric N_2O and O_3. *Trans.* B 277, 159–181 (1977).

McElwain, D.L.S. *See* Hogarth & McElwain.

Macer, R.C.F. The resistance of cereals to yellow rust and its exploitation by plant breeding (Discussion). *Proc.* B 181, 281–301 (1972).

McEwan, J. *See* Chisholm & McEwan.

McEwen, K.A. The Fermi surface of barium. *Proc.* A 322, 509–521 (1971).

McFadden, B.A. *See* Lord, McFadden & Kornberg.

McFarland, D.J. & Sibly, R.M. The behavioural final common path. *Trans.* B 270, 265–293 (1975).

Macfarlane, P.W. & Lawrie, T.D.V. Automated e.c.g. interpretation in the National Health Service (Discussion). *Proc.* B 184, 455–471 (1973).

Macfarlane, R.G. Montague Maizels. *Biogr. Mem.* 23, 345–366 (1977).

McFarlane, I.D. Multiple conducting systems and the control of behaviour in the brain coral *Meandrina meandrites* (L.). *Proc.* B 200, 193–216 (1978).

   See also Shelton (G.A.B.) & McFarlane.

McGarry, M.G. Appropriate technologies for environmental hygiene (Discussion).
   Proc. B 209, 37–46 (1980).

McGeehin, P., Henderson, B. & Benson, P.C. Magnetic resonance studies of $(ReO_4)^{2-}$ in calcium tungstate. *Proc.* A 346, 497–513 (1975).

McGlynn, J.C. *See* Irving (E.) & McGlynn.

McGowan, L.M. *See* Burgis, Darlington, Dunn, Ganf, Gwahaba & McGowan.

McGrath, W.L. The controlled environment (Discussion). *Trans.* A 272, 603–609 (1972).

Macgregor, H.C. Some trends in the evolution of very large chromosomes (Discussion).
   Trans. B 283, 309–318 (1978).

McGregor, V.R. The early Precambrian gneisses of the Godthåb district, West Greenland (Discussion).
   Trans. A 273, 343–358 (1973).

McIlroy, D.K. *See* Mason (D.P.) & McIlroy; Mason (D.P.), McIlroy & Wright.

McIntosh, A.C. *See* Clarke (J.F.) & McIntosh.

MacIntosh, F.C. & Paton, W.D.M. George Lindor Brown. *Biogr. Mem.* 20, 41 (1974).

McIntyre, A. *See* Ruddiman, Sancetta & McIntyre.

McIntyre, J.A. *See* Faulk (W. Page), Yeager, McIntyre & Ueda.

McIntyre, M.E. Towards a Lagrangian-mean description of stratospheric circulations and chemical transports (Discussion). *Trans.* A 296, 129–148 (1980).

Mackay, S. *See* Williams (A.) & Mackay.

McKeag, R.J. *See* Miller (D.S.), Baker, Bowden, Evans, Holt and others.

McKeever, S.W.S. *See* Durrani, Bull & McKeever.

McKeith, C.D. *See* Bates (B.), Bradley, McBride and others; Bates, (B.), Carson, Dufton, McKeith and others; Boksenberg, Kirkham, Michelson, Pettini and others.

McKeith, N.E. *See* Bates (B.), Bradley, McBride and others.

McKellar, A.R.W. & Welsh, H.L. Collision-induced spectra of hydrogen in the first and second overtone regions with applications to planetary atmospheres. *Proc.* A 322, 421–434 (1971).
   Errata. *Proc.* A 324, 506 (1971).

McKellar, J.F. *See* Davies (A.K.), McKellar & Phillips.

MacKenzie, A.P. Non-equilibrium freezing behaviour of aqueous systems (Discussion).
   Trans. B 278, 167–189 (1977).

McKenzie, D.R. *See* McPhedran & McKenzie; Perrins, McKenzie & McPhedran.

McKenzie, D.R., McPhedran, R.C. & Derrick, G.H. The conductivity of lattices of spheres. II. The body centred and face centred cubic lattices. *Proc.* A 362, 211–232 (1978).

McKenzie, G.H. *See* Johnson (P.) & McKenzie.

McKenzie, J.F. *See* Mekki & McKenzie.

McKenzie, K.G. Entomostraca of Aldabra, with special reference to the genus *Heterocypris* (Crustacea, Ostracoda). *Trans.* B **260**, 257—297 (1971).
Note on evidence of human interference on South Island, Aldabra (Discussion).
Trans. B **260**, 629—630 (1971).

McKenzie, P. *See* Hughes (D.E.) & McKenzie.

McKibbin, C.S. *See* Bates (D.R.) & McKibbin.

MacKie, E.W. Archaeological tests on supposed prehistoric astronomical sites in Scotland (Discussion) (with an appendix by J.S. Bibby). *Trans.* A **276**, 169—194 (1974).

Mackie, G.O. & Bone, Q. Luminescence and associated effector activity in *Pyrosoma* (Tunicata: Pyrosomida). *Proc.* B **202**, 483—495 (1978).

Mackie, G.O., Paul, D.H., Singla, G.M., Sleigh, M.A. & Williams, D.E. Branchial innervation and ciliary control in the ascidian *Corella*. *Proc.* B **187**, 1—35 (1974).

MacKinlay, R.R. *See* Franklin (R.N.), MacKinlay, Edgley & Wall.

MacKinnon, D.I. The formation of muscle scars in articulate brachiopods.
Trans. B **280**, 1—27 (1977).

Mackintosh, Anne H. *See* Bond & Mackintosh.

MacKinven, R. *See* Gray (P.), Jones & MacKinven.

Mackley, M.R. *See* Berry (M.V.) & Mackley; Knott & Mackley.

Mackley, M.R. & Keller, A. Flow induced polymer chain extension and its relation to fibrous crystallization. *Trans.* A **278**, 29—66 (1975).

McKnight, A.T., Hughes, J. & Kosterlitz, H.W. Synthesis of enkephalins by guinea-pig striatum *in vitro*.
Proc. B **205**, 199—207 (1979).

McLachlan, D.S. *See* Bibby, Nabarro, McLachlan & Stephen.

McLafferty, F.W. Collisional activation mass spectra (Discussion). *Trans.* A **293**, 93—102 (1979).

McLaren, Anne. *See* Grüneberg, Cattanach, McLaren, Wolfe & Bowman; Grüneberg & McLaren.

McLauchlan, K.A. *See* Chapman (G.E.), Danyluk & McLauchlan.

McLaughlin, A.C. *See* Dwek, Jones, Marsh, McLaughlin, Press and others.

McLaughlin, E. & Pittman, J.F.T. Determination of the thermal conductivity of toluene — a proposed data standard — from 180 to 400 K under saturation pressure by the transient hot-wire method.
I. The theory of the technique. *Trans.* A **270**, 557—578 (1971).
II. New measurements and a discussion of existing data. *Trans.* A **270**, 579—602 (1971).

McLaughlin, R. *See* Laws (N.) & McLaughlin.

McLaurin, G.E. *See* Kell, McLaurin & Whalley.

McLean, A.E.M. Hazards from chemicals: scientific questions and conflicts of interest (Discussion).
Proc. B **205**, 179—197 (1979).

McLean, D. Concluding remarks to a Discussion. *Trans.* A **295**, 337—341 (1980).

Maclean, I.G. *See* Boylett & Maclean.

McLean, M. Friction stress and recovery during high-temperature creep: interpretation of creep transients following a stress reduction. *Proc.* A **371**, 279—294 (1980).
The contribution of friction stress to the creep behaviour of the nickel-base *in situ* composite, γ-γ'-Cr_3C_2. *Proc.* A **373**, 93—109 (1980).
See also Bullock (E.), Lea & McLean; Horridge, McLean, Stange & Lillywhite.

McLean, Miriam. *See* Horridge & McLean.

McLean, R.F. *See* Polach, McLean, Caldwell & Thom; Scoffin & McLean; Stoddart (D.R.), McLean & Hopley; Stoddart (D.R.), McLean, Scoffin & Gibbs; Stoddart (D.R.), McLean, Scoffin, Thom & Hopley; Scoffin, Stoddart, McLean & Flood.

McLean, R.F. & Stoddart, D.R. Reef island sediments of the northern Great Barrier Reef (Discussion). *Trans.* A **291**, 101—117 (1978).

McLean, R.F., Stoddart, D.R., Hopley, D. & Polach, H.A. Sea level change in the Holocene on the northern Great Barrier Reef (Discussion). *Trans.* A **291**, 167—186 (1978).

McLenaghan, R.G. & Leroy, J. Complex recurrent space-times. *Proc.* A **327**, 229—249 (1972).

McLeod, J.B. The existence of axially symmetric flow above a rotating disk.
Proc. A **324**, 391—414 (1971).

MacLeod, N.K. *See* Dyer (R.G.), MacLeod & Ellendorff.

McLeod, R.J.Y. The Steiner surface revisited. *Proc.* A **369**, 157–174 (1979).

McLintock, D.N. *See* Ashkenazi, McLintock & Sykes.

McMahan, U.J. *See* Muller (K.J.) & McMahan; Peper & McMahan.

McMahan, U.J. & Kuffler, S.W. Visual identification of synaptic boutons on living ganglion cells and of varicosities in postganglionic axons in the heart of the frog. *Proc.* B **177**, 485–508 (1971).

McMahan, U.J., Spitzer, N.C. & Peper, K. Visual identification of nerve terminals in living isolated skeletal muscle. *Proc.* B **181**, 421–430 (1972).

McMahon, C.J., Jr, & Yu, J. Compositional factors that enhance or retard temper embrittlement of alloy steels [abstract] (Discussion). *Trans.* A **295**, 299–300 (1980).

Macmillan, D.L. A physiological analysis of walking in the American lobster (*Homarus americanus*). *Trans.* B **270**, 1–59 (1975).

See also Laverack, Macmillan & Neil; Neil, Macmillan, Robertson & Laverack.

Macmillan, D.L., Neil, D.M. & Laverack, M.S. A quantitative analysis of exopodite beating in the larvae of the lobster *Homarus gammarus* (L.). *Trans.* B **274**, 69–85 (1976).

Macmillan, N.H. & Kelly, A. The mechanical properties of perfect crystals.

I. The ideal strength. *Proc.* A **330**, 291–308 (1972).

II. The stability and mode of fracture of highly stressed ideal crystals. *Proc.* A **330**, 309–317 (1972).

McMillan, Nora F. *See* Mitchell (G.F.), Catt, Weir, McMillan, Margerel & Whatley.

McMorris, T.C. Antheridiol and the oogoniols, steroid hormones which control sexual reproduction in *Achlya* (Discussion). *Trans.* B **284**, 459–470 (1978).

Macnae, W. Mangroves on Aldabra (Discussion). *Trans.* B **260**, 237–247 (1971).

Macnair, M.R. *See* Murray (N.D.), Bishop & Macnair.

McNamara, J.M. Instability of black hole inner horizons. *Proc.* A **358**, 499–517 (1978).

Behaviour of scalar perturbations of a Reissner–Nordström black hole inside the event horizon. *Proc.* A **364**, 121–134 (1978).

McNaughton, P.A. *See* Brown (Hilary F.), McNaughton, Noble & Noble.

McPeters, R. *See* Krueger, Guenther, Fleig, Heath, Hilsenrath and others.

McPhedran, R.C. *See* McKenzie (D.R.), McPhedran & Derrick; Perrins, McKenzie & McPhedran.

McPhedran, R.C. & McKenzie, D.R. The conductivity of lattices of spheres. I. The simple cubic lattice. *Proc.* A **359**, 45–63 (1978).

Macpherson, A.K. Thermal stability in a rotating spherical fluid shell with non-uniform heating. *Proc.* A **353**, 349–362 (1977).

On the solution of the Landau equation. *Proc.* A **366**, 47–61 (1979).

The effect of non-Markovian terms on the Landau equation. *Proc.* A **371**, 381–392 (1980).

MacPherson, G.G. Development of megakaryocytes in bone marrow of the rat: an analysis by electron microscopy and high resolution autoradiography. *Proc.* B **177**, 265–274 (1971).

Macpherson, I.A. Reversion in cells transformed by tumour viruses (Discussion). *Proc.* B **177**, 41–48 (1971).

See also Robbins (P.W.) & Macpherson.

McPherson, R. *See* Presser & McPherson.

MacQueen, R.M. Coronal transients: a summary (Discussion). *Trans.* A **297**, 605–620 (1980).

MacQueen, R.M., Gosling, J.T., Hildner, E., Munro, R.H., Poland, A.I. & Ross, C.L. Initial results from the High Altitude Observatory white light coronagraph on Skylab – a progress report (Discussion). *Trans.* A **281**, 405–414 (1976).

MacRobbie, E.A.C. Ion transport in plant cells (Discussion). *Trans.* B **262**, 333–342 (1971).

McWeeny, R. *See* Moores (W.H.) & McWeeny.

McWhirter, R.W.P. The energy budget in and out of coronal holes (Discussion). *Trans.* A **297**, 531–540 (1980).

McWhirter, R.W.P. & Wilson, R. The energy and pressure balance in the corona (Discussion). *Trans.* A **281**, 331–337 (1976).

McWilliam, I.G. *See* Bolton (H.C.), Grant, McWilliam, Nicholson & Swingler; Bolton (H.C.) & McWilliam.

Madden, P.A. Light scattering studies of the dynamics and structure of liquids (Discussion). *Trans.* A **293**, 419–428 (1979).

Maddock, Sir Ieuan. The Seventh Royal Society Technology Lecture. Science, technology, and industry. *Proc.* A **345**, 295–326 (1975).

Maddrell, S.H.P. Fluid secretion by the Malpighian tubules of insects (Discussion).
Trans. B **262**, 197–208 (1971).

Madin, A.B. *See* Bailey (P.), Madin & Preston.

Maetz, J. Fish gills: mechanisms of salt transfer in fresh water and sea water (Discussion).
Trans. B **262**, 209–249 (1971).

Magat, E.E. Fibres from extended chain aromatic polyamides (Discussion).
Trans. A **294**, 463–472 (1980).

Magnin, P. *See* Béné, Borcard, Hiltbrand & Magnin.

Magnon, Anne. *See* Ashtekar & Magnon.

Maguire, Marjorie P. Homologous chromosome pairing (Discussion). *Trans.* B **277**, 245–258 (1977).

Mahadevan, P. Merging beams study of ionization of positive ions by electrons.
Proc. A **327**, 317–328 (1972).

Mahalingam, S. *See* Bishop (R.E.D.) & Mahalingam.

Mahmud, Khalida. *See* Chapman (J.A.), Grant, Taylor, Mahmud, Sardar-ul-Mulk & Shahid.

Mahoney, R.T. *See* Perkin (G.W.), Duncan, Mahoney & Smith.

Mahony, J.J. Validity of averaging methods for certain systems with periodic solutions.
Proc. A **330**, 349–371 (1972).
See also Barnard (B.J.S.), Mahony & Pritchard; Benjamin (T.B.), Bona & Mahony.

Mahy, B.W.J., Barrett, I., Briedis, D.J., Brownson, J.M. & Wolstenholme, A.J. Influence of the host cell on influenza virus replication (Discussion). *Trans.* B **288**, 349–357 (1980).

Mai, Y.W. *See* Gurney (C.), Mai & Owen.

Main, P. *See* Bernstein (J.), Regev, Herbstein, Main and others.

Maino, V.C. *See* Crumpton, Allen, Auger, Green & Maino.

Mair, W.A. & Maull, D.J. Aerodynamic behaviour of bodies in the wakes of other bodies (Discussion).
Trans. A **269**, 425–437 (1971).

Majumdar, S.R. *See* Chee-Seng, Majumdar & Westbrook.

Majumdar, S.R. & Michael, D.H. The equilibrium and stability of two dimensional pendent drops.
Proc. A **351**, 89–115 (1976).

Mäkelä, P.H. *See* Stocker & Mäkelä.

Malaisse, F. *See* Brooks (R.R.), McCleave & Malaisse.

Malaviya, V. *See* Bates (D.R.), Malaviya & Young.

Malchow, D. *See* Gerisch, Hülser, Malchow & Wick.

Malcolm, D.B. *See* Sommerville, Malcolm & Callan.

Maldonado, A. *See* Sibuet, Ryan, Arthur, Barnes, Blechsmidt and others.

Malhotra, M.S. *See* Cotes, Dabbs, Hall, Lakhera, Saunders & Malhotra.

Malhotra, S.K. & Tewari, J.P. Molecular alterations in the plasma membrane of sporangiospores of *Phycomyces* related to germination. *Proc.* B **184**, 207–216 (1973).

Malhotra, S.K. & Tipnis, U. Alterations in the structure of sarcolemma in a denervated skeletal muscle. *Proc.* B **203**, 59–68 (1978).

Malik, Renuka. *See* Verma, Malik & Dhir.

Malin, S.R.C. Worldwide distribution of geomagnetic tides. *Trans.* A **274**, 551–594 (1973).
See also Chapman (S.), Gupta & Malin.

Mallaby, R. *See* Cornforth, Clifford, Mallaby & Phillips.

Mallard, J., Hutchison, J.M.S., Edelstein, W.A., Ling, C.R., Foster, M.A. & Johnson, G. *In vivo* n.m.r. imaging in medicine: the Aberdeen approach, both physical and biological (Discussion).
Trans. B **289**, 519–533 (1980).

Mallick, D.I.J. Development of the New Hebrides archipelago (Discussion).
Trans. B **272**, 277–285 (1975).

Mallinson, J.R. & Landsberg, P.T. Meteorological effects on solar cells. *Proc.* A **355**, 115–130 (1977).

Mallinson, L.G., Jefferson, D.A., Thomas, J.M. & Hutchison, J.L. The internal structure of nephrite: experimental and computational evidence for the coexistence of multiple-chain silicates within an amphibole host. *Trans.* A **295**, 537–552 (1980).

Mallion, R.B. Some graph-theoretical aspects of simple ring current calculations on conjugated systems. *Proc.* A **341**, 429–449 (1975).

Malmejac, Y. The proposed French metallurgy programme for *Spacelab* (Discussion).
Proc. A **361**, 165–174 (1978).

Maloiy, G.M.O. The water metabolism of a small East African antelope: the dik-dik.
Proc. B **184**, 167–178 (1973).

Malpas, J. Magma generation in the upper mantle, field evidence from ophiolite suites, and application to the generation of oceanic lithosphere (Discussion). *Trans.* A **288**, 527–546 (1978).

von der Malsburg, C. *See* Willshaw & von der Malsburg.

Malvern, A.R., Pinder, A.C., Stacey, D.N. & Thompson, R.C. Self-broadening in singlet spectral lines of helium. *Proc.* A **371**, 259–278 (1980).

Mandel, P. & Chryssostomidis, C. A design methodology for ships and other complex systems (Discussion). *Trans.* A **273**, 85–98 (1972).

Mandelstam, J. The Leeuwenhoek Lecture, 1975. Bacterial sporulation: a problem in the biochemistry and genetics of a primitive developmental system. *Proc.* B **193**, 89–106 (1976).

Mandelstam, S. New results of X-ray flare studies (Discussion). *Trans.* A **270**, 135–142 (1971).

Mangion, M.M. *See* DeLucia, Kelly, Mangion, Moews & Knox.

Mani, G.S. A theoretical study of morph ratio clines with special reference to melanism in moths.
Proc. B **210**, 299–316 (1980).

Mani, R. Sound propagation in parallel sheared flows in ducts: the mode estimation problem.
Proc. A **371**, 393–412 (1980).

Mann, A. *See* Atalay, Mann & Peierls.

Mann, D.W. Scientific aspects of well logging in deep ocean drill sites on Leg 48 – Biscay and Rockall areas (Discussion). *Trans.* A **294**, 105–119 (1980).

Mann, T. Review Lecture. Relevance of physiological and biochemical research to problems in animal fertility. *Proc.* B **193**, 1–15 (1976).
See also Brooks (D.E.), Gaughwin & Mann; Brooks (D.E.), Lutwak-Mann, Mann & Martin; Brooks (D.E.), Mann & Martin; Jones (R.) & Mann; Jones (R.). Mann & Sherins; Martin (A.W.), Jones & Mann; Tash & Mann.

Manning, G. *See* Fender, Hobbis & Manning.

Mansfield, E.H. Large-deflexion torsion and flexure of initially curved strips.
Proc. A **334**, 279–298 (1973).
Analysis of wrinkled membranes with anisotropic and nonlinear elastic properties.
Proc. A **353**, 475–498 (1977).
On the deflexion of an anisotropic cantilever plate with variable rigidity.
Proc. A **366**, 491–515 (1979).

Mansfield, E.H. & Young, D.H. Stephen Prokofievitch Timoshenko.
Biogr. Mem. **19**, 679–694 (1973).

Mansfield, M.W.D. An improved analysis of singly ionized potassium (K II) by means of Hartree-Fock calculations. *Proc.* A **341**, 277–283 (1974).
The K I absorption spectrum in the vacuum ultraviolet: 3p-subshell excitation.
Proc. A **346**, 539–553 (1975).
The K I absorption spectrum in the vacuum ultraviolet: 2p-subshell excitation.
Proc. A **346**, 555–563 (1975).
The Ca I absorption spectrum in the extreme ultraviolet: excitation of the 2p subshell.
Proc. A **348**, 143–151 (1976).
Excitation of the 3p-subshell in chromium vapour. *Proc.* A **358**, 253–265 (1977).
The simultaneous excitation of two electrons in atomic cadmium. *Proc.* A **362**, 129–144 (1978).
A new interpretation of the Rb I 4p subshell excitation spectrum between 15 and 19 eV.
Proc. A **364**, 135–144 (1978).
See also Connerade, Baig, Mansfield & Radtke; Connerade, Drerup & Mansfield; Connerade, Garton, Mansfield & Martin; Connerade & Mansfield; Connerade, Mansfield & Martin; Connerade, Mansfield, Newsom, Tracy, Baig & Thimm; Connerade, Mansfield & Thimm.

Mansfield, M.W.D. & Connerade, J.P. The absorption spectrum of Sr I between 40 and 95Å.
Proc. A **342**, 421–430 (1975).

Potential barrier effect and discrete structure between 40 and 120 Å in the Rb I absorption spectrum. *Proc.* A **344**, 303–309 (1975).

Observation of 4d→f transitions in europium vapour. *Proc.* A **352**, 125–139 (1976).

On the simultaneous excitation of two electrons in neutral atomic zinc. *Proc.* A **359**, 389–410 (1978).

Mansfield, M.W.D. & Newsom, G.H. The Ca I absorption spectrum in the vacuum ultraviolet: excitation of the 3p-subshell. *Proc.* A **357**, 77–102 (1977).

Mansfield, M.W.D. & Ottley, T.W. The identification of low energy K and Ca$^+$ autoionizing levels observed in electron impact experiments. *Proc.* A **365**, 413–424 (1979).

Mansfield, P., Morris, P.G., Ordidge, R.J., Pykett, I.L., Bangert, V. & Coupland, R.E. Human whole body imaging and detection of breast tumours by n.m.r. (Discussion). *Trans.* B **289**, 503–510 (1980).

Mansfield, T.A. Chemical control of stomatal movements (Discussion). *Trans.* B **273**, 541–550 (1976).

Mansfield, T.A., Wellburn, A.R. & Moreira, T.J.S. The role of abscisic acid and farnesol in the alleviation of water stress (Discussion). *Trans.* B **284**, 471–482 (1978).

Manson, A.J. *See* Runcorn, Collinson, O'Reilly, Stephenson, Battey, and others.

Manton, Irene & Leadbeater, B.S.C. Some critical qualitative details of lorica construction in the type species of *Calliacantha* Leadbeater (Choanoflagellata). *Proc.* B **203**, 49–57 (1978).

Manton, Irene & Oates, K. Further observations on *Calliacantha* Leadbeater (Choanoflagellata), with special reference to *C. simplex* sp. nov. from many parts of the world. *Proc.* B **204**, 287–300 (1979).

Manton, Irene, Sutherland, Joan & Leadbeater, B.S.C. Four new species of choanoflagellates from Arctic Canada. *Proc.* B **189**, 15–27 (1975).

Manton, Irene, Sutherland, Joan & Oates, K. Arctic coccolithophorids: two species of *Turrisphaera* gen. nov. from West Greenland, Alaska and the Northwest Passage. *Proc.* B **194**, 179–194 (1976).

A reinvestigation of collared flagellates in the genus *Bicosta* Leadbeater with special reference to correlations with climate. *Trans.* B **290**, 431–447 (1980).

Manton, I., Sutherland, J., & Oates, K. Arctic coccolithophorids: *Wigwamma artica* gen. et sp.nov. from Greenland and arctic Canada. *W. annulifera* sp.nov. from South Africa and S. Alaska and Calciarcus alaskensis gen. et sp.nov. from S. Alaska. *Proc.* B **197**, 145–168 (1977).

Manton, Sidnie M. *See* Anderson (D.T.) & Manton.

Manzoni, G. & Marchesini, C. A 60 m laser strainmeter (Discussion). *Trans.* A **274**, 285 (1973).

Maples, Joanne. *See* Blest & Maples.

Maranci, A. *See* Doyle (M.J.), Maranci, Orowan & Stork.

Marble, F.E. *See* Cumpsty & Marble.

March, N.H. *See* Beattie, Stoddart & March; Parrinello, Tosi & March.

March, N.H. & Tosi, M.P. Plasmons propagating in periodic lattices and their dispersion. *Proc.* A **330**, 373–387 (1972).

Marchesini, C. *See* Manzoni & Marchesini.

Marchiafava, P.L. & Weiler, R. Intracellular analysis and structural correlates of the organization of inputs to ganglion cells in the retina of the turtle. *Proc.* B **208**, 103–113 (1980).

Marder, L. Gravitational waves in general relativity. XII. Correspondence between toroidal and cylindrical waves. *Proc.* A **327**, 123–130 (1972).

Margerel, J.P. *See* Mitchell (G.F.), Catt, Weir, McMillan, Margerel & Whatley.

Margulis, Lynn, Chase, D. & To, L.P. Possible evolutionary significance of spirochaetes. (Discussion). *Proc.* B **204**, 189–198 (1979).

Mariani, J., Crepel, F., Mikoshiba, K., Changeux, J.P. & Sotelo, C. Anatomical, physiological and biochemical studies of the cerebellum from *reeler* mutant mouse. *Trans.* B **281**, 1–28 (1977).

Mark, R.F. *See* Cass & Mark; Genat & Mark; Johnston, Schrameck & Mark; Rogers (L.J.), Oettinger, Szer & Mark; Watts (M.E.) & Mark.

Mark, R.F. & Watts, M.E. Drug inhibition of memory formation in chickens. I. Long-term memory. *Proc.* B **178**, 439–454 (1971).

Markham, R., Horne, R.W. & Hicks, R. Marian. Introductory remarks to a Discussion. *Trans.* B **268**, 3 (1974).

Markoff, L.J. *See* Murphy (B.R.), Markoff, Chanock, Spring and others.
Markushin, S.G. *See* Ghendon & Markushin.
Marlow, Jean. *See* Goldsmith & Marlow.
Marmion, B.P. Prospects for new viral vaccines (Discussion). *Trans.* B **290**, 395–407 (1980).
Maroudas, Alice. Glycosaminoglycan turn-over in articular cartilage (Discussion).
  Trans. B **271**, 293–313 (1975).
Marr, D. Simple memory: a theory for archicortex. *Trans.* B **262**, 23–81 (1971).
  Early processing of visual information. *Trans.* B **275**, 483–519 (1976).
  Analysis of occluding contour. *Proc.* B **197**, 441–475 (1977).
  Visual information processing: the structure and creation of visual representations (Discussion).
    Trans. B **290**, 199–218 (1980).
Marr, D. & Hildreth, E. Theory of edge detection. *Proc.* B **207**, 187–217 (1980).
Marr, D. & Nishihara, H.K. Representation and recognition of the spatial organization of three-dimensional shapes. *Proc.* B **200**, 269–294 (1978).
Marr, D. & Poggio, T. A computational theory of human stereo vision. *Proc.* B **204**, 301–328 (1979).
Marr, G.V. *See* West (J.B.) & Marr; Woodruff (Pamela R.) & Marr.
Marriner, G.F. *See* Morrison (M. Ann), Thompson, Gibson & Marriner; Norry, Truckle, Lippard, Hawkesworth, Weaver & Marriner.
Marriott, J.B. *See* Llewellyn (D.T.), Marriott, Naylor & Thewlis.
Marsh, B.D. On the cooling of ascending andesitic magma (Discussion).
  Trans. A **288**, 611–625 (1978).
Marsh, D. *See* Baker (J.M.) & Marsh; Dwek, Jones, Marsh, McLaughlin, Press and others.
Marshall, A.G. *See* Medway & Marshall.
Marshall, G.P. *See* Williams (J.G.) & Marshall.
Marshall, Jean M. & Kroeger, E.A. Adrenergic influences on uterine smooth muscle (Discussion).
  Trans. B **265**, 135–148 (1973).
Marshall, N.B. Francis Charles Fraser. *Biogr. Mem.* **25**, 287–317 (1979).
Marshall, R.D. *See* Dunstan, Grant, Marshall & Neuberger.
Marshall, R.M. *See* Bull (K.R.), Marshall & Purnell; Lexton, Marshall & Purnell.
Marshall, R.M., Purnell, J.H. & Storey, P.D. The mechanism of the pyrolysis of 2, 2, 3, 3-tetramethylbutane. *Proc.* A **363**, 503–523 (1978).
Marthy, H.-J. *See* Froesch & Marthy.
Martin, A. Aspects of the fundamental theory of proton–proton scattering (Discussion).
  Proc. A **335**, 503–507 (1973).
Martin, A.W. *See* Brooks (D.E.), Lutwak-Mann, Mann & Martin; Brooks (D.E.), Mann & Martin.
Martin, A.W., Jones, R. & Mann, T. D(–)Lactic acid formation and D(–)lactate dehydrogenase in octopus spermatozoa. *Proc.* B **193**, 235–243 (1976).
Martin, B.G.H. *See* Logan (A.G.), Tenyi, Peart, Breathnach & Martin.
Martin, B.W. *See* Fargie & Martin; Hasoon & Martin.
Martin, B.W. & Payne, A. Tangential flow development for laminar axial flow in an annulus with a rotating inner cylinder. *Proc.* A **328**, 123–141 (1972).
Martin, H.H., Tonn-Ehlers, Margrit & Schilf, W. Cooperation of benzyl penicillin and cefotoxin in bacterial growth inhibition [synopsis] (Discussion). *Trans.* B **289**, 365–367 (1980).
Martin, J.A. *See* Schroepfer, Lutsky, Martin, Huntoon and others.
Martin, M.A.P. *See* Connerade, Garton, Mansfield & Martin; Connerade, Mansfield & Martin; Connerade & Martin.
Martin, R. & Barlow, J.J. Changes in glial cells of the octopus brain after 6-hydroxydopamine administration. *Proc.* B **196**, 431–441 (1977).
Martin, R. & Miledi, R. A structural study of the squid synapse after intraaxonal injection of calcium.
  Proc. B **201**, 317–333 (1978).
Martin, R.D. Review Lecture. Adaptive radiation and behaviour of the Malagasy lemurs.
  Trans. B **264**, 295–352 (1972).
Martin, R.L. *See* Gregson, Martin & Mitra.
de Martínez, Clara Verde. *See* Whittembury, de Martínez, Linares & Paz-Aliaga.

Martinussen-Runde, O.J., Melrose, M.P. & Derrick, P.J. Molecular orbital theory in mass spectrometry: *ab initio* calculations on 2-methylpropene and the 2-methylpropene radical-cation.
Proc. A **350**, 553–564 (1976).

Martodipoero, Soebagio. *See* King (M.H.), King & Martodipoera.

Maruno, T. *See* Ise, Maruno & Okubo; Okubo, Maruno & Ise.

Marvin, D.A. & Wachtel, E.J. Structure and assembly of filamentous bacterial viruses (Discussion).
Trans. B **276**, 81–98 (1976).

Marwick, A.D. *See* Harries (D.R.) & Marwick.

Masironi, R. Geochemistry and cardiovascular diseases (Discussion). *Trans.* B **288**, 193–203 (1979).

Maskell, E.C. *See* Owen (P.R.) & Maskell.

Maslin, N.M. Theory of energy flux and polarization changes of a radio wave with two magnetoionic components undergoing self demodulation in the ionosphere. *Proc.* A **341**, 361–381 (1974).
Theory of the modifications imposed on the ionospheric plasma by a powerful radio wave reflected in the D or E region. *Proc.* A **343**, 109–131 (1975).
Theory of heating in the lower ionsphere by obliquely incident waves from a powerful point transmitter. *Proc.* A **346**, 37–57 (1975).
The effect of a time varying collison frequency on a radio wave obliquely incident on the lower ionosphere. *Proc.* A **348**, 245–263 (1976).
Theory of the modulation imposed on an obliquely incident radio wave reflected in a disturbed region of the lower ionosphere. *Proc.* A **349**, 555–570 (1976).
Estimating ionospheric cross-modulation. *Proc.* A **351**, 277–293 (1976).

Mason, B.J. The Bakerian Lecture, 1971. The physics of the thunderstorm.
Proc. A **327**, 433–466 (1972).
Review Lecture. Recent advances in the numerical prediction of weather and climate.
Proc. A **363**, 297–333 (1978).

Mason, D.P. & McIlroy, D.K. A perturbation solution to the problem of Wien dissociation in weak electrolytes. *Proc.* A **359**, 303–317 (1978).

Mason, D.P., McIlroy, D.K. & Wright, C.J. Ion trajectories in weak electrolytes.
Proc. A **371**, 413–427 (1980).

Mason, P.J. Baroclinic waves in a container with sloping end walls. *Trans.* A **278**, 397–445 (1975).

Mason, R. The Limpopo mobile belt – southern Africa (Discussion). *Trans.* A **273**, 463–485 (1973).
See also Brown (P. Jane), Forsyth & Mason; Clarke (T.A.), Mason & Tescari; Elder (M.), Hitchcock, Mason & Shipley; Gay (I.D.), Textor, Mason & Iwasawa; Mason (Sir Ronald); Textor, Gay & Mason; Varghese & Mason.

Mason, R. & Textor, M. The chemisorption of simple- and halogeno-substituted unsaturated hydrocarbons on the α-Fe(111) single crystal surface: photoelectron spectroscopic studies.
Proc. A **356**, 47–60 (1977).

Mason, R., Textor, M., Iwasawa, Y. & Gay, I.D. Photoelectron spectroscopic studies of the chemisorption of ethylene and halo-substituted alkenes on the (100) and (111) crystal surfaces of platinum and a general model for the dissociation of unsaturated molecules at surfaces.
Proc. A **354**, 171–196 (1977).

Mason, S.A. *See* Bentley (G.A.) & Mason.

Mason, S.G. *See* Torza, Cox & Mason.

Mason, Sir Ronald. *See* Mason (R.).

Massalski, T.B. *See* Mizutani & Massalski.

Massalski, T.B. & Mizutani, U. Electronic band structure of h.c.p. electron phases based on the noble metals. *Proc.* A **351**, 423–436 (1976).

Massalski, T.B., Mizutani, U. & Noguchi, S. Low temperature specific heats of zinc alloyed with silver. *Proc.* A **343**, 363–374 (1975).

Massebeuf, M. *See* Spizzichino & Massebeuf.

Massey, Sir Harrie. David Forbes Martyn. *Biogr. Mem.* **17**, 497–510 (1971).
Atomic energy and the development of large teams and organizations (Discussion).
Proc. A **342**, 491–497 (1975).

Massey, Sir Harrie & Feather, N. James Chadwick. *Biogr. Mem.* **22**, 11–70 (1976).

Massey, Sir Harrie & McCrea, W.H. Preface to a symposium. *Proc.* A **336**, 3–4 (1974).
Massey, Sir Harrie & Thompson, Sir Harold. David Christie Martin. *Biogr. Mem.* **24**, 391–407 (1978).
Mather, K. *See* Caligari & Mather; Hanks & Mather; Mather (Sir Kenneth).
Mather, Sir Kenneth. *See* Caligari & Mather; Mather (K.).
Mathew, P. *See* Hastings, Mathew & Oxley.
Matlin, S.A. *See* Jackson (A.H.), Sancovich, Ferramola, Evans and others.
Matsumiya, T. *See* Yamaguchi, Kobayashi, Matsumiya & Hayami.
Matthews, D.H. Altered basalts from Swallow Bank, an abyssal hill in the NE Atlantic, and from a nearby seamount (Discussion). *Trans.* A **268**, 551–571 (1971).
    An account of the meeting for informal discussion held on Friday 14 November 1969 (Discussion). *Trans.* A **268**, 733–736 (1971).
Matthews, J.D. Prospects for improvement by site amelioration, breeding, and protection (Discussion). *Trans.* B **271**, 115–138 (1975).
Matthews, Margaret R. An ultrastructural study of axonal changes following constriction of post-ganglionic branches of the superior cervical ganglion in the rat. *Trans.* B **264**, 479–505 (1973).
Matthews, Margaret R. & Raisman, G. A light and electron microscopic study of the cellular response to axonal injury in the superior cervical ganglion of the rat. *Proc.* B **181**, 43–79 (1972).
Matthews, P.T. *See* Feldman (G.) & Matthews.
Matus, A.I. & Taff-Jones, D.H. Morphology and molecular composition of isolated postsynaptic junctional structures. *Proc.* B **203**, 135–151 (1978).
Matzner, R.A. & Nutku, Y. On the method of virtual quanta and gravitational radiation. *Proc.* A **336**, 285–305 (1974).
Matzner, R.A. & Zamorano, N.A. Imaginary-frequency interior modes of black holes. *Proc.* A **373**, 223–233 (1980).
Maugin, G.A. Relation between wavespeeds in the crust of dense magnetic stars. *Proc.* A **364**, 537–552 (1978).
Maull, D.J. *See* Mair & Maull.
Maurer, P. *See* Geiss, Eberhardt, Grögler, Guggisberg, Maurer & Stettler.
Maurette, M. *See* Dran, Duraud, Klossa, Langevin & Maurette.
Mauzerall, D. Chlorophyll and photosynthesis (Discussion). *Trans.* B **273**, 287–294 (1976).
Maxwell, J.R. *See* Eglinton, HajIbrahim, Maxwell, Quirke, Shaw and others.
Maxwell, W.L. Spermiogenesis of *Eledone cirrhosa* Lamarck (Cephalopoda, Octopoda). *Proc.* B **186**, 181–190 (1974).
    Spermiogenesis of *Eusepia officinalis* (L.), *Loligo forbesi* (Steenstrup) and *Alloteuthis subulata* (L.) (Cephalopoda, Decapoda). *Proc.* B **191**, 527–535 (1975).
May, J.W. Heterogeneous catalysis at line defects.
  I. Active sites in periodic adlayers. *Proc.* A **331**, 185–193 (1972).
 II. Simple reactions. *Proc.* A **331**, 195–202 (1972).
May, M.J., Gladman, T. & Walker, E.F. Recent developments in ultra high strength steels and their applications (Symposium). *Trans.* A **282**, 377–387 (1976).
May, R. *See* Dainton, May, Morrow, Salmon & Thompson; O'Neill (P.), Salmon & May.
Mayaud, P.N. Comparison of magnetic observations in the Northern and Southern Hemispheres [abstract] (Discussion). *Trans.* B **279**, 273 (1977).
Mayers, D.F. Appendix to a paper by A.M. Binnie. *Proc.* A **339**, 447–449 (1974).
Mayes, I.W. & Dowker, J.S. Canonical functional integrals in general coordinates. *Proc.* A **327**, 131–135 (1972).
Mayhew, D.F. Reinterpretation of the extinct beaver *Trogontherium* (Mammalia, Rodentia). *Trans.* B **281**, 407–438 (1978).
    The vertebrate fauna at Bramerton (appendix to a paper by Funnell, Naton & West). *Trans.* B **287**, 531–534 (1979).
Mayhew, J.E.W. *See* Frisby & Mayhew.
Maynard, D.M. & Dando, M.R. The structure of the stomatogastric neuromuscular system in *Callinectes sapidus, Homarus americanus* and *Panulirus argus* (Decapoda Crustacea). *Trans.* B **268**, 161–220 (1974).

Mayneord, W.V. John Alfred Valentine Butler. *Biogr. Mem.* **25**, 145–178 (1979).

Mead, D.G. & Wilkinson, G.R. Far infrared emission of alkali halide crystals and melts. *Proc.* A **354**, 349–366 (1977).

Meade, C.J. *See* Ashwell (Margaret), Meade, Medawar & Sowter.

Meade, C.J., Ashwell, Margaret & Sowter, C. Is genetically transmitted obesity due to an adipose tissue defect? *Proc.* B **205**, 395–410 (1979).

Meadows, A.J. *See* Hughes (S.) & Meadows.

Meares, P. The mechanism of water transport in membranes (Discussion). *Trans.* B **278**, 113–150 (1977).
See also Foley, Klinowski & Meares.

Meares, P. & Page, K.R. Rapid force-flux transitions in highly porous membranes. *Trans.* A **272**, 1–46 (1972).
Oscillatory fluxes in highly porous membranes. *Proc.* A **339**, 513–532 (1974).

Mearns, C. *See* Coles (E.C.), Beilin, Bulpitt, Dollery, Johnson and others.

Measures, J.C. *See* Gould (G.W.) & Measures.

Medawar, Sir Peter. Michael Abercrombie. *Biogr. Mem.* **26**, 1–15 (1980).
See also Ashwell (Margaret), Meade, Medawar & Sowter.

Medawar, Sir Peter, Hunt, Ruth & Mertin, J. An influence of diet on transplantation immunity. *Proc.* B **206**, 265–280 (1979).

Medway, Lord & Marshall, A.G. Terrestrial vertebrates of the New Hebrides: origin and distribution (Discussion). *Trans.* B **272**, 423–465 (1975).

Mee, M.S.R. *See* Tait (J.F.), Tait, Gould & Mee.

Meek, M. The designer's response from the owner's side (Discussion). *Trans.* A **273**, 45–59 (1972).

Mehlum, E. & Sørensen, P.F. Example of an existing system in the ship-building industry: the Auto-kon system (Discussion). *Proc.* A **321**, 219–233 (1971).

Mehra, J. Satyendra Nath Bose. *Biogr. Mem.* **21**, 117–154 (1975).

Mehrishi, J.N. *See* Donner & Mehrishi.

Meinertzhagen, I. *See* Miller (D.S.), Baker, Bowden, Evans, Holt and others.

Meinertzhagen, I.A. The organization of perpendicular fibre pathways in the insect optic lobe. *Trans.* B **274**, 555–594 (1976).
See also Armett-Kibel, Meinertzhagen & Dowling.

Meissner, R. Lunar viscosity models (Discussion). *Trans.* A **285**, 463–467 (1977).

Mekki, O.M. & McKenzie, J.F. The propagation of atmospheric Rossby gravity waves in latitudinally sheared zonal flows. *Trans.* A **287**, 115–143 (1977).

Melchior, P. Some problems of tilt and strain measurements (Discussion). *Trans.* A **274**, 203–208 (1973).

Meldolesi, J. Dynamics of cytoplasmic membranes in pancreatic acinar cells (Discussion). *Trans.* B **268**, 39–53 (1974).

Meldrum, D.T. *See* Robin, Drewry & Meldrum.

Meleka, A.H. Component manufacture and joining techniques (Discussion). *Trans.* A **275**, 373–379 (1973).

Melford, D.A. The influence of residual and trace elements on hot shortness and high temperature embrittlement. *Trans.* A **295**, 89–103 (1980).

Mellema, J.E. *See* Driedonks, Krijgsman & Mellema.

Melrose, M.P. *See* Martinussen-Runde, Melrose & Derrick.

Melson, W.G. & Thompson, G. Petrology of a transform fault and adjacent ridge segments (Discussion). *Trans.* A **268**, 423–441 (1971).

Mendaš, I. *See* Bates (D.R.) & Mendaš.

Mendell, Nancy R. *See* Roberts (R.C.) & Mendell.

de Mendonca, F. The Brazilian project (Discussion). *Proc.* A **345**, 449–458 (1975).

Mengel, L. *See* Benedek (G.B.), Clark, Serrallach, Young and others.

Menon, M.G.K. *See* Krishnaswamy (M.R.), Menon, Narasimham, Hinotani, Ito and others.

Menter, Sir James. The future role of physical metallurgy in relation to engineering practice (Symposium). *Trans.* A **282**, 451–460 (1976).

Menzel, N. *See* Brimicombe, Stacey, Stacey, Hühnermann & Menzel.
Mercer, W.L. Materials requirements for pipeline construction (Symposium).
    Trans. A **282**, 41–51 (1976).
Merchant, M.E. The future of batch manufacture (Discussion). *Trans.* A **275**, 357–372 (1973).
Merkine, L. *See* Liu & Merkine.
Merminod, A. *See* Cooper (G.A.), Berlie & Merminod.
Merritt, R.R., Hyde, B.G. and, in part, Bursill, L.A. & Philp, D.K. The thermodynamics of the titan-
    ium+oxygen system: an isothermal gravimetric study of the composition range Ti_3O_5 to TiO_2
    at 1304 K. *Trans.* A **274**, 627–661 (1973).
Mertin, J. *See* Medawar, Hunt & Mertin.
Merton, L.F.H. *See* Hnatiuk (R.J.) & Merton.
Messenger, J.B. The nervous system of *Loligo*. IV. The peduncle and olfactory lobes.
    Trans. B **285**, 275–309 (1979).
Metafora, S., Felsani, A., Cotrufo, R., Tajana, G.F., Del Rio, A., De Prisco, P.P., Rutigliano, B. &
    Esposito, V. Neural control of gene expression in the skeletal muscle fibre: changes in the muscu-
    lar mRNA population following denervation. *Proc.* B **209**, 257–273 (1980).
Metafora, S., Felsani, A., Cotrufo, R., Tajana, G.F., Di Iorio, G., Del Rio, A., De Prisco, P.P. &
    Esposito, V. Neural control of gene expression in the skeletal muscle fibre: the nature of the
    lesion in the muscular protein-synthesizing machinery following denervation.
    Proc. B **209**, 239–255 (1980).
Metcalfe, E. *See* Bradley (J.N.) & Metcalfe.
Metcalfe, J.C. *See* Lee (A.G.), Birdsall, Metcalfe, Warren & Roberts.
Metiu, H. *See* Dewar & Metiu.
Metz, F., Howard, W.E., Wunsch, L., Neusser, H.J. & Schlag, E.W. The theory of molecular two-
    photon spectroscopy in the gas phase. *Proc.* A **363**, 381–401 (1978).
Meunier, F.M. *See* Israel (M.), Lesbats, Meunier & Stinnakre.
Meurling, P. *See* Knowles (Sir Francis), Vollrath & Meurling.
Meves, H. Calcium currents in squid giant axon (Discussion). *Trans.* B **270**, 377–387 (1975).
  Asymmetry currents in intracellularly perfused squid giant axons (Discussion).
    Trans. B **270**, 493–500 (1975).
  See also Levinson & Meves.
Mews, Lynne K. The green hydra symbiosis. III. The biotrophic transport of carbohydrate from alga
    to animal. *Proc.* B **209**, 377–401 (1980).
Meyer, P. Composition and spectra of primary cosmic-ray electrons and nuclei above 10^{10} eV (Dis-
    cussion). *Trans.* A **277**, 349–363 (1974).
Meyer-Rochow, V.B. The dioptric system of the eye of *Cybister* (Dytiscidae: Coleoptera).
    Proc. B **183**, 159–178 (1973).
Meyer-Rochow, V.B. & Horridge, G.A. The eye of *Anoplognathus* (Coleoptera, Scarabaeidae).
    Proc. B **188**, 1–30 (1975).
Meyer-Rochow, V.B. & Tiang, K.M. The effects of light and temperature on the structural organiza-
    tion of the eye of the Antarctic amphipod *Orchomene plebs* (Crustacea).
    Proc. B **206**, 353–368 (1979).
Mian, Salma, Bond, G. & Rodriguez-Barrueco, C. Effective and ineffective root nodules in *Myrica faya*.
    Proc. B **194**, 285–293 (1976).
Michael, D.H. *See* Majumdar & Michael.
Michael, D.H. & O'Neill, M.E. Two-dimensional problems of electrohydrostatic stability.
    Trans. A **272**, 331–359 (1972).
  The bursting of a charged cylindrical film. *Proc.* A **328**, 529–539 (1972).
Michael, D.H. & Williams, P.G. The equilibrium and stability of axisymmetric pendent drops.
    Proc. A **351**, 117–127 (1976).
  The equilibrium and stability of sessile drops. *Proc.* A **354**, 127–136 (1977).
Michal, F. *See* Boisseau, Lorient, Born & Michal.
Michels, H.J. *See* Duruz, Michels & Ubbelohde.
Michels, H.J. & Ubbelohde, A.R. Horizon formation by molten organic salts as a test of their fluidity.
    Proc. A **338**, 447–457 (1974).

Michelson, E. *See* Boksenberg, Kirkham, Michelson, Pettini and others.

Middendorf, H.D. & Randall, Sir John. Molecular dynamics of hydrated proteins (Discussion).
Trans. B **290**, 639–655 (1980).

Middlehurst, Barbara M. Transient lunar phenomena, deep moonquakes, and high-frequency tele-seismic events: possible connections (Discussion). *Trans.* A **285**, 485–487 (1977).

Middleton, A.J. & Smith, E.B. General anaesthetics and bacterial luminescence.
I. The effect of diethyl ether on the *in vivo* light emission of *Vibrio fischeri*.
Proc. B **193**, 159–171 (1976).
II. The effect of diethyl ether on the *in vitro* light emission of *Vibrio fischeri*.
Proc. B **193**, 173–190 (1976).

Middleton, C.J. Cavitation control in steels of high residual element content [abstract] (Discussion).
Trans. A **295**, 305 (1980).

Migus, A. *See* Ippen, Shank, Wiesenfeld & Migus; Shank, Ippen, Fork, Migus & Kobayashi.

Mikhail, F.I. & Wanas, M.I. A generalized field theory. I. Field equations.
Proc. A **356**, 471–481 (1977).

Mikoshiba, K. *See* Mariani, Crepel, Mikoshiba, Changeux & Sotelo.

Miledi, R. Transmitter release induced by injection of calcium ions into nerve terminals.
Proc. B **183**, 421–425 (1973).
Intracellular calcium and desensitization of acetylcholine receptors. *Proc.* B **209**, 447–452 (1980).
See also Abe, Alemá & Miledi; Abe, Limbrick & Miledi; Abe & Miledi; Bevan (S.), Grampp & Miledi;
Bevan (S.J.), Katz & Miledi; Bregestovski, Miledi & Parker; Cull-Candy, Miledi, Nakajima &
Uchitel; Fulton, Miledi & Szczepaniak; Fulton, Miledi & Takahashi; Green (D.P.L.), Ito, Miledi
& Vincent; Green (D.P.L.), Miledi, de la Mora & Vincent; Green (D.P.L.), Miledi & Vincent;
Heuser, Katz & Miledi; Heuser & Miledi; Ito (Y.) & Miledi; Ito (Y.), Miledi, Molenaar, Vincent
and others; Ito (Y.), Miledi & Vincent; Katz (Sir Bernard) & Miledi; Kusano, Miledi & Stinnakre;
Martin (R.) & Miledi; de Santis, Eusebi & Miledi.

Miledi, R., Molenaar, P.C. & Polak, R.L. An analysis of acetylcholine in frog muscle by mass frag-mentography. *Proc.* B **197**, 285–297 (1977).

Miledi, R., Nakajima, S. & Parker, I. Endplate currents in sucrose solution.
Proc. B **211**, 135–141 (1980).

Miledi, R. & Parker, I. Blocking of acetylcholine-induced channels by extracellular or intracellular application of D600. *Proc.* B **211**, 143–150 (1980).

Miledi, R., Parker, I. & Schalow, G. Measurement of calcium transients in frog muscle by the use of arsenazo. III. *Proc.* B **198**, 201–210 (1977).

Miledi, R. & Szczepaniak, Anna C. Effect of *Dendroaspis* neurotoxins on synaptic transmission in the spinal cord of the frog. *Proc.* B **190**, 267–274 (1975).

Miles, A.E.W. *See* Shellis & Miles.

Miles, J.L. Solid-state anodization of aluminium, beryllium, and zirconium.
Proc. A **321**, 503–514 (1971).

Miles, J.W. Asymptotic eigensolutions of Laplace's tidal equation. *Proc.* A **353**, 377–400 (1977).
On the second Painlevé transcendent. *Proc.* A **361**, 277–291 (1978).

Miles, R.S. The Holonematidae (placoderm fishes), a review based on new specimens of *Holonema* from the Upper Devonian of Western Australia. *Trans.* B **263**, 101–234 (1971).

Miles, Sir Ashley. *See* Bishop (Ann) & Miles.

Mill, P.J. *See* Lowe (D.A.), Mill & Knapp.

Mill, P.J. & Lowe, D.A. The fine structure of the PD proprioceptor of *Cancer pagurus*. I. The receptor strand and the movement sensitive cells. *Proc.* B **184**, 179–197 (1973).

Milledge, H. Judith. *See* Rood, Emerson & Milledge.

Millen, D.J. *See* Bevan (J.W.), Kisiel, Legon, Millen & Rogers; Bevan (J.W.), Legon & Millen; Bevan (J.W.), Legon, Millen & Rogers; Georgiou, Legon & Millen; Legon, Millen & Rogers.

Miller, A. *See* Doyle (B.B.), Hulmes, Miller, Parry, Piez & Woodhead-Galloway.

Miller, D.E., Brownscombe, J.L., Carruthers, G.P., Pick, D.R. & Stewart, K.H. Operational tempera-ture sounding of the stratosphere (Discussion). *Trans.* A **296**, 65–71 (1980).

Miller, D.S. *See* Stock & Miller.

Miller, D.S., Baker, J., Bowden, M., Evans, E., Holt, J., McKeag, R.J., Meinertzhagen, I., Mumford, P.M., Oddy, D.J., Rivers, J.P.W.R., Sevenhuysen, G., Stock, M.J., Watts, M., Kebede, A., Wolde-Gabriel, Y. & Wolde-Gabriel, Z. The Ethiopia applied nutrition project (Discussion). *Proc.* B **194**, 23–48 (1976).

Miller, Elaine K. *See* Kornberg & Miller.

Miller, G.F. *See* Burton (A.J.) & Miller.

Miller, G.J. *See* Edwards (R.H.T.), Miller, Hearn & Cotes.

Miller, G.H. *See* Welberry, Miller & Pickard.

Miller, J. & Clarkson, E.N.K. The post-ecdysial development of the cuticle and the eye of the Devonian trilobite *Phacops rana milleri* Stewart 1927. *Trans.* B **288**, 461–480 (1980).

Miller, J.C.P. On rotational tessellations and copses. *Trans.* A **293**, 599–641 (1980).

Miller, K.M. *See* Lowndes, Miller, Poulsen & Springford.

Miller, K.W. *See* Griffiths (H.B.), Miller, Paton & Smith.

Miller, M.K. *See* Beaven (P.A.), Miller, Williams, Delargy & Smith.

Miller, Sir Stephen. *See* Lyle, Miller & Ashton.

Millington, D.S. *See* Craig (R.D.), Bateman, Green & Millington.

Mills, A.A. *See* Geake, Walker, Telfer & Mills.

Mills, B. Effect of residual elements on the machinability of leaded free machining steels (Discussion). *Trans.* A **295**, 87–88 (1980).

Mills, C.F. Trace elements in animals (Discussion). *Trans.* B **288**, 51–63 (1979).

Mills, P.F. *See* Waugh, Mills & Southon.

Mimura, K. *See* Horridge & Mimura; Horridge, Mimura & Hardie; Horridge, Mimura & Tsukahara.

Mines, G.W. & Thomas, R.K. The photoelectron spectrum of sulphur trioxide: Jahn–Teller distortion in SO_3^+. *Proc.* A **336**, 355–364 (1974).

Mines, G.W., Thomas, R.K. & Thompson, Sir Harold. Photoelectron spectra of compounds containing thionyl and sulphuryl groups. *Proc.* A **329**, 275–282 (1972).
The photoelectron spectra of thiocarbonyl fluoride and thiocarbonyl chloride. *Proc.* A **333**, 171–181 (1973).

Mines, G.W. & Thompson, Sir Harold. Infrared and photoelectron spectra, and keto–enol tautomerism of acetylacetones and acetoacetic esters. *Proc.* A **342**, 327–339 (1975).

Mingins, J., Zobel, F.G.R., Pethica, B.A. & Smart, C. Potential differences due to spread monolayers at the polar oil/water interface. *Proc.* A **324**, 99–116 (1971).

Minster, J.-F. *See* Allègre, Brévart, Dupré & Minster.

Mintz, B. & Arrowsmith, J.M. The hot ductility behaviour of C–Mn–Nb–Al steels and its relation to crack propagation during the straightening of concast strand [abstract] (Discussion). *Trans.* A **295**, 124 (1980).

Mirault, M.-E. *See* Moran, Mirault, Arrigo, Goldschmidt-Clermont & Tissières.

Mirza, M.Y. & Duley, W.W. Two photon laser spectroscopy of indium. *Proc.* A **364**, 255–263 (1978).

Mita, K. *See* Shikata, Kim, Mita, Ise & Kunugi.

Mitchell, A.R. Variational principles and the finite-element method in partial differential equations (Discussion). *Proc.* A **323**, 211–217 (1971).

Mitchell, D. & Hellon, R.F. Neuronal and behavioural responses in rats during noxious stimulation of the tail. *Proc.* B **197**, 169–194 (1977).

Mitchell, E.W.J. & Stewart, R.J. Diffuse neutron scattering from crystal imperfections (Discussion). *Trans.* B **290**, 511–525 (1980).

Mitchell, F.L. Equipment for clinical chemistry and its evaluation (Discussion). *Proc.* B **184**, 351–359 (1973).

Mitchell, G.F. Periglacial Ireland (Discussion). *Trans.* B **280**, 199–209 (1977).

Mitchell, G.F., with contributions by Catt, J.A. & Weir, A.H., McMillan, Nora F., Margerel, J.P. and Whatley, R.C. The Late Pliocene marine formation at St Erth, Cornwall. *Trans.* B **266**, 1–37 (1973).

Mitchell, J.W. Dislocations in crystals of silver halides (Symposium). *Proc.* A **371**, 149–159 (1980).

Mitchell, P.I. *See* Carroll & Mitchell.

Mitchell, R.L. & Burridge, J.C. Trace elements in soils and crops (Discussion). *Trans.* B **288**, 15–24 (1979).

Mitchinson, C. *See* Carey, Mitchinson, Pain & Virden.

Mitchison, G.J. A model for vein formation in higher plants. *Proc.* B **207**, 79–109 (1980).
The dynamics of auxin transport. *Proc.* B **209**, 489–511 (1980).

Mitchison, T. *See* Sambrook, Botchan, Hu, Mitchison & Stringer.

Mitchum, Jr, R.M. *See* Vail (P.R.), Mitchum, Shipley & Buffler.

Mitra, S. *See* Gregson, Martin & Mitra.

Mittwoch, Ursula. *See* Clarke (Sir Cyril), Mittwoch & Traut.

Miyake, S. *See* Krishnaswamy (M.R.), Menon, Narasimham, Hinotani, Ito and others.

Miyashiro, A., Shido, F. & Ewing, M. Metamorphism in the Mid-Atlantic Ridge near 24° and 30°N (Discussion). *Trans.* A **268**, 589–602 (1971).

Miyazaki, S. & Nicholls, J.G. The properties and connections of nerve cells in leech ganglia maintained in culture. *Proc.* B **194**, 295–311 (1976).

Mizrahi, Rachel. *See* Shaltiel, Mizrahi & Sela.

Mizushina, T., Ogino, F., Ueda, H. & Komori, S. Application of laser Doppler velocimetry to turbulence measurement in non-isothermal flow. *Proc.* A **366**, 63–79 (1979).

Mizutani, U. *See* Massalski & Mizutani; Massalski, Mizutani & Noguchi.

Mizutani, U. & Massalski, T.B. Experimental evidence for the Fermi surface overlap effect in ϵ-phase Ag–Zn alloys, obtained from low temperature specific heats. *Proc.* A **343**, 375–387 (1975).

Modarress, D. *See* Holt (M.) & Modarress.

Moens, P.B., Mowat, M., Esposito, M.S. & Esposito, R.E. Meiosis in a temperature-sensitive DNA-synthesis mutant and in an apomictic yeast strain (*Saccharomyces cerevisiae*) (Discussion). *Trans.* B **277**, 351–358 (1977).

Moews, P.C. *See* Bunn, Moews & Baumber; DeLucia, Kelly, Mangion, Moews & Knox.

Molenaar, P.C. *See* Ito (Y.), Miledi, Molenaar, Vincent, Polak and others; Miledi, Molenaar & Polak.

Møllgård, K. & Saunders, N.R. A possible transepithelial pathway via endoplasmic reticulum in foetal sheep choroid plexus. *Proc.* B **199**, 321–326 (1977).

Mollon, J.D. *See* Polden & Mollon.

Mollon, J.D. & Polden, P.G. An anomaly in the response of the eye to light of short wavelengths. (With an appendix by W.S. Stiles). *Trans.* B **278**, 207–240 (1977).

Mollowney, B.M. *See* Barrett (M.J.) & Mollowney.

Molyneux, D.H. *See* Killick-Kendrick, Leaney, Ready & Molyneux; Killick-Kendrick, Molyneux & Ashford; Killick-Kendrick, Molyneux, Hommel, Leaney & Robertson.

Molyneux, D.H., Killick-Kendrick, R. & Ashford, R.W. *Leishmania* in phlebotomid sandflies. III. The ultrastructure of *Leishmania mexicana amazonensis* in the midgut and pharynx of *Lutzomyia longipalpis*. *Proc.* B **190**, 341–357 (1975).

Monciardini, C. *See* Andreieff, Bouysse, Curry, Fletcher and others.

Monfils, A. *See* Nandy, Thompson, Jamar, Monfils & Wilson.

Monfils, A. & Macau-Hercot, D. Preparation of a u.v. spectrophotometric catalogue of bright stars (Discussion). *Trans.* A **279**, 405–411 (1975).

Monge, C. *See* Zeuthen (T.) & Monge.

Monneron, Ariane. One-step isolation and characterization of nuclear membranes (Discussion). *Trans.* B **268**, 101–108 (1974).

Montadert, L. *See* Roberts (D.G.) & Montadert; Roberts (D.G.), Montadert, Thompson, Auffret and others; Sibuet, Ryan, Arthur, Barnes, Blechsmidt and others.

Montazemi, K. *See* Lehmann (H.), Ala, Hedeyat, Montazemi and others.

Monteith, J.L. Closing remarks to a Discussion. *Trans.* B **273**, 611–613 (1976).
Climate and the efficiency of crop production in Britain (Discussion).
Trans. B **281**, 277–294 (1977).
See also Cena & Monteith; McArthur & Monteith.

Montgomery, H.B.S. & Posnette, A.F. Franklin Kidd. *Biogr. Mem.* **21**, 407–430 (1975).

Montgomery, R.F. *See* Ratter, Askew, Montgomery & Gifford.

Moody, M.F. Application of optical diffraction to helical structures in the bacteriophage tail (Discussion). *Trans.* B **261**, 181–195 (1971).

Moon, P.B. George Paget Thomson. *Biogr. Mem.* **23**, 529–556 (1977).

Rutherford Memorial Lecture, 1975. Yarns and spinners: recollections of Rutherford and applications of swift rotation. *Proc.* A **360**, 303–315 (1978).

Moor, H. Recent progress in the freeze-etching technique (Discussion).
Trans. B **261**, 121–131 (1971).
See also Pfenninger, Akert, Moor & Sandri.

Moorbath, S. Age and isotope evidence for the evolution of continental crust (Discussion).
Trans. A **288**, 401–413 (1978).

Moore, A.J. *See* Hirst & Moore.

Moore, D.G. *See* Sibuet, Ryan, Arthur, Barnes, Blechsmidt and others.

Moore, D.W. The rolling up of a semi-infinite vortex sheet. *Proc.* A **345**, 417–430 (1975).
The spontaneous appearance of a singularity in the shape of an evolving vortex sheet.
Proc. A **365**, 105–119 (1979).
The velocity of a vortex ring with a thin core of elliptical cross section.
Proc. A **370**, 407–415 (1980).
See also Broadbent (E.G.) & Moore.

Moore, D.W. & Saffman, P.G. The motion of a vortex filament with axial flow.
Trans. A **272**, 403–429 (1972).
Axial flow in laminar trailing vortices. *Proc.* A **333**, 491–508 (1973).
A note on the stability of a vortex ring of small cross-section. *Proc.* A **338**, 535–537 (1974).
The instability of a straight vortex filament in a strain field. *Proc.* A **346**, 413–425 (1975).

Moore, F.H. *See* Venkatesan, Dale, Hodgkin, Nockolds, Moore & O'Connor.

Moore, J.G. *See* Sato & Moore.

Moore, Katharine. *See* Johnson (F.A.) & Moore.

Moore, M.N. *See* Bayne, Moore, Widdows, Livingstone & Salkeld.

Moore, P. The observation of transient lunar phenomena (Discussion). *Trans.* A **285**, 481–483 (1977).

Moore, P.J. *See* Plant & Moore.

Moore, S.L. *See* Hamilton (G.M.) & Moore.

Moore, W.S. & Holland, G.N. Experimental considerations in implementing a whole body multiple sensitive point nuclear magnetic resonance imaging system (Discussion).
Trans. B **289**, 511–518 (1980).

Moores, E.M. & Vine, F.J. The Troodos Massif, Cyprus and other ophiolites as oceanic crust: evaluation and implications (Discussion). *Trans.* A **268**, 443–466 (1971).

Moores, W.H. & McWeeny, R. The calculation of spin-orbit splitting and g tensors for small molecules and radicals. *Proc.* A **332**, 365–384 (1973).

Moorhouse, S.R. *See* Baños, Daniel, Moorhouse & Pratt.

Moos, H. Warren. *See* Fastie, Moos, Henry & Feldman.

de la Mora, M. Perez. *See* Green (D.P.L.), Miledi, de la Mora & Vincent.

Moran, L., Mirault, M.-E., Arrigo, A.P., Goldschmidt-Clermont, M. & Tissières, A. Heat shock of *Drosophila melanogaster* induces the synthesis of new messenger RNAs and proteins (Discussion).
Trans. B **283**, 391–406 (1978).

Moreira, T.J.S. *See* Mansfield (T.A.), Wellburn & Moreira.

Moreton, R. *See* Reynolds & Moreton.

Morgan, A.V. The Pleistocene geology of the area north and west of Wolverhampton, Staffordshire, England. *Trans.* B **265**, 233–297 (1973).

Morgan, D.A.O. Metalliferous potential of the United Kingdom (Discussion).
Proc. A **339**, 289–297 (1974).

Morgan, D.M.L. & Neuberger, A. Mutarotation of some biologically important 2-substituted hexoses.
Proc. A **337**, 317–332 (1974).

Morgan, D.V. & Wood, D.R. Surface studies by means of α particles of high energy (2–4 MeV).
Proc. A **335**, 509–523 (1973).

Morgan, G. *See* Tucker (E.M.), Ellory, Wooding, Morgan & Herbert.

Morgan, G.E. *See* Hailwood, Hamilton & Morgan; Sibuet, Ryan, Arthur, Barnes, Blechsmidt and others.

Morgan, J.D. The interaction of sound with a subsonic cylindrical vortex layer.

Proc. A **344**, 341–362 (1975).

See also Jones (D.S.) & Morgan.

Morgan, M.J. Analogue models of motion perception (Discussion). *Trans.* B **290**, 117–135 (1980).

Morgan, R.P. *See* Beynon (J.H.), Morgan & Brenton; Szulejko, Amaya, Morgan, Brenton & Beynon.

Morgan, Sir Morien. *See* Hall (Sir Arnold) & Morgan.

Moriarty, Christine M. *See* Moriarty (D.J.W.), Darlington, Dunn, Moriarty & Tevlin.

Moriarty, D.J.W., Darlington, Johanna P.E.C., Dunn, I.G., Moriarty, Christine M. & Tevlin, M.P. Feeding and grazing in Lake George, Uganda (Discussion). *Proc.* B **184**, 299–319 (1973).

Morita, A. *See* Calderwood, Coffey, Morita & Walker.

Morita, K., Nomura, H., Numata, M., Ochiai, M. & Yoneda, M. An approach to broad-spectrum cephalosporins (Discussion). *Trans.* B **289**, 181–190 (1980).

Morland, L.W. & Boulton, G.S. Stress in an elastic hump: the effects of glacier flow over elastic bedrock. *Proc.* A **344**, 157–173 (1975).

Morley, D. Organization of paediatric care (Discussion). *Proc.* B **199**, 161–168 (1977).

Morley, J.G. *See* McColl (I.R.) & Morley.

Morley, L.S.D. Finite element solution of boundary-value problems with non-removable singularities. *Trans.* A **275**, 463–488 (1973).

Analysis of developable shells with special reference to the finite element method and circular cylinders. *Trans.* A **281**, 113–170 (1976).

Moroz, I.M. *See* Gibbon, James & Moroz.

Morris, H.R. Research on peptides and glycopeptides (Discussion). *Trans.* A **293**, 39–51 (1979).

Morris, N.J. The infaunal descendants of the Cycloconchidae: an outline of the evolutionary history and taxonomy of the Heteroconchia, superfamilies Cycloconchacea to Chamacea (Discussion). *Trans.* B **284**, 259–275 (1978).

Morris, P. *See* Hall (D.O.), Adams, Morris & Rao.

Morris, P.G. *See* Mansfield (P.), Morris, Ordidge, Pykett, Bangert & Coupland.

Morris, P.L. *See* Dillamore, Morris, Smith & Hutchinson.

Morris, S. Conway. Middle Cambrian polychaetes from the Burgess Shale of British Columbia. *Trans.* B **285**, 227–274 (1979).

Morris, V.J. & Jennings, B.R. Light scattering by bacteria. I. Angular dependence of the scattered intensity. *Proc.* A **338**, 197–208 (1974).

Morrison, C.J. *See* Hayhurst (D.R.), Leckie & Morrison; Leckie, Hayhurst & Morrison.

Morrison, D.A. *See* Crozaz, Poupeau, Walker, Zinner & Morrison.

Morrison, D.A. & Zinner, E. Distribution and flux of micrometeoroids (Discussion). *Trans.* A **285**, 379–384 (1977).

Morrison, D.R.O. Review of inelastic proton–proton reactions (Discussion). *Proc.* A **335**, 461–483 (1973).

Morrison, M.Ann, Thompson, R.N., Gibson, I.L. & Marriner, G.F. Lateral chemical heterogeneity in the Palaeocene upper mantle beneath the Scottish Hebrides (Discussion). *Trans.* A **297**, 229–244 (1980).

Morrison, R.S. *See* Brooks (R.R.), Morrison, Reeves, Dudley & Akman.

Morrison, W.B. & Chapman, J.A. Controlled rolling (Symposium). *Trans.* A **282**, 289–303 (1976).

Morrow, T. *See* Bradley (D.J.), Hutchinson, Koetser, Morrow, New & Petty; Dainton, May, Morrow, Salmon & Thompson; Dainton, Morrow, Salmon & Thompson; Frankevich, Morrow & Salmon.

Morse, J.D. *See* Friedman (M.), Teufel & Morse.

Morse, R.D. *See* Basco & Morse.

Mortell, M.P. & Seymour, B.R. Nonlinear forced oscillations in a closed tube: continuous solutions of a functional equation. *Proc.* A **367**, 253–270 (1979).

Mortensen, C.E. *See* Allen (R.V.), Wood & Mortensen.

Mortimer, C.H. & Fee, E.J. Free surface oscillations and tides of Lakes Michigan and Superior (with a supplementary note and figure by C.H. Mortimer, D.B. Rao and D.J. Schwab). *Trans.* A **281**, 1–61 (1976).

Mortimer, C.H., Rao, D.B. & Schwab, D.J. Supplementary note and figure to a paper by C.H. Mortimer & E.J. Fee. *Trans.* A **281**, 58–60 (1976).

Mortimer, M.J. *See* Gordon (J.E.), Hall, Lee & Mortimer.

Morton, D.C. Observations of the interstellar gas with the Copernicus satellite (Discussion).
  Trans. A **279**, 299–302 (1975).
  P-Cygni profiles in hot stars [abstract] (Discussion). *Trans.* A **279**, 443 (1975).

Morton, H., Hegh, V. & Clunie, G.J.A. Studies of the rosette inhibition test in pregnant mice: evidence of immunosuppression? *Proc.* B **193**, 413–419 (1976).

Morton, J.E. The intertidal ecology of the British Solomon Islands. I. The zonation patterns of the weather coasts. *Trans.* B **265**, 491–537 (1973).

Morton, K.W. Stability and convergence in fluid flow problems. (Discussion).
  Proc. A **323**, 237–253 (1971).

Morton, P.H. Titanium alloys for engineering structures (Symposium).
  Trans. A **282**, 401–411 (1976).

Morzadec-Kerfourn, M.T. Appendix to a paper by Destombes, Shephard-Thorn & Redding (Discussion). *Trans.* A **279**, 253–255 (1975).

Moses, R.W., Jr. Aberration correction for high-voltage electron microscopy.
  Proc. A **339**, 483–512 (1974).

Moss, C.J. Versatility versus specialization in cultivation and harvesting for crops and in livestock production. *Trans.* B **267**, 61–70 (1973).

Moss, R.L. *See* Julian, Sollins & Moss.

Motherwell, W.D.S. *See* Kennard, Isaacs, Motherwell, Coppola and others.

Motley, R.W. *See* Franklin (R.N.), Edgley, Hamberger & Motley.

Mott, N.F. Introduction to a Symposium. *Proc.* A **371**, 3–7 (1980).
  Memories of early days in solid state physics (Symposium). *Proc.* A **371**, 56–66 (1980).
  See also Mott, Sir Nevill.

Mott, R.L. *See* Steward, Israel, Mott, Wilson & Krikorian.

Mott, Sir Nevill & Peierls, Sir Rudolf. Werner Heisenberg. *Biogr. Mem.* **23**, 213–251 (1977).

Mott, Sir Nevill, Pepper, M., Pollitt, S., Wallis, R.H. & Adkins, C.J. The Anderson transition.
  Proc. A **345**, 169–205 (1975).

Moulden, B. After-effects and the integration of patterns of neural activity within a channel (Discussion). *Trans.* B **290**, 39–55 (1980).

Moulder, J.W. The cell as an extreme environment (Discussion). *Proc.* B **204**, 199–210 (1979).

Moult, J. *See* Brehm & Moult.

Mountain, G. *See* Sibuet, Ryan, Arthur, Barnes, Blechsmidt and others.

Moura, Teresa. *See* Lew, Ferreira & Moura.

Mourant, A.E. The hereditary blood factors of the peoples of New Guinea and the surrounding regions (Discussion). *Trans.* B **268**, 251–255 (1974).
  See also Edholm (O.G.), Samueloff, Mourant, Fox and others; Godber, Kopeć, Mourant, Tills & Lehmann; Lehmann (H.), Ala, Hedeyat, Montazemi and others.

Mourant, A.E., Godber, Marilyn J., Kopeć, Ada C., Tills, D. & Woodhead, Bridget G. Genetical studies at high and low altitudes in Ethiopia (Discussion). *Proc.* B **194**, 17–22 (1976).

Mouravieff, I. Activity of solutions of methanol, ethanol, and *n*-butanol on stomatal opening in presence or absence of carbon dioxide (Discussion). *Trans.* B **273**, 561–564 (1976).

Mowat, M. *See* Moens, Mowat, Esposito & Esposito.

Moynahan, E.J. Trace elements in man (Discussion). *Trans.* B **288**, 65–79 (1979).

Muddle, B.C. & Edmonds, D.V. Interfacial segregation and embrittlement in liquid-phase sintered tungsten alloys [abstract] (Discussion). *Trans.* A **295**, 129 (1980).

Muggleton, J. *See* Bishop (J.A.), Cook & Muggleton.

Mühl, S. *See* Collins (R.A.), Mühl & Dearnaley.

Muir, Helen. *See* Wiebkin & Muir.

Muir, T.C. *See* Gillespie (J.S.), Creed & Muir.

Mulholland, H.P. *See* Klein (G.) & Mulholland.

Müller, Carla. *See* Roberts (D.G.), Montadert, Thompson, Auffret and others.

Muller, H.G. Long-period meteor wind oscillations (Discussion). *Trans.* A **271**, 585–598 (1972).

Muller, H.L. Flowline manufacture (Discussion). *Trans.* A **275**, 381–389 (1973).

Muller, K.J. & McMahan, U.J. The shapes of sensory and motor neurones and the distribution of their synapses in ganglia of the leech: a study using intracellular injection of horseradish peroxidase. *Proc.* B **194**, 481–499 (1976).

Müller, Ulrike. *See* Zentgraf, Keller & Müller.

Müller, W.A. Studies of the ecoclimatological optimum of the African Migratory Locust *Locusta migratoria migratorioides* [abstract] (Discussion). *Trans.* B **287**, 357 (1979).

Mullik, S.U. & Norrish, R.G.W. The photolysis of polyvinyl bromide. *Proc.* A **344**, 1–19 (1975).

Mullinger, A.M. *See* Johnson (R.T.), Mullinger & Skaer.

Mullins, L.J. *See* Beaugé & Mullins.

Mumford, P.M. *See* Miller (D.S.), Baker, Bowden, Evans, Holt and others.

Mundhenke, R. *See* Bartel, Duinker, Heintze, Heinzelmann and others.

Munro, A.J. Interactions between cells of a syngeneic immune system involving products of the major histocompatibility complex (Discussion). *Proc.* B **202**, 177–189 (1978).

Munro, E., Siegel, R.L., Craig, I.W. & Sly, W.S. Cytoplasmic transfer of a determinant for chloramphenicol resistance between mammalian cell lines. *Proc.* B **201**, 73–85 (1978).

Munro, R.H. *See* MacQueen, Gosling, Hildner, Munro, Poland & Ross.

Munro-Faure, A.D. *See* Coles (E.C.), Beilin, Bulpitt, Dollery, Johnson and others.

Munton, R. Future propulsion systems for merchant ships (Discussion). *Trans.* A **273**, 137–150 (1972).

von Muralt, A. The optical spike (Discussion). *Trans.* B **270**, 411–423 (1975).

Murdmaa, I.O. *See* Chernysheva & Murdmaa.

Murdock, J.W. The generation of a Tollmien–Schlichting wave by a sound wave. *Proc.* A **372**, 517–534 (1980).

Murgatroyd, R.J. & O'Neill, A. Interaction between the troposphere and stratosphere (Discussion). *Trans.* A **296**, 87–102 (1980).

Murphy, B.R., Markoff, L.J., Chanock, R.M., Spring, Susan B., Maassab, H.F., Kendal, A.P., Cox, Nancy J., Levine, M.M., Douglas, Jr, R.G., Betts, R.F., Couch, R.B. & Cate, Jr, T.R. Genetic approaches to attenuation of influenza A viruses for man (Discussion). *Trans.* B **288**, 401–415 (1980).

Murphy, J.O. & Steiner, J.M. The effect of a magnetic field and rotation acting simultaneously on the onset of stationary convection in a field layer with rigid boundaries. *Proc.* A **347**, 85–98 (1975).

Murphy, M.C. *See* Batte & Murphy.

Murray, C.A. & Yallop, B.D. Lunar laser ranging and fundamental astrometry (Discussion). *Trans.* A **284**, 507–514 (1977).

Murray, C.L. *See* Ash, Barrer, Clint, Dolphin & Murray.

Murray, J. & Clarke, B. The genus *Partula* on Moorea: speciation in progress. *Proc.* B **211**, 83–117 (1980).

Murray, J.D. On the molecular mechanism of facilitated oxygen diffusion by haemoglobin and myoglobin. *Proc.* B **178**, 95–110 (1971).

Murray, K. Genetic engineering: possibilities and prospects for its application in industrial microbiology (Discussion). *Trans.* B **290**, 369–386 (1980).

Murray, M.A.F. *See* Brown-Grant, Murray, Raisman & Sood.

Murray, M.J. Fracture of WC-Co alloys: an example of spatially constrained crack tip opening displacement. *Proc.* A **356**, 483–508 (1977).

See also Hannink, Kohlstedt & Murray.

Murray, N.D., Bishop, J.A. & Macnair, M.R. Melanism and predation by birds in the moths *Biston betularia* and *Phigalia pilosaria*. *Proc.* B **210**, 277–283 (1980).

Murthy, V. Rama. Lunar evolution: is there a global radioactive crust on the Moon? (Discussion). *Trans.* A **285**, 127–136 (1977).

Murthy, V.S. *See* Ramasastry, Reddy & Murthy.

Muscatine, L., Boyle, J. Elizabeth & Smith, D.C. Symbiosis of the acoel flatworm *Convoluta roscoffensis* with the alga *Platymonas convolutae*. *Proc.* B **187**, 221–234 (1974).

Muscatine, L. & Pool, R.R. Regulation of numbers of intracellular algae. (Discussion). *Proc.* B **204**, 131–139 (1979).

Musselec, P. *See* Boillot & Musselec.

Myatt, G. Observations of $(\mu^- e^+)$ pairs associated with strange particles in neutrino interactions in Gargamelle (Discussion). *Proc.* A **355**, 581–583 (1977).

Myers, J. The influence of impurity and alloy content on stress relief cracking in CrMoV steels [abstract] (Discussion). *Trans.* A **295**, 289–290 (1980).

See also Batte, Brear, Holdsworth, Myers & Reynolds.

Nabarro, F.R.N. Recollections of the early days of dislocation physics (Symposium).
    Proc. A **371**, 131–135 (1980).

See also Bibby, Nabarro, McLachlan & Stephen.

Nabarro, F.R.N. & Bibby, B. Rothberg. A possible hypo-critical point in the phase diagram of a moderately small superconductor in a magnetic field. *Trans.* A **278**, 343–349 (1975).

Nachman, A. & Talliaferro, S. Mass transfer into boundary layers for power law fluids.
    Proc. A **365**, 313–326 (1979).

Nag, B.R. *See* Chattopadhyay & Nag.

Nagasawa, H., Cox, A.B. & Lett, J.T. The radiation responses of synchronous ·L5178Y S/S cells and their significance for radiobiological theory. *Proc.* B **211**, 25–49 (1980).

Nagata, T. *See* Fisher (R.M.), Huffman, Nagata & Schwerer.

Nagata, W. Straightforward synthesis of 7α-methoxy-l-oxacephems from penicillins (Discussion).
    Trans. B **289**, 225–230 (1980).

Nagenthiram, P. *See* Cullen (A.L.), Nagenthiram & Williams.

Naghdi, P.M. *See* Green (A.E.), Laws & Naghdi; Green (A.E.) & Naghdi; Green (A.E.), Naghdi & Wenner.

Naghdi, P.M. & Tang, P.Y. Large deformation possible in every isotropic elastic membrane.
    Trans. A **287**, 145–187 (1977).

Nagle, J.F. Statistical mechanics of the melting transition in lattice models of polymers.
    Proc. A **337**, 569–589 (1974).

Nakajima, S. *See* Miledi, Nakajima & Parker.

Nakajima, Y. *See* Cull-Candy, Miledi, Nakajima & Uchitel.

Nakanishi, H., Jones, W., Thomas, J.M., Hasegawa, M. & Rees, W.L. Topochemically controlled solid-state polymerization. *Proc.* A **369**, 307–325 (1980).

Nakanishi, K. & Occolowitz, J.L. Applications of mass spectrometry in structural studies of bioactive compounds (Discussion). *Trans.* A **293**, 3–11 (1979).

Nambudripad, N. *See* Allsop, Bleaney, Bowden, Nambudripad, Stone & Suzuki.

Nandy, K. *See* Evans (R.G.), Nandy & Wilson.

Nandy, K., Thompson, G.I., Jamar, C., Monfils, A. & Wilson, R. Preliminary results of ultraviolet interstellar extinction from TD1 satellite observations (Discussion).
    Trans. A **279**, 337–343 (1975).

Narasimham, V.S. *See* Krishnaswamy (M.R.), Menon, Narasimham, Hinotani, Ito and others.

Nash, C.H. *See* Neuss, Nash, Lemke & Grutzner.

Nasir, M.J. *See* Axon, Nasir & Knowles.

Nason, R.D. *See* King (C-Y.), Nason & Tocher.

Nayler, J.H.C. Structure—activity relationships in semi-synthetic penicillins (Discussion).
    Proc. B **179**, 357–367 (1971).

Naylor, D.J. *See* Llewellyn (D.T.), Marriott, Naylor & Thewlis.

Naylor, E. *See* Aréchiga, Huberman & Naylor.

Nayudu, Y.R. Petrology of submarine volcanics from the NE Pacific [abstract] (Discussion).
    Trans. A **268**, 651 (1971).

Nazaré, Maria H. *See* Davies (G.) & Nazaré; Davies (G.), Nazaré & Hamer.

Ndiaye, A. *See* Abdallahi, Skaf, Castel & Ndiaye.

Neaves, W.B. The annual testicular cycle in an equatorial colony of lesser rock hyrax, *Heterohyrax brucei. Proc.* B **206**, 183–189 (1979).

Needham, Dorothy M. Introductory remarks to a Discussion. *Trans.* B **265**, 167 (1973).

Needham, J. Astronomy in ancient and medieval China (Discussion). *Trans.* A **276**, 67—82 (1974).

Needham, N.G. & Orr, J. The effect of residuals on the elevated temperature properties of some creep resistant steels (Discussion). *Trans.* A **295**, 279—288 (1980).

Neerhoff, F.L. Scattering of SH-waves by an irregularity at the mass-loaded boundary of a semi-infinite elastic medium. *Proc.* A **342**, 237—257 (1975).

Neidle, S., Rogers, D. & Hursthouse, M.B. The crystal and molecular structure of streptomycin oxime selenate tetrahydrate. *Proc.* A **359**, 365—388 (1978).

Neil, D.M. *See* Laverack, Neil & Robertson; Laverack, Macmillan & Neil; Macmillan, Neil & Laverack.

Neil, D.M., Macmillan, D.L., Robertson, R.M. & Laverack, M.S. The structure and function of thoracic exopodites in the larvae of the lobster *Homarus gammarus* (L.). *Trans.* B **274**, 53—68 (1976).

Neill, D.W. Automated clinical chemistry and haematology in a hospital region (Discussion). *Proc.* B **184**, 361—368 (1973).

Nejad, H. Karini. *See* Lehmann (H.), Ala, Hedeyat, Montazemi and others.

Nelson, A.C. *See* Boland (B.J.), Brown, Carrington & Nelson.

Nelson, L. *See* Daneholt, Case, Lamb, Nelson & Wieslander.

Nelson, S.A. *See* Carmichael, Nicholls, Spera, Wood & Nelson.

Nelson-Smith, A. Effects of the oil industry on shore life in estuaries (Discussion). *Proc.* B **180**, 487—496 (1972).

Nemet, A. Non-invasive and non-destructive techniques in medicine and industry (Discussion). *Trans.* A **292**, 137—146 (1979).

Nesbitt, R.W. & Sun, S.-S. Geochemical features of some Archaean and post-Archaean high-magnesian—low-alkali liquids (Discussion). *Trans.* A **297**, 365—381 (1980).

Nesheim, M.C. *See* Parshad, Crompton & Nesheim.

Nesheim, M.C., Crompton, D.W.T., Arnold, Susan & Barnard, D. Dietary relations between *Moniliformis* (Acanthocephala) and laboratory rats. *Proc.* B **197**, 363—383 (1977).
  Host dietary starch and *Moniliformis* (Acanthocephala) in growing rats. *Proc.* B **202**, 399—408 (1978).

Nesterov, V.P. *See* Andreeva, Katasyev, Uvarov, Nesterov & Chasovitin.

de Nettancourt, D., Devreux, M., Carluccio, F., Laneri, U., Cresti, M., Pacini, E., Sarfatti, G. & van Gastel, A.J.G. Facts and hypotheses on the origin of S mutations and on the function of the S gene in *Nicotiana alata* and *Lycopersicum peruvianum* (Discussion). *Proc.* B **188**, 345—360 (1975).

Nettleship, R. *See* Woodgate (B.E.), Knight, Uribe, Sheather, Bowles & Nettleship.

Nettleton, M.A. & Stirling, R. The combustion of clouds of coal particles in shock-heated mixtures of oxygen and nitrogen. *Proc.* A **322**, 207—221 (1971).

Neuberger, A. James Norman Davidson. *Biogr. Mem.* **19**, 281—303 (1973).
  Introductory remarks to a Discussion. *Trans.* B **273**, 77—78 (1976).
  Charles Enrique Dent. *Biogr. Mem.* **24**, 15—31 (1978).
  See also Dunstan, Grant, Marshall & Neuberger; Morgan (D.M.L.) & Neuberger; Wider de Xifra, Sandy, Davies & Neuberger.

Neukum, G. Lunar cratering (Discussion). *Trans.* A **285**, 267—272 (1977).

Neupert, W.M. Solar flare X-ray spectra (Discussion). *Trans.* A **270**, 143—155 (1971).

Neuss, N., Nash, C.H., Lemke, P.A. & Grutzner, J.B. The use of ^{13}C n.m.r. (c.m.r.) spectroscopy in biosynthetic studies of β-lactam antibiotics. I. The incorporation of $[1\text{-}^{13}$C]- and $[2\text{-}^{13}$C]sodium acetate, and DL-$[1\text{-}^{13}$C]- and DL-$[2\text{-}^{13}$C]valine into cephalosporin C (Discussion). *Proc.* B **179**, 335—344 (1971).

Neusser, H.J. *See* Metz, Howard, Wunsch, Neusser & Schlag.

Nevell, T.G. *See* Cullis & Nevell.

Neville, C. *See* Rothschild (Miriam), Schlein, Parker, Neville & Sternberg.

New, G.H.C. Mode-locked laser systems: theoretical models (Discussion). *Trans.* A **298**, 247—256 (1980).
  See also Bradley (D.J.), Hutchinson, Koetser, Morrow, New & Petty.

Newcombe, G. *See* Chubb (J.P.), Billingham, Hancock, Dimbylow & Newcombe.

Newell, A.C. The general structure of integrable evolution equations. *Proc.* A **365**, 283–311 (1979).
See also Kaup & Newell.

Newell, N.D. & Boyd, D.W. A palaeontologist's view of bivalve phylogeny (Discussion).
Trans. B **284**, 203–215 (1978).

Newland, D.E. Buckling and rupture of the double bellows expansion joint assembly at Flixborough.
Proc. A **351**, 525–549 (1976).

Newman, B.G. Shape of a towed boom of logs. *Proc.* A **346**, 329–348 (1975).

Newman, E.I. Water movement through root systems (Discussion). *Trans.* B **273**, 463–478 (1976).

Newman, E.T. *See* Hansen (R.O.), Newman, Penrose & Tod.

Newsom, G.H. *See* Baig, Connerade & Newsom; Connerade, Baig, Garton & Newsom; Connerade,
Mansfield, Newsom, Tracy, Baig & Thimm; Mansfield (M.W.D.) & Newsom.

Newson, M.D. *See* Clarke (R.T.) & Newson.

Newton, D. *See* Chamberlain, Clough, Heard, Newton, Stott & Wells.

Newton, G., Andrews, D.A. & Unsworth, P.J. A precision determination of the Lamb shift in hydrogen. *Trans.* A **290**, 373–404 (1979).

Newton, R.R. Introduction to some basic astronomical concepts (Discussion).
Trans. A **276**, 5–20 (1974).

Two uses of ancient astronomy (Discussion). *Trans.* A **276**, 99–116 (1974).

Ng, B.S. *See* Lakin, Ng & Reid.

Ngan, K.M. *See* Gurney (C.) & Ngan.

Nibbering, N.M.M. Mechanistic studies (Discussion). *Trans.* A **293**, 103–115 (1979).

Nicholes, K.R.K. *See* Fife & Nicholes.

Nicholls, G.D. & Islam, M.R. Geochemical investigations of basalts and associated rocks from the
ocean floor and their implications (Discussion). *Trans.* A **268**, 469–486 (1971).

Nicholls, J. *See* Carmichael, Nicholls, Spera, Wood & Nelson.

Nicholls, J.G. *See* Miyazaki & Nicholls; Wallace (B.G.), Adal & Nicholls.

Nicholls, R.W. *See* Creek & Nicholls; Danylewych & Nicholls.

Nicholson, A.J.C. *See* Bolton (H.C.), Grant, McWilliam, Nicholson & Swingler.

Nicholson, R.B. Nickel base materials developments for high temperatures (Symposium).
Trans. A **282**, 389–399 (1976).

Nickel, E. & Potter, L.T. Synaptic vesicles in freeze-etched electric tissue of *Torpedo* (Discussion).
Trans. B **261**, 383–385 (1971).

Nicklas, R.B. Chromosome distribution: experiments on cell hybrids and *in vitro* (Discussion).
Trans. B **277**, 267–276 (1977).

Nicol, J.A. *See* Zyznar & Nicol.

Nicol, J.A.C. *See* Arnott, Best, Ito & Nicol; Arnott, Nicol & Querfeld; Fineran & Nicol; Ito (S.) &
Nicol; Ito (S.), Thurston & Nicol; Wang (R.T.), Nicol, Thurston & McCants; Watson (M.),
Thurston & Nicol; Zyznar, Cross & Nicol.

Nicol, J.A.C. & Arnott, H.J. Tapeta lucida in the eyes of goatsuckers (Caprimulgidae).
Proc. B **187**, 349–352 (1974).

Nicolas, A. Stress estimates from structural studies in some mantle peridotites (Discussion).
Trans. A **288**, 49–57 (1978).

Nicoll, R.A. *See* Allen (G.I.), Eccles, Nicoll, Oshima & Rubia; Eccles, Nicoll, Oshima & Rubia.

Nicoll, R.A., Alger, B.E. & Jahr, C.E. Peptides as putative excitatory neurotransmitters: carnosine,
enkephalin, substance P and TRH (Discussion). *Proc.* B **210**, 133–149 (1980).

Niebel, K.F. & Venables, J.A. An explanation of the crystal structure of the rare gas solids.
Proc. A **336**, 365–377 (1974).

Niedergerke, R. *See* Gadsby, Niedergerke & Ogden; Lammel, Niedergerke & Page.

Niedergerke, R. & Page, Sally. Analysis of catecholamine effects in single atrial trabeculae of the frog
heart. *Proc.* B **197**, 333–362 (1977).

Niekerke, J. *See* Hessberg, Niekerke & Stephan.

Nielsen, D. *See* Steiner (D.F.), Patzelt, Chan, Quinn, Tager; Nielsen and others.

Nielsen, J.B.K. *See* Lampen, Nielsen, Izui & Caulfield.

Nieto, M. *See* Ghuysen, Frère, Leyh-Bouille, Perkins & Nieto.

Nieto-Vesperinas, M. *See* Ross (G.), Fiddy, Nieto-Vesperinas & Wheeler.
Ninham, B.W. *See* Horridge, Ninham & Diesendorf.
Nisbet, J.D. *See* Kemball, Nisbet, Robertson & Scurrell.
Nishihara, H.K. *See* Marr (D.) & Nishihara.
Nishiyama, A. & Petersen, O.H. Biphasic membrane potential changes in pancreatic acinar cells following short pulses of acetylcholine stimulation. *Proc.* B **191**, 549–553 (1975).
Nitzan, A., Jortner, J. & Rentzepis, P.M. Intermediate level structure in highly excited electronic states of large molecules. *Proc.* A **327**, 367–391 (1972).
Nixon, P.E. *See* Lee (D.L.), Nixon & North.
Nixon, W.C. The general principles of scanning electron microscopy (Discussion). *Trans.* B **261**, 45–50 (1971).
Noble, B. *See* Sewell & Noble.
Noble, D. *See* Brown (Hilary F.), McNaughton, Noble & Noble.
Noble, Susan J. *See* Brown (Hilary F.), McNaughton, Noble & Noble.
Noci, G. *See* Boland (B.C.), Jones, Wilson, Engstrom & Noci.
Nockolds, C.E. *See* Deer & Nockolds; Venkatesan, Dale, Hodgkin, Nockolds, Moore & O'Connor.
Noda, M.T. *See* Araki, Yano, Ueda & Noda.
Noguchi, S. *See* Massalaki, Mizutani & Noguchi.
Nolan, J. Determination of specific component activities in tholeiitic melts: principles and technique (Discussion). *Trans.* A **286**, 343–351 (1977).
Nomura, H. *See* Morita (K.), Nomura, Numata, Ochiai & Yoneda.
Nonner, W., Rojas, E. & Stämpfli, R. Gating currents in the node of Ranvier: voltage and time dependence (Discussion). *Trans.* B **270**, 483–492 (1975).
Norberg, R.Å. Occurrence and independent evolution of bilateral ear asymmetry in owls and implications on owl taxonomy. *Trans.* B **280**, 375–408 (1977).
Skull asymmetry, ear structure and function, and auditory localization in Tengmalm's owl, *Aegolius funereus* (Linné). *Trans.* B **282**, 325–410 (1978).
Norberg, Ulla M. Morphology of the wings, legs and tail of three coniferous forest tits, the goldcrest, and the treecreeper in relation to locomotor pattern and feeding station selection. *Trans.* B **287**, 131–165 (1979).
Norbury, J. *See* Keady & Norbury.
Norgan, N.G., Ferro-Luzzi, A. & Durnin, J.V.G.A. The energy and nutrient intake and the energy expenditure of 204 New Guinean adults (Discussion). *Trans.* B **268**, 309–348 (1974).
Norris, J.R. *See* Katz (J.J.), Oettmeier & Norris.
Norris, P.R. *See* Padalia, Lang, Norris, Watson & Fabian.
Norrish, R.G.W. *See* Mullik & Norrish.
Norry, M.J., Truckle, P.H., Lippard, S.J., Hawkesworth, C.J., Weaver, S.D. & Marriner, G.F. Isotope and trace element evidence from lavas, bearing on mantle heterogeneity beneath Kenya (Discussion). *Trans.* A **297**, 259–271 (1980).
North, A.C.T. *See* Lee (D.L.), Nixon & North.
North, R.A. *See* Katayama, North & Williams.
Northcote, D.H. Membrane systems of plant cells (Discussion). *Trans.* B **268**, 119–128 (1974).
Norton, P.E.P. *See* Funnell, Norton & West; West (R.G.) & Norton.
Norton, P.R. *See* Batt, Eley & Norton.
Norton, T.A. *See* Goss-Custard, Jones, Kitching & Norton.
Norwood, F.W. The satellite technology demonstration and health-education telecommunications experiments on ATS-6 (Discussion). *Proc.* A **345**, 541–557 (1975).
Nouri-Moghadam, M. & Taylor, J.G. One-loop divergencies for the Einstein-charged meson system. *Proc.* A **344**, 87–99 (1975).
Non-renormalizability of Einstein-spontaneous-symmetry breaking vector and scalar meson interaction. *Proc.* A **351**, 197–208 (1976).
Novak, Ulrike. *See* Griffin (Beverly E.), Dilworth, Ito & Novak.
Noy, D.J.M. *See* Girdler, Brown, Noy & Styles.
Noyes, B.E. *See* Steiner (D.F.), Patzelt, Chan, Quinn, Tager, Nielson and others.

Nozik, A.J. Photoelectrochemical cells (Discussion). *Trans.* A **295**, 453–470 (1980).
Nugaliyadde, L. Appendix to a paper by H. Grüneberg. *Trans.* B **275**, 436–437 (1976).
Nugroho, G. & Elliott, Katherine. The Dana Sehat programme in Solo, Indonesia (Discussion). *Proc.* B **199**, 145–150 (1977).
Numata, M. *See* Morita (K.), Nomura, Numata, Ochiai & Yoneda.
Nunnally, R.L. *See* Hollis & Nunnally.
Nur, A. Role of pore fluids in faulting (Discussion). *Trans.* A **274**, 297–304 (1973).
Nurse, R.W. The contribution of the materials scientist (Discussion). *Trans.* A **272**, 585–593 (1972).
Nutku, Y. *See* Matzner & Nutku.
Nutman, P.S. Henry Gerard Thornton. *Biogr. Mem.* **23**, 557–574 (1977).
Nyburg, S.C. & Wong-Ng, W. Anisotropic atom–atom forces and the space group of solid chlorine. *Proc.* A **367**, 29–45 (1979).
Nye, J.F. Optical caustics in the near field from liquid drops. *Proc.* A **361**, 21–41 (1978).
Optical caustics from liquid drops under gravity: observations of the parabolic and symbolic umbilics. *Trans.* A **292**, 25–44 (1979).
See also Berry (M.V.), Nye & Wright; Evans (D.C.B.), Nye & Cheeseman.
Nye, J.F. & Berry, M.V. Dislocations in wave trains. *Proc.* A **336**, 165–190 (1974).

Oakley, B. *See* Poulsen (J.H.) & Oakley.
Oakley, C.L. Leonard Colebrook. *Biogr. Mem.* **17**, 91–138 (1971).
Francis William Rogers Brambell. *Biogr. Mem.* **19**, 129–171 (1973).
Carl Hamilton Browning. *Biogr. Mem.* **19**, 173–215 (1973).
Oates, G. *See* Coffey (J.M.), Oates & Whittle.
Oates, K. *See* Hunt (S.) & Oates; Manton (Irene), Sutherland & Oates; Manton (Irene) & Oates.
O'Brien, E.J., Bennett, Pauline M. & Hanson, Jean. Optical diffraction studies of myofibrillar structure (Discussion). *Trans.* B **261**, 201–208 (1971).
O'Brien, M.C.M. *See* Gee (J.V.), Hayes & O'Brien.
O'Brien, R.W. *See* Batchelor (G.K.) & O'Brien.
O'Callaghan, Cynthia H. Structure–activity relations and β-lactamase resistance (Discussion). *Trans.* B **289**, 197–205 (1980).
O'Callaghan, J.R. Problems of power: modification versus innovation (Discussion). *Trans.* B **267**, 51–59 (1973).
Occolowitz, J.L. *See* Nakanishi (K.) & Occolowitz.
Ochiai, M. *See* Morita (K.), Nomura, Numata, Ochiai & Yoneda.
Ockendon, J.R. & Tayler, A.B. The dynamics of a current collection system for an electric locomotive. *Proc.* A **322**, 447–468 (1971).
Ockleford, C.D. *See* Clint (Jane M.), Wakely & Ockleford.
O'Connor, A.J. *See* Gates, O'Connor & Westcott.
O'Connor, B.H. *See* Venkatesan, Dale, Hodgkin, Nockolds, Moore & O'Connor.
O'Connor, J.J. *See* Johnson (K.L.), O'Connor & Woodward.
Oddy, D.J. *See* Miller (D.S.), Baker, Bowden, Evans, Holt and others.
Odiyo, P.O. Forecasting infestations of a migrant pest: the African armyworm *Spodoptera exempta* (Walk.) (Discussion). *Trans.* B **287**, 403–413 (1979).
Oele, E. *See* Kirby & Oele.
Oertel, G. *See* Wood (D.S.), Oertel, Singh & Bennett.
Oettinger, R. *See* Rogers (L.J.), Oettinger, Szer & Mark.
Oettmeier, W. *See* Katz (J.J.), Oettmeier & Norris.
Offer, G. The antigenicity of myosin and C-protein. *Proc.* B **192**, 439–449 (1976).
See also Craig (R.) & Offer; Elliott (A.), Offer & Burridge.
Offermann, D. A winter anomaly campaign in Western Europe (Discussion). *Trans.* A **296**, 261–268 (1980).
Ogden, D.C. *See* Gadsby, Niedergerke & Ogden.

Ogden, R.W. Large deformation isotropic elasticity – on the correlation of theory and experiment for incompressible rubberlike solids. *Proc.* A **326**, 565–584 (1972).

Large deformation isotropic elasticity: on the correlation of theory and experiment for compressible rubberlike solids. *Proc.* A **328**, 567–583 (1972).

Ogino, F. *See* Mizushina, Ogino, Ueda & Komori.

Ogston, A.G. Appendix to a paper by Ogston, Preston & Wells. *Proc.* A **333**, 310 (1973).

Harold Brewer Hartley. *Biogr. Mem.* **19**, 349–373 (1973).

Corrigendum. *Biogr. Mem.* **19**, 695 (1973).

Ogston, A.G., Preston, B.N. & Wells, J.D. On the transport of compact particles through solutions of chain-polymers (With appendices by A.G. Ogston, by B.N. Preston & J. McK. Snowden, and by J.D. Wells). *Proc.* A **333**, 297–316 (1973).

Oh, K.P. A diffusion model for fatigue crack growth. *Proc.* A **367**, 47–58 (1979).

See also Rohde & Oh.

O'Hara, M.J. A mechanism for ocean-floor spreading (Discussion). *Trans.* A **268**, 731 (1971).

Thermal history of magmas; the low pressure reference point (Discussion).
Trans. A **288**, 627–629 (1978).

Nonlinear nature of the unavoidable long-lived isotopic, trace and major element contamination of a developing magma chamber (Discussion). *Trans.* A **297**, 215–227 (1980).

See also Brown (G.M.), O'Hara & Oxburgh; Ford (C.E.), O'Hara & Spencer; Howells (Susan) & O'Hara.

O'Hara, M.J. & Humphries, D.J. Gravitational separation of quenching crystals: a cause of chemical differentiation in lunar basalts (Discussion). *Trans.* A **285**, 177–192 (1977).

Problems of iron gain and loss during experimentation on natural rocks: the experimental crystallization of five lunar basalts at low pressures (Discussion). *Trans.* A **286**, 313–330 (1977).

O'Hara, M.J. & Yarwood, G. High pressure–temperature point on an Archaean geotherm, implied magma genesis by crustal anatexis, and consequences for garnet–pyroxene thermometry and barometry (Discussion). *Trans.* A **288**, 441–456 (1978).

Okada, Sharon. *See* Dulbecco & Okada.

Okamoto, K. & Quastel, J.H. Spontaneous action potentials in isolated guinea-pig cerebellar slices: effects of amino acids and conditions affecting sodium and water uptake.
Proc. B **184**, 83–90 (1973).

O'Keefe, J.A. & Urey, H.C. The deficiency of siderophile elements in the Moon (Discussion).
Trans. A **285**, 569–575 (1977).

O'Keefe, M. & Hyde, B.G. Plane nets in crystal chemistry. *Trans.* A **295**, 553–618 (1980).

Okubo, T. *See* Ise, Maruno & Okubo.

Okubo, T. & Ise, N. Catalysis of the ammonium cyanate–urea conversion by polyelectrolytes.
Proc. A **327**, 413–424 (1972).

Okubo, T., Kitano, H., Ishiwatari, T. & Ise, N. Conductance stopped-flow study on the micellar equilibria of ionic surfactants. *Proc.* A **366**, 81–90 (1979).

Okubo, T., Maruno, T. & Ise, N. Role of solvation and desolvation in polymer 'catalysis'. II. The influence of high pressures on the alkaline hydrolyses of $Co(NH_3)_5Br^{2+}$ calalysed by macro-ions. *Proc.* A **370**, 501–508 (1980).

Olek, A. *See* Robbins (N.), Olek, Kelly, Takach & Christopher.

Olhoeft, G.R. *See* Strangway & Olhoeft.

d'Olier, B. Subsidence and sea-level rise in the Thomas Estuary (Discussion).
Trans. A **272**, 121–130 (1972).

Oliver, G. & Allen, J.A. The functional and adaptive morphology of the deep-sea species of the Arcacea (Mollusca: Bivalvia) from the Atlantic. *Trans.* B **291**, 45–76 (1980).

The functional and adaptive morphology of the deep-sea species of the family Limopsidae (Bivalvia: Arcoida) from the Atlantic. *Trans.* B **291**, 77–125 (1980).

Olsson, J.E. *See* Bartel, Duinker, Heintze, Heinzelmann and others.

Olunloyo, V.O.S. *See* Hutter & Olunloyo.

Olver, F.W.J. Second-order linear differential equations with two turning points.
Trans. A **278**, 137–174 (1975).

Legendre functions with both parameters large. *Trans.* A **278**, 175–185 (1975).

General connection formulae for Liouville–Green approximations in the complex plane. *Trans.* A **289**, 501–548 (1978).

Corrigendum. *Trans.* A **290**, 686 (1979).

O'Neill, A. *See* Murgatroyd & O'Neill.

O'Neill, J.B. *See* Brookes (C.A.), O'Neill & Redfern.

O'Neill, M.E. *See* Michael & O'Neill.

O'Neill, P., Salmon, G.A. & May, R. The radiation-induced formation of excited states of aromatic hydrocarbons in alicyclic hydrocarbons. I. Yields of excited singlet and triplet state solute molecules. *Proc.* A **347**, 61–73 (1975).

O'Nions, R.K., Evensen, N.M. & Hamilton, P.J. Differentiation and evolution of the mantle (Discussion). *Trans.* A **297**, 479–493 (1980).

O'Nions, R.K., Evensen, N.M., Hamilton, P.J. & Carter, S.R. Melting of the mantle past and present: isotope and trace element evidence (Discussion). *Trans.* A **288**, 547–559 (1978).

Onishi, T. *See* Watanabe (K.), Kondow, Soma, Onishi & Tamaru.

Onomakpome, N. *See* Fynn, Onomakpome & Peart.

Orchard, A.F. *See* Bird (B.D.), Cooke, Day & Orchard.

Orchard, S.W. & Thrush, B.A. Photochemical studies of unimolecular processes.

IV. Quenching of vibrationally excited cycloheptatriene by added gases. *Proc.* A **329**, 233–240 (1972).

V. The photolysis and quenching of *cis*- and *trans*-hexa-1,3,5-triene and of cyclohexa-1,3-diene. *Proc.* A **337**, 243–256 (1974).

VI. The unimolecular reactions of C_6H_8 isomers and the interpretation of their photolyses. *Proc.* A **337**, 257–274 (1974).

Ordidge, R.J. *See* Mansfield (P.), Morris, Ordidge, Pykett, Bangert & Coupland.

O'Reilly, W. *See* Runcorn, Collinson, O'Reilly, Stephenson, Battey, Manson & Readman.

Orgel, L.E. Selection *in vitro* (Discussion). *Proc.* B **205**, 435–442 (1979).

Orger, B.H. *See* Costa (S.M. de B.), Froines, Harris, Leblanc, Orger & Porter.

Orlyansky, A.D. *See* Lysenko, Orlyansky & Portnyagin.

Orme, G.R. *See* Flood, Orme & Scoffin; Thom (B.G.), Orme & Polach.

Orme, G.R., Flood, P.G. & Sargent, G.E.G. Sedimentation trends in the lee of outer (ribbon) reefs, Northern Region of the Great Barrier Reef Province (Discussion). *Trans.* A **291**, 85–99 (1978).

Orme, G.R., Webb, J.P., Kelland, N.C. & Sargent, G.E.G. Aspects of the geological history and structure of the northern Great Barrier Reef (Discussion). *Trans.* A **291**, 23–35 (1978).

Ormerod, M.G. *See* Lenherr & Ormerod.

Orowan, E. Origin of the surface features of the Moon. *Proc.* A **336**, 141–163 (1974).

See also Doyle (M.J.), Maranci, Orowan & Stork.

Orr, D. *See* Kaiser (T.R.), Orr & Smith.

Orr, D.C. *See* Curtis (N.A.C.), Orr & Boulton.

Orr, J. *See* Needham (N.G.) & Orr.

Osborn, J.W. The ontogeny of tooth succession in *Lacerta vivipara* Jacquin (1787). *Proc.* B **179**, 261–289 (1971).

Osborne, Daphne J. & Wright M. Gravity-induced cell elongation (Discussion). *Proc.* B **199**, 551–564 (1977).

Osborne, J.L. *See* Krishnaswamy (M.R.), Menon, Narasimham, Hinotani, Ito and others.

Osborne, P.J. Insect faunas of Late Devensian and Flandrian age from Church Stretton, Shropshire. *Trans.* B **263**, 327–367 (1972).

The insect fauna of the organic deposits at Sugworth and its environmental and stratigraphic implications. *Trans.* B **289**, 119–133 (1980).

Oshima, T. *See* Allen (G.I.), Eccles, Nicoll, Oshima & Rubia; Eccles, Nicoll, Oshima & Rubia.

Osmaston, H.A. *See* Shotton (F.W.), Goudie, Briggs & Osmaston.

Ostle, D. *See* Simpson (P.R.), Brown, Plant & Ostle.

O'Sullivan, A.J. Ecological effects of sewage discharge in the marine environment (Discussion). *Proc.* B **177**, 331–351 (1971).

O'Sullivan, J., Huddleston, J.A. & Abraham, E.P. Biosynthesis of penicillins and cephalosporins in cell-free systems [synopsis] (Discussion). *Trans.* B 289, 363–365 (1980).

Ottley, T.W. *See* Mansfield (M.W.D.) & Ottley.

Oudet, P., Germond, J.E., Bellard, M., Spadafora, C. & Chambon, P. Nucleosome structure (Discussion). *Trans.* B 283, 241–258 (1978).

Ourisson, G. *See* Anding, Brandt, Ourisson, Pryce & Rohmer.

Owen, D.R.J. *See* Haward & Owen.

Owen, G. The fine structure and histochemistry of the digestive diverticula of the protobranchiate bivalve *Nucula sulcata*. *Proc.* B 183, 249–264 (1973).

Studies on the gill of *Mytilus edulis:* the eu-latero-frontal cirri. *Proc.* B 187, 83–91 (1974).

Classification and the bivalve gill (Discussion). *Trans.* B 284, 377–385 (1978).

Owen, G. & McCrae, J.M. Further studies on the latero-frontal tracts of bivalves. *Proc.* B 194, 527–544 (1976).

Sensory cell/gland cell complexes associated with the pallial tentacles of the bivalve *Lima hians* (Gmelin), with a note on specialized cilia on the pallial curtains. *Trans.* B 287, 45–62 (1979).

Owen, H.G. Continental displacement and expansion of the Earth during the Mesozoic and Cenozoic. *Trans.* A 281, 223–291 (1976).

Owen, J.M., Haynes, C.M. & Bayley, F.J. Heat transfer from an air-cooled rotating disk. *Proc.* A 336, 453–473 (1974).

Owen, M.J. Fatigue processes in fibre reinforced plastics (Discussion). *Trans.* A 294, 535–543 (1980).

Owen, P.J. *See* Andrews (E.H.), Owen & Singh.

Owen, P.R. The prospect ahead (Discussion). *Trans.* A 269, 545–554 (1971).

Owen, P.R. & Maskell, E.C. Dietrich Küchemann. *Biogr. Mem.* 26, 305–326 (1980).

Owen, R.C. *See* Gurney (C.), Mai & Owen.

Owen, W.S. *See* Cohen (M.) & Owen.

Owens, W.H. & Bamford, D. Magnetic, seismic, and other anisotropic properties of rock fabrics (Discussion). *Trans.* A 283, 55–68 (1976).

Oxburgh, E.R. The structure of the oceanic lithosphere [abstract] (Discussion). *Trans.* A 268, 619 (1971).

See also Brown (G.M.), O'Hara & Oxburgh; Turcotte & Oxburgh.

Oxburgh, E.R. & Parmentier, E.M. Thermal processes in the formation of continental lithosphere (Discussion). *Trans.* A 288, 415–429 (1978).

Oxley, P.L.B. *See* Hastings, Mathew & Oxley.

Oxley, P.L.B. & Hastings, W.F. Minimum work as a possible criterion for determining the frictional conditions at the tool/chip interface in machining. *Trans.* A 282, 565–584 (1976).

Predicting the strain rate in the zone of intense shear in which the chip is formed in machining from the dynamic flow stress properties of the work material and the cutting conditions. *Proc.* A 356, 395–410 (1977).

Oyster, C.W. & Takahashi, Ellen S. Interplexiform cells in rabbit retina. *Proc.* B 197, 477–484 (1977).

Ozbekhan, H. The future of Paris: a systems study in strategic urban planning (Discussion). *Trans.* A 287, 523–544 (1977).

Pace, N.G. & Saunders, G.A. Ultrasonic study of lattice stability in indium+thallium alloys. *Proc.* A 326, 521–533 (1972).

Pacini, E. *See* de Nettancourt, Devreux, Carluccio and others.

Pack, D.C. *See* Cole (R.J.) & Pack.

Packer, K.J. The dynamics of water in heterogeneous systems (Discussion). *Trans.* B 278, 59–87 (1977).

Padalia, B.D., Lang, W.C., Norris, P.R., Watson, L.M. & Fabian, D.J. X-ray photoelectron core-level studies of the heavy rare-earth metals and their oxides. *Proc.* A 354, 269–290 (1977).

Padday, J.F. The profiles of axially symmetric menisci. *Trans.* A **269**, 265–293 (1971).
  Sessile drop profiles: corrected methods for surface tension and spreading coefficients.
    Proc. A **330**, 561–572 (1972).
Padday, J.F. & Pitt, A.R. Surface and interfacial tensions from the profile of a sessile drop.
    Proc. A **329**, 421–431 (1972).
  The stability of axisymmetric menisci. *Trans.* A **275**, 489–528 (1973).
Padgham, R.C. *See* Parkin (D.W.) & Padgham.
Padley, P.J. *See* Bulewicz & Padley; Kelly (R.) & Padley.
Paffett, J.A.H. Hydrodynamics and ship performance (Discussion). *Trans.* A **273**, 77–84 (1972).
Page, J.K., Rodgers, G.G. & Souster, C.G. Systematic design assessment techniques for solar buildings
    (Discussion). *Trans.* A **295**, 379–401 (1980).
Page, K.R. *See* Meares & Page.
Page, Sally. *See* Lammel, Niedergerke & Page; Niedergerke & Page.
Page, T.F. & Ralph, B. Field-ion microscopy of dilute substitutional solid solutions.
    Proc. A **339**, 223–243 (1974).
Pain, R.H. *See* Carrey, Mitchinson, Pain & Virden.
Painter, H.A. Biodegradability (Discussion). *Proc.* B **185**, 149–158 (1974).
Pal, Y. Some experiences in preparing for a satellite television experiment for rural India (Discussion).
    Proc. A **345**, 437–447 (1975).
Palese, P., Racaniello, V.R., Desselberger, U., Young, J. & Baez, M. Genetic structure and genetic
    variation of influenza viruses (Discussion). *Trans.* B **288**, 299–305 (1980).
Palin, C.J. Amplitude of the de Haas—van Alphen effect in mercury. *Proc.* A **329**, 17–34 (1972).
Palka, J. *See* Edwards (J.S.) & Palka.
Palka, J. & Edwards, J.S. The cerci and abdominal giant fibres of the house cricket, *Acheta domesti-*
    cus. II. Regeneration and effects of chronic deprivation. *Proc.* B **185**, 105–121 (1974).
Palme, H. *See* Wänke, Palme, Baddenhausen, Dreibus, Kruse & Spettel.
Palme, H. & Wänke, H. Lunar differentiation processes as characterized by trace element abundances
    (Discussion). *Trans.* A **285**, 199–205 (1977).
Palmer, A.C. & Rice, J.R. The growth of slip surfaces in the progressive failure of over-consolidated
    clay. *Proc.* A **332**, 527–548 (1973).
Palmer, F. Wood supply and demand – the medium term prospects for Britain (Discussion).
    Trans. B **271**, 81–89 (1975).
Palmer, H.P. & Anderson, B. Useful geodetic measurements with radio interferometers? (Discussion).
    Trans. A **274**, 195–197 (1973).
Palmer, S.B. & Lee, E.W. The elastic constants of dysprosium and holmium.
    Proc. A **327**, 519–543 (1972).
Palmer, S.B., Lee, E.W. & Islam, M.N. The elastic constants of gadolinium, terbium and erbium.
    Proc. A **338**, 341–357 (1974).
Palmer, T.F. *See* Archer (A.S.), Cundall, Evans & Palmer; Archer (A.S.), Cundall & Palmer.
Palmork, K.H. *See* Jensen (S.), Lange, Berge, Palmork & Renberg.
Pancharatnam, S. On the magnetic resonance of an alined spin-assembly.
    Proc. A **330**, 265–270 (1972).
  The ellipsoid of alinement and its precessional motion in magnetic resonance.
    Proc. A **330**, 271–280 (1972).
  Theory of dispersion in relation to light shifts. *Proc.* A **330**, 281–289 (1972).
Panchen, A.L. The skull and skeleton of *Eogyrinus attheyi* Watson (Amphibia: Labyrinthodontia).
    Trans. B **263**, 279–326 (1972).
  A new genus and species of anthracosaur amphibian from the Lower Carboniferous of Scotland
    and the status of *Pholidogaster pisciformis* Huxley. *Trans.* B **269**, 581–637 (1975).
  On *Anthracosaurus russelli* Huxley (Amphibia: Labyrinthodontia) and the family Anthracosauridae.
    Trans. B **279**, 447–512 (1977).
Pandey, Dhananjai, Lele, Shrikant & Krishna, P. X-ray diffraction from one-dimensionally disordered
    2H crystals undergoing solid state transformation to the 6H structure.
  I. The layer displacement mechanism. *Proc.* A **369**, 435–449 (1980).

II. The deformation mechanism. *Proc.* A **369**, 451–461 (1980).

III. Comparison with experimental observations on SiC. *Proc.* A **369**, 463–477 (1980).

Pankhurst, R.C. *See* Young (A.D.), Pankhurst & Schultz.

Pant, D.D. & Khare, P.K. *Damudopteris* gen.nov. – a new genus of ferns from the Lower Gondwanas of the Raniganj coalfield, India. *Proc.* B **186**, 121–135 (1974).

Paoletti, E. Grossi. *See* Fiecchi, Kienle, Scala, Galli and others.

Paoletti, R. *See* Fiecchi, Kienle, Scala, Galli and others.

Papadopoulos, G.J. Functional integrals for Fermi systems without quaternions. *Proc.* A **350**, 547–552 (1976).

Papadopoulos, R. *See* Barrer & Papadopoulos; Barrer, Papadopoulos & Ramsay.

Papanastassiou, D.A. *See* Wasserburg, Papanastassiou, Tera & Huneke.

Páquet, P. Variations of Doppler results with software and time (Discussion). *Trans.* A **294**, 237–244 (1980).

du Parcq, R.P. *See* Barrow (R.F.) & du Parcq.

Pardoe, G.K.C. Design and operational considerations of educational satellite systems (Discussion). *Proc.* A **345**, 477–491 (1975).

Pardon, J.F. *See* Richards (B.M.), Pardon, Lilley, Cotter *et al.*

Parinello, M., Tosi, M.P. & March, N.H. Partial structure factors and atomic dynamics in conformal solutions. *Proc.* A **341**, 91–104 (1974).

Paris, R.B. The asymptotic behaviour of solutions of the differential equation

$$\frac{d^4 u}{dz^4} + \left[z^2 \frac{d^2 u}{dz^2} + az \frac{du}{dz} + bu \right] = 0. \ Proc. \ A \ 346, \ 171–207 \ (1975).$$

Paris, R.B. & Wood, A.D. The asymptotic expansion of solutions of the differential equation $u^{iv} + \lambda^2 \left[(z^2 + c) u'' + azu' + bu \right] = 0$ for large $|z|$. *Trans.* A **293**, 511–533 (1980).

Parish, C.R. & Hayward, J.A. The lymphocyte surface.

I. Relation between Fc receptors, $C'3$ receptors and surface immunoglobulin. *Proc.* B **187**, 47–63 (1974).

II. Separation of Fc receptor, $C'3$ receptor and surface immunoglobulin-bearing lymphocytes. *Proc.* B **187**, 65–81 (1974).

III. Function of Fc receptor, $C'3$ receptor and surface Ig bearing lymphocytes: identification of a radioresistant B cell. *Proc.* B **187**, 379–395 (1974).

Park, D. *See* Landsberg & Park.

Parker, A.B. *See* Bhasavanich & Parker; Johnson (P.C.) & Parker.

Parker, A.B. & Johnson, P.C. The dielectric breakdown of low density gases. I. Theoretical. *Proc.* A **325**, 511–527 (1971).

Parker, B.A. The effect of minor element concentrations on the strain rate sensitivity and ductility of commercial purity aluminium sheet [abstract] (Discussion). *Trans.* A **295**, 130 (1980).

Parker, D.F. The decay of sawtooth solutions to the Burgers equation. *Proc.* A **369**, 409–424 (1980).

Parker, G.A. *See* Baker (R.R.) & Parker.

Parker, I. *See* Bregestovski, Miledi & Parker; Miledi, Nakajima & Parker; Miledi & Parker; Miledi, Parker & Schalow.

Parker, K. *See* Rothschild (Miriam), Schlein, Parker, Neville & Sternberg.

Parker, R.A. Ancient Egyptian astronomy (Discussion). *Trans.* A **276**, 51–65 (1974).

Parkes, E.W. The expanding and contracting hinge in a rapidly heated rigid-plastic beam. *Proc.* A **337**, 351–364 (1974).

Parkes, E.W. & Carter, G.A. Dynamic thermal stresses in a pulsed reactor. *Trans.* A **270**, 325–347 (1971).

Parkes, Sir Alan. Sydney John Folley. *Biogr. Mem.* **18**, 241–265 (1972). Corrigendum. *Biogr. Mem.* **19**, 695 (1973).

Parkin, A. *See* Heddle, Keesing & Parkin.

Parkin, D.W. Trade-winds during the glacial cycles. *Proc.* A **337**, 73–100 (1974).

Parkin, D.W. & Padgham, R.C. Further studies on trade winds during the glacial cycles. *Proc.* A **346**, 245–260 (1975).

Parkin, D.W., Sullivan, R.A.L. & Andrews, J.N. Further studies on cosmic spherules from deep-sea sediments. *Trans.* A **297**, 495–518 (1980).

Parkin, P.H. Acoustics of concert and multi-purpose halls (Discussion).
Trans. A **272**, 621–625 (1972).

Parkinson, G.V. Wind-induced instability of structures (Discussion). *Trans.* A **269**, 395–413 (1971).

Parkinson, J.H. High resolution X-ray spectra of the Sun (Discussion). *Trans.* A **281**, 375–382 (1976). *See also* Herring (J.R.H.), Glencross, Parkinson & Pounds.

Parkinson, W.H. *See* Garton & Parkinson.

Parkinson, W.H. & Reeves, E.M. Autoionization resonances and configuration-mixing in the emission spectra of ZnI and CdI. *Proc.* A **331**, 237–247 (1972).

Parkinson, W.H., Reeves, E.M. & Tomkins, F.S. Measurements of Sc I *gf*-values.
Proc. A **351**, 569–579 (1976).

Parmentier, E.M. *See* Oxburgh & Parmentier.

Parrinello, M., Tosi, M.P. & March, N.H. Partial structure factors and atomic dynamics in conformal solutions. *Proc.* A **341**, 91–104 (1974).

Parrington, F.R. On the Upper Triassic mammals. *Trans.* B **261**, 231–272 (1971).
A further account of the Triassic mammals. *Trans.* B **282**, 177–204 (1978).
See also Jenkins (F.A.) & Parrington; Westoll & Parrington.

Parrington, F.R. & Westoll, T.S. David Meredith Seares Watson. *Biogr. Mem.* **20**, 483–504 (1974).

Parry, D.A.D. *See* Doyle (B.B.), Hulmes, Miller, Parry, Piez & Woodhead-Galloway.

Parry, D.A.D., Barnes, G.R.G. & Craig, A.S. A comparison of the size distribution of collagen fibrils in connective tissues as a function of age and a possible relation between fibril size distribution and mechanical properties. *Proc.* B **203**, 305–321 (1978).

Parry, D.A.D., Craig, A.S. & Barnes, G.R.G. Tendon and ligament from the horse: an ultrastructural study of collagen fibrils and elastic fibres as a function of age. *Proc.* B **203**, 293–303 (1978).

Parshad, V.R., Crompton, D.W.T. & Nesheim, M.C. The growth of *Moniliformis* (Acanthocephala) in rats fed on various monosaccharides and disaccharides. *Proc.* B **209**, 299–315 (1980).

Parsons, A.J. *See* Krishnaswamy (M.R.), Menon, Narasimham, Hinotani, Ito and others.

Parsons, B.J. Spectroscopic mode Grüneisen parameters for diamond. *Proc.* A **352**, 397–417 (1977).

Partington, G.A. *See* De Robertis, Partington & Gurdon.

Partridge, S. Miles & Blombäck, Birger. Pehr Victor Edman. *Biogr. Mem.* **25**, 241–265 (1979).

Parvizi, A. *See* Bailey (J.E.), Curtis & Parvizi.

Pasquill, F. Wind structure in the atmospheric boundary layer (Discussion).
Trans. A **269**, 439–456 (1971).
See also Sutcliffe (R.C.) & Pasquill.

Pasquill, F., Sheppard, P.A. & Sutcliffe, R.C. Oliver Graham Sutton.
Biogr. Mem. **24**, 529–546 (1978).

Passatore, M. *See* Gallagher, Kent, Passatore, Phipps & Richardson; Phipps, Richardson, Corfield, Gallagher and others.

Passmore, R. *See* McCance, El Neil, El Din, Widdowson, Southgate and others.

Paster, T.P. *See* Watkins (N.D.) & Paster.

Patel, C.K.N. High resolution spectroscopy and atmospheric and stratospheric detection of minor constituents with the use of tunable spin flip Raman lasers (Discussion).
Trans. A **293**, 257–275 (1979).

Patel, M.H. On laminar boundary layers in oscillatory flow. *Proc.* A **347**, 99–123 (1975).
Corrigenda. *Proc.* A **348**, 557 (1976).
On turbulent boundary layers in oscillatory flow. *Proc.* A **353**, 121–144 (1977).

Paterson, A.B. Animal health (Discussion). *Trans.* B **267**, 113–130 (1973).

Paterson, M.S. Some current aspects of experimental rock deformation (Discussion).
Trans. A **283**, 163–172 (1976).

Paterson, S.J. *See* Kosterlitz & Paterson.

Patey, A.L. *See* Hemmings (C.), Hemmings, Patey & Wood.

Paton, Sir Angus. William Hudson. *Biogr. Mem.* **25**, 319–335 (1979).
The Eighth Royal Society Technology Lecture. Dams and their interfaces.

Proc. A **351**, 1–17 (1976).

See also Thomas (A.R.) & Paton.

Paton, W.D.M. *See* Griffiths (H.B.), Miller, Paton & Smith; MacIntosh (F.C.) & Paton.

Patrick, G. & Stirling, Christine. The retention of particles in large airways of the respiratory tract.
Proc. B **198**, 455–462 (1977).

Patrick, J.M. *See* Cotes, Anderson & Patrick.

Patrick, J.M. & Cotes, J.E. Anthropometric and other factors affecting respiratory responses to carbon dioxide in New Guineans (Discussion). *Trans.* B **268**, 363–373 (1974).

Patterson, C. The braincase of pholidophorid and leptolepid fishes with a review of the actinopterygian braincase. *Trans.* B **269**, 275–579 (1975).

Patterson, L.K. *See* Kelly (Angela R.) & Patterson.

Patterson, S.J. Diophantine approximation in Fuchsian groups. *Trans.* A **282**, 527–563 (1976).

Patzelt, C. *See* Steiner (D.F.), Patzelt, Chan, Quinn, Tager, Nielsen and others.

Pau, R.N., Brunet, P.C.J. & Williams, Monica J. The isolation and characterization of proteins from the left colleterial gland of the cockroach, *Periplaneta americana* (L.).
Proc. B **177**, 565–579 (1971).

Paul, D.H. *See* Mackie (G.O.), Paul, Singla, Sleigh & Williams.

Paul, D.H. & Roberts, B.L. Studies on a primitive cerebellar cortex.
 I. The anatomy of the lateral-line lobes of the dogfish, *Scyliorhinus canicula.*
Proc. B **195**, 453–466 (1977).
 II. The projection of the posterior lateral-line nerve to the lateral-line lobes of the dogfish brain.
Proc. B **195**, 467–478 (1977).
 III. The projection of the anterior lateral-line nerve to the lateral-line lobes of the dogfish brain.
Proc. B **195**, 479–496 (1977).

Paul, D.K. *See* Harris (P.G.), Hutchison & Paul.

Paul, G.R. *See* Duckworth (R.F.), Paul & Cameron.

Paul, G.R. & Cameron, A. An absolute high-pressure microviscometer based on refractive index.
Proc. A **331**, 171–184 (1972).
The ultimate shear stress of fluids at high pressures measured by a modified impact microviscometer.
Proc. A **365**, 31–41 (1979).

Paul, J. Introductory remarks (Discussion). *Trans.* B **283**, 331–332 (1978).

Paul, J.P. Force actions transmitted by joints in the human body (Discussion).
Proc. B **192**, 163–172 (1976).

Pauling, L. The theory of resonance in chemistry. *Proc.* A **356**, 433–441 (1977).

Paulitschke, W. *See* Grabke, Petersen & Paulitschke.

Pawley, G.S. *See* Dolling, Pawley & Powell.

Pawson, I.G. Growth and development in high altitude populations: a review of Ethiopian, Peruvian, and Nepalese studies (Discussion). *Proc.* B **194**, 83–98 (1976).

See also Clegg (E.J.), Pawson, Ashton & Flinn.

Paxton, H.J.B. *See* Bates (B.), Bradley, McBride & others.

Paxton, J.D. *See* Howlett, Knox, Paxton & Heslop-Harrison.

Payne, A. *See* Martin (B.W.) & Payne.

Payne, D.N. *See* Gambling & Payne.

Payne, M.R. Growth of a fur seal population (Discussion). *Trans.* B **279**, 67–79 (1977).

Paz-Aliaga, A. *See* Whittembury, de Martínez, Linares & Paz-Aliaga.

Peachey, J.E. Environmental information and data handling (Discussion).
Proc. B **185**, 209–219 (1974).

Peacock, M.A. *See* Smith (M.W.) & Peacock.

Peake, J.F. The evolution of terrestrial faunas in the western Indian Ocean (Discussion).
Trans. B **260**, 581–610 (1971).

See also Stoddart (D.R.) & Peake; Taylor (J.D.), Braithwaite, Peake & Arnold.

Peaker, M. Avian salt glands (Discussion). *Trans.* B **262**, 289–300 (1971).

Pearlman, M.R., Lanham, N.W., Lehr, C.G. & Wohn, J. Smithsonian Astrophysical Observatory laser tracking systems (Discussion). *Trans.* A **284**, 431–442 (1977).

Pearse, A.G.E. The phylogeny of the common peptides (Chairman's address) (Discussion).
Proc. B **210**, 61–62 (1980).

Pearson, C.R. & McConnell, G. Chlorinated C_1 and C_2 hydrocarbons in the marine environment (Discussion). *Proc.* B **189**, 305–332 (1975).

Pearson, J.C. A revision of the subfamily Haplorchinae Looss, 1899 (Trematoda: Heterophyidae). II. Genus *Galactosomum. Trans.* B **266**, 341–447 (1973).

Pearson, Lucy. The corpora pedunculata of *Sphinx ligustri* L. and other Lepidoptera: an anatomical study. *Trans.* B **259**, 477–516 (1971).

Pearson, R.C.A. *See* Shanks, Pearson & Powell.

Pearson, R.C.A. & Powell, T.P.S. The cortico-cortical connections to area 5 of the parietal lobe from the primary somatic sensory cortex of the monkey. *Proc.* B **200**, 103–108 (1978).

Pearson, T.H. The effect of industrial effluent from pulp and paper mills on the marine benthic environment (Discussion). *Proc.* B **180**, 469–485 (1972).

Pearson, W.B. The stability of metallic phases and structures: phases with the AIB_2 and related structures. *Proc.* A **365**, 523–535 (1979).
Dimensional analysis of the crystal structures of intermetallic phases.
Trans. A **298**, 415–449 (1980).

Peart, W.S. *See* Fynn, Onomakpome & Peart; Logan (A.G.), Tenyi, Peart, Breathnach & Martin; Pessina, Hulme & Peart; Pessina & Peart.

Peck, D.R. *See* Dobbs, Lea & Peck; Lea (M.J.), Llewellyn, Peck & Dobbs.

Peckham, G. *See* Ellis (P.), Holah, Houghton, Jones, Peckham and others.

Peckham, G.E. *See* Chapman (W.A.), Cross, Flower, Peckham & Smith; Chapman (W.A.) & Peckham.

Peckover, R.S. *See* Gimblett & Peckover.

Pedersen, K.O. *See* Claesson & Pedersen.

Pedersen, Kai O. *See* Kekwick & Pedersen.

Pedgley, D.E. Weather during Desert Locust plague upsurges (Discussion).
Trans. B **287**, 387–391 (1979).

Pedley, M.D. *See* Borrell (P.), Borrell, Pedley & Grant.

Pedley, T.J. *See* Springer (S.G.) & Pedley.

Pegg, D.T. & Series, G.W. On the reduction of a problem in magnetic resonance.
Proc. A **332**, 281–289 (1973).

Peglar, Sylvia. *See* Birks (H.J.B.), Deacon & Peglar.

Peichl, L. *See* Boycott, Peichl & Wässle; Wässle, Boycott & Peichl; Wässle, Peichl & Boycott.

Peierls, Sir Rudolf. Perturbation theory for projected states. *Proc.* A **333**, 157–170 (1973).
The momentum of light in a refracting medium. *Proc.* A **347**, 475–491 (1976).
The momentum of light in a refracting medium. II. Generalization. Application to oblique reflexion.
Proc. A **355**, 141–151 (1977).
Recollections of early solid state physics (Symposium). *Proc.* A **371**, 28–38 (1980).
See also Atalay, Mann & Peierls; Burt (M.G.) & Peierls; Mott (Sir Nevill) & Peierls.

Peirson, D.H. & Cawse, P.A. Trace elements in the atmosphere (Discussion).
Trans. B **288**, 41–49 (1979).

Pekeris, C.L. A derivation of Laplace's tidal equation from the theory of inertial oscillations.
Proc. A **344**, 81–86 (1975).
Relativistic axially symmetric flows of a perfect fluid. *Proc.* A **355**, 53–60 (1977).
Oscillations of a gravitating and rotating uniform liquid sphere. *Proc.* A **366**, 143–154 (1979).
See also Accad & Pekeris.

Pekeris, C.L. & Accad, Y. Dynamics of the liquid core of the Earth. *Trans.* A **273**, 237–260 (1972).

Pekeris, C.L., Accad, Y. & Shkoller, B. Kinematic dynamos and the Earth's magnetic field.
Trans. A **275**, 425–461 (1973). Corrigendum. *Trans.* A **275**, 647 (1974).

Pelizzari, C.A. *See* Sköld & Pelizzari.

Penasse, L. *See* Bucourt, Heymès, Lutz, Penasse & Perronnet.

Penchina, C.M. *See* Davies (G.) & Penchina.

Pendlebury, J.M. & Smith K. The electric and magnetic moments of the neutron (Discussion).
Trans. B **290**, 617–626 (1980).

Pennington, Winifred (Mrs T.G. Tutin). The Late Devensian flora and vegetation of Britain (Discussion). *Trans.* B **280**, 247–271 (1977).

Pennington, Winifred (Mrs T.G. Tutin), Haworth, E.Y., Bonny, A.P. & Lishman, J.P. Lake sediments in northern Scotland (with a contribution by J. Johansen). *Trans.* B **264**, 191–294 (1972).

Pennock, J.F. *See* Glover, Pennock, Pitt & Goodwin.

Penny, F.D. State of the art (Discussion). *Proc.* A **321**, 147–155 (1971).

Penny, M.J. Migrant waders at Aldabra, September 1967–March 1968 (Discussion). *Trans.* B **260**, 549–559 (1971).

See also Benson (C.W.) & Penny.

Penny, M.J. & Diamond, A.W. The White-throated Rail *Dryolimnas cuvieri* on Aldabra (Discussion). *Trans.* B **260**, 529–548 (1971).

Penrose, R. General introduction to a Discussion. *Proc.* A **368**, 3–4 (1979).

Global General Relativity [abstract] (Discussion). *Proc.* A **368**, 19–21 (1979).

See also Geroch, Kronheimer & Penrose; Hansen (R.O.), Newman, Penrose & Tod.

Pentecost, A. Blue–green algae and freshwater carbonate deposits. *Proc.* B **200**, 43–61 (1978).

Peper, K. *See* McMahan (U.J.), Spitzer & Peper.

Peper, K. & McMahan, U.J. Distribution of acetylcholine receptors in the vicinity of nerve terminals on skeletal muscle of the frog. *Proc.* B **181**, 431–440 (1972).

Pepper, D.C. *See* Lorimer & Pepper.

Pepper, M. The Anderson transition in silicon inversion layers: the origin of the random field and the effect of substrate bias. *Proc.* A **353**, 225–246 (1977).

See also Mott (Sir Nevill), Pepper, Pollitt, Wallis & Adkins.

Peppiatt, S.J. The melting of small particles. II. Bismuth. *Proc.* A **345**, 401–412 (1975).

Peppiatt, S.J. & Sambles, J.R. The melting of small particles. I. Lead. *Proc.* A **345**, 387–399 (1975).

Percival, Elizabeth. *See* Stacey (M.) & Percival.

Percival, I.C. Planetary atoms. *Proc.* A **353**, 289–297 (1977).

Stochastic metastability and Hamiltonian dynamics. *Proc.* A **366**, 129–141 (1979).

Peregrine, D.H. & Smith, R. Nonlinear effects upon waves near caustics. *Trans.* A **292**, 341–370 (1979).

Peregrine, D.H. & Thomas, G.P. Finite-amplitude deep-water waves on currents. *Trans.* A **292**, 371–390 (1979).

Pereira, H.C. Land-use in semi-arid southern Africa (Discussion). *Trans.* B **278**, 555–563 (1977).

Concluding remarks to a Discussion. *Trans.* B **281**, 299–301 (1977).

See also Pereira (Sir Charles).

Pereira, H.G. *See* Tyrrell & Pereira.

Pereira, Marguerite S. The effects of shifts and drifts on the epidemiology of influenza in man (Discussion). *Trans.* B **288**, 423–432 (1980).

Pereira, Sir Charles. Opening remarks to a Discussion. *Proc.* B **199**, 3–4 (1977).

Concluding remarks to a Discussion. *Proc.* A **363**, 131–133 (1978).

See also Pereira (H.C.).

Peretz, B. *See* Shimahara & Peretz.

Perham, R.N. Self-assembly of biological macromolecules (Discussion). *Trans.* B **272**, 123–136 (1975).

Perkin, G.W., Duncan, G.W., Mahoney, R.T. & Smith, R.H. Contraceptive development for developing countries: unmet needs (Discussion). *Proc.* B **195**, 187–198 (1976).

Perkins, D.F. *See* Heal & Perkins.

Perkins, D.H. *See* Frank (F.C.) & Perkins.

Perkins, E.J. The need for sublethal studies (Discussion). *Trans.* B **286**, 425–442 (1979).

Perkins, F.T. Technology for prophylactic immunization (Discussion). *Proc.* B **199**, 99–107 (1977).

Perkins, H.R. *See* Ghuysen, Frère, Leyh-Bouille, Perkins & Nieto.

Perlman, S. *See* Bishop (J.O.), Beckmann, Campo, Hastie & others.

Pernow, B. *See* Hökfelt, Lundberg, Schultzberg, Johansson, Skirboll, Ånggård and others.

Perram, J.W. *See* Barouch, Perram & Smith; de Leeuw, Perram & Smith.

Perram, J.W. & Smith, E.R. Competitive adsorption via the Percus–Yevick approximation. *Proc.* A **353**, 193–220 (1977).

Perram, J.W. & Stiles, P.J. On the application of ellipsoidal harmonics to potential problems in molecular electrostatics and magnetostatics. *Proc.* A **349**, 125–139 (1976).

Perrella, M. *See* Johnson (P.) & Perrella.

Perrin, R.M.S., Rose, J. & Davies, H. The distribution, variation and origins of pre-Devensian tills in eastern England. *Trans.* B **287**, 535–570 (1979).

Perrins, W.T., McKenzie, D.R. & McPhedran, R.C. Transport properties of regular arrays of cylinders. *Proc.* A **369**, 207–225 (1979).

Perronet, J. *See* Bucourt, Heymès, Lutz, Penasse & Perronnet.

Perry, G.E. *See* Wood (C.D.) & Perry.

Perry, J.S. Implantation, foetal membranes and early placentation of the African elephant, *Loxodonta africana. Trans.* B **269**, 109–135 (1974).

  See also Amoroso & Perry.

Perry, M.J. *See* Gibbons (G.W.) & Perry.

Perry, P.E. *See* Callan & Perry.

Perry, R.A. The evaluation and exploitation of semi-arid lands: Australian experience (Discussion). *Trans.* B **278**, 493–505 (1977).

Perry, S.J. *See* Double, Hellawell & Perry.

Perry, V.H. The ganglion cell layer of the retina of the rat: a Golgi study. *Proc.* B **204**, 363–375 (1979).

Perry, V.H. & Walker, M. Amacrine cells, displaced amacrine cells and interplexiform cells in the retina of the rat. *Proc.* B **208**, 415–431 (1980).

  Morphology of cells in the ganglion cell layer during development of the rat retina. *Proc.* B **208**, 433–446 (1980).

Personne, P. *See* Thieffry, Bruner & Personne.

Perutz, M.F. Review Lecture. Stereochemical mechanism of oxygen transport by haemoglobin. *Proc.* B **208**, 135–162 (1980).

Peskett, G.D. *See* Curtis (P.D.), Houghton, Peskett & Rodgers; Drummond (J.R.), Houghton, Peskett, Rodgers, Wale and others; Ellis (P.), Holah, Houghton, Jones, Peckham and others.

Pessina, A.C., Hulme, B. & Peart, W.S. Renin induced proteinuria and the effects of adrenalectomy. II. Morphology in relation to function. *Proc.* B **180**, 61–71 (1972).

Pessina, A.C. & Peart, W.S. Renin induced proteinuria and the effects of adrenalectomy. I. Haemodynamic changes in relation to function. *Proc.* B **180**, 43–60 (1972).

  The effects of renin and adrenalectomy on blood distribution and capillary permeability. *Proc.* B **180**, 73–85 (1972).

Petch, N.J. & Wright, E. The plasticity and cleavage of polycrystalline beryllium.

  I. Yield and flow stresses. *Proc.* A **370**, 17–27 (1980).

  II. The cleavage strength and ductility transition temperature. *Proc.* A **370**, 29–39 (1980).

Peters, Josephine. *See* Berry (R.J.) & Peters.

Peters, Sir Rudolph, Shorthouse, M., Ward, P.F.V. & McDowell, E.M. Observations upon the metabolism of fluorocitrate in rats. *Proc.* B **182**, 1–8 (1972).

Peters, W., Garnham, P.C.C., Killick-Kendrick, R., Rajapaksa, N., Cheong, W.H. & Cadigan, F.C. Malaria of the orang-utan (*Pongo pygmaeus*) in Borneo. *Trans.* B **275**, 439–482 (1976).

Petersen, E.M. *See* Grabke, Petersen & Paulitschke.

Petersen, O.H. Initiation of salt and water transport in mammalian salivary glands by acetylcholine (Discussion). *Trans.* B **262**, 307–314 (1971).

  Electrogenic sodium pump in pancreatic acinar cells. *Proc.* B **184**, 115–119 (1973).

  See also Graf (J.) & Petersen; Nishiyama & Petersen.

Petersen, R.L. *See* Symons & Petersen.

Peterson, P.J. Geochemistry and ecology (Discussion). *Trans.* B **288**, 169–177 (1979).

Pethica, B.A. *See* Hall (D.G.) & Pethica; Mingins, Zobel, Pethica & Smart.

Peto, R. Detection of risk of cancer to man (Discussion). *Proc.* B **205**, 111–120 (1979).

Petrie, Isobel A. *See* Hill (P.), Campbell & Petrie.

Petrie, W.L. *See* Cannon & Petrie.

Petrini, D. *See* Asaad & Petrini.

Pettigrew, J.D. Binocular visual processing in the owl's telencephalon (Discussion).
  Proc. B **204**, 435–454 (1979).
Pettini, M. *See* Bates (B.), Carson, Dufton, McKeith and others; Boksenberg, Kirkham, Michelson, Pettini and others.
Pettitt, J.M. The megaspore wall in gymnosperms: ultrastructure in some zooidogamous forms.
  Proc. B **195**, 497–515. (1977).
Petty, H.C. *See* Kelsey (T.) & Petty.
Petty, J.A. The aspiration of bordered pits in conifer wood. *Proc.* B **181**, 395–406 (1972).
Petty, M.S. *See* Bradley (D.J.), Hutchinson, Koetser, Morrow, New & Petty.
Pfaelzer, P.F. *See* Gane, Pfaelzer & Tabor.
Pfeffer, R. *See* Leichtberg, Weinbaum, Pfeffer & Gluckman.
Pfeiffer, P. *See* Jacrot, Pfeiffer & Witz.
Pfeiffer, P., Herzog, M. & Hirth, L. Stabilization of brome mosaic virus (Discussion).
  Trans. B **276**, 99–107 (1976).
Pfenninger, K., Akert, K., Moor, H. & Sandri, C. Freeze-fracturing of presynaptic membranes in the central nervous system (Discussion). *Trans.* B **261**, 387 (1971).
Phelps, A.D.R. & Allen, J.E. A floating electrostatic sheath in a thermally produced plasma.
  Proc. A **348**, 221–233 (1976).
  High frequency oscillations in a bounded thermally produced plasma.
  Proc. A **360**, 541–555 (1978).
Philippon, B. *See* Le Berre, Garms, Davies, Walsh & Philippon.
Phillips, A. *See* Allum, McClintock, Phillips & Bowley.
Phillips, A. & McClintock, P.V.E. Field emission and field ionization in liquid ^{4}He.
  Trans. A **278**, 271–310 (1975).
Phillips, C.G. Francis Martin Rouse Walshe. *Biogr. Mem.* **20**, 457–481 (1974).
  On integration and teleonomy (Discussion). *Proc.* B **199**, 415–424 (1977).
  See also Andersen, Hagan, Phillips & Powell.
Phillips, G.O. *See* Davies (A.K.), McKellar & Phillips.
Phillips, G.T. *See* Cornforth, Clifford, Mallaby & Phillips.
Phillips, Linda. Pleistocene vegetational history and geology in Norfolk (with an appendix by B.W. Sparks). *Trans.* B **275**, 215–286 (1976).
Phillips, M.C., Barlow, A.J. & Lamb, J. Relaxation in liquids: a defect-diffusion model of visco-elasticity. *Proc.* A **329**, 193–218 (1972).
Phillips, M.J. *See* Whitehouse & Phillips.
Phillips, R.J.N. Theoretical models of proton—proton scattering (Discussion).
  Proc. A **335**, 485–501 (1973).
Phillips, Sir David. William Lawrence Bragg. *Biogr. Mem.* **25**, 75–143 (1979).
Phillips, V. *See* Gillett, Roman & Phillips.
Philp, D.K. *See* Merritt, Hyde, Bursill & Philp.
Philp, J. McL. A multi-national company, the public and the environment (Discussion).
  Proc. B **185**, 199–208 (1974).
Phipps, R.J. *See* Gallagher, Kent, Passatore, Phipps & Richardson.
Phipps, R.J., Richardson, P.S., Corfield, A., Gallagher, J.T., Jeffery, P.K., Kent, P.W. & Passatore, M.
  A physiological, biochemical and histological study of goose tracheal mucin and its secretion.
  Trans. B **279**, 513–540 (1977).
Phizackerley, R.P. *See* Poljak, Amzel, Chen, Phizackerley & Saul.
Piccolino, M. *See* Gerschenfeld & Piccolino.
Piccolino, M. & Gerschenfeld, H.M. Activation of a regenerative calcium conductance in turtle cones by peripheral stimulation. *Proc.* B **201**, 309–315 (1978).
  Characteristics and ionic processes involved in feedback spikes of turtle cones.
  Proc. B **206**, 439–463 (1980).
Pick, D.R. *See* Ellis (P.), Holah, Houghton, Jones, Peckham and others; Miller (D.E.), Brownscombe, Carruthers, Pick & Stewart.
Pick, Monique. The radioemission from flares: impulsive phase and type III bursts (Discussion).
  Trans. A **297**, 587–593 (1980).

Pickard, D.K. *See* Welberry, Miller & Pickard.

Pickering, **Sir George**. Max Leonard Rosenheim (Baron Rosenheim of Camden).
Biogr. Mem. **20**, 349–358 (1974).

Pickering, W.M. *See* Sozou & Pickering.

Pickersgill, **Barbara** & **Heiser, C.B., Jr.** Cytogenetics and evolutionary change under domestication
(Discussion). *Trans.* B **275**, 55–69 (1976).

Pickup, B.T. The symmetric group and the method of diatomics in molecules: an application to small
lithium clusters. *Proc.* A **333**, 69–87 (1973).

Piddington, R.W. & Sattelle, D.B. Motion in nerve ganglia detected by light-beating spectroscopy.
Proc. B **190**, 415–420 (1975).

Pidgeon, C.R. *See* Dennis (R.B.), Pidgeon, Smith, Wherrett & Wood.

Pierre, J. *See* Rothschild (Miriam), von Euw, Reichstein, Smith & Pierre.

Piez, K.A. *See* Doyle (B.B.), Hulmes, Miller, Parry, Piez & Woodhead-Galloway.

Piggott, L.S. *See* Kilburn & Piggott.

Piggott, S. Concluding remarks to a Discussion. *Trans.* A **276**, 275–276 (1974).
Robert Eric Mortimer Wheeler. *Biogr. Mem.* **23**, 623–642 (1977).
See also Kendall (D.G.) & Piggott.

Piggott, W.R. The importance of the Antarctic in atmospheric sciences (Discussion).
Trans. B **279**, 275–285 (1977).

Pigott, C.D. Natural regeneration of *Tilia cordata* in relation to forest-structure in the forest of
Białowieża, Poland. *Trans.* B **270**, 151–179 (1975).
The scientific basis of practical conservation: aims and methods of conservation (Discussion).
Proc. B **197**, 59–68 (1977).

Pigott, C.D. & Wilson, Joan F. The vegetation of North Fen at Esthwaite in 1967–9.
Proc. B **200**, 331–351 (1978).

Pigott, R.W. The development of endoscopy of the palatopharyngeal isthmus (Discussion).
Proc. B **195**, 269–275 (1977).

Pike, E.R. Light scattering in the study of dynamical properties (Discussion).
Trans. A **293**, 349–358 (1979).

Pilcher, J.R. & Smith, A.G. Palaeoecological investigations at Ballynagilly, a Neolithic and Bronze
Age settlement in County Tyrone, Northern Ireland. *Trans.* B **286**, 345–369 (1979).

Pillinger, C.T. & Eglinton, G. The chemistry of carbon in the lunar regolith (Discussion).
Trans. **285**, 369–377 (1977).

Pillinger, C.T., Eglinton, G., Gowar, A.P. & Jull, A.J.T. Carbon chemistry of the Luna 16 and 20
samples. *Trans.* A **284**, 145–150 (1977).

Pillinger, C.T. & Fabian, D.M. The separation and distribution of some Luna 24 core materials.
Trans. A **297**, 1–6 (1980).

Pillinger, C.T. & Gowar, A.P. The separation and subdivision of two 0.5 g samples of lunar soil
collected by the Luna 16 and 20 missions. *Trans.* A **284**, 137–143 (1977).

Pinder, A.C. *See* Malvern, Pinder, Stacey & Thompson.

Pineau, A. *See* Lemblé, Pineau, Catagné, Dumoulin & Guttmann.

Piper, J.D.A. Palaeomagnetic evidence for a Proterozoic super-continent (Discussion).
Trans. A **280**, 469–490 (1976).
See also Gibson (I.L.) & Piper.

Pippard, A.B. *See* Harding (G.L.), Pippard & Tomlinson; Pippard, (Sir Brian).

Pippard, A.B., Shepherd, J.G. & Tindall, D.A. Resistance of superconducting–normal interfaces.
Proc. A **324**, 17–35 (1971).

Pippard, **Sir Brian**. The magnetoresistance of polycrystalline copper. *Trans.* A **291**, 569–598 (1979).
See also Pippard (A.B.).

Pirie, N.W. Introductory remarks to a Discussion. *Proc.* B **179**, 173–175 (1971).
Frederick Charles Bawden. *Biogr. Mem.* **19**, 19–63 (1973).
Fixation of nucleic acid by leaf fibre and calcium phosphate. *Proc.* B **185**, 343–356 (1974).
Introductory remarks to a Discussion. *Proc.* B **189**, 139–141 (1975).
Food protein sources (Discussion). *Trans.* B **274**, 489–498 (1976).

The extended use of fractionation processes (Discussion). *Trans.* B **281**, 139–151 (1977).
Concluding remarks to a Discussion. *Proc.* B **199**, 565–566 (1979).
See also Bawden & Pirie.
Pitt, A.R. *See* Padday & Pitt.
Pitt, G.A.J. *See* Glover, Pennock, Pitt & Goodwin.
Pittman, J.F.T. *See* McLaughlin (E.) & Pittman.
Pitt-Rivers, Rosalind. *See* Himsworth & Pitt-Rivers.
Pitts, J.N. Photochemical and biological implications of the atmospheric reactions of amines and benzo(a)pyrene (Discussion). *Trans.* A **290**, 551–576 (1979).
Plant, Jane. *See* Simpson (P.R.), Brown, Plant & Ostle; Watson (Janet V.) & Plant.
Plant, Jane & Moore, P.J. Regional geochemical mapping and interpretation in Britain (Discussion). *Trans.* B **288**, 95–112 (1979).
Plaskett, J.S. & Barton, G. Analysis of the adsorption potential in the hydrodynamic model of metals. *Proc.* A **372**, 415–439 (1980).
Platt, C.M. *See* Boyce (A.J.), Harrison, Platt, Hornabrook and others.
Player, M.A. Dispersion and the transverse aether drag. *Proc.* A **345**, 343–344 (1975).
On the dragging of the plane of polarization of light propagating in a rotating medium.
Proc. A **349**, 441–445 (1976).
Plotch, S.J. *See* Krug, Bouloy & Plotch.
Poggio, G.F. *See* Fischer & Poggio.
Poggio, T. *See* Marr (D.) & Poggio; Torre & Poggio.
Pojeta, J., Jr. The origin and early taxonomic diversification of pelecypods (Discussion).
Trans. B **284**, 225–246 (1978).
Polach, H.A. *See* McLean (R.F.), Stoddart, Hopley & Polach; Thom (B.G.), Orme & Polach.
Polach, H.A., McLean, R.F., Caldwell, J.R. & Thom, B.G. Radiocarbon ages from the northern Great Barrier Reef (Discussion). *Trans.* A **291**, 139–158 (1978).
Polak, R.L. *See* Ito (Y.), Miledi, Molenaar, Vincent, Polak and others; Miledi, Molenaar & Polak.
Poland, A.I. *See* MacQueen, Gosling, Hildner, Munro, Poland & Ross.
Polani, P.E. *See* Embury, Seller, Adinolfi & Polani; Seller, Embury, Polani & Adinolfi.
Polden, P.G. *See* Mollon & Polden.
Polden, P.G. & Mollon, J.D. Reversed effect of adapting stimuli on visual sensitivity.
Proc. B **210**, 235–272 (1980).
Poldy, F. *See* Batley, Bramley, Poldy & Robinson.
Poljak, R.J., Amzel, L.M., Chen, B.L., Phizackerley, R.P. & Saul, F. Structure and specificity of antibody molecules (Discussion). *Trans.* B **272**, 43–51 (1975).
Pollack, H.N. *See* Gass, Chapman, Pollack & Thorpe.
Pollack, J.B. *See* Greeley, Iversen, Pollack, Udovich & White.
Pollak, M. A dielectric theory for amorphous semiconductors. *Proc.* A **325**, 383–400 (1971).
Pollard, R.T. *See* Barat, Cullis & Pollard.
Pollitt, S. *See* Mott (Sir Nevill), Pepper, Pollitt, Wallis & Adkins.
Pollock, M.D. The interaction between a weak magnetic field and a slowly rotating black hole.
Proc. A **350**, 239–252 (1976).
The interaction between a rotating, spherical shell of matter and a central black hole.
Proc. A **362**, 469–492 (1978).
Pollock, M.D. & Brinkmann, W.P. The interaction between a weak magnetic field and a black hole.
Proc. A **356**, 351–362 (1977).
Pollock, M.R. The function and evolution of penicillinase (Discussion).
Proc. B **179**, 385–401 (1971).
Pollock, M.R. & Abraham, E.P. Concluding remarks to a Discussion. *Trans.* B **289**, 377–378 (1980).
Pond, R.C. Periodic grain boundary structures in aluminium. II. A geometrical method for analysing periodic grain boundary structure and some related transmission electron microscope observations. *Proc.* A **357**, 471–483 (1977).
Pond, R.C. & Bollmann, W. The symmetry and interfacial structure of bicrystals.
Trans. A **292**, 449–472 (1979).

Pond, R.C. & Smith, D.A. Plasticity of grain boundaries [abstract] (Discussion). *Trans.* A **295**, 166—167 (1980).

Pond, R.C. & Vitek, V. Periodic grain boundary structures in aluminium. I. A combined experimental and theoretical investigation of coincidence grain boundary structure in aluminium. *Proc.* A **357**, 453—470 (1977).

Pontecorvo, G. & Bokhari, M. Hedge-like habit of *Juniperus excelsa* at high altitude on the Southern Zagros Mountains in Iran. *Proc.* B **188**, 507—508 (1975).

Pool, R.R. *See* Muscatine & Pool.

Pool, R.R., Jr. *See* Trench, Pool, Logan & Engelland.

Pooley, Christine M. & Tabor, D. Friction and molecular structure: the behaviour of some thermoplastics. *Proc.* A **329**, 251—274 (1972).

Pooley, D. *See* Hobbs (L.W.), Hughes & Pooley.

Pooley, F.D. The use of an analytical electron microscope in the analysis of mineral dusts (Discussion). *Trans.* A **286**, 625—638 (1977).

Poore, M.E.D. A conservation viewpoint (Discussion). *Proc.* A **339**, 395—410 (1974).

Pope, R. *See* Buller & Pope.

Pope, R.L. & Tassie, L.J. Recombination and dissociation of diatomic molecules. *Proc.* A **320**, 487—503 (1971).

Pope, S.B. The statistical theory of turbulent flames. *Trans.* A **291**, 529—568 (1979).

Popov, G.B. *See* Hemming (C.F.), Popov, Roffey & Waloff.

Porter, A. Some proposed experiments with the Communications Technology Satellite (Discussion). *Proc.* A **345**, 459—475 (1975).

Porter, D.A. & Edington, J.W. Microanalysis and cell boundary velocity measurements for the cellular reaction in a Mg—9 % Al alloy. *Proc.* A **358**, 335—350 (1978).

Porter, G. *See* Ashpole, Formosinho & Porter; Costa (S.M.de B.), Froines, Harris, Leblanc, Orger & Porter; Kemp (D.R.) & Porter; Porter (Sir George).

Porter, Helen K. & Ranson, S.L. Meirion Thomas. *Biogr. Mem.* **24**, 547—568 (1978).

Porter, J. *See* Cowey & Porter.

Porter, R. *See* Lemon, Hanby & Porter; Lemon & Porter.

Porter, R.R. The Croonian Lecture, 1980. The complex proteases of the complement system. *Proc.* B **210**, 477—498 (1980).

Porter, Sir George. The Bakerian Lecture, 1977. *In vitro* models for photosynthesis. *Proc.* A **362**, 281—303 (1978). Corrigenda. *Proc.* A **362**, 572 (1978).

[Abstract.] *Proc.* B **202**, 539—540 (1978).

Chairman's comment (Discussion). *Trans.* A **295**, 471—472 (1980).

See also Beddard, Fleming, Gijzeman & Porter; Beddard, Fleming, Porter & Robbins; Beddard, Porter & Weese; Costa (Silvia M. de B.) & Porter; Creed (D.), Hales & Porter; Darwent, Kalyanasundaram & Porter; Formosinho, Porter & West; Kalyanasundaram & Porter; Porter (G.).

Portman, J.E. The bioaccumulation and effects of organochlorine pesticides in marine animals (Discussion). *Proc.* B **189**, 291—304 (1975).

Portmann, R. *See* Birnstiel, Kressmann, Schaffner, Portmann & Busslinger.

Portnyagin, Yu. I. *See* Lysenko, Orlyansky & Portnyagin.

Posner, A.M. *See* Atkinson (R.J.), Posner & Quirk.

Posnette, A.F. *See* Montgomery & Posnette.

Possingham, J.V. & Rose, R.J. Chloroplast replication and chloroplast DNA synthesis in spinach leaves. *Proc.* B **193**, 295—305 (1976).

Postgate, J.R. Possibilities for the enhancement of biological nitrogen fixation (Discussion). *Trans.* B **281**, 249—260 (1977).

Photoelectron spectra and valence shell orbital structures of groups V and VI hydrides. *Proc.* A **326**, 181—197 (1972). Corrigenda. *Proc.* A **328**, 585 (1972).

Posthuma, H. *See* Bovée, Creyghton, Getreuer, Korbee, Lobregt and others.

Potter, A.J.B. *See* Edmunds, Potter & Stuart.

Potter, L.T. *See* Nickel & Potter.

Potter, R.T. On the mechanism of tensile fracture in notched fibre reinforced plastics.

Proc. A **361**, 325–341 (1978).

See also Ewins & Potter.

de Potter, W.P. Noradrenaline storage particles in splenic nerve (Discussion).
Trans. B **261**, 313–317 (1971).

Potts, A.W. *See* Lee (E.P.F.) & Potts.

Potts, A.W. & Price, W.C. The photoelectron spectra of methane, silane, germane and stannane.
Proc. A **326**, 165–179 (1972).

Photoelectron spectra and valance shell orbital structures of groups V and VI hydrides.
Proc. A **326**, 181–197 (1972). Corrigenda. *Proc.* A **328**, 585 (1972).

Potts, A.W., Williams, T.A. & Price, W.C. Photoelectron spectra and electronic structure of diatomic alkali halides. *Proc.* A **341**, 147–161 (1974).

Potts, D.M. The implementation of family planning programmes (Discussion).
Proc. B **195**, 213–224 (1976).

Family spacing and limitation: acceptable and effective techniques – still in the future? (Discussion).
Proc. B **199**, 129–144 (1977).

Potts, M. & Whitton, B.A. Vegetation of the intertidal zone of the lagoon of Aldabra, with particular reference to the photosynthetic prokaryotic communities. *Proc.* B **208**, 13–55 (1980).

Poulain, D.A., Wakerley, J.B. & Dyball, R.E.J. Electrophysiological differentiation of oxytocin- and vasopressin-secreting neurones. *Proc.* B **196**, 367–384 (1977).

Poulsen, F.M. *See* Costa (J.L.), Dobson, Kirk, Poulsen, Valeri & Vecchione.

Poulsen, F.R. *See* Lowy, Vibert, Haselgrove & Poulsen.

Poulsen, J.H. & Oakley, B, II. Intracellular potassium ion activity in resting and stimulated mouse pancreas and submandibular gland. *Proc.* B **204**, 99–104 (1979).

Poulsen, R.G. *See* Lowndes, Miller, Poulsen & Springford.

Pounds, K.A. Observations of binary X-ray sources with Ariel 5 (Discussion).
Proc. A **350**, 441–461 (1976).

The 2A catalogue and supplementary results from the Ariel 5 sky survey (Discussion).
Proc. A **366**, 375–390 (1979).

See also Herring (J.R.H.), Glencross, Parkinson & Pounds.

Poupeau, G. *See* Crozaz, Poupeau, Walker, Zinner & Morrison.

Powell, B.M. *See* Dolling, Pawley & Powell.

Powell, G. *See* Block (H.), Ions, Powell, Singh & Walker.

Powell, H.M. *See* Blundell & Powell.

Powell, J.R. *See* Dobzhansky & Powell.

Powell, R. The thermodynamics of pyroxene geotherms (Discussion). *Trans.* A **288**, 457–469 (1978).

Powell, T.P.S. *See* Andersen, Hagan, Phillips & Powell; Fisken, Garey & Powell; Garey & Powell; Harding (B.N.) & Powell; Kemp (Janet) & Powell; Pearson (R.C.A.) & Powell; Shanks, Pearson & Powell; Sloper, Hiorns & Powell; Sloper & Powell; Winfield, Hiorns & Powell; Winfield & Powell.

Power, E.A. *See* Babiker, Power & Thirunamachandran; Craig (D.P.), Power & Thirunamachandran.

Power, E.A. & Thirunamachandran, T. The multipolar Hamiltonian in radiation theory.
Proc. A **372**, 265–273 (1980).

Power, J.D. Fixed nuclei two-centre problem in quantum mechanics. *Trans.* A **274**, 663–697 (1973).

Power, R.J.D. & Longuet-Higgins, H.C. Learning to count: a computational model of language acquisition [abstract]. *Proc.* A **360**, 301 (1978).

Learning to count: a computational model of language acquisition. *Proc.* B **200**, 391–417 (1978).

Prabhakara, C. *See* Krueger, Guenther, Fleig, Heath, Hilsenrath and others.

Prandle, D. Storm surges in the southern North Sea and River Thames. *Proc.* A **344**, 509–539 (1975).

Residual flows and elevations in the southern North Sea. *Proc.* A **359**, 189–228 (1978).

Prasad, Phoolan. *See* Bhatnagar (P.L.) & Prasad.

Prather, M.J. *See* Logan (Jennifer A.), Prather, Wofsy & McElroy.

Pratt, K.C. & Wakeham, W.A. The mutual diffusion coefficient of ethanol–water mixtures: determination by a rapid, new method. *Proc.* A **336**, 393–406 (1974).

The mutual diffusion coefficient for binary mixtures of water and the isomers of propanol.
Proc. A **342**, 401–419 (1975).

Pratt, K.F. *See* Jenkins (H.D.B.) & Pratt.

Pratt, O.E. *See* Bachelard, Daniel, Love & Pratt; Baños, Daniel, Moorhouse & Pratt; Daniel, Love & Pratt; Daniel, Pratt & Spargo; Daniel, Pratt & Wilson.

Prazdny, K. *See* Longuet-Higgins (H.C.) & Prazdny.

Preece, R.C. *See* Kerney, Preece & Turner.

Prelog, V. & Jeger, O. Leopold Ruzicka. *Biogr. Mem.* **26**, 411–501 (1980).

Prentice, J.E. Sedimentation in the inner estuary of the Thames, and its relation to the regional subsidence (Discussion). *Trans.* A **272**, 115–119 (1972).

Press, Elizabeth M. *See* Dwek, Jones, Marsh, McLaughlin, Press and others.

Presser, R.I. & McPherson, R. Prior austenite grain boundary embrittlement of low alloy steel by boron [abstract] (Discussion). *Trans.* A **295**, 298 (1980).

Pressouyre, G.M. & Bernstein, I.M. Titanium: a hydrogen trap in iron [abstract] (Discussion). *Trans.* A **295**, 304 (1980).

Prestige, M.C. & Willshaw, D.J. On a role for competition in the formation of patterned neural connexions. *Proc.* B **190**, 77–98 (1975).

Preston, A. Artificial radioactivity in freshwater and estuarine systems (Discussion). *Proc.* B **180**, 421–436 (1972).

Standards and environmental criteria: the practical application of the results of laboratory experiments and field trials to pollution control (Discussion). *Trans.* B **286**, 611–624 (1979).

Preston, A. & Wood, P.C. Monitoring the marine environment (Discussion). *Proc.* B **177**, 451–462 (1971).

Preston, B.N. *See* Ogston, Preston & Wells.

Preston, B.N. & Snowden, J.McK. Appendix to a paper by Ogston, Preston & Wells. *Proc.* A **333**, 311–313 (1973).

Preston, L.L. *See* Bailey (P.), Madin & Preston.

Prestt, I. Techniques for assessment of pollution effects on seabirds (Discussion). *Proc.* B **177**, 287–294 (1971).

Prewett, P.D. & Allen, J.E. The double sheath associated with a hot cathode. *Proc.* A **348**, 435–446 (1976).

Prewo, K.M. The importance of fibres in achieving impact tolerant composites (Discussion). *Trans.* A **294**, 551–558 (1980).

Price, B.T. Airflow problems related to surface transport systems (Discussion). *Trans.* A **269**, 327–333 (1971).

Price, C. Thermodynamics of rubber elasticity (Discussion). *Proc.* A **351**, 331–350 (1976).

Price, D.H.A. Summarizing remarks (Discussion). *Proc.* B **180**, 535–536 (1972).

Price, J.H. The shallow sublittoral marine ecology of Aldabra (Discussion). *Trans.* B **260**, 123–171 (1971).

Price, N.C. *See* Dwek, Jones, Marsh, McLaughlin, Press and others.

Price, W.C. Arthur Donald Walsh. *Biogr. Mem.* **24**, 569–582 (1980).

See also Potts (A.W.) & Price; Potts, (A.W.), Williams & Price.

Price, W.G. *See* Bishop (R.E.D.), Burcher & Price; Bishop (R.E.D.) & Price.

Pridor, A. *See* DiPrima & Pridor.

Priest, E.R. Current sheets (Discussion). *Trans.* A **281**, 497–505 (1976).

See also Soward & Priest.

Prince, R.H. *See* Sayer (R.J.), Prince & Duley.

Prince, W.T. *See* Berridge & Prince.

Pringle, J.E. The nature of transient X-ray sources (Discussion). *Proc.* A **350**, 481–486 (1976).

Pringle, J.W.S. Effects of World War II on the development of knowledge in the biological sciences (Discussion). *Proc.* A **342**, 537–548 (1975).

The Croonian Lecture, 1977. Stretch activation of muscle: function and mechanism. *Proc.* B **201**, 107–130 (1978).

Prior, C.R. Angular momentum in general relativity.

I. Definition and asymptotic behaviour. *Proc.* A **354**, 379–405 (1977).

II. Perturbations of a rotating black hole. *Proc.* A **355**, 1–29 (1977).

Pritchard, H.O. *See* Yau & Pritchard.
Pritchard, R.H. Review Lecture. On the growth and form of a bacterial cell.
Trans. B 267, 303—336 (1974).
Pritchard, W.G. Measurement of the viscometric functions for a fluid in steady shear flows.
Trans. A 270, 507—556 (1971).
See also Barnard (B.J.S.), Mahony & Pritchard; Jean & Pritchard.
Proctor, B.A. & Yale, B. Glass fibres for cement reinforcement (Discussion).
Trans. A 294, 427—436 (1980).
Prosser, R.F. A survey of design and operational problems — environmental and regulatory aspects
(Discussion). *Trans.* A 273, 35—43 (1972).
Prostka, H.J. *See* Lipman, Prostka & Christiansen.
Prothero, A. *See* Halstead, Kirsch, Prothero & Quinn; Halstead, Prothero & Quinn.
Prunell, Ariel & Kornberg, R.D. Relation of nucleosomes to nucleotide sequences in the rat (Discussion). *Trans.* B 283, 269—273 (1978).
Prutton, M. *See* Ahmad, Prutton & Whiting; Browning (R.), Bassett, El Gomati & Prutton.
Pryce, R.J. *See* Anding, Brandt, Ourisson, Pryce & Rohmer.
Prŷs-Jones, R.P. The ecology and conservation of the Aldabran brush warbler, *Nesillas aldabranus*
(Discussion). *Trans.* B 286, 211—224 (1979).
Pugh, J.R. *See* Drever (R.W.P.), Hough, Pugh, Edelstein and others.
Pugh, K.B. *See* Gibbs (C.F.), Pugh & Andrews.
Pugsley, Sir Alfred. Andrew Robertson. *Biogr. Mem.* 24, 515—528 (1978).
Pullen, J. & Williamson, J.B.P. On the plastic contact of rough surfaces.
Proc. A 327, 159—173 (1972).
Pulsford, A. *See* Bone, Anderson & Pulsford.
Pulvertaft, T.C.R. Recumbent folding and flat-lying structure in the Precambrian of northern West
Greenland (Discussion). *Trans.* A 273, 535—545 (1973).
Purchio, A.F. *See* Erikson (R.L.), Collett, Erikson & Purchio.
Purchon, R.D. An analytical approach to a classification of the Bivalvia (with an appendix by G.
Clarke) (Discussion). *Trans.* B 284, 425—436 (1978).
Purnell, J.H. *See* Bowrey & Purnell; Bull (K.R.), Marshall & Purnell; Lexton, Marshall & Purnell;
Marshall (R.M.), Purnell & Storey.
Pursley, W.C. *See* Barlow (A.J.), Harrison, Irving, Kim, Lamb & Pursley.
Pusey, P.N. The study of Brownian motion by intensity fluctuation spectroscopy (Discussion).
Trans. A 293, 429—439 (1979).
Pykett, I.L. *See* Mansfield (P.), Morris, Ordidge, Pykett, Bangert & Coupland.
Pynn, R. & Squires, G.L. Measurements of the normal-mode frequencies of magnesium.
Proc. A 326, 347—360 (1972).
Pyott, G.A.D. *See* Crapper, Dombrowski & Pyott.
Pyper, N.C. & Gerratt, J. Spin-coupled theory of molecular wavefunctions: applications to the structure and properties of $LiH(X^1\Sigma^+)$, $BH(X^1\Sigma^+)$, $Li_2(X^1\Sigma_g^+)$ and $HF(X^1\Sigma^+)$.
Proc. A 355, 407—439 (1977).
Pyper, N.C. & Grant, I.P. The relation between successive atomic ionization potentials.
Proc. A 359, 525—543 (1978). Corrigendum. *Proc.* A 361, 527 (1978).

de Quadros, C.A. More effective immunization (Discussion). *Proc.* B 209, 111—118 (1980).
Quantin, P. Soils of the New Hebrides islands (Discussion). *Trans.* B 272, 287—292 (1975).
Quastel, J.H. *See* Okamoto & Quastel.
Quenby, J.J. High energy solar particles (Discussion). *Trans.* A 281, 491—496 (1976).
Propagation of solar particles in the interplanetary medium (Discussion).
Trans. A 297, 629—640 (1980).
See also Carpenter (G.F.), Coe, Engel & Quenby.
Quenby, J.J., Coe, M.J. & Engel, A.R. Ariel 5 hard X-ray observations (Discussion).
Proc. A 366, 295—310 (1979).

Querfeld, C.W. *See* Arnott, Nicol & Querfeld.
Quinn, C.P. *See* Halstead, Kirsch, Prothero & Quinn; Halstead, Prothero & Quinn.
Quinn, P.S. *See* Steiner (D.F.), Patzelt, Chan, Quinn, Tager, Nielsen and others.
Quinn, T.J. *See* Colclough, Quinn & Chandler.
Quinn, T.J., Colclough, A.R. & Chandler, T.R.D. A new determination of the gas constant by an acoustical method. *Trans.* A **283**, 367–420 (1976).
Quintana, H. *See* Carter (B.) & Quintana.
Quirk, J.P. *See* Atkinson (R.J.), Posner & Quirk.
Quirke, J.M.E. *See* Eglinton, HajIbrahim, Maxwell, Quirke, Shaw and others.

Raab, R.E. *See* Buckingham (A.D.) & Raab; de Figueiredo & Raab.
Rabiyah, Y. Abu. *See* McCance, Abu Rabiyah, Beer, Edholm, Even-Paz and others.
Racaniello, V.R. *See* Palese, Racaniello, Desselberger, Young & Baez.
Rackham, G.M. *See* Buxton (B.F.), Eades, Steeds & Rackham; Jones (P.M.), Rackham & Steeds.
von Rad, U. & Einsele, G. Mesozoic-Cainozoic subsidence history and palaeobathymetry of the north-west African continental margin (Aaiun Basin to D.S.D.P. Site 397) (Discussion). *Trans.* A **294**, 37–50 (1980).
Radda, G.K. Fluorescent probes in membrane studies (Discussion). *Trans.* B **270**, 539–549 (1975). The dynamic properties of biological membranes (Discussion). *Trans.* B **272**, 159–171 (1975). *See also* Ackerman, Bore, Gadian, Grove & Radda.
Radford, C.C. *See* Brooks (R.R.) & Radford.
Radford, C.J. *See* Bryant, Hagston & Radford.
Radford, H.E. *See* Davies (P.B.), Russell, Thrush & Radford.
Rado, R. Selective families of sets. *Proc.* A **372**, 307–315 (1980).
Radom, L. *See* Craig (D.P.), Radom & Stiles.
Radtke, E. *See* Connerade, Baig, Mansfield & Radtke.
Rafique, S. *See* Collins (A.T.) & Rafique.
Rager, G. Morphogenesis and physiogenesis of the retino-tectal connection in the chicken.
  I. The retinal ganglion cells and their axons. *Proc.* B **192**, 331–352 (1976). Erratum. *Proc.* B **199**, 587 (1977).
 II. The retino-tectal synapses. *Proc.* B **192**, 353–370 (1976).
Rahimtula, A.D. *See* Akhtar, Wilton, Watkinson & Rahimtula.
Raimondi, M. *See* Gerratt & Raimondi.
Rainey, P. *See* Evans (T.) & Rainey.
Rainey, R.C. Rainfall: scarce resource in 'opportunity country' (Discussion). *Trans.* B **278**, 439–455 (1977).
Rainey, R.C. & Betts, Elizabeth. Continuity in major populations of migrant pests: the Desert Locust and the African armyworm (Discussion). *Trans.* B **287**, 359–374 (1979).
Rainey, R.C., Betts, Elizabeth & Lumley, Ann. The decline of the Desert Locust plague in the 1960s: control operations or natural causes? (Discussion). *Trans.* B **287**, 315–344 (1979).
Rainey, R.C. & Gunn, D.L. Introductory remarks to a Discussion. *Trans.* B **287**, 249 (1979).
Rainis, A., Tung, R. & Szwarc, M. Kinetics of protonation of Li^+, Na^+ and K^+ salts of anthracenide radical ions in DME and THF by methanol and *tert*-butanol; the significant contribution of the encounter complex to the protonation. *Proc.* A **339**, 417–433 (1974).
Raisman, G. Formation of synapses in the adult rat after injury: similarities and differences between a peripheral and a central nervous site (Discussion). *Trans.* B **278**, 349–359 (1977). *See also* Brown-Grant, Murray, Raisman & Sood; Brown-Grant & Raisman; Matthews (Margaret R.) & Raisman.
Raisman, G. & Brown-Grant, K. The 'suprachiasmatic syndrome': endocrine and behavioural abnormalities following lesions of the suprachiasmatic nuclei in the female rat. *Proc.* B **198**, 297–314 (1977).
Rajagopal, K.R. *See* Fosdick & Rajagopal.
Rajapaksa, N. *See* Peters (W.), Garnham, Killick-Kendrick, Rajapaksa and others.

Rajewsky, K. The carrier effect and cellular cooperation in the induction of antibodies (Discussion). *Proc.* B **176**, 385–392 (1971).

Rakic, P. Prenatal development of the visual system in rhesus monkey (Discussion). *Trans.* B **278**, 245–260 (1977).

Rakowski, R.F. *See* Chandler (W.K.), Schneider, Rakowski & Adrian.

Ralph, B. *See* Page (T.F.) & Ralph; Jones (A.R.), Ralph & Hansen.

Ramachandra, R. *See* Gopal, Sekhar, Ananthakrishna and others.

Ramachandran, K. *See* Cassell, Henderson & Ramachandran.

Ramachandran, V.S. *See* Clarke (P.G.H.), Ramachandran & Whitteridge.

Ramalingaswami, V. The people (Discussion). *Proc.* B **209**, 83–88 (1980).

Ramasastry, C. & Reddy, K. Viswanatha. Point defects in sodium chlorate crystals. *Proc.* A **335**, 1–14 (1973).

Ramasastry, C., Reddy, K. Viswanatha & Murthy, V.S. Electrical conductivity in sodium chlorate crystals. *Proc.* A **325**, 347–361 (1971).

Ramírez, G. *See* Viñuela, Camacho, Jiménez, Carrascosa, Ramírez & Salas.

Rämme, G. *See* Fisher (M.), Rämme, Claesson & Szwarc.

Rämme, G., Fisher, M., Claesson, S. & Szwarc, M. Kinetics of electron transfer processes involving free ions and ion pairs studied by flash photolysis. *Proc.* A **327**, 467–479 (1972).

Ramsay, D.A. *See* Bolman, Brown, Carrington, Kopp & Ramsay.

Ramsay, G. *See* Hayman (M.J.), Ramsay, Kitchener, Graf, Beug and others.

Ramsay, J.A. Insect rectum (Discussion). *Trans.* B **262**, 251–260 (1971).
The rectal complex in the larvae of Lepidoptera. *Trans.* B **274**, 203–226 (1976).

Ramsay, J.D.F. *See* Barrer, Papadopoulos & Ramsay.

Ramsay, J.G. Displacement and strain (Discussion). *Trans.* A **283**, 3–25 (1976).

Ramsden, S.A. Future developments in lunar and satellite laser ranging (Discussion). *Trans.* A **284**, 457–460 (1977).

Ramshaw, J.A.M. *See* Boulter, Ramshaw, Thompson, Richardson & Brown.

Ramussen, S.W. Meiosis in *Bombyx mori* females (Discussion). *Trans.* B **277**, 343–350 (1977).

Randall, Sir John. Concluding remarks to a Discussion. *Trans.* B **268**, 155–159 (1974).
Emmeline Jean Hanson. *Biogr. Mem.* **21**, 313–344 (1975).
See also Middendorf & Randall.

Randall, Sir John & Vaughan, J.M. Brillouin scattering in systems of biological significance (Discussion). *Trans.* A **293**, 341–347 (1979).

Randles, D.L. De Haas–van Alphen measurements of the conduction electron g-factor in the noble metals and potassium. *Proc.* A **331**, 85–101 (1972).

Rangaswami, S. *See* Baker (Wilson) & Rangaswami.

Ranger, A.P., Ettles, C.M.M. & Cameron, A. The solution of the point contact elasto-hydrodynamic problem. *Proc.* A **346**, 227–244 (1975).

Ranson, S.L. *See* Porter (Helen K.) & Ranson.

Rao, A.N. *See* Lewis (D.) & Rao.

Rao, C.N.R. *See* Hutchison (J.L.), Anderson & Rao.

Rao, C.N.R., Sarma, D.D. & Hegde, M.S. A novel approach to the study of surface oxidation states and oxidation of transition metals by Auger electron spectroscopy. *Proc.* A **370**, 269–280 (1980).

Rao, C.N.R., Sarma, D.D., Vasudevan, S. & Hegde, M.S. Study of transition metal oxides by photoelectron spectroscopy. *Proc.* A **367**, 239–252 (1979).

Rao, C.R. Prasantha Chandra Mahalanobis. *Biogr. Mem.* **19**, 455–492 (1973).
Corrigenda. *Biogr. Mem.* **20**, 505 (1974). Corrigendum. *Biogr. Mem.* **21**, 585 (1975).

Rao, C.R.A. & Goda, M.A.A. Generalization of Lamb's problem to a class of inhomogeneous elastic halfspaces. *Proc.* A **359**, 93–110 (1978).

Rao, D.B. *See* Mortimer (C.H.), Rao & Schwab.

Rao, D.B. & Schwab, D.J. Two dimensional normal modes in arbitrary enclosed basins on a rotating earth: application to Lakes Ontario and Superior. *Trans.* A **281**, 63–96 (1976).

Rao, K.K. *See* Hall (D.O.), Àdams, Morris & Rao.

Rao, M.K. *See* Bennett (M.D.), Finch, Smith & Rao; Bennett (M.D.), Rao, Smith & Bayliss.

Rapp, U.R. *See* Todaro, Callahan, Rapp & De Larco.

Rappaport, L. & Adams, D. Gibberellins: synthesis, compartmentation and physiological process (Discussion). *Trans.* B 284, 521—539 (1978).

Raschke, K. How stomata resolve the dilemma of opposing priorities (Discussion).
   Trans. B 273, 551—560 (1976).

Ratcliffe, D.A. Ecological effects of mineral exploitation in the United Kingdom and their significance to nature conservation (Discussion). *Proc.* A 339, 355—372 (1974).

   Conservation of terrestrial communities (Discussion). *Trans.* B 274, 417—435 (1976).

   Nature conservation: aims, methods and achievements (Discussion). *Proc.* B 197, 11—29 (1977).

Ratcliffe, J.A. William Henry Eccles. *Biogr. Mem.* 17, 195—214 (1971).

   Physics in a university laboratory before and after World War II (Discussion).
   Proc. A 342, 457—464 (1975).

   The early ionosphere investigations of Appleton and his colleagues (Discussion).
   Trans. A 280, 3—9 (1975).

   Robert Alexander Watson-Watt. *Biogr. Mem.* 21, 549—568 (1975).

Ratcliffe, R.A.S. Meteorological aspects of the 1975—76 drought (Discussion).
   Proc. A 363, 3—20 (1978).

Ratcliffe, R.W. *See* Salzmann, Ratcliffe, Bouffard & Christensen.

Ratter, J.A., Askew, G.P., Montgomery, R.F. & Gifford, D.R. Observations on the vegetation of northeastern Mato Grosso. II. Forests and soils of the Rio Suiá—Missu area.
   Proc. B 203, 191—208 (1978).

Ratter, J.A., Richards, P.W., Argent, G. & Gifford, D.R. Observations on the vegetation of northeastern Mato Grosso. I. The woody vegetation types of the Xavantina—Cachimbo Expedition area. *Trans.* B 266, 449—492 (1973).

Raudkivi, A.J. & Hutchison, D.L. Erosion of kaolinite clay by flowing water.
   Proc. A 337, 537—554 (1974).

Ravetz, J.R. Nicolaus Copernicus (1473—1543) (Symposium). *Proc.* A 336, 5—9 (1974).

Raviart, P.A. Pseudo-viscosity methods and nonlinear hyperbolic equations (Discussion).
   Proc. A 323, 277—283 (1971).

Rawcliffe, G.H. The Clifford Paterson Lecture, 1977. Induction motors: old and new.
   Proc. A 362, 145—178 (1978).

Rawlins, A.D. The solution of a mixed boundary value problem in the theory of diffraction by a semi-infinite plane. *Proc.* A 346, 469—484 (1975).

   Radiation of sound from an unflanged rigid cylindrical duct with an acoustically absorbing internal surface. *Proc.* A 361, 65—91 (1978).

Ray, D.K. & Kajzar, F. Studies on the interaction between two correlated electronic bands. I. Phase transition in NiS. *Proc.* A 373, 253—268 (1980).

Ray, I.L.F. & Cockayne, D.J.H. The dissociation of dislocations in silicon.
   Proc. A 325, 543—554 (1971).

Rayment, T. *See* Bomchil, Hüller, Rayment, Roser, Smalley, Thomas & White.

Raymont, J.E.G. Some aspects of pollution in Southampton Water (Discussion).
   Proc. B 180, 451—468 (1972).

Raymont, J.E.G., Krishnaswamy, S., Woodhouse, M.A. & Griffin, R.L. Studies on the fine structure of Copepoda. Observations on *Calanus finmarchicus* (Gunnerus).
   Proc. B 185, 409—424 (1974).

Raynor, G.V. Alan Richard Powell. *Biogr. Mem.* 22, 307—318 (1976).

Rayns, D.G. Freeze-etching studies on muscle (Discussion). *Trans.* B 261, 139—142 (1971).

Readman, P.W. *See* Runcorn, Collinson, O'Reilly, Stephenson, Battey and others.

Ready, P.D. *See* Killick-Kendrick, Leaney, Ready & Molyneux; Lainson, Ready & Shaw.

Reay, J.S.S. The philosophy of monitoring (Discussion). *Trans.* A 290, 609—623 (1979).

Reay, N.A. *See* Hunt (G.W.), Reay & Yoshimura.

Redding, J.H. *See* Destombes, Shephard-Thorn & Redding; Kellaway, Redding, Shephard-Thorn & Destombes.

Reddy, K. Viswanatha. *See* Ramasastry & Reddy; Ramasastry, Reddy & Murthy.

Redfern, B.A.W. *See* Bröokes (C.A.), O'Neill & Redfern.

Reece, M.P. A review of the development of the vacuum interrupter (Discussion). *Trans.* A **275**, 121–129 (1973).

Reed, P.E. The influence of crystalline texture on the tensile properties of natural rubber. I. *Proc.* A **338**, 459–478 (1974).

Rees, D. Winds and temperatures in the auroral zone and their relations to geomagnetic activity (Discussion). *Trans.* A **271**, 563–575 (1972).

Rees, D., Roper, R.G., Lloyd, K.H. & Low, C.H. Determination of the structure of the atmosphere between 90 and 250 km by means of contaminant releases at Woomera, May 1968. *Trans.* A **271**, 631–663 (1972).

Rees, F. Gwendolen. Studies on the pigmented and unpigmented photoreceptors of the cercaria of *Cryptocotyle lingua* (Creplin) from *Littorina littorea* (L.). *Proc.* B **188**, 121–138 (1975).
The arrangement and ultrastructure of the musculature, nerves and epidermis, in the tail of the cercaria of *Cryptocotyle lingua* (Creplin) from *Littorina littorea* (L.). *Proc.* B **190**, 165–186 (1975).
The development of the tail and the excretory system in the cercaria of *Cryptocotyle lingua* (Creplin) (Digenea: Heterophyidae from *Littorina littorea* (L.)). *Proc.* B **195**, 425–452 (1977).
The ultrastructure, development and mode of operation of ventrogenital complex of *Cryptocotyle lingua* (Creplin) (Digenea: Heterophyidae). *Proc.* B **200**, 245–267 (1978).

Rees, F. Gwendolen & Day, M.F. The origin and development of the epidermis and associated structures in the cercaria of *Cryptocotyle lingua* (Creplin) (Digenea: Heterophyidae) from *Littorina littorea* (L.). *Proc.* B **192**, 299–321 (1976).

Rees, G.R. & Satow, P.F.C. Navigation on the Thames (Discussion). *Trans.* A **272**, 201–212 (1972).

Rees, H. *See* Teoh & Rees.

Rees, H., Shaw, D.D. & Wilkinson, P. Nuclear DNA variation among acridid grasshoppers. *Proc.* B **202**, 517–525 (1978).

Rees, M.J. Observational status of black holes [abstract] (Discussion). *Proc.* A **368**, 27–32 (1979). Concluding remarks to a Discussion. *Trans.* A **296**, 431–435 (1980).

Rees, W.L. *See* Nakanishi (H.), Jones, Thomas, Hasegawa & Rees.

Reeves, D. *See* Gillett, Cole & Reeves.

Reeves, E.M. *See* Garton, Reeves & Tomkins; Garton, Reeves, Tomkins & Ercoli; Parkinson (W.H.) & Reeves; Parkinson (W.H.), Reeves & Tomkins.

Reeves, E.M., Vernazza, J.E. & Withbroe, G.L. The quiet Sun in the extreme ultraviolet (Discussion). *Trans.* A **281**, 319–329 (1976).

Reeves, H. *See* David & Reeves.

Reeves, R.D. *See* Brooks (R.R.), Morrison, Reeves, Dudley & Akman; Jaffré, Kersten, Brooks & Reeves.

Regan, D., Beverley, K.I. & Cynader, M. Steroeoscopic subsystems for position in depth and for motion in depth (Discussion). *Proc.* B **204**, 485–501 (1979).

Regev, H. *See* Bernstein (J.), Regev, Herbstein, Main, Rizvi and others.

Rehault, J.P. *See* Sibuet, Ryan, Arthur, Barnes, Blechsmidt and others.

Rehfeld, J. *See* Hökfelt, Lundberg, Schultzberg, Johansson, Skirboll, Anggård and others.

Reichstein, T. *See* Rothschild (Miriam), von Euw & Reichstein; Rothschild (Miriam), von Euw, Reichstein, Smith & Pierre.

Reid, A.A.L. New telecommunications services and their social implications (Discussion). *Trans.* A **289**, 175–184 (1978).

Reid, R.L. *See* Goldstein (M.E.) & Reid.

Reid, W.H. *See* Lakin, Ng & Reid.

Reidenberg, M. *See* James (Margaret O.), Smith, Williams & Reidenberg.

Reintjes, J. Extreme ultraviolet picosecond pulses (Discussion). *Trans.* A **298**, 273–280 (1980).

Renberg, L. *See* Jensen (S.), Lange, Berge, Palmork & Renberg.

Rendall, M. *See* McColl (L.) & Rendall.

Rentzepis, P.M. Picosecond spectroscopic studies of biological systems (Discussion).

Trans. A **293**, 455–468 (1979).
See also Nitzan, Jortner & Rentzepis.
Renvoize, S.A. The origin and distribution of the flora of Aldabra (Discussion).
Trans. B **260**, 227–236 (1971).
Requena, J. *See* Brooks (D.E.), Levine, Requena & Haydon.
Requena, J., Billett, D.F. & Haydon, D.A. Van der Waals forces in oil–water systems from the study of thin lipid films. I. Measurement of the contact angle and the estimation of the van der Waals free energy of thinning of a film. *Proc.* A **347**, 141–159 (1975).
Requena, J. & Haydon, D.A. Van der Waals forces in oil–water systems from the study of thin lipid films. II. The dependence of the van der Waals free energy of thinning on film composition and structure. *Proc.* A **347**, 161–177 (1975).
Reuter, H., Blaustein, M.P. & Haeusler, G. Na–Ca exchange and tension development in arterial smooth muscle (Discussion). *Trans.* B **265**, 87–94 (1973).
Rex, D.C. *See* Baker (P.E.), Buckley & Rex; Briden, Rex, Faller & Tomblin.
Rey, L. Freezing and freeze-drying (Discussion). *Proc.* B **191**, 9–19 (1975).
Reyment, R.A. & Tait, E.A. Biostratigraphical dating of the early history of the South Atlantic Ocean. *Trans.* B **264**, 55–95 (1972).
Reynaud, S. *See* Cohen-Tannoudji & Reynaud.
Reynolds, C.S. Growth and buoyancy of *Microcystis aeruginosa* Kütz. emend. Elenkin in a shallow eutrophic lake. *Proc.* B **184**, 29–50 (1973).
Reynolds, D.R. *See* Riley (J.R.) & Reynolds.
Reynolds, P.E. *See* Batte, Brear, Holdsworth, Myers & Reynolds.
Reynolds, W.N. & Moreton, R. Some factors affecting the strengths of carbon fibres (Discussion). *Trans.* A **294**, 451–461 (1980).
Rhee, Hyun-Ku, Aris, R. & Amundson, N.R. Multicomponent adsorption in continuous counter-current exchangers. *Trans.* A **269**, 187–215 (1971).
Rhines, P. A comment on the *Aries* observations (Discussion). *Trans.* A **270**, 461–463 (1971).
Rhodes, D.R. A reactance theorem. *Proc.* A **353**, 1–10 (1977).
Rhodes, J.M. Some compositional aspects of lunar regolith evolution (Discussion). *Trans.* A **285**, 293–301 (1977).
Ribeiro, M.M. & Whitelaw, J.H. The structure of turbulent jets. *Proc.* A **370**, 281–301 (1980).
Rice, J.R. *See* Palmer (A.C.) & Rice.
Rice, R.V. *See* Somlyo (A.P.), Devine, Somlyo & Rice.
Richards, B.M., Pardon, J.F., Lilley, D.M.J., Cotter, Rosalind I., Wooley, J.C. & Worcester, D.L. Nucleosome sub-structure during transcription and replication (Discussion). *Trans.* B **283**, 287–289 (1978).
Richards, D.J.W. *See* Rowbottom & Richards.
Richards, H.J. *See* Gibb (O.) & Richards.
Richards, K.E. *See* Guilley, Jonard, Richards & Hirth.
Richards, P.W. *See* Ratter, Richards, Argent & Gifford.
Richards, R.E. Introductory remarks to a Discussion. *Proc.* A **345**, 3 (1975).
See also Brown (F.F.), Halsey & Richards; Hall (C.), Richards & Sharp; Hoult & Richards.
Richards, W.G. *See* Gold (Elizabeth), Hammersley & Richards; Hinkley, Walker & Richards.
Richards, W.G., Clarkson, R. & Ganellin, C.R. Molecule–receptor specificity (Discussion). *Trans.* B **272**, 75–85 (1975).
Richards, W.G. & Wallis, Jenifer. The distribution of electronic charge in some biologically active amines. *Proc.* B **199**, 291–307 (1977).
Richardson, C.A., Crisp, D.J. & Runham, N.W. Factors influencing shell growth in *Cerastoderma edule*. *Proc.* B **210**, 513–531 (1980).
Richardson, Elspeth. Deformation and haemolysis of red cells in shear flow. *Proc.* A **338**, 129–253 (1974).
Richardson, M. *See* Boulter, Ramshaw, Thompson, Richardson & Brown.
Richardson, P.S. *See* Gallagher, Kent, Passatore, Phipps & Richardson; Phipps, Richardson, Corfield, Gallagher and others.

Richardson, R.M. *See* Frost (J.C.), Leadbetter & Richardson.

Richelmi, J. *See* Leroy (V.), Richelmi & Graas.

Richmond, M.H. 'Cells' and 'organisms' as a habitat for DNA (Discussion).
  Proc. B **204**, 235–250 (1979).

Richmond, M.H., Bennett, P.M., Choi, C.-L., Brown, N., Brunton, J., Grinsted, J. & Wallace, L. The genetic basis of the spread of β-lactamase synthesis among plasmid-carrying bacteria (Discussion). *Trans.* B **289**, 349–359 (1980).

Richmond, P. *See* Chan (D.) & Richmond.

Richter, B. The production of new particles and muon-electron events in e^+-e^- annihilation at SPEAR (Discussion). *Proc.* A **355**, 447–480 (1977).

Rickards, R.B., Hyde, P.J.W. & Krinsley, D.H. Periderm ultrastructure of a species of *Monograptus* (Phylum Hemichordata). *Proc.* B **178**, 347–356 (1971).

Ridgeley, A. *See* Burton (W.M.), Jordan, Ridgeley & Wilson.

Riding, G. *See* Treloar & Riding.

Ridley, W.I. Some petrological aspects of Imbrium stratigraphy (Discussion).
  Trans. A **285**, 105–114 (1977).

Riecke, W.D. Prospects for high resolution electron microscopy (Discussion).
  Trans. B **261**, 15–34 (1971).

Riemann, H.J. *See* Wässle & Riemann.

Ries, A.C. & Shackleton, R.M. Patterns of strain variation in arcuate fold belts (Discussion).
  Trans. A **283**, 281–288 (1976).

Rieseberg, H. *See* Bartel, Duinker, Heintze, Heinzelmann and others.

Rigby, P.W.J., Chia, W., Clayton, Christine E. & Lovett, M. The structure and expression of the integrated viral DNA in mouse cells transformed by simian virus 40 (Discussion). *Proc.* B **210**, 437–450 (1980).

Righelato, R.C. Microbial production of energy sources from biomass (Discussion).
  Trans. A **295**, 491–500 (1980).
  Anaerobic fermentation: alcohol production. *Trans.* B **290**, 303–312 (1980).

Rijks, D. The conservation and utilization of water (Discussion). *Trans.* B **278**, 583–592 (1977).

Riley, J.R. & Reynolds, D.R. Radar-based studies of the migratory flight of grasshoppers in the middle Niger area of Mali. *Proc.* B **204**, 67–82 (1979).
  [Abstract] (Discussion). *Trans.* B **287**, 457 (1979).

Riley, R. Concluding remarks – the evolution of crops and of agriculture (Discussion).
  Trans. B **275**, 209–213 (1976).
  See also Bennett (M.D.), Chapman & Riley; Bennett (M.D.), Dover & Riley; Dover & Riley.

Riley, R. & Flavell, R.B. A first view of the meiotic process (Discussion).
  Trans. B **277**, 191–199 (1977).

Rimington, C. & Gray, C.H. Max Rudolf Lemberg. *Biogr. Mem.* **22**, 257–294 (1976).

Rimmer, J. *See* Kenner, Rimmer, Smith & Unsworth.

Ringwood, A.E. Mare basalt petrogenesis and the composition of the lunar interior (Discussion).
  Trans. A **285**, 577–586 (1977).

Riordan, Claudia & Kornberg, H.L. Location of *galP*, a gene which specifies galactose permease activity, on the *Escherichia coli* linkage map. *Proc.* B **198**, 401–410 (1977).

Rishbeth, J. Resistance to fungal pathogens of tree roots (Symposium). *Proc.* B **181**, 333–351 (1972).

Ritchie, J.M. Binding of tetrodotoxin and saxitoxin to sodium channels (Discussion).
  Trans. B **270**, 319–336 (1975).
  See also Howarth (J.V.), Ritchie & Stagg.

Rittmann, A. Structure and evolution of Mount Etna (Symposium). *Trans.* A **274**, 5–16 (1973).

Rivers, J.P.W.R. *See* Miller (D.S.), Baker, Bowden, Evans, Holt and others.

Rivett, B.H.P. Policy selection by structural mapping (with an appendix by D.G. Kendall).
  Proc. A **354**, 407–423 (1977).

Rivière, J.C. *See* Coad & Rivière.

Rizvi, S.H. *See* Bernstein (J.), Regev, Herbstein, Main and others.

Robbins, A.R. Introduction to a Discussion. *Trans.* A **294**, 211–215 (1980).

Robbins, N., Olek, A., Kelly, S.S., Takach, P. & Christopher, M. Quantitative study of motor end-plates in muscle fibres dissociated by a simple procedure. *Proc.* B **209**, 555–562 (1980).

Robbins, P.W. & Macpherson, I.A. Glycolipid synthesis in normal and transformed animal cells (Discussion). *Proc.* B **177**, 49–58 (1971).

Robbins, R.J. *See* Beddard, Fleming, Porter & Robbins.

Robbins, W.E. *See* Thompson (M.J.), Svoboda, Kaplanis & Robbins.

Roberts, A. & Blight, A.R. Anatomy, physiology and behavioural rôle of sensory nerve endings in the cement gland of embryonic *Xenopus*. *Proc.* B **192**, 111–127 (1975).

Roberts, A.D. *See* Johnson (K.L.), Kendall & Roberts; Johnson (K.L.) & Roberts.

Roberts, A.D. & Tabor, D. The extrusion of liquids between highly elastic solids.
Proc. A **325**, 323–345 (1971).

Roberts, Alan & Hayes, B.P. The anatomy and function of 'free' nerve endings in an amphibian skin sensory system. *Proc.* B **196**, 415–429 (1977).

Roberts, B.B. Conservation in the Antarctic (Discussion). *Trans.* B **279**, 97–104 (1977).

Roberts, B.L. *See* Paul (D.H.) & Roberts; Williamson (R.M.) & Roberts; Witkovsky & Roberts.

Roberts, B.L. & Ryan, K.P. The fine structure of the lateral-line sense organs of dogfish.
Proc. B **179**, 157–169 (1971).

Roberts, B.L. & Witkovsky, P. A functional analysis of the mesencephalic nucleus of the fifth nerve in the selachian brain. *Proc.* B **190**, 473–495 (1975).

Roberts, D.E. *See* Chisholm & Roberts.

Roberts, D.G. Marine geology of the Rockall Plateau and Trough. *Trans.* A **278**, 447–509 (1975).
See also Laughton & Roberts.

Roberts, D.G. & Montadert, L. Contrasts in the structure of the passive margins of the Bay of Biscay and Rockall Plateau (Discussion). *Trans.* A **294**, 97–103 (1980).

Roberts, D.G., Montadert, L., Thompson, R.W., Auffret, G.A., Lumsden, D.N., Kagami, H., Timofeev, P.P., Müller, Carla, Bock, W.D., DuPeuble, P.A., Schnitker, D., Hailwood, E.A., Harrison, W. & Thompson, T.L. Geological setting and principal results of drilling on the margins of the Bay of Biscay and Rockall Plateau during Leg 48 (Discussion). *Trans.* A **294**, 65–75 (1980).

Roberts, D.G.M. Public health engineering in the external environment: water supply and re-use, waste disposal and pollution control (Discussion). *Trans.* A **272**, 639–650 (1972).

Roberts, D.H. Microtechnology (Discussion). *Trans.* A **289**, 93–101 (1978).

Roberts, G.C.K. *See* Birdsall (B.), Griffiths, Roberts, Feeney & Burgen; Feeney, Roberts, Birdsall, Griffiths and others; Lee (A.G.), Birdsall, Metcalfe, Warren & Roberts.

Roberts, G.O. Dynamo action of fluid motions with two-dimensional periodicity.
Trans. A **271**, 411–454 (1972).

Roberts, J.B. *See* Gaster & Roberts.

Roberts, J.B. & Gaster, M. On the estimation of spectra from randomly sampled signals: a method of reducing variability. *Proc.* A **371**, 235–258 (1980).

Roberts, J.E. Development of normalized structural steels (Symposium).
Trans. A **282**, 277–287 (1976).

Roberts, K. Crystalline glycoprotein cell walls of algae: their structure, composition and assembly (Discussion). *Trans.* B **268**, 129–146 (1974).

Roberts, K.V. An objective interpretation of Lagrangian quantum mechanics.
Proc. A **360**, 135–160 (1978).

Roberts, M.W. *See* Brundle & Roberts; Carley & Roberts; Joyner, Kishi & Roberts; Joyner & Roberts; Kishi & Roberts.

Roberts, P.H. Kinematic dynamo models. *Trans.* A **272**, 663–698 (1972).
See also Baldwin (P.) & Roberts; Binnie & Roberts; Braginsky & Roberts; Donnelly & Roberts; Kumar (S.) & Roberts.

Roberts, P.H. & Soward, A.M. Stellar winds and breezes. *Proc.* A **328**, 185–215 (1972).

Roberts, P.H. & Stewartson, K. On finite amplitude convection in a rotating magnetic system.
Trans. A **277**, 287–315 (1974).

Roberts, P.O. Food transport in the 1980s (Discussion). *Proc.* B **191**, 155–168 (1975).

Roberts, R.C. & Mendell, Nancy R. A case of polydactyly with multiple thresholds in the mouse.
Proc. B **191**, 427–444 (1975).

Robertson, A.J.B. *See* Derrick (P.J.) & Robertson; Hibbert (D.B.) & Robertson.

Robertson, Alan. Conrad Hal Waddington. *Biogr. Mem.* **23**, 575–622 (1977).

Robertson, E.F. *See* Campbell (C.M.), Coxeter & Robertson.

Robertson, E.S. *See* Killick-Kendrick, Molyneux, Hommel, Leaney & Robertson.

Robertson, G.N. A theory of a broadening of the infrared absorption spectra of hydrogen-bonded species. III. The kinematic and electronic coupling mechanisms.
Trans. A **286**, 25–53 (1977).
See also Coulson & Robertson.

Robertson, J. & Webb, G. Catalysis by supported group VIII metal compounds. I. The interaction of *n*-butene with hydrogen over silica-supported ruthenium carbonyl catalysts.
Proc. A **341**, 383–398 (1974).

Robertson, J.M. James Wilfred Cook. *Biogr. Mem.* **22**, 71–103 (1976).

Robertson, J.S. Nucleotide sequences from the terminal regions of fowl plague virus genome RNA (Discussion). *Trans.* B **288**, 371–374 (1980).

Robertson, N.A. *See* Drever (R.W.P.), Hough, Pugh, Edelstein and others.

Robertson, P.J. *See* Kemball, Nisbet, Robertson & Scurrell.

Robertson, R.M. *See* Laverack, Neil & Robertson; Neil, Macmillan, Robertson & Laverack.

Robertson, R.M. & Laverack, M.S. The structure and function of the labrum in the lobster *Homarus gammarus* (L.). *Proc.* B **206**, 209–233 (1979).
Oesophageal sensors and their modulatory influence on oesophageal peristalsis in the lobster, *Homarus gammarus*. *Proc.* B **206**, 235–263 (1979).

Robertson, V.C. Experience in the Middle East (Discussion). *Trans.* B **278**, 525–535 (1977).

Robin, G.de Q. Ice cores and climatic change (Discussion). *Trans.* B **280**, 143–168 (1977).

Robin, G.de Q., Drewry, D.J. & Meldrum, D.T. International studies of ice sheet and bedrock (Discussion). *Trans.* B **279**, 185–196 (1977).

Robinson, A.R. The Gulf Stream (Discussion). *Trans.* A **270**, 351–370 (1971).

Robinson, C. *See* Boddington, Gray & Robinson.

Robinson, C.L. & Cameron, A. Studies in hydrodynamic thrust bearings.
I. Theory considering thermal and elastic distortions. *Trans.* A **278**, 351–366 (1975).
II. Comparison of calculated and measured performance of tilting pads by means of interferometry. *Trans.* A **278**, 367–384 (1975).
III. The parallel surface bearing. *Trans.* A **278**, 385–395 (1975).

Robinson, F.N.H. *See* Bleaney, Robinson & Wells.

Robinson, I.S. A theoretical analysis of the use of submarine cables as electromagnetic oceanographic flowmeters. *Trans.* A **280**, 355–396 (1976).

Robinson, J.E. The ostracod fauna of the interglacial deposits at Sugworth, Oxfordshire.
Trans. B **289**, 99–106 (1980).

Robinson, J.L. & Scott, M.H. Liquation cracking during the welding of austenitic stainless steels and nickel alloys (Discussion). *Trans.* A **295**, 105–117 (1980).

Robinson, K. *See* Batley, Bramley, Poldy & Robinson; Batley, Bramley & Robinson.

Robinson, P.D. *See* Barnsley & Robinson.

Robson, J.N. Storage and shelf life (Discussion). *Proc.* B **191**, 185–191 (1975).

Robson, Linda E. & Kosterlitz, H.W. Specific protection of the binding sites of D-Ala^2-D-Leu^5-enkephalin (δ-receptors) and dihydromorphine (μ-receptors). *Proc.* B **205**, 425–432 (1979).

Rocha, A. & Acrivos, A. Experiments on the effective conductivity of dilute dispersions containing highly conducting slender inclusions. *Proc.* A **337**, 123–133 (1974).

Rochester, C.H. *See* Eley, Rochester & Scurrell.

Rochester, G.D. *See* Ditchburn & Rochester.

Rochester, G.D. & Wolfendale, A.W. Introductory remarks to a Discussion.
Trans. A **277**, 318 (1974).

Rodda, J.C. *See* Day (J.B.W.) & Rodda.

Rodgers, C.D. *See* Curtis (P.D.), Houghton, Peskett & Rodgers; Drummond (J.R.), Houghton, Peskett, Rodgers, Wale and others; Ellis (P.), Holah, Houghton, Jones, Peckham and others.

Rodgers, G.G. *See* Page (J.K.), Rodgers & Souster.

Rodgers, M.J. *See* Dyson (B.F.), Loveday & Rodgers.

Rodriguez-Barrueco, C. *See* Mian, Bond & Rodriguez-Barrueco.

Roffey, J. *See* Hemming (C.F.), Popov, Roffey & Waloff.

Rogers, A.J. The electrogyration effect in crystalline quartz. *Proc.* A 353, 177–192 (1977).

Rogers, A.W. Recent developments in the use of autoradiographic techniques with electron microscopy (Discussion). *Trans.* B 261, 159–171 (1971).

Rogers, C.A., Burgess, D.A., Halberstam, H. & Birch, B.J. Harold Davenport.
Biogr. Mem. 17, 159 (1971).

Rogers, D. *See* Neidle, Rogers & Hursthouse.

Rogers, E.W.E. *See* Scruton (C.) & Rogers.

Rogers, G.L. The presence of a dispersion term in the 'transverse Fresnel aether drag' experiment.
Proc. A 345, 345–349 (1975).

Rogers, L.J., Oettinger, R., Szer, J. & Mark, R.F. Separate chemical inhibitors of long-term and short-term memory: contrasting effects of cycloheximide, ouabain and ethacrynic acid on various learning tasks in chickens. *Proc.* B 196, 171–195 (1977).

Rogers, S.C. *See* Bevan (J.W.), Kisiel, Legon, Millen & Rogers; Bevan (J.W.), Legon, Millen & Rogers; Legon, Millen & Rogers.

Rohde, S.M. & McAllister, G.T. On the optimization of fluid film bearings.
Proc. A 351, 481–497 (1976).

Rohde, S.M. & Oh, K.P. A unified treatment of thick and thin film elastohydrodynamic problems by using higher order element methods. *Proc.* A 343, 315–331 (1975).

Rohl, A.N. *See* Bowes, Langer & Rohl.

Rohmer, M. *See* Anding, Brandt, Ourisson, Pryce & Rohmer.

Rojas, E. *See* Keynes, Bezanilla, Rojas & Taylor; Nonner, Rojas & Stämpfli.

Rojas, E. & Keynes, R.D. On the relation between displacement currents and activation of the sodium conductance in the squid giant axon (Discussion). *Trans.* B 270, 459–482 (1975).

Rolinson, G.N. Bacterial resistance to penicillins and cephalosporins (Discussion).
Proc. B 179, 403–410 (1971).

Roman, E.A. *See* Gillett, Roman & Phillips.

Romano, R. & Sturiale, C. Some considerations on the magma of the 1971 eruption (Symposium).
Trans. A 274, 37–43 (1973).

Romano, S. *See* Luckhurst & Romano.

Romero-Herrera, A.E. *See* Dene, Goodman & Romero-Herrera.

Romero-Herrera, A.E. & Lehmann, H. The amino acid sequence of human myoglobin and its minor fractions. *Proc.* B 186, 249–279 (1974).

Romero-Herrera, A.E., Lehmann, H., Joysey, K.A. & Friday, A.E. On the evolution of myglobin.
Trans. B 283, 61–163 (1978).

Ronay, Maria. Determination of the dynamic surface tension of liquids from the instability of excited capillary jets and from the oscillation frequency of drops issued from such jets.
Proc. A 361, 181–206 (1978).

Ronzio, G.S. *See* Trench & Ronzio.

Rood, A.P., Emerson, D. & Milledge, H. Judith. Photodimerization of anthracene single crystals *in situ* in the electron microscope. *Proc.* A 324, 37–43 (1971).

Rooke, D.E. Future trends in gas production and transmission (Discussion).
Trans. A 276, 547–558 (1974).

Roper, R.G. *See* Rees (D.), Roper, Lloyd & Low.

Rosales, R.R. The similarity solution for the Korteweg–de Vries equation and the related Painlevé transcendent. *Proc.* A 361, 265–275 (1978).

Roscoe, H. *See* Ellis (P.), Holah, Houghton, Jones, Peckham and others.

Roscoe, H.K. *See* Chaloner, Drummond, Houghton, Jarnot & Roscoe.

Rose, D.J.W. The significance of low-density populations of the African armyworm *Spodoptera exempta* (Walk.) (Discussion). *Trans.* B 287, 393–402 (1979).

Rose, F.L. To what extent can molecular structure point to potential hazards? (Discussion).
Proc. B 185, 159–164 (1974).

See also Schild & Rose.

Rose, G., Widdel, H.U., Azcárraga, A. & Sanchez, L. A payload for small sounding rockets for wind finding and density measurements in the height region between 95 and 75 km (Discussion). *Trans.* A **271**, 509–528 (1972).

Results of an experimental investigation of correlations between D-region neutral gas winds, density changes and short-wave radio-wave absorption (Discussion). *Trans.* A **271**, 529–545 (1972).

Rose, J. *See* Perrin, Rose & Davies.

Rose, J.D. Ronald Holroyd. *Biogr. Mem.* **20**, 235–245 (1974).

Rose, L.R.F. An approximate (Wiener—Hopf) kernel for dynamic crack problems in linear elasticity and viscoelasticity. *Proc.* A **349**, 497–521 (1976).

Rose, R.J. *See* Possingham & Rose.

Rose, S.J., Grant, I.P. & Connerade, J.P. A study of 5p excitation in atomic barium. II. A fully relativistic analysis of 5p excitation in atomic barium. *Trans.* A **296**, 527–544 (1980).

Rose, S.P.R. Early visual experience, learning, and neurochemical plasticity in the rat and the chick (Discussion). *Trans.* B **278**, 307–318 (1977).

Rosen, B.R. Determination of a collection of coral microatoll specimens from the northern Great Barrier Reef (appendix to a paper by Scoffin & Stoddart) (Discussion).
Trans. B **284**, 115–122 (1978).

Rosenberg, H. Solar radio observations and interpretations (Discussion).
Trans. A **281**, 461–471 (1976).

See also de Jager, Kuperus & Rosenberg.

Rosenberg, H.M. The production of powder-filled/metal composites in space (Discussion).
Proc. A **361**, 175–178 (1978).

See also Bleaney, Loftus & Rosenberg.

Rosenblatt, M. *See* Evans (A.G.), Gulden & Rosenblatt.

Rosendorff, S. A modified approach to Glauber approximation for particle—atom collisions.
Proc. A **353**, 11–34 (1977).

Rosenfeld, J.L.J. *See* Graham (S.C.), Homer & Rosenfeld; Jones (A.), & Rosenfeld.

Rosensteel, G. *See* Ihrig, Rosensteel, Chow & Trainor.

Rosensteel, G., Ihrig, E. & Trainor, L.E.H. Group theory and many body diagrams. I. Classification and structure of diagrams. *Proc.* A **344**, 387–401 (1975).

Rosenthal, D., McEachran, R.P. & Cohen, M. Sum rules for electric quadrupole transitions.
Proc. A **337**, 365–378 (1974).

Roser, S.J. *See* Bomchil, Hüller, Rayment, Roser, Smalley, Thomas & White.

Roskams, Mary. *See* Frankel & Roskams.

Ross, C.L. *See* MacQueen, Gosling, Hildner, Munro, Poland & Ross.

Ross, D.I. *See* Aumento, Longcarevic & Ross.

Ross, F.P. *See* Cornforth & Ross.

Ross, G. *See* Burge, Fiddy, Greenaway & Ross.

Ross, G., Fiddy, M.A., Nieto-Vesperinas, M. & Wheeler, M.W.L. The phase problem in scattering phenomena: the zeros of entire functions and their significance. *Proc.* A **360**, 25–45 (1978).

Ross, R. Connective tissue cells, cell proliferation and synthesis of extracellular matrix – a review (Discussion). *Trans.* B **271**, 247–259 (1975).

Rossiter, J.R. Sea-level observations and their secular variation (Discussion).
Trans. A **272**, 131–139 (1972).

Rossiter, R.J. *See* Barr (M.L.) & Rossiter.

Rostas, J. *See* Duxbury, Horani & Rostas.

Roth, M. *See* Finch (J.T.), Lewit-Bentley, Bentley, Roth & Timmins.

Rother, J.A. & Fay, P. Sporulation and the development of planktonic blue-green algae in two Salopian meres. *Proc.* B **196**, 317–332 (1977).

Rotheram, Susan. The surface of the egg of a parasitic insect.

I. The surface of the egg and first-instar larva of *Nemeritis*. *Proc.* B **183**, 179–194 (1973).

II. The ultrastructure of the particulate coat on the egg of *Nemeritis*. *Proc.* B **183**, 195–204 (1973).

Rotherham, L. Hans Kronberger. *Biogr. Mem.* **18**, 413–426 (1972).

Rothmayr, W.W. Food process engineering (Discussion). *Proc.* B **191**, 71–86 (1975).

Rothschild, Lord. The Fourth Royal Society Technology Lecture. Petrol and pollution. *Proc.* A **322**, 147–163 (1971).

Rothschild, Miriam, von Euw, J. & Reichstein, T. Cardiac glycosides (heart poisons) in the polka-dot moth *Syntomeida epilais* Walk. (Ctenuchidae: Lep.) with some observations on the toxic qualities of *Amata* (=*Syntomis*) *phegea* (L.). *Proc.* B **183**, 227–247 (1973).

Rothschild, Miriam, von Euw, J., Reichstein, T., Smith, D.A.S. & Pierre, J. Cardenolide storage in *Danaus chrysippus* (L.) with additional notes on *D. plexippus* (L.). *Proc.* B **190**, 1–31 (1975).

Rothschild, Miriam & Schlein, J. The jumping mechanism of *Xenopsylla cheopis*. I. Exoskeletal structures and musculature. *Trans.* B **271**, 457–490 (1975).

Rothschild, Miriam, Schlein, J., Parker, K., Neville, C. & Sternberg, S. The jumping mechanism of *Xenopsylla cheopis*. III. Execution of the jump and activity. *Trans.* B **271**, 499–515 (1975).

Rott, R. Genetic determinants for infectivity and pathogenicity of influenza viruses (Discussion). *Trans.* B **288**, 393–399 (1980).

Roufosse, A. *See* Herzfeld, Roufosse, Haberkorn, Griffin & Glimcher.

Round, G.F. *See* Chan (K.W.), Baird & Round.

Roussel, Martine. *See* Hayman (M.J.), Ramsay, Kitchener, Graf, Beug and others.

Rowbottom, M.D. & Richards, D.J.W. Mechanical and aerodynamic problems associated with future overhead lines (Discussion). *Trans.* A **275**, 181–188 (1973).

Rowell, P. *See* Apte, Rowell & Stewart.

Rowlands, G. *See* Infeld & Rowlands.

Rowlands, P.C. *See* Ferguson, Hicks, Hoaksey, Rowlands & Lloyd.

Rowley, W.R.C. *See* Blaney, Bradley, Edwards, Jolliffe and others.

Rowlinson, J.S. *See* Harrington & Rowlinson; Leng, Rowlinson & Thompson.

Rowlinson, J.S. & Tildesley, D.J. The determination of the gas constant from the speed of sound. *Proc.* A **358**, 281–286 (1978).

Rowntree, Sir Norman. Water resources management – England and Wales (Discussion). *Proc.* B **180**, 367–369 (1972).

Rowold, A. *See* Franck, Rowold, Wegner & Eckert.

Roxburgh, I.W. Testing relativity and gravitational theories by radar ranging to a heliocentric satellite (Discussion). *Trans.* A **284**, 589–593 (1977).
 See also Schwartz (S.J.) & Roxburgh.

Roy, A.D. Surgical care in the village (Discussion). *Proc.* B **209**, 147–151 (1980).

Roy, J. Decisive steps towards control of the Desert Locust 1952–62 (Discussion). *Trans.* B **287**, 301–304 (1979).

Roy-Chowdhury, T. *See* Jain (S.L.), Kutty, Roy-Chowdhury & Chatterjee.

Rubenstein, A.H. *See* Steiner (D.F.), Patzelt, Chan, Quinn, Tager, Nielsen and others.

Rubia, F.J. *See* Allen (G.I.), Eccles, Nicoll, Oshima & Rubia; Eccles, Nicoll, Oshima & Rubia.

Rubincam, D.P. *See* Kolenkiewicz, Smith, Rubincam, Dunn & Torrence.

Rubinson, K.A. & Baker, P.F. The flow properties of axoplasm in a defined chemical environment: influence of anions and calcium. *Proc.* B **205**, 323–345 (1979).

Rubinstein, I. *See* Goad, Rubinstein & Smith.

Rubio, J. *See* García-Moliner & Rubio.

Ruddiman, W.F., Sancetta, C.D. & McIntyre, A. Glacial/Interglacial response rate of subpolar North Atlantic waters to climatic change: the record in oceanic sediments (Discussion). *Trans.* B **280**, 119–142 (1977).

Rudge, M.R.H. On the scattering of electrons by atomic hydrogen. *Proc.* A **327**, 425–431 (1972).
 See also McCavert & Rudge.

Rudraiah, N. & Srimani, P.K. Finite-amplitude cellular convection in a fluid-saturated porous layer. *Proc.* A **373**, 199–222 (1980).

Rudy, B. Sodium gating currents in *Myxicola* giant axons. *Proc.* B **193**, 469–475 (1976).
 See also Tsien, Green, Levinson, Rudy & Sanders.

Rumsby, P.L. Faulting in the Kent Coalfield (Discussion). *Trans.* A **110**, 111–113 (1972).

Runciman, W.A. *See* Judd & Runciman.

Runcorn, S.K. Some aspects of the physics of the Moon (Symposium). *Proc.* A **336**, 11–33 (1974).

Interpretation of lunar potential fields (Discussion). *Trans.* A **285**, 507–516 (1977).

See also Collinson, Stephensen & Runcorn; Stephenson (A.), Collinson & Runcorn.

Runcorn, S.K., Collinson, D.W., O'Reilly, W., Stephenson, A., Battey, M.H., Manson, A.J. & Readman, P.W. Magnetic properties of Apollo 12 lunar samples. *Proc.* A **325**, 157–174 (1971).

Rundle, C.C. & Snelling, N.J. The geochronology of uraniferous minerals in the Witwatersrand Triad; an interpretation of new and existing U–Pb age data on rocks and minerals from the Dominion Reef, Witwatersrand and Ventersdorp Supergroups (Discussion). *Trans.* A **286**, 567–583 (1977).

Runham, N.W. *See* Richardson (C.A.), Crisp & Runham.

Runnegar, B. Origin and evolution of the Class Rostroconchia (Discussion). *Trans.* B **284**, 319–331 (1978).

Rush, R.A. *See* Livett, Geffen & Rush.

Rushton, W.A.H. Hamilton Hartridge. *Biogr. Mem.* **23**, 193–211 (1977).

Rushworth, A.J. *See* Elliott (R.J.), Hayes, Kleppmann, Rushworth & Ryan.

Ruskell, L.E.C. Reynolds equation and elastohydrodynamic lubrication in metal seals. *Proc.* A **349**, 383–396 (1976).

Russell, C.T. *See* Coleman (P.J.) & Russell.

Russell, D.K. *See* Davies (P.B.), Russell, Thrush & Radford.

Russell, E.W. The role of organic matter in soil fertility (Discussion). *Trans.* B **281**, 209–219 (1977).

Russell, G.E. Inherited resistance to virus yellows in sugar beet (Discussion). *Proc.* B **181**, 267–279 (1972).

Russell, J.M., III. *See* Gille, Bailey & Russell.

Russell, P.J. *See* Ledwith, Russell & Sutcliffe.

Russell, R. Scott. Improvement in crop production (Discussion). *Trans.* B **199**, 17–31 (1977).

Russell, S.H. *See* Eley & Russell.

Russell, Sir Frederick. Introductory remarks to a Discussion. *Proc.* B **180**, 365 (1972).

Sheina Macalister Marshall. *Biogr. Mem.* **24**, 369–389 (1978).

Rust, D.M. Optical and magnetic measurements of the photosphere and low chromosphere (Discussion). *Trans.* A **281**, 353–358 (1976).

Observations of flare-associated magnetic field changes (Discussion). *Trans.* A **281**, 427–433 (1976).

Rutigliano, B. *See* Metafora, Felsani, Cotrufo, Tajana, Del Rio, De Prisco and others.

Rutovitz, D. Pattern recognition by computer (Discussion). *Proc.* B **184**, 441–454 (1973).

Rutter, E.H. The kinetics of rock deformation by pressure solution (Discussion). *Trans.* A **283**, 203–219 (1976).

Ruzicka, L. Arthur Stoll. *Biogr. Mem.* **18**, 567–593 (1972).

Ryan, G.R. The genesis of Proterozoic uranium deposits in Australia (Discussion). *Trans.* A **291**, 339–353 (1979).

Ryan, J.F. *See* Elliott (R.J.), Hayes, Kleppmann, Rushworth & Ryan.

Ryan, K.P. *See* Bone & Ryan; Roberts (B.L.) & Ryan.

Ryan, M.P., Jr. *See* Breuer, Ryan & Waller.

Ryan, W.B.F. *See* Sibuet, Ryan, Arthur, Barnes, Blechsmidt and others.

Ryan, W.B.F. & Fox, P.J. Rifting and its geological environment [abstract] (Discussion). *Trans.* A **294**, 121–122 (1980).

Ryhming, I., Cooper, G.A. & Berlie, J. A novel concept for a rock-breaking machine. I. Theoretical consideration and model experiments. *Proc.* A **373**, 331–351 (1980).

de Sa, E.S. & Davies, G. Uniaxial stress studies of the 2.498 eV (H4), 2.417 eV and 2.536 eV vibronic bands in diamond. *Proc.* A **357**, 231–251 (1977).

Sabath, L.D. Achievements and problems from the view of a physician (Discussion). *Trans.* B **289**, 251–256 (1980).

Sabatier, P.C. Remarks on approximate methods in geophysical inverse problems. *Proc.* A **337**, 49–71 (1974).

Sabin, M.A. An existing system in the aircraft industry. The British Aircraft Corporation Numerical Master Geometry system (Discussion). *Proc.* A **321**, 197–205 (1971).

Sabina, F.J. *See* Burridge (R.) & Sabina.
Sabine, P.A. Metamorphic processes at high temperature and low pressure: the petrogenesis of the metasomatized and assimilated rocks of Carneal, Co. Antrim (with X-ray studies by B.R. Young). *Trans.* A **280**, 225–269 (1975).
See also Sutton (J.), Sabine & Skelhorn.
Sachs, A. Babylonian observational astronomy (Discussion). *Trans.* A **276**, 43–50 (1974).
Sachs, K. The role of residuals in engineering steels [abstract] (Discussion). *Trans.* A **295**, 119–120 (1980).
Sadler, P.J. *See* Barry (C.D.), Hill, Sadler & Williams.
Saffman, P.G. *See* Goodstein & Saffman; Moore (D.W.) & Saffman.
Sagan, C. The recognition of extraterrestrial intelligence (Discussion). *Proc.* B **189**, 143–153 (1975).
Sai, F.T. The needs of the developing world (Discussion). *Proc.* B **195**, 57–68 (1976).
Saka, H. *See* Cherns, Hirsch & Saka.
Sakore, T.D. *See* Sobell, Tsai, Jain & Sakore.
Salahub, D.R. *See* English (C.A.), Venables & Salahub.
Salam, Abdus. The unconfined unstable quark (predictions from a unified gauge theory of strong, weak and electromagnetic interactions) (Discussion). *Proc.* A **355**, 515–538 (1977).
Salas, Margarita. *See* Viñuela, Camacho, Jiménez, Carrascosa, Ramírez & Salas.
Saldin, D.K., Stathopoulos, A.Y. & Whelan, M.J. Electron microscope image contrast of small dislocation loops and stacking-fault tetrahedra. *Trans.* A **292**, 523–537 (1979).
Saldin, D.K. & Whelan, M.J. The construction of displacement fields of dislocation loops and stacking-fault tetrahedra from angular dislocation segments. *Trans.* A **292**, 513–521 (1979).
Salisbury, Sir Edward. The organization of the ranunculaceous flower with especial regard to the correlated variations of its constituent members. *Proc.* B **183**, 205–225 (1973).
The variations in the reproductive organs of *Stellaria media (sensu stricto)* and allied species with special regard to their relative frequency and prevalent modes of pollination. *Proc.* B **185**, 331–342 (1974).
Seed size and mass in relation to environment. *Proc.* B **186**, 83–88 (1974).
The floral morphology of *Ranunculus tripartitus* var. *terrestris (R. lutarius)* and comparison with related taxa. *Proc.* B **186**, 89–97 (1974).
The survival value of modes of dispersal. *Proc.* B **188**, 183–188 (1975).
A note on shade tolerance and vegetative propagation of woodland species. *Proc.* B **192**, 257–258 (1976).
Seed output and the efficacy of dispersal by wind. *Proc.* B **192**, 323–329 (1976).
Exceptional fruitfulness and its biological significance. *Proc.* B **193**, 455–460 (1976).
A note on seed production and frequency. *Proc.* B **200**, 485–487 (1978).
Salkeld, P. *See* Bayne, Moore, Widdows, Livingstone & Salkeld.
Salmon, G.A. *See* Dainton, Janovský & Salmon; Dainton, May, Morrow, Salmon & Thompson; Dainton, Morrow, Salmon & Thompson; Dainton, Salmon & Zucker; Ellison, Salmon & Wilkinson; Frankevich, Morrow & Salmon; O'Neill (P.), Salmon & May.
Salmon, P.R. New clinical procedures in endoscopy of the digestive tract (Discussion). *Proc.* B **195**, 243–249 (1977).
Salt, G. Experimental studies in insect parasitism. XVI. The mechanism of the resistance of *Nemeritis* to defence reactions. *Proc.* B **183**, 337–350 (1973).
Howard Everest Hinton. *Biogr. Mem.* **24**, 151–182 (1978).
A note on the resistance of two parasitoids to the defence reactions of their insect hosts. *Proc.* B **207**, 351–353 (1980).
Salzmann, T.N., Ratcliffe, R.W., Bouffard, F.A. & Christensen, B.G. A stereocontrolled, enantiomerically specific total synthesis of thienamycin (Discussion). *Trans.* B **289**, 191–195 (1980).
Sambles, J.R. An electron microscope study of evaporating gold particles: the Kelvin equation for liquid gold and the lowering of the melting point of solid gold particles. *Proc.* A **324**, 339–351 (1971).
See also Peppiatt & Sambles.
Sambrook, J., Botchan, M., Hu, S.-L., Mitchison, T. & Stringer, J. Integration of viral DNA sequences

in cells transformed by adenovirus 2 or SV40 (Discussion). *Proc.* B **210**, 423–435 (1980).

Sammes, P.G. *See* Barton (D.H.R.) & Sammes; Hill (H.A.O.), Sammes & Waley.

Samueloff, S. *See* Edholm (O.G.) & Samueloff; Edholm (O.G.), Samueloff, Mourant and others; Lehmann (E.E.), Gadoth & Samueloff; McCance, Abu Rabiyah, Beer, Edholm, Even-Paz, Luff & Samueloff.

Samueloff, S., Davies, C.T.M. & Shvartz, E. Biological studies of Yemenite and Kurdish Jews in Israel and other groups in southwest Asia. VII. The physical working capacity of Yemenite and Kurdish Jews in Israel. *Trans.* B **266**, 141–147 (1973).

Sancetta, C.D. *See* Ruddiman, Sancetta & McIntyre.

Sanchez, L. *See* Rose (G.), Widdel, Azcárraga & Sanchez.

Sancovich, H.A. *See* Jackson (A.H.), Sancovich, Ferramola, Evans and others.

Sandars, P.G.H. *See* Angel, Sandars & Woodgate.

Sandeman, D.R. *See* Zeki (S.M.) & Sandeman.

Sandeman, R.J. *See* Huber & Sandeman; Huber, Sandeman & Tubbs.

Sanders, G.D. & Young, J.Z. Reappearance of specific colour patterns after nerve regeneration in *Octopus*. *Proc.* B **186**, 1–11 (1974).

Sanders, J.K.M. *See* Tsien, Green, Levinson, Rudy & Sanders.

Sanders, P.G. *See* Wilkie (N.M.), Eglin, Sanders & Clements.

Sanderson, C.J. The mechanism of T cell mediated cytotoxicity.
 I. The release of different cell components. *Proc.* B **192**, 221–239 (1976).
 II. Morphological studies of cell death by time-lapse microcinematography.
  Proc. B **192**, 241–255 (1976).

Sanderson, C.J. & Glauert, Audrey M. The mechanism of T cell mediated cytotoxicity. V. Morphological studies by electron microscopy. *Proc.* B **198**, 315–323 (1977).

Sanderson, C.J., Hall, P.J. & Thomas, Jennifer A. The mechanism of T cell mediated cytotoxicity. IV. Studies on communicating junctions between cells in contact. *Proc.* B **196**, 73–84 (1977).

Sanderson, C.J. & Thomas, Jennifer A. The mechanism of T cell mediated cytotoxicity. III. Changes in target cell susceptibility during the cell cycle. *Proc.* B **194**, 417–429 (1976).
 The mechanism of K cell (antibody-dependent) cell mediated cytotoxicity.
 I. The release of different cell components. *Proc.* B **197**, 407–415 (1977).
 II. Characteristics of the effector cell and morphological changes in the target cell.
  Proc. B **197**, 417–424 (1977).

Sandler, L. *See* Lindsley & Sandler.

Sandri, C. *See* Pfenninger, Akert, Moor & Sandri.

Sandwell, R. *See* Ellis (P.), Holah, Houghton, Jones, Peckham and others.

Sandy, J.D. *See* Wider de Xifra, Sandy, Davies & Neuberger.

Sanford, P.A. *See* Browne (J.L.), Sanford & Smyth.

Sanford, P.W. X-ray observations of variable sources (Discussion). *Proc.* A **340**, 411–422 (1974).
 Observations of compact X-ray sources (Discussion). *Proc.* A **366**, 281–293 (1979).

Sanford, P.W. & Ives, J.C. Ariel results on extragalactic X-ray sources (Discussion).
 Proc. A **350**, 491–503 (1976).

Sanger, F. The Croonian Lecture, 1975. Nucleotide sequences in DNA. *Proc.* B **191**, 317–333 (1975).

Sankar, P.N. On the aerodynamic performance of a class of vertical shaft windmills.
 Proc. A **349**, 35–51 (1976).

Sankarasubramanian, R. *See* Gill (W.N.) & Sankarasubramanian.

Sankarasubramanian, R. & Gill, W.N. Dispersion from a prescribed concentration distribution in time variable flow. *Proc.* A **329**, 479–492 (1972).
 Unsteady convective diffusion with interphase mass transfer. *Proc.* A **333**, 115–132 (1973).
 Correction to 'Unsteady convective diffusion with interphase mass transfer'.
  Proc. A **341**, 407–408 (1974).

Sansom, Janet M. *See* Barnes (D.W.H.), Loutit & Sansom.

de Santis, A., Eusebi, F. & Miledi, R. Kainic acid and synaptic transmission in the stellate ganglion of the squid. *Proc.* B **202**, 527–532 (1978).

Saraph, Hannelore E. & Seaton, M.J. The calculation of energy levels for atoms in configurations

$1s^2 2s^2 2p^q nl$. *Trans.* A **271**, 1–39 (1971).

Sardar-ul-Mulk. *See* Chapman (J.A.), Grant, Taylor, Mahmud, Sardar-ul-Mulk & Shahid.

Sarfatti, G. *See* de Nettancourt, Devreux, Carluccio and others.

Sargent, G.E.G. *See* Orme, Flood & Sargent; Orme, Webb, Kelland & Sargent.

Sargent, J.A., Ingram, D.S. & Tommerup, I.C. Oospore development in *Bremia lactucae* Regel.: an ultrastructural study. *Proc.* B **198**, 129–138 (1977).

Sarkar, S.C. *See* Barnes (K.J.), Dondi & Sarkar.

Sarma, D.D. *See* Rao (C.N.R.), Sarma & Hegde; Rao (C.N.R.), Sarma, Vasudevan & Hegde.

Sasaki, R. Geometric approach to soliton equations. *Proc.* A **373**, 373–384 (1980).

Sastri, K.S. *See* Vajravelu & Sastri.

Sasvari, K. *See* Bernstein (J.), Regev, Herbstein, Main and others.

Sato, M. & Moore, J.G. Oxygen and sulphur fugacities of magmatic gases directly measured in active vents of Mount Etna (Symposium). *Trans.* A **274**, 137–146 (1973).

Satow, P.F.C. *See* Rees (G.R.) & Satow.

Sattelle, D.B. *See* Piddington & Sattelle.

Sattinger, D.H. On the free surface of a viscous fluid motion. *Proc.* A **349**, 183–204 (1976).

Sauke, T. *See* Benedek (G.B.), Clark, Serrallach, Young and others.

Saul, F. *See* Poljak, Amzel, Chen, Phizackerley & Saul.

Saule, S. *See* Hayman (M.J.), Ramsay, Kitchener, Graf, Beug and others.

Saunders, A.D. *See* Tarney, Wood, Saunders, Cann & Varet.

Saunders, C.P.R. *See* Brazier-Smith, Brook, Latham, Saunders & Smith.

Saunders, G.A. *See* Gunton & Saunders; Pace & Saunders.

Saunders, G.A. & Sümengen, Z. Frozen-in defects in bismuth in relation to its magnetoresistivity and thermoelectric power. *Proc.* A **329**, 453–466 (1972).

Saunders, M.J. *See* Cotes, Dabbs, Hall, Lakhera, Saunders & Malhotra.

Saunders, N.R. *See* Møllgård & Saunders.

Saunders, R.D. *See* Lowenstein & Saunders.

Saunders, S.R.J. *See* Lloyd (G.O.), Kent, Saunders & Lea.

Saunders, V.R. *See* DeKock, Lloyd, Hillier & Saunders; Guest (M.F.), Hillier, Saunders & Wood.

Savage, R.J.G. *See* Lloyd (A.J.), Savage, Stride & Donovan.

Saville, B.P. *See* Dobb, Johnson & Saville.

Savkoor, A.R. & Briggs, G.A.D. The effect of tangential force on the contact of elastic solids in adhesion. *Proc.* A **356**, 103–114 (1977).

Sawle, R. *See* Lea (C.), Sawle & Sellars.

Sawyer, B.C. *See* Bullock (G.), Eakins, Sawyer & Slater.

Sawyer, J.S. Introductory remarks to a Discussion. *Trans.* B **280**, 105 (1977).

Sayer, P. The long-wave behaviour of the virtual mass in water of finite depth. *Proc.* A **372**, 65–91 (1980).

An integral-equation method for determining the fluid motion due to a cylinder heaving on water of finite depth. *Proc.* A **372**, 93–110 (1980).

Sayer, R.J., Prince, R.H. & Duley, W.W. Luminescence of N atoms in solid N_2 stimulated by low energy electrons. *Proc.* A **365**, 235–251 (1979).

Scaife, B.K.P. *See* Calderwood & Scaife.

Scaife, W.G.S. & Lyons, C.G.R. Dielectric permittivity and pvT data of some n-alkanes. *Proc.* A **370**, 193–211 (1980).

Scala, A. *See* Fiecchi, Kienle, Scala, Galli and others.

Scali, V. *See* Clarke (Sir Cyril), Sheppard & Scali.

Scaramuzzi, R.J. *See* Baird (D.T.), Land, Scaramuzzi & Wheeler.

Scarlato, O.A. & Starobogatov, Y.I. Phylogenetic relations and the early evolution of the class Bivalvia (Discussion). *Trans.* B **284**, 217–224 (1978).

Scarlett, B., Buxton, R.E. & Faulkner, R.G. Formation of glass spheres on the lunar surface (Discussion). *Trans.* A **285**, 279–284 (1977).

Scarrott, G.G. The Clifford Paterson Lecture, 1979. From computing slave to knowledgeable servant: the evolution of computers. *Proc.* A **369**, 1–30 (1979).

Schaefer, G.W. An airborne radar technique for the investigation and control of migrating pest insects (Discussion). *Trans.* B 287, 459–465 (1979).

Schäefer, R. *See* Franklin (R.M.), Hinnen, Schäefer & Tsukagoshi.

Schaeffer, O.A. Lunar chronology as determined from the radiometric ages of returned lunar samples (Discussion). *Trans.* A 285, 137–143 (1977).

Schaffner, W. *See* Birnstiel, Kressmann, Schaffner, Portmann & Busslinger.

Schafir, R.L. Bondi's vector, and the Seliger & Whitham Variational Principle for matter fields. *Proc.* A 368, 59–73 (1979).

Schagen, P. Review Lecture. Electronic aids to night vision. *Trans.* A 269, 233–263 (1971).
  X-ray image intensifiers: design and future possibilities (Discussion). *Trans.* A 292, 265–272 (1979).

Schalow, G. *See* Miledi, Parker & Schalow.

Schalow, G. & Schmidt, H. Local development of action potentials in slow muscle fibres after complete or partial denervation. *Proc.* B 203, 445–457 (1979).

Scheer, U. *See* Franke & Scheer.

Scheffler, T.B. Analyticity of the eigenvalues and eigenfunctions of an ordinary differential operator with respect to a parameter. *Proc.* A 336, 475–486 (1974).

Schell, J. *See* Van Montagu, Holsters, Zambryski, Hernalsteens, Depicker and others.

Schell, J., Van Montagu, M., De Beuckeleer, M., De Block, M., Depicker, A., De Wilde, M., Engler, G., Genetello, C., Hernalsteens, J.P., Holsters, M., Seurinck, J., Silva, B., Van Vliet, F. & Villarroel, R. Interactions and DNA transfer between *Agrobacterium tumefaciens*, the Ti-plasmid and the plant host. *Proc.* B 204, 251–266 (1979).

Schild, H.O. & Rose, F.L. Harry Raymond Ing. *Biogr. Mem.* 22, 239–255 (1976).

Schilf, W. *See* Martin (H.H.), Tonn-Ehlers & Schilf.

Schilling, J.-G. Sea-floor evolution: rare-earth evidence (Discussion). *Trans.* A 268, 663–706 (1971).

Schilling, J.-G., Bergeron, M.B. & Evans, R. Halogens in the mantle beneath the North Atlantic (Discussion). *Trans.* A 297, 147–178 (1980).

Schindler, P. Enzyme inhibitors of microbial origin (Discussion). *Trans.* B 290, 291–301 (1980).

Schipper, P.E. *See* Craig (D.P.) & Schipper.

Schipper, P.E. & Walmsley, S.H. On the rôle of exciton trapping at extended defects in the photodimerization of anthracene. *Proc.* A 348, 203–219 (1976).

Schlag, E.W. *See* Metz, Howard, Wunsch, Neusser & Schlag.

Schlapp, R. *See* Kemmer & Schlapp.

Schlein, J. *See* Rothschild (Miriam) & Schlein; Rothschild (Miriam), Schlein, Parker, Neville & Sternberg.

Schmatz, W. Magnetic diffuse and small-angle scattering (Discussion). *Trans.* B 290, 527–535 (1980).

Schmid, M. La flore et la végétation de la partie méridionale de l'Archipel des Nouvelles Hébrides (Discussion). *Trans.* B 272, 329–342 (1975).

Schmidbauer, M. The Symphonie project (Discussion). *Proc.* A 345, 559–565 (1975).

Schmidt, B.G. & Stewart, J.M. The scalar wave equation in a Schwarzschild space–time. *Proc.* A 367, 503–525 (1979).

Schmidt, D.L. *See* Greenwood (W.R.), Hadley, Anderson, Fleck & Schmidt.

Schmidt, H. *See* Lehouelleur & Schmidt; Schalow & Schmidt.

Schmidt, H. & Tong, E.Y. Inhibition by actinomycin D of the denervation-induced action potential in frog slow muscle fibres. *Proc.* B 184, 91–95 (1973).

Schmidt, J.T. The laminar organization of optic nerve fibres in the tectum of goldfish. *Proc.* B 205, 287–306 (1979).

Schmidt-Nielsen, K. The physiology of wild animals (Discussion). *Proc.* B 199, 345–360 (1977).

Schneider, M.F. *See* Chandler (W.K.), Schneider, Rakowski & Adrian.

Schneider, S. Flashlamp-pumped mode-locked dye lasers (Discussion). *Trans.* A 298, 233–245 (1980).

Schneider, W. *See* Lucas (M.L.), Schneider, Haberich & Blair.

Schnitker, D. *See* Roberts (D.G.), Montadert, Thompson, Auffret and others.

Schnurmann, R. & Davies, P.A. Admissible Reynolds numbers for Poiseuille flow in jet viscometer orifices. *Proc.* A 347, 47–60 (1975).

Schnurmann, R. & Lardge, M.G.C. Electric climbing of dielectric liquids.
 Proc. A **327**, 393–402 (1972).
 Enhanced heat flux in non-uniform electric fields. *Proc.* A **334**, 71–82 (1973).
Schoenberg, D.A. & Trench, R.K. Genetic variation in *Symbiodinium* (= *Gymnodinium*) *micro-adriaticum* Freudenthal, and specificity in its symbiosis with marine invertebrates.
 I. Isoenzyme and soluble protein patterns of axenic cultures of *Symbiodinium microadriaticum.*
  Proc. B **207**, 405–427 (1980).
 II. Morphological variation in *Symbiodinium microadriaticum. Proc.* B **207**, 429–444 (1980).
 III. Specificity and infectivity of *Symbiodinium microadriaticum. Proc.* B **207**, 445–460 (1980).
Scholes, J.H. Colour receptors, and their synaptic connexions, in the retina of a cyprinid fish.
 Trans. B **270**, 61–118 (1975).
Schöll, E. Formal conditions for non-equilibrium phase transitions in semiconductors.
 Proc. A **365**, 511–521 (1979).
Schöll, E. & Landsberg, P.T. Semiconductor models for first and second order non-equilibrium phase transitions. *Proc.* A **365**, 495–510 (1979).
Scholtissek, C. Evolution of pandemic influenza virus strains (Discussion).
 Trans. B **288**, 307–312 (1980).
Schrameck, Joan E. *See* Johnston, Schrameck & Mark.
Schroepfer, G.J., Jr, Lutsky, B.N., Martin, J.A., Huntoon, S., Fourcans, B., Lee, W.-H. & Vermilion, J. Recent investigations on the nature of sterol intermediates in the biosynthesis of cholesterol (Discussion). *Proc.* B **180**, 125–146 (1972).
Schroter, R.C. *See* Caro, Fitz-Gerald & Schroter.
Schubert, G., Young, R.E. & Cassen, P. Solid state convection models of the lunar internal temperature (Discussion). *Trans.* A **285**, 523–536 (1977).
Schulman, H.M. *See* Beringer, Brewin, Johnston, Schulman & Hopwood.
Schultz, D.L. *See* Young (A.D.), Pankhurst & Schultz.
Schultzberg, M. *See* Hökfelt, Lundberg, Schultzberg, Johansson, Skirboll, Anggård and others.
Schürlein, B. *See* Bartel, Duinker, Heintze, Heinzelmann and others.
Schurr, S.H. *See* Darmstadter & Schurr.
Schutz, B.F. *See* Comins & Schutz; Dyson (J.) & Schutz.
Schwab, D.J. *See* Mortimer (C.H.), Rao & Schwab; Rao & Schwab.
Schwar, M.J.R. *See* Jones (A.R.), Schwar & Weinberg.
Schwartz, L.W. *See* Vanden Broeck, Schwartz & Tuck.
Schwartz, S.J. & Roxburgh, I.W. Instabilities in the solar wind (Discussion).
 Trans. A **297**, 555–563 (1980).
Schweiger, H.-G. *See* Bannwarth, Ikehara & Schweiger; Bannwarth & Schweiger.
Schwerer, F.C. *See* Fisher (R.M.), Huffman, Nagata & Schwerer.
Schwyzer, R. Structure and function in neuropeptides (Discussion). *Proc.* B **210**, 5–20 (1980).
Sciama, D.W. Cosmological implications of the 3 K background [abstract] (Discussion).
 Proc. B **368**, 17–18 (1979).
Scoffin, T.P. *See* Flood, Orme & Scoffin; Flood & Scoffin; Stoddart (D.R.), McLean, Scoffin & Gibbs; Stoddart (D.R.), McLean, Scoffin, Thom & Hopley.
Scoffin, T.P. & McLean, R.F. Exposed limestones of the Northern Province of the Great Barrier Reef (Discussion). *Trans.* A **291**, 119–138 (1978).
Scoffin, T.P. & Stoddart, D.R. Nature and significance of microatolls (with an appendix by B.R. Rosen) (Discussion). *Trans.* B **284**, 99–122 (1978).
Scoffin, T.P., Stoddart, D.R., McLean, R.F. & Flood, P.G. The Recent development of reefs in the Northern Province of the Great Barrier Reef (Discussion). *Trans.* B **284**, 129–139 (1978).
Scott, A.F.D. Mesospheric temperatures and winds during a stratospheric warming (Discussion).
 Trans. A **271**, 547–557 (1972).
Scott, A.I. The biosynthesis of vitamin B_{12} (Discussion). *Trans.* B **273**, 303–318 (1976).
Scott, J.E. Physiological function and chemical composition of pericellular proteoglycan (an evolutionary view) (Discussion). *Trans.* B **271**, 235–242 (1975).
Scott, J.F. *See* Crighton & Scott; Huerre & Scott.

Scott, M.H. *See* Robinson (J.L.) & Scott.

Scott, R.L. *See* van Konynenburg & Scott.

Scott, S.C. The geology of Longonot volcano, Central Kenya: a question of volumes.
    Trans. A **296**, 437–465 (1980).

Scriven, L.E. *See* Brown (R.A.) & Scriven.

Scruton, B. *See* Briscoe, Scruton & Willis.

Scruton, C. & Rogers, E.W.E. Steady and unsteady wind loading of buildings and structures (Discussion). *Trans.* A **269**, 353–383 (1971).

Scudder, P. *See* Feeney, Roberts, Birdsall, Griffiths and others.

Scurrell, M.S. *See* Eley, Rochester & Scurrell; Kemball, Nisbet, Robertson & Scurrell.

Seah, M.P. Segregation and the strength of grain boundaries. *Proc.* A **349**, 535–554 (1976).
  Impurities, segregation and creep embrittlement (Discussion). *Trans.* A **295**, 265–278 (1980).
  See also Lea (C.), Seah & Hondros.

Seah, M.P. & Hondros, E.D. Grain boundary segregation. *Proc.* A **335**, 191–212 (1973).

Seah, M.P., Spencer, P.J. & Hondros, E.D. Investigation of an additive remedy for temper brittleness [abstract] (Discussion). *Trans.* A **295**, 301 (1980).

Seaton, M.J. *See* Saraph & Seaton.

Seeger, A.K. Some recollections of the radiation damage work of the 1950s (Symposium).
    Proc. A **371**, 165–172 (1980).
  Early work on imperfections in crystals, and forerunners of dislocation theory (Symposium).
    Proc. A **371**, 173–177 (1980).

Segal, S.J. *See* Harkavy, Jaffe, Koblinsky & Segal.

Segoufin, J. Structure du plateau continental armoricain (Discussion). *Trans.* A **279**, 109–121 (1975).

Seibert, G. *Spacelab* and material processing facilities and experiments (Discussion).
    Proc. A **361**, 131–142 (1978).

Seifert, F.A. Reconstruction of rock cooling paths from kinetic data on the Fe^{2+}–Mg exchange reaction in anthophyllite (Discussion). *Trans.* A **286**, 303–311 (1977).

Seitz, F. Biographical notes (Symposium). *Proc.* A **371**, 84–99 (1980).

Sekerka, R.F. *See* Foglio, Sekerka & Van Vleck.

Sekhar, P. Chandra. *See* Gopal, Sekhar, Ananthakrishna and others.

Sela, M. *See* Shaltiel, Mizrahi & Sela.

Self, S. *See* Booth (B.) & Self.

Selikoff, I.J. Polybrominated biphenyls in Michigan [abstract] (Discussion).
    Proc. B **205**, 153–156 (1979).

Sellars, C.M. Recrystallization of metals during hot deformation (Discussion).
    Trans. A **288**, 147–158 (1978).
  See also Lea (C.), Sawle & Sellars.

Seller, Mary J. *See* Embury, Seller, Adinolfi & Polani.

Seller, Mary J., Embury, Susan, Polani, P.E. & Adinolfi, M. Neural tube defects in curly-tail mice. II. Effect of maternal administration of vitamin A. *Proc.* B **206**, 95–107 (1979).

Seminara, G. & Hall, P. Linear stability of slowly varying unsteady flows in a curved channel.
    Proc. A **346**, 279–303 (1975).
  Centrifugal instability of a Stokes layer: linear theory. *Proc.* A **350**, 299–316 (1976).
  The centrifugal instability of a Stokes layer: nonlinear theory. *Proc.* A **354**, 119–126 (1977).

Senanayake, Pramilla. Applying family planning in rural communities (Discussion).
    Proc. B **199**, 115–127 (1977).

Sengupta, S. *See* Ghosh, Basu & Sengupta.

Sequeira, L. *See* Kelman & Sequeira.

Series, G.W. Introductory remarks to a Discussion. *Trans.* A **293**, 211 (1979).
  See also Pegg & Series.

Serjeantson, S. *See* Boyce (A.J.), Harrison, Platt, Hornabrook and others.

Serrallach, E.N. *See* Benedek (G.B.), Clark, Serrallach, Young and others.

Serrin, J. The swirling vortex. *Trans.* A **271**, 325–360 (1972).
  See also Fosdick & Serrin.

Seth, B.B. *See* Hattangadi, Wagner & Seth.

Seurinck, J. *See* Schell, Van Montagu, De Beuckeleer, De Block and others.

Sevenhuysen, G. *See* Miller (D.S.), Baker, Bowden, Evans, Holt and others.

Severn, R.T. Numerical methods for calculation of stress and strain (Discussion).
  Trans. A **274**, 339–350 (1973).

Sewell, M.J. & Noble, B. General estimates for linear functionals in nonlinear problems.
  Proc. A **361**, 293–324 (1978).

Seymour, B.R. *See* Mortell & Seymour.

Shackleton, N.J. The oxygen isotope stratigraphic record of the Late Pleistocene (Discussion).
  Trans. B **280**, 169–182 (1977).

Shackleton, R.M. Problems of the evolution of the continental crust (Discussion).
  Trans. A **273**, 317–320 (1973).
  Pan-African structures (Discussion). *Trans.* A **280**, 491–497 (1976).
  See also Ries & Shackleton.

Shadrake, L.G. *See* Guiu & Shadrake.

Shaffer, J.C. *See* Cogan, Hutson & Schaffer.

Shahid, M.A. *See* Chapman (J.A.), Grant, Taylor, Mahmud, Sardar-ul-Mulk & Shahid.

Shalaby, A.H. *See* Havner & Shalaby.

Shaltiel, S., Mizrahi, Rachel & Sela, M. On the immunological properties of penicillins (Discussion).
  Proc. B **179**, 411–432 (1971).

Shank, C.V. *See* Ippen, Shank, Wiesenfeld & Migus.

Shank, C.V., Ippen, E.P., Fork, R.L., Migus, A. & Kobayashi, T. Application of subpicosecond optical techniques to molecular dynamics (Discussion). *Trans.* A **298**, 303–308 (1980).

Shankar, P.N. On the aerodynamic performance of a class of vertical shaft windmills.
  Proc. A **349**, 35–51 (1976).

Shanks, M.F., Pearson, R.C.A. & Powell, T.P.S. The intrinsic connections of the primary somatic sensory cortex of the monkey. *Proc.* B **200**, 95–101 (1978).

Shaper, A.G. Cardiovascular disease and trace metals (Discussion). *Proc.* B **205**, 135–143 (1979).
  Epidemiology for geochemists (Discussion). *Trans.* B **288**, 127–136 (1979).

Shapiro, B.I. *See* Hille, Woodhull & Shapiro.

Shapiro, M.M. & Silberberg, R. Cosmic-ray nuclei up to 10^{10} eV/u in the Galaxy (Discussion).
  Trans. A **277**, 319–348 (1974).

Shapley, R. & Gordon, J. The visual sensitivity of the retina of the conger eel.
  Proc. B **209**, 317–330 (1980).

Sharp, G.W.G. *See* Leaf & Sharp.

Sharp, R.R. *See* Hall (C.), Richards & Sharp.

Sharpe, R.S. Current limitations of non-destructive testing in engineering (Discussion).
  Trans. A **292**, 163–174 (1979).

Shaw, D.D. *See* Rees (H.), Shaw & Wilkinson.

Shaw, G.J. *See* Eglinton, HajIbrahim, Maxwell, Quirke, Shaw and others.

Shaw, J.J. *See* Lainson, Ready & Shaw; Lainson, Ward & Shaw.

Shaw, Linda & Spencer, A.J.M. Transverse impact of ideal fibre-reinforced rigid–plastic plates.
  Proc. A **361**, 43–64 (1978).

Sheather, P. *See* Woodgate (B.E.), Knight, Uribe, Sheather, Bowles & Nettleship.

Sheehan, J.C. Introductory remarks to a Discussion. *Trans.* B **289**, 167 (1980).

Sheffield, Elizabeth & Bell, P.R. Phytoferritin in the reproductive cells of a fern, *Pteridium aquilinum* (L.) Kuhn. *Proc.* B **202**, 297–306 (1978).

Sheldon, J. Freezing technology in the market place (Discussion). *Proc.* B **191**, 111–129 (1975).

Sheldrick, G.M. *See* Bonnett (R.), Davies, Hursthouse & Sheldrick.

Shellis, R.P. & Miles, A.E.W. Autoradiographic study of the formation of enameloid and dentine matrices in teleost fishes using tritiated amino acids. *Proc.* B **185**, 51–72 (1974).
  Observations with the electron microscope on enameloid formation in the common eel (*Anguilla anguilla*; Teleostei). *Proc.* B **194**, 253–269 (1976).

Shelton, G.A.B. Colonial behaviour and electrical activity in the Hexacorallia.
  Proc. B **190**, 239–256 (1975).

Shelton, G.A.B. & McFarlane, I.D. Electrophysiology of two parallel conducting systems in the colonial Hexacorallia. *Proc.* B **193**, 77—87 (1976).

Shelton, R.G.J. Effects of oil and oil dispersants on the marine environment (Discussion). *Proc.* B **177**, 411—422 (1971).

Shemin, D. δ-Aminolaevulinic acid dehydratase: structure, function, and mechanism (Discussion). *Trans.* B **273**, 109—115 (1976).

Shenton, D.B. *See* Bates (B.), Bradley, McBride and others.

Shephard-Thorn, E.R. *See* Destombes, Shephard-Thorn & Redding; Kellaway, Redding, Shephard-Thorn & Destombes.

Shephard-Thorn, E.R., Lake, R.D. & Atitullah, E.A. Basement control of structures in the Mesozoic rocks of the Strait of Dover region, and its reflexion in certain features of the present land and submarine topography (with an appendix by P.L. Rumsby) (Discussion). *Trans.* A **272**, 99—113 (1972).

Shepherd, J.G. Supercurrents through thick, clean S—N—S sandwiches. *Proc.* A **326**, 421—430 (1972). *See also* Pippard, Shepherd & Tindall.

Shepherd, J.V. *See* Cullis, Hucknall & Shepherd.

Shepherd, T.J. Fluid inclusion study of the Witwatersrand gold—uranium ores (Discussion). *Trans.* A **286**, 549—565 (1977).

Sheppard, B.L. Platelet adhesion in the rabbit abdominal aorta following the removal of endothelium with EDTA. *Proc.* B **182**, 103—108 (1972).

Sheppard, B.L. & French, J.E. Platelet adhesion in the rabbit abdominal aorta following the removal of the endothelium: a scanning and transmission electron microscopical study. *Proc.* B **176**, 427—432 (1971).

Sheppard, C.J.R. & Wilson, T. Fourier imaging of phase information in scanning and conventional optical microscopes. *Trans.* A **295**, 513—536 (1980).

Sheppard, N. *See* Smart (R.StC.) & Sheppard.

Sheppard, P.A. *See* Pasquill, Sheppard & Sutcliffe.

Sheppard, P.M. *See* Allen (W.R.) & Sheppard; Brown (K.S.), Sheppard & Turner; Clarke (C.A.) & Sheppard; Clarke (Sir Cyril) & Sheppard; Clarke (Sir Cyril), Sheppard & Scali; Whittle, Clarke, Sheppard & Bishop.

Sheppard, S.M.F. *See* Easton, Hamilton, Kempe & Sheppard.

Shepstone, B.J. Diagnostic visualization (Discussion). *Trans.* A **292**, 177—185 (1979).

Sherif, M.A. *See* Bostrom & Sherif.

Sherins, R.J. *See* Jones (R.), Mann & Sherins.

Sherman, P. Textural properties and food acceptability (Discussion). *Proc.* B **191**, 131—144 (1975).

Shido, F. *See* Miyashiro, Shido & Ewing.

Shikata, M., Kim, S., Mita, K., Ise, N. & Kunugi, S. 'Catalysis' by anionic and cationic polyelectrolytes on reactions between cationic species: outer and inner sphere electron transfer reactions of Co complexes. *Proc.* A **351**, 233—243 (1976).

Shimahara, T. & Peretz, B. Quantal transmitter release in an identified inhibitory cholinergic synapse of *Aplysia*. *Proc.* B **206**, 403—409 (1980).

Shimizu, N. *See* Allègre, Shimizu & Treuil.

Shipley, G.G. *See* Elder (M.), Hitchcock, Mason & Shipley.

Shipley, T.H. *See* Vail (P.R.), Mitchum, Shipley & Buffler.

Shirling, D. *See* McCance, El Neil, El Din, Widdowson, Southgate and others.

Shkoller, B. *See* Pekeris, Accad & Shkoller.

Shmatchenko, V.V. *See* Varshavsky, Bakayev, Bakayeva, Chumackov *et al.*

Shoenberg, Catherine F. The influence of temperature on the thick filaments of vertebrate smooth muscle (Discussion). *Trans.* B **265**, 197—202 (1973).

Shoenberg, D. Heinz London. *Biogr. Mem.* **17**, 441—461 (1971).

Shoppee, C.W. Christopher Kelk Ingold. *Biogr. Mem.* **18**, 349—411 (1972). Corrigenda. *Biogr. Mem.* **19**, 695 (1973).

Short, L. The practical evaluation of multivariate approximants with branch points. *Proc.* A **362**, 57—69 (1978).

Short, R.V. The evolution of human reproduction (Discussion). *Proc.* B **195**, 3—24 (1976).

Shorthouse, M. *See* Peters, Shorthouse, Ward & McDowell.

Shortt, H.E. & Garnham, P.C.C. Samuel Rickard Christophers. *Biogr. Mem.* **25**, 179—207 (1979).

Shotton, F.W. The Devensian Stage: its development, limits and substages (Discussion). *Trans.* B **280**, 107—118 (1977).

See also Colhoun, Dickson, McCabe & Shotton.

Shotton, F.W., Goudie, A.S., Briggs, D.J. & Osmaston, H.A. Cromerian interglacial deposits at Sugworth, near Oxford, England, and their relation to the Plateau Drift of the Cotswolds and the terrace sequence of the Upper and Middle Thames. *Trans.* B **289**, 55—86 (1980).

Shotton, K.C. *See* Blaney, Bradley, Edwards, Jolliffe and others; Till (Susan M.), Jones & Shotton.

Shreve, R.L. *See* Drake (L.D.) & Shreve.

Shulman, R.G. *See* Cohen (S.M.) & Shulman.

Shum, W.K. *See* Cuthbert & Shum.

Shvartz, E. *See* Samueloff, Davies & Shvartz.

Sibbett, W. *See* Adams (M.C.), Bradley, Sibbett & Taylor; Bradley (D.J.), Jones & Sibbett.

Sibly, R.M. *See* McFarland & Sibly.

Sibuet, J.C. *See* Foucher & Sibuet.

Sibuet, J.C., Ryan, W.B.F., Arthur, M., Barnes, R., Blechsmidt, G., De Charpal, O., De Gracianský, P.C., Habib, D., Iaccarino, S., Johnson, D., Lopatin, B.G., Maldonado, A., Montadert, L., Moore, D.G., Morgan, G.E., Mountain, G., Rehault, J.P., Sigal, J. & Williams, C.A. Deep drilling results of Leg 47b (Galicia Bank area) in the framework of the early evolution of the North Atlantic Ocean (Discussion). *Trans.* A **294**, 51—61 (1980).

Siddans, A.W.B. Deformed rocks and their textures (Discussion). *Trans.* A **283**, 43—54 (1976).

Sidia, Abdallahi Ould M. *See* Abdallahi Ould M. Sidia and others.

Siebens, H.C. *See* Trench & Siebens.

Siebens, H.C. & Trench, R.K. Aspects of the relation between *Cyanophora paradoxa* (Korschikoff) and its endosymbiotic cyanelles *Cyanocyta korschikoffiana* (Hall & Claus). III. Characterization of ribosomal ribonucleic acids. *Proc.* B **202**, 463—472 (1978).

Siegel, B.M. Current and future prospects in electron microscopy for observations on biomolecular structure (Discussion). *Trans.* B **261**, 5—14 (1971).

Siegel, R.L. *See* Munro (E.), Siegel, Craig & Sly.

Siegenthaler, A. *See* Henderson (L.F.) & Siegenthaler.

Sievers, A. & Volkmann, D. Ultrastructure of gravity-perceiving cells in plant roots (Discussion). *Proc.* B **199**, 525—536 (1977).

Sigal, J. *See* Sibuet, Ryan, Arthur, Barnes, Blechsmidt and others.

Sigel, R. Optical diagnostics of laser-produced plasmas with ultra-short laser pulses (Discussion). *Trans.* A **298**, 407—414 (1980).

Signer, P., Baur, H., Etique, Philippe, Frick, U. & Funk, H. On the question of the [40]Ar excess in lunar soils (Discussion). *Trans.* A **285**, 385—390 (1977).

Silberberg, R. *See* Shapiro (M.M.) & Silberberg.

Silva, B. *See* Schell, Van Montagu, De Beuckeleer, De Block and others.

Silver, D.M. *See* Wilson (S.), Silver & Farrell.

Silver, I.A. Measurement of pH and ionic composition of pericellular sites (Discussion). *Trans.* B **271**, 261—272 (1975).

Silver, J.D. & Stacey, D.N. Isotope shift and hyperfine structure in the atomic spectrum of tin. *Proc.* A **332**, 129—138 (1973).

Isotope effects in the nuclear charge distribution in tin. *Proc.* A **332**, 139—150 (1973).

Silvester, W.B. *See* Lex, Silvester & Stewart.

Simkin, T. *See* Cann & Simkin.

Simmie, J.M. *See* Simpson (C.J.S.M.), Gait & Simmie; Simpson (C.J.S.M.) & Simmie.

Simmonds, N.W. Plant breeding (Discussion). *Trans.* B **267**, 145—156 (1973).

Simmons, M.D. *See* White (G.W.T.) & Simmons.

Simoneit, B.R.T. *See* Eglinton, Simoneit & Zoro.

Simons, J.P. *See* Boxall & Simons.

Simpkins, P.G. & Krause, J.T. Dynamic response of glass fibres during tensile fracture.
Proc. A **350**, 253—265 (1976).

Simpson, C.J.S.M., Gait, P.D. & Simmie, J.M. Vibration—rotation energy exchange in carbon dioxide—hydrogen mixtures.
II. *Proc.* A **348**, 57—72 (1976).
III. Comparison with theory. *Proc.* A **348**, 73—80 (1976).

Simpson, C.J.S.M. & Simmie, J.M. A study of vibrational—rotational energy exchange in a shock tube.
Proc. A **325**, 197—206 (1971).

Simpson, G.G. Review of fossil penguins from Seymour Island. *Proc.* B **178**, 357—387 (1971).

Simpson, P.R. & Bowles, J.F.W. Uranium mineralization of the Witwatersrand and Dominion Reef systems (with a preface by S.H.U. Bowie) (Discussion). *Trans.* A **286**, 527—548 (1977).

Simpson, P.R., Brown, G.C., Plant, Jane & Ostle, D. Uranium mineralization and granite magmatism in the British Isles (Discussion). *Trans.* A **291**, 385—412 (1979).

Sinclair, A.T. *See* Wilkins (G.A.) & Sinclair.

Sinclair, J.E. & Lawn, B.R. An atomistic study of cracks in diamond-structure crystals.
Proc. A **329**, 83—103 (1972).

Sinclair, R. *See* Driver (J.H.), Sinclair & Jack.

Sinden, R.E., Canning, Elizabeth U., Bray, R.S. & Smalley, M.E. Gametocyte and gamete development in *Plasmodium falciparum. Proc.* B **201**, 375—399 (1978).

Sinden, R.E., Canning, Elizabeth U. & Spain, Barbara. Gametogenesis and fertilization in *Plasmodium yoelii nigeriensis:* a transmission electron microscope study. *Proc.* B **193**, 55—76 (1976).

Singer, I.M. *See* Atiyah, Hitchin & Singer.

Singer, K. *See* Adams (Eveline M.), McDonald & Singer.

Singer, K.E. *See* Lee (A.E.) & Singer.

Singh, A. *See* Andrews (E.H.), Owen & Singh.

Singh, A.B. *See* Singh (G.), Joshi, Chopra & Singh.

Singh, G., Joshi, R.D., Chopra, S.K. & Singh, A.B. Late Quaternary history of vegetation and climate of the Rajasthan Desert, India. *Trans.* B **267**, 467—501 (1974).

Singh, J. *See* Wood (D.S.), Oertel, Singh & Bennett.

Singh, P. *See* Baldwin (J.E.), Jung, Singh, Wan, Haber and others.

Singh, R.K. & Gupta, H.N. An extended three-body force shell model for the lattice-dynamics of ionic crystals. *Proc.* A **349**, 289—308 (1976).

Singh, R.P. *See* Block (H.), Ions, Powell, Singh & Walker.

Singh, S.J., Ben-Menahem, A. & Vered, M. A unified approach to the representation of seismic sources. *Proc.* A **331**, 525—551 (1973).

Singla, C.M. *See* Mackie (G.O.), Paul, Singla, Sleigh & Williams.

Singleton, T. Coordination and management of resources for learning and their application to satellite systems for education (Discussion). *Proc.* A **345**, 531—539 (1975).

Sivan, J.P. *See* Courtès, Laget, Sivan, Viton and others.

Sjogren, W.L. Lunar gravity determinations and their implications (Discussion).
Trans. A **285**, 219—226 (1977).

Skaer, R.J. Transcription in the interbands of *Drosophila* (Discussion).
Trans. B **283**, 411—413 (1978).
See also Johnson (R.T.), Mullinger & Skaer.

Skaf, R. *See* Abdallahi, Skaf, Castel & Ndiaye.

Skehel, J.J. *See* Hay, Skehel & McCauley.

Skehel, J.J., Waterfield, M.D., McCauley, J.W., Elder, K. & Wiley, D.C. Studies on the structure of the haemagglutinin (Discussion). *Trans.* B **288**, 335—339 (1980).

Skeldon, P., Calvert, J.M. & Lees, D.G. A study of the oxidation mechanisms of some austenitic stainless steels in carbon dioxide at 1123 K by means of charged-particle nuclear techniques. I, II, III. *Trans.* A **296**, 545—580 (1980).

Skelhorn, R.R. *See* Sutton (J.), Sabine & Skelhorn.

Skelton, P.W. The evolution of functional design in rudists (Hippuritacea) and its taxonomic implications (Discussion). *Trans.* B **284**, 305—318 (1978).

Skempton, A.W. Introductory remarks to a Discussion. *Trans.* A **283**, 423–426 (1976).
See also Chandler (R.J.), Kellaway, Skempton & Wyatt.

Skempton, A.W. & Weeks, A.G. The Quaternary history of the Lower Greensand escarpment and Weald Clay vale near Sevenoaks, Kent (Discussion). *Trans.* A **283**, 493–526 (1976).

Skene, D.S. Chloroplast structure in mature apple leaves grown under different levels of illumination and their response to changed illumination. *Proc.* B **186**, 75–78 (1974).

Sketch, H.J.H. The Prospero satellite (Discussion). *Proc.* A **343**, 265–275 (1975).

Skews, B.W. *See* Cruickshank & Skews.

Skilling, J. The complete set of uniform polyhedra. *Trans.* A **278**, 111–135 (1975).

Skinner, A.E. *See* Bishop (A.W.) & Skinner.

Skinner, G.K. The galactic centre (Discussion). *Proc.* A **366**, 345–355 (1979).

Skirboll, L. *See* Hökfelt, Lundberg, Schultzberg, Johansson, Skirboll, Änggård and others.

Sköld, K. & Pelizzari, C.A. Elementary excitations in liquid ^{3}He (Discussion).
Trans. B **290**, 605–616 (1980).

Slack, G.A. *See* Challis (L.J.), de Goër, Guckelsberger & Slack.

Slade, Carole T. *See* Benjamin (P.R.), Slade & Soffe.

Slater, T.F. *See* Bullock (G.), Eakins, Sawyer & Slater.

Sleigh, M.A. *See* Mackie (G.O.), Paul, Singla, Sleigh & Williams.

Sleytr, Uwe B. *See* Thornley, Glauert & Sleytr.

Sloper, J.J., Hiorns, R.W. & Powell, T.P.S. A qualitative and quantitative electron microscopic study of the neurons in the primate motor and somatic sensory cortices.
Trans. B **285**, 141–171 (1979).

Sloper, J.J. & Powell, T.P.S. Dendro-dendritic and reciprocal synapses in the primate motor cortex.
Proc. B **203**, 23–38 (1978).
Gap junctions between dendrites and somata of neurons in the primate sensori-motor cortex.
Proc. B **203**, 39–47 (1978).
Ultrastructural features of the sensori-motor cortex of the primate. *Trans.* B **285**, 123–139 (1979).
A study of the axon initial segment and proximal axon of neurons in the primate motor and somatic sensory cortices. *Trans.* B **285**, 173–197 (1979).
An experimental electron microscopic study of afferent connections to the primate motor and somatic sensory cortices. *Trans.* B **285**, 199–226 (1979).

Sly, W.S. *See* Munro, Siegel, Craig & Sly.

Smale-Adams, K.B. & Jackson, G.O. Manganese nodule mining (Discussion).
Trans. A **290**, 125–133 (1978).

Small, J.V. *See* Sobieszek & Small.

Smalley, M.E. *See* Sinden, Canning, Bray & Smalley.

Smalley, M.V. *See* Bomchil, Hüller, Rayment, Roser, Smalley, Thomas & White.

Smarr, L. Numerical construction of space–time [abstract] (Discussion).
Proc. A **363**, 15–16 (1979).

Smart, C. *See* Mingins, Zobel, Pethica & Smart.

Smart, R.StC. & Sheppard, N. Infrared spectroscopic studies of adsorption on alkali halide surfaces.
I. HCl, HBr and HI on fluorides, chlorides, bromides and iodides. *Proc.* A **320**, 417–436 (1971).

Smid, L. *See* Bovée, Creyghton, Getreuer, Korbee, Lobregt and others.

Smidt, J. *See* Bovée, Creyghton, Getreuer, Korbee, Lobregt and others.

Smith, A.B. The structure and arrangement of echinoid tubercles. *Trans.* B **289**, 1–54 (1980).

Smith, A.D. Secretion of proteins (chromogranin A and dopamine β-hydroxylase) from a sympathetic neuron (Discussion). *Trans.* B **261**, 363–370 (1971).
Summing up: some implications of the neuron as a secreting cell (Discussion).
Trans. B **261**, 423–437 (1971).
See also Chubb (I.W.) & Smith; Somogyi, Chubb & Smith.

Smith, A.G. Environmental factors influencing pupal colour determination in Lepidoptera.
I. Experiments with *Papilio polytes, Papilio demoleus* and *Papilio polyxenes*.
Proc. B **200**, 295–329 (1978).
II. Experiments with *Pieris rapae, Pieris napi* and *Pieris brassicae*. *Proc.* B **207**, 163–186 (1980).

Smi

See also Goad, Rubinstein & Smith; Pilcher & Smith.

Smith, A.J. *See* Andreieff, Bouysse, Curry, Fletcher and others; Curry (D.) & Smith; Hamilton (D.), Hommeril, Larsonneur & Smith; Kaiser (T.R.), Orr & Smith.

Smith, A.J. & Cameron, A. Rigid surface films. *Proc.* A **328**, 541–560 (1972).

Smith, A.J. & Curry, D. The structure and geological evolution of the English Channel (Discussion). *Trans.* A **279**, 3–20 (1975).

Smith, A.M. A study of interstellar lines in the rocket spectrum of δ Scorpii (Discussion). *Trans.* A **279**, 317–322 (1975).

Smith, C.H. Llewellyn. Heavy leptons (Discussion). *Proc.* A **355**, 585–599 (1977).

Smith, C.J.E. *See* Dillamore, Morris, Smith & Hutchinson.

Smith, D.A. *See* Pond & Smith.

Smith, D.A.S. *See* Rothschild (Miriam), von Euw, Reichstein, Smith & Pierre.

Smith, D.C. From extracellular to intracellular: the establishment of a symbiosis (Discussion). *Proc.* B **204**, 115–130 (1979).

See also Boyle (J. Elizabeth) & Smith; Gallop, Bartrop & Smith; Jolley & Smith; Lewis (D.H.) & Smith; Muscatine, Boyle & Smith; Trench, Boyle & Smith.

Smith, D.D. The development of international legal principles to facilitate educational satellite telecommunications: a humanistic approach (Discussion). *Proc.* A **345**, 575–590 (1975).

Smith, D.E. *See* Kolenkiewicz, Smith, Rubincam, Dunn & Torrence.

Smith, D.E., Kolenkiewicz, R., Wyatt, G.H., Dunn, P.J. & Torrence, M.H. Geodetic applications of laser ranging (Discussion). *Trans.* A **284**, 529–536 (1977).

Smith, D.M. Stanley Fabes Dorey. *Biogr. Mem.* **19**, 305–316 (1973).

Smith, D.N.R. & Banks, W.B. The mechanism of flow of gases through coniferous wood. *Proc.* B **177**, 197–223 (1971).

Smith, D.S. On the significance of cross-bridges between microtubules and synaptic vesicles (Discussion). *Trans.* B **261**, 395–405 (1971).

Smith, E. Planar distributions of dislocations. III. The exact positions of important dislocations when the number in an array is large. *Proc.* A **320**, 527–535 (1971).

Smith, E.B. *See* Griffiths (H.B.), Miller, Paton & Smith; Middleton (A.J.) & Smith.

Smith, E.R. *See* Barouch, Perram & Smith; de Leeuw, Perram & Smith; Perram & Smith.

Smith, F.B. & Hunt, R.D. The dispersion of sulphur pollutants over western Europe (Discussion). *Trans.* A **290**, 523–542 (1979).

Smith, F.T. Three dimensional stagnation point flow into a corner. *Proc.* A **344**, 489–507 (1975).

Steady motion within a curved pipe. *Proc.* A **347**, 345–370 (1976).

Fluid flow into a curved pipe. *Proc.* A **351**, 71–87 (1976).

Steady motion through a branching tube. *Proc.* A **355**, 167–187 (1977).

The laminar separation of an incompressible fluid streaming past a smooth surface. *Proc.* A **356**, 443–463 (1977).

On the non-parallel flow stability of the Blasius boundary layer. *Proc.* A **366**, 91–109 (1979).

Nonlinear stability of boundary layers for disturbances of various sizes. *Proc.* A **368**, 573–589 (1979).

Corrections to 'Nonlinear stability of boundary layers for disturbances of various sizes'. *Proc.* A **371**, 439–440 (1980).

See also Dennis (S.C.R.) & Smith.

Smith, F.T. & Stewartson, K. On slot injection into a supersonic laminar boundary layer. *Proc.* A **332**, 1–22 (1973).

Smith, G. & Tomkins, F.S. Autoionization resonances in the Eu I absorption spectrum and a new determination of the ionization potential. *Proc.* A **342**, 149–156 (1975).

The absorption spectrum of europium. *Trans.* A **283**, 345–365 (1976).

Smith, G.D.W. *See* Beaven (P.A.), Miller, Williams, Delargy & Smith.

Smith, G.J. *See* Dixon (H.E.), Earnshaw, Hook, Hough, Smith, Stephenson & Turver; Franklin (R.N.), Hamberger, Lampis & Smith.

Smith, I.R. *See* Viner & Smith.

Smith, J.B. *See* Bennett (M.D.), Finch, Smith & Rao; Bennett (M.D.), Rao, Smith & Bayliss;

206

Bennett (M.D.) & Smith; Bennett (M.D.), Smith & Barclay.

Smith, J.F. & Courtier, G.M. The Ariel 5 programme (Discussion). *Proc.* A **350**, 421–439 (1976).

Smith, J.M. *See* Blow & Smith.

Smith, J. Maynard. Game theory and the evolution of behaviour (Discussion).
  Proc. B **205**, 475–488 (1979).

Smith, J.V. Mineralogy of the planets: a voyage in space and time (summary of the centenary Halli-
  mond Lecture) (Discussion). *Trans.* A **286**, 433–437 (1977).

The relation of mantle heterogeneity to the bulk composition and origin of the Earth (Discussion).
  Trans. A **297**, 139–146 (1980).

See also Dawson (J.B.), Smith & Hervig.

Smith, K. *See* Pendlebury & Smith.

Smith, K.M. *See* Kenner, Rimmer, Smith & Unsworth.

Smith, M.H. *See* Brazier-Smith, Brook, Latham, Saunders & Smith.

Smith, M.L. *See* Dahlen & Smith.

Smith, Moya M. The microstructure of the dentition and dermal ornament of three dipnoans from
  the Devonian of Western Australia: a contribution towards dipnoan interrelations, and morphogenesis, growth and adaptation of the skeletal tissues. *Trans.* B **281**, 29–72 (1977).

Structure and histogenesis of tooth plates in *Sagenodus inaequalis* Owen considered in relation to
  the phylogeny of post-Devonian dipnoans. *Proc.* B **204**, 15–39 (1979).

Smith, M.S. Phase integrals and coupling for radio waves in the ionosphere.
  Proc. A **335**, 213–233 (1973).

Coupling points of the Booker quartic equation for radio wave propagation in the ionosphere.
  Proc. A **336**, 229–250 (1974).

Theory of the reflexion of low-frequency radio waves obliquely incident on the ionosphere.
  Proc. A **336**, 525–542 (1974).

Theory of the reflexion in the ionosphere of radio waves with frequencies near the electron gyro-
  frequency. *Proc.* A **343**, 133–153 (1975).

Phase memory in W.K.B. and phase integral solutions of ionospheric propagation problems.
  Proc. A **346**, 59–79 (1975).

See also Budden & Smith.

Smith, M.W. & Ellory, J.C. Sodium–amino acid interactions in the intestinal epithelium (Discussion).
  Trans. B **262**, 131–140 (1971).

Smith, M.W. & Jarvis, L.G. Growth and cell replacement in the new-born pig intestine.
  Proc. B **203**, 69–89 (1978).

Smith, M.W. & Peacock, M.A. Anomalous replacement of foetal enterocytes in the neonatal pig.
  Proc. B **206**, 411–420 (1980).

Smith, P. Some applications of extremum principles to magnetohydrodynamic pipe flow.
  Proc. A **336**, 211–222 (1974).

Smith, P.A. Cerenkov relations between r.f. noise and energetic particles on Ariel 4 (Discussion).
  Proc. A **343**, 241–255 (1975).

Smith, P.L. & Kühne, M. Oscillator strengths of neutral titanium from hook method measurements in
  a furnace. I. Lines from the $a^3F_{2,3 \text{ and } 4}$ levels at 0, 0.021, and 0.048 eV.
  Proc. A **362**, 263–279 (1978).

Smith, R. Buoyancy effects upon longitudinal dispersion in wide well-mixed estuaries.
  Trans. A **296**, 467–496 (1980).

See also Bona (J.L.) & Smith; Peregrine & Smith.

Smith, R.A. Review Lecture. Lasers and light scattering. *Proc.* A **323**, 305–320 (1971).

Excitation of transitions between atomic or molecular energy levels by monochromatic laser radiation.

I. Excitation under conditions of strictly homogeneous line broadening.
  Proc. A **362**, 1–12 (1978).

II. Excitation under conditions of strictly inhomogeneous line broadening and mixed homogeneous
  and inhomogeneous broadening. *Proc.* A **362**, 13–25 (1978).

III. Effect of Doppler broadening when phase-destroying collisions predominate.
  Proc. A **368**, 163–175 (1979).

IV. Transient excitation of Doppler-broadened transitions — effect of detuning.
   Proc. A **371**, 319—329 (1980).

Smith, R.E. Diatomic hydride and deuteride spectra of the second row transition metals.
   Proc. A **332**, 113—127 (1973).

Smith, R.H. *See* Perkin (G.W.), Duncan, Mahoney & Smith.

Smith, R.I.L. *See* Callaghan, Smith & Walton.

Smith, R.L. A probability model for fibrous composites with local load sharing.
   Proc. A **372**, 539—553 (1980).
   See also James (Margaret O.), Smith, Williams & Reidenberg.

Smith, R.W. *See* Croker, Fidler & Smith.

Smith, S.D. *See* Butcher, Dennis & Smith; Chapman (W.A.), Cross, Flower, Peckham & Smith; Dennis (R.B.), Pidgeon, Smith, Wherrett & Wood; Dennis (R.B.), Smith & Summers; Ellis (P.), Holah, Houghton, Jones, Peckham and others.

Smith, S.G. *See* Jackson (A.H.), Sancovich, Ferramola, Evans and others.

Smith, S.R.L. Single cell protein (Discussion). *Trans.* B **290**, 341—354 (1980).

Smith, S.R.P. *See* Elliott (R.J.), Harley, Hayes & Smith.

Smith, Una. Uptake of ferritin into neurosecretory terminals (Discussion).
   Trans. B **261**, 391—394 (1971).

Smoluchowski, R. Random comments on the early days of solid state physics (Symposium).
   Proc. A **371**, 100—101 (1980).

Smyth, D.H. Sodium—hexose interactions (Discussion). *Trans.* B **262**, 121—130 (1971).
   See also Browne (J.L.), Sanford & Smyth; Cheeseman (C.I.) & Smyth.

Snary, D. *See* Crumpton, Snary, Walsh, Barnstable and others.

Snelling, N.J. *See* Rundle & Snelling.

Snow, M.H.L. & Ansell, J.D. The chromosomes of giant trophoblast cells of the mouse.
   Proc. B **187**, 93—98 (1974).

Snow, P., Jr. *See* Gull, York, Snow & Henize.

Snow, P.J. The motor innervation and the musculature of the antennule of the Australian mud crab, *Scylla serrata* (Forskål) (Portunoidea, Brachyura). *Proc.* B **207**, 219—237 (1980).

Snowden, J.McK. *See* Preston (B.N.) & Snowden.

Snyder, A.W. *See* Diesendorf, Stange & Snyder.

Snyder, S.H. *See* Gorenstein & Snyder.

Sobell, H.M. *See* Tsai (C.-C.), Jain & Sobell.

Sobell, H.M., Tsai, C.-C., Jain, S.C. & Sakore, T.D. Conformational flexibility in DNA structure and its implications in understanding the organization of DNA in chromatin (Discussion).
   Trans. B **283**, 295—298 (1978).

Sobieszek, A. & Small, J.V. The assembly of ribbon-shaped structures in low ionic strength extracts obtained from vertebrate smooth muscle (Discussion). *Trans.* B **265**, 203—212 (1973).

Soffe, S.R. *See* Benjamin (P.R.), Slade & Soffe.

Soh, W.K. *See* Fink & Soh.

Sollins, M.R. *See* Julian, Sollins & Moss.

Soma, M. *See* Watanabe (K.), Kondow, Soma, Onishi & Tamaru.

Somers, M.L. Appendix to a paper by Stride, Belderson & Kenyon. *Trans.* A **284**, 281—285 (1977).

Somlyo, A.P. *See* Devine, Somlyo & Somlyo.

Somlyo, A.P., Devine, C.E., Somlyo, Avril V. & Rice, R.V. Filament organization in vertebrate smooth muscle (Discussion). *Trans.* B **265**, 223—229 (1973).

Somlyo, Avril V. *See* Devine, Somlyo & Somlyo; Somlyo (A.P.), Devine, Somlyo & Rice.

Sommers, P. Properties of shear-free congruences of null geodesics. *Proc.* A **349**, 309—318 (1976).

Sommerville, J., Malcolm, D.B. & Callan, H.G. The organization of transcription on lampbrush chromosomes (Discussion). *Trans.* B **283**, 359—366 (1978).

Somogyi, P., Chubb, I.W. & Smith, A.D. A possible structural basis for the extracellular release of acetylcholinesterase. *Proc.* B **191**, 271—283 (1975).

Somorjai, G.A., Joyner, R.W. & Lang, B. The reactivity of low index [(111) and (100)] and stepped platinum single crystal surfaces (Discussion). *Proc.* A **331**, 335—346 (1972).

Sonnett, C.P. Some consequences of solar wind induction in the Moon (Discussion).
  Trans. A **285**, 537–547 (1977). Erratum. *Trans.* A **288**, 644 (1978).
Sood, M.C. *See* Brown-Grant, Murray, Raisman & Sood.
Sørensen, P.F. *See* Mehlum & Sørensen.
Sotelo, C. *See* Mariani, Crepel, Mikoshiba, Changeux & Sotelo.
Sotelo, C. & Beaudet, A. Influence of experimentally induced agranularity on the synaptogenesis of serotonin nerve terminals in rat cerebellar cortex. *Proc.* B **206**, 133–138 (1979).
Soundalgekar, V.M. Free convection effects on the oscillatory flow past an infinite, vertical, porous plate with constant suction.
  I. *Proc.* A **333**, 25–36 (1973).
  II. *Proc.* A **333**, 37–50 (1973).
Souriau, M. *See* Froidevaux & Souriau.
Souster, C.G. *See* Page (J.K.), Rodgers & Souster.
South, G.P. & Grant, E.H. Dielectric dispersion and dipole moment of myoglobin in water.
  Proc. A **328**, 371–387 (1972).
Southall, D.M. *See* Jackson (W.J.) & Southall.
Southgate, D.A.T. *See* McCance, El Neil, El Din, Widdowson, Southgate and others.
Southgate, Eileen. *See* White (J.G.), Southgate, Thomson & Brenner.
Southon, M.J. *See* Waugh, Mills & Southon.
Southwood, D.J. *See* Dungey (J.W.) & Southwood.
Soward, A.M. A kinematic theory of large magnetic Reynolds number dynamos.
  Trans. A **272**, 431–462 (1972).
  A convection-driven dynamo. I. The weak field case. *Trans.* A **275**, 611–646 (1974).
  A unified approach to Stefan's problem for spheres and cylinders. *Proc.* A **373**, 131–147 (1980).
  See also Roberts (P.H.) & Soward.
Soward, A.M. & Priest, E.R. Fast magnetic field line reconnection. *Trans.* A **284**, 369–417 (1977).
Sowter, C. *See* Ashwell (Margaret), Meade, Medawar & Sowter; Meade, Ashwell & Sowter.
Sozou, C. Electrohydrodynamics of a liquid drop: the time-dependent problem.
  Proc. A **331**, 263–272 (1972).
  Electrohydrodynamics of a liquid drop: the development of the flow field.
  Proc. A **334**, 343–356 (1973).
Sozou, C. & English, H. Fluid motions induced by an electric current discharge.
  Proc. A **329**, 71–81 (1972).
Sozou, C. & Hewson-Browne, R.C. On two dimensional electrohydrostatic stability.
  Proc. A **349**, 231–243 (1976).
Sozou, C. & Pickering, W.M. Magnetohydrodynamic flow in a container due to the discharge of an electric current from a finite size electrode. *Proc.* A **362**, 509–523 (1978).
Spadafora, C. *See* Oudet, Germond, Bellard, Spadafora & Chambon.
Spain, Barbara. *See* Sinden, Canning & Spain.
Spalding, D.B. & Stephenson, P.L. Laminar flame propagation in hydrogen + bromine mixtures.
  Proc. A **324**, 315–337 (1971).
Spargo, E. *See* Daniel, Pratt & Spargo.
Sparks, B.W. The non-marine Mollusca from Swanton Morley sample D (appendix to a paper by Linda Phillips). *Trans.* B **275**, 278–281 (1976).
  See also West (R.G.), Dickson, Catt, Weir & Sparks.
Sparks, B.W., Williams, R.B.G. & Bell, F.G. Presumed ground-ice depressions in East Anglia.
  Proc. A **327**, 329–343 (1972).
Spassky, B. *See* Dobzhansky, Levene & Spassky.
Spaull, V.W. Distribution of soil and litter arthropods on Aldabra Atoll (Discussion).
  Trans. B **286**, 109–117 (1979).
Speakman, J.C. *See* Kerr (K.Ann), Ashmore & Speakman.
Speedy, Gillian. *See* Woodruff (Sir Michael) & Speedy.
Speer, R.J. *See* Franks (A.), Lindsey, Bennett, Speer, Turner & Hunt.
Spence, D.A. & Turcotte, D.L. An elastostatic model of stress accumulation on the San Andreas fault.

Proc. A **349**, 319—341 (1976).
Viscoelastic relaxation of cyclic displacements on the San Andreas Fault.
Proc. A **365**, 121—144 (1979).
Spencer, A.J.M. *See* Shaw (Linda) & Spencer.
Spencer, P.J. *See* Seah, Spencer & Hondros.
Spencer, P.N. *See* Ford (C.E.), O'Hara & Spencer.
Spencer, R. *See* Cartwright (D.E.), Edden, Spencer & Vassie.
Spera, F.J. *See* Carmichael, Nicholls, Spera, Wood & Nelson.
Spettel, B. *See* Dreibus, Spettel & Wänke; Wänke, Palme, Baddenhausen, Dreibus, Kruse & Spettel.
Spicer, A. Contributions of processing preservation towards the food technologies of the 1980s (Discussion). *Proc.* B **191**, 3—7 (1975).
Spiegelman, S. DNA and the RNA viruses (Discussion). *Proc.* B **177**, 87—108 (1971).
Spikes, H.A. & Cameron, A. A comparison of adsorption and boundary lubricant failure.
Proc. A **336**, 407—419 (1974).
Spinks, A. *See* Day (R.L.) & Spinks.
Spiro, T.G. Biological applications of resonance Raman spectroscopy: haem proteins (Discussion). *Trans.* A **345**, 89—105 (1975).
Spitzer, N.C. *See* McMahan (U.J.), Spitzer & Peper.
Spizzichino, A. & Glass, M. C.w. radar continuous wind measurements over France [abstract] (Discussion). *Trans.* A **271**, 599 (1972).
Spizzichino, A. & Massebeuf, M. Atomspheric waves observed over Garchy (47° N) [abstract] (Discussion). *Trans.* A **271**, 599 (1972).
Spracklen, C.T. *See* Dudeney, Jones, Kressman & Spracklen.
Spratt, B.G. Biochemical and genetical approaches to the mechanism of action of penicillin (Discussion). *Trans.* B **289**, 273—283 (1980).
Sprenger, K. & Lysenko, I.A. The significance and interpretation of ionospheric drift measurements in the low-frequency range (Discussion). *Trans.* A **271**, 473—484 (1972).
Spring, Susan B. *See* Murphy (B.R.), Markoff, Chanock, Spring and others.
Springer, C. *See* Jorna & Springer.
Springer, S.G. The solution of heat-transfer problems by the Wiener—Hopf technique. II. Trailing edge of a hot film. *Proc.* A **337**, 395—412 (1974). Corrigenda. *Proc.* A **338**, 544 (1974).
Springer, S.G. & Pedley, T.J. The solution of heat transfer problems by the Wiener—Hopf technique. I. Leading edge of a hot film. *Proc.* A **333**, 347—362 (1973).
Springer, T. Developments in experimental neutron physics at the Institut Laue—Langevin (Discussion). *Trans.* B **290**, 673—681 (1980).
Springford, M. *See* Lowndes, Miller, Poulsen & Springford.
Squires, G.L. *See* Pynn & Squires.
Sreekantan, B.V. Blackett Memorial Lecture, 1978. Fundamental research in India in the area of the physical sciences. *Proc.* A **365**, 145—160 (1979).
Srimani, P.K. *See* Rudraiah & Srimani.
Sriram, G. *See* Webster (R.G.), Hinshaw, Bean & Sriram.
Srivastava, R.S. *See* Chopra (M.G.) & Srivastava.
Stacey, D.N. *See* Baird (P.E.G.) & Stacey; Baird (P.E.G.), Brambley, Burnett, Stacey, Warrington & Woodgate; Brimicombe, Stacey, Stacey, Hühnermann & Menzel; Kuhn, Baird, Brimicombe, Stacey & Stacey; Malvern, Pinder, Stacey & Thompson; Silver (J.D.) & Stacey.
Stacey, M. John Kenyon Netherton Jones. *Biogr. Mem.* **25**, 365—389 (1979).
Stacey, M. & Percival, Elizabeth. Edmund Langley Hirst. *Biogr. Mem.* **22**, 137—168 (1976).
Stacey, V. *See* Brimicombe, Stacey, Stacey, Hühnermann & Menzel; Kuhn, Baird, Brimicombe, Stacey & Stacey.
Stacey, Virginia & Kuhn, H.G. Isotope shifts and hyperfine structure in the atomic spectrum of platinum. *Proc.* A **322**, 301—311 (1971).
Stagg, D. *See* Howarth (J.V.), Ritchie & Stagg.
Stammers, Judith R. *See* Davies (D.S.) & Stammers.
Stämpfli, R. *See* Nonner, Rojas & Stämpfli.

Standley, C.C. *See* Kessler (A.) & Standley.

Stange, G. *See* Diesendorf, Stange & Snyder; Horridge, Giddings & Stange; Horridge, McLean, Stange & Lillywhite.

Stanhope, J.M. *See* Hornabrook, Crane & Stanhope.

Stanley, D.J. Low-temperature magnetothermopower of silver. *Proc.* A **339**, 97—117 (1974).

Stanley, P.I. & Bunyan, P.J. Hazards to wintering geese and other wildlife from the use of dieldrin, chlorfenvinphos and carbophenothion as wheat seed treatments (Discussion). *Proc.* B **205**, 31—45 (1979).

Stanley, S.M. Aspects of the adaptive morphology and evolution of the Trigoniidae (Discussion). *Trans.* B **284**, 247—258 (1978).

Stanway, R. *See* Burrows (C.R.) & Stanway.

Stark, J.P.W. The X-ray spectrum and variability of active galaxies (Discussion). *Proc.* A **366**, 435—448 (1979).

Starkel, L. The palaeogeography of mid- and east Europe during the last cold stage, with west European comparisons (Discussion). *Trans.* B **280**, 351—372 (1977).

Starobogatov, Y.I. *See* Scarlato & Starobogatov.

Stathopoulos, A.Y. *See* Saldin, Stathopoulos & Whelan.

Staveley, L.A.K. *See* Callanan, Weir & Staveley.

Stean, J.P.B. *See* Burns (B. Delisle), Stean & Webb.

Stebbing, A.R.D. An experimental approach to the determinants of biological water quality (Discussion). *Trans.* B **286**, 465—481 (1979).

Stebbing, N., Lindley, I.J.D. & Eaton, M.A.W. The direct anti-viral activity of single stranded poly-ribonucleotides.
  I. Potentiation of activity by mixtures of polymers which do not anneal.
     Proc. B **198**, 411—428 (1977).
 II. The effect of molecular size and the involvement of cellular receptors.
     Proc. B **198**, 429—437 (1977).

Stebbins, R.T. *See* Levine (J.) & Stebbins.

Steeds, J.W. *See* Buxton (B.F.), Eades, Steeds & Rackham; Jones (P.M.), Rackham & Steeds.

Steele, J.H. Comparative studies of beaches (Discussion). *Trans.* B **274**, 401—415 (1976).
  The uses of experimental ecosystems (Discussion). *Trans.* B **286**, 583—595 (1979).

Steele, J.H. & Frost, B.W. The structure of plankton communities. *Trans.* B **280**, 485—534 (1977).

Steele, R.C. Forests and wildlife (Discussion). *Trans.* B **271**, 163—178 (1975).

Steensberg, A. The husbandry of food production (Discussion). *Trans.* B **275**, 43—54 (1976).

Steers, J.A. Concluding remarks to a Discussion. *Trans.* A **291**, 195—196 (1978).
  Concluding remarks to a Discussion. *Trans.* B **284**, 161—162 (1978).

Stefani, E. *See* Beaty & Stefani.

Steffen, P. *See* Bartel, Duinker, Heintze, Heinzelmann and others.

Stehbens, W.E. Haemodynamic production of lipid deposition, intimal tears, mural dissection and thrombosis in the blood vessel wall. *Proc.* B **185**, 357—373 (1974).

Stehelin, Dominique. *See* Hayman (M.J.), Ramsay, Kitchener, Graf, Beug and others.

Steinberg, R.H. & Wood, I. Pigment epithelial cell ensheathment of cone outer segments in the retina of the domestic cat. *Proc.* B **187**, 461—478 (1974).

Steinberg, R.H., Wood, I. & Hogan, M.J. Pigment epithelial ensheathment and phagocytosis of extrafoveal cones in human retina. *Trans.* B **277**, 459—471 (1977).

Steiner, D.F., Patzelt, C., Chan, S.J., Quinn, P.S., Tager, H.S., Nielsen, D., Lernmark, Å., Noyes, B.E., Agarwal, K.L., Gabbay, K.H. & Rubenstein, A.H. Formation of biologically active peptides (Discussion). *Proc.* B **210**, 45—59 (1980).

Steiner, J.M. *See* Murphy (J.O.) & Steiner.

Steinmetz, M., Streeck, R.E. & Zachau, H.G. Reconstituted histone—DNA complexes (Discussion). *Trans.* B **283**, 259—268 (1978).

Stengelin, R. *See* von Engelhardt & Stengelin.

Stephan, K.-H. *See* Hessberg, Niekerke & Stephan.

Stephansson, O. Finite element analysis of folds (Discussion). *Trans.* A **283**, 153—161 (1976).

Stephen, M.J. *See* Bibby, Nabarro, McLachlan & Stephen.

Stephens, R.W.B. *See* Kumar (Ram) & Stephens.

Stephenson, A. *See* Collinson, Stephenson & Runcorn; Runcorn, Collinson, O'Reilly, Stephenson, Battey, Manson & Readman.

Stephenson, A., Collinson, D.W. & Runcorn, S.K. Magnetic characteristics of Luna 16 and 20 samples. *Trans.* A 284, 151–156 (1977).

Stephenson, P.L. *See* Spalding & Stephenson.

Stephenson, R.F. *See* Brozel, Evans & Stephenson.

Stephenson, W. *See* Dixon (H.E.), Earnshaw, Hook, Hough, Smith, Stephenson & Turver.

de Sterke, A. Advances in the technology of mechanized ultrasonic testing (Discussion). *Trans.* A 292, 207–221 (1979).

Stern, H. Concluding remarks to a Discussion. *Trans.* B 277, 371–376 (1977).
See also Bennett (M.D.) & Stern.

Stern, H. & Hotta, Y. Biochemistry of meiosis (Discussion). *Trans.* B 277, 277–294 (1977).

Sternberg, S. *See* Rothschild (Miriam), Schlein, Parker, Neville & Sternberg.

Stettler, A. *See* Geiss, Eberhardt, Grögler, Guggisberg, Maurer & Stettler.

Steven, J.H. *See* Charlesby, Folland & Steven.

Steward, F.C., Israel, H.W., Mott, R.L., Wilson, H.J. & Krikorian, A.D. Observations on growth and morphogenesis in cultured cells of carrot (*Daucus carota* L.). *Trans.* B 273, 33–53 (1975).

Stewart, J.M. Global solutions of linear ordinary differential equations with polynomial coefficients. *Proc.* A 334, 51–64 (1975).
On the stability of Kerr's space-time. *Proc.* A 344, 65–79 (1975).
On transient relativistic thermodynamics and kinetic theory. *Proc.* A 357, 59–75 (1977).
Hertz–Bromwich–Debye–Whittaker–Penrose potentials in general relativity. *Proc.* A 367, 527–538 (1979).
See also Israel (W.) & Stewart; Schmidt & Stewart.

Stewart, J.M. & Walker, M. Perturbations of space-times in general relativity. *Proc.* A 341, 49–74 (1974).

Stewart, K.H. *See* Miller (D.E.), Brownscombe, Carruthers, Pick & Stewart.

Stewart, Murray. The location of the troponin binding site on tropomyosin. *Proc.* B 190, 257–266 (1975).

Stewart, P.A.E. General discussion: close range X-ray photogrammetry at Rolls-Royce (Discussion). *Trans.* A 292, 175–176 (1979).

Stewart, R.J. *See* Mitchell (E.W.J.) & Stewart.

Stewart, R.W. & Thomson, R.E. Re-examination of vorticity transfer theory. *Proc.* A 354, 1–8 (1977).

Stewart, W.D.P. Nitrogen fixation (Discussion). *Trans.* B 274, 341–358 (1976).
See also Apte, Rowell & Stewart; Lex, Silvester & Stewart; Tel-Or & Stewart.

Stewartson, K. *See* Brown (Susan N.) & Stewartson; Davey & Stewartson; Hocking & Stewartson; Roberts (P.H.) & Stewartson; Smith (F.T.) & Stewartson.

Stewartson, K. & Waechter, R.T. On Stefan's problem for spheres. *Proc.* A 348, 415–426 (1976).

Stewartson, K. & Walton, I.C. On waves in a thin shell of stratified rotating fluid. *Proc.* A 349, 141–156 (1976).

Stiles, P.J. Multipolar expansions for magnetic shielding near anisotropic molecules. *Proc.* A 336, 251–256 (1974).
'Ring-current' and 'pseudo-contact' magnetic shielding by non-spherical molecules. *Proc.* A 346, 209–225 (1975).
See also Craig (D.P.), Radom & Stiles; Perram & Stiles.

Stiles, W.S. Appendix to a paper by J.D. Mollon & P.G. Polden. *Trans.* B 278, 233–240 (1977).

Stillinger, F.H. Theoretical approaches to the intermolecular nature of water (Discussion). *Trans.* B 278, 97–112 (1977).

Stimson, I.L. & Fisher, R. Design and engineering of carbon brakes (Discussion). *Trans.* A 294, 583–590 (1980).

Stinnakre, J. *See* Israel (M.), Lesbats, Meunier & Stinnakre; Kusano, Miledi & Stinnakre.

Stirling, Christine. *See* Patrick (G.) & Stirling.

Stirling, R. *See* Nettleton & Stirling.

Stock, M.J. *See* Miller (D.S.), Baker, Bowden, Evans, Holt and others.

Stock, M.J. & Miller, D.S. Dietary-induced thermogenesis at high and low altitudes (Discussion).
Proc. B **194**, 57–62 (1976).

Stocker, B.A.D. & Mäkelä, P.H. Genetics of the (gram-negative) bacterial surface (Discussion).
Proc. B **202**, 5–30 (1978).

Stockmayer, W.H. *See* Edwards (S.F.) & Stockmayer.

Stockwell, R.A. Structural and histochemical aspects of the pericellular environment in cartilage (Discussion). *Trans.* B **271**, 243–245 (1975).

Stoddart, C.T.H. & Hunt, C.P. Alloy surface composition resulting from fabrication, as determined by s.i.m.s. (secondary ion mass spectrometry) [abstract] (Discussion).
Trans. A **295**, 134–135 (1980).

Stoddart, D.R. Scientific studies at Aldabra and neighbouring islands (Discussion).
Trans. B **260**, 5–29 (1971).
Settlement, development and conservation of Aldabra (Discussion). *Trans.* B **260**, 611–628 (1971).
Place names of Aldabra (Discussion). *Trans.* B **260**, 631–632 (1971).
The Great Barrier Reef and the Great Barrier Reef Expedition 1973 (Discussion).
Trans. A **291**, 5–22 (1978).
Aldabra and the Aldabra Research Station (Discussion). *Trans.* B **286**, 3–10 (1979).
See also McLean (R.F.) & Stoddart; McLean (R.F.), Stoddart, Hopley & Polach; Scoffin & Stoddart; Scoffin, Stoddart, McLean & Flood.

Stoddart, D.R., McLean, R.F. & Hopley, D. Geomorphology of reef islands, northern Great Barrier Reef (Discussion). *Trans.* B **284**, 39–61 (1978).

Stoddart, D.R., McLean, R.F., Scoffin, T.P. & Gibbs, P.E. Forty-five years of change on low wooded islands, Great Barrier Reef (Discussion). *Trans.* B **284**, 63–80 (1978).

Stoddart, D.R., McLean, R.F., Scoffin, T.P., Thom, B.G. & Hopley, D. Evolution of reefs and islands, northern Great Barrier Reef: synthesis and interpretation (Discussion).
Trans. B **284**, 149–159 (1978).

Stoddart, D.R. & Peake, J.F. Historical records of Indian Ocean giant tortoise populations (with appendixes by C. Gordon and R. Burleigh) (Discussion). *Trans.* B **286**, 147–161 (1979).

Stoddart, D.R., Taylor, J.D., Fosberg, F.R. & Farrow, G.E. Geomorphology of Aldabra Atoll (Discussion). *Trans.* B **260**, 31–66 (1971).

Stoddart, D.R. & Walsh, R.P.D. Long-term climatic change in the western Indian Ocean (Discussion).
Trans. B **286**, 11–23 (1979).

Stoddart, J.C. *See* Beattie, Stoddart & March.

Stoicheff, B.P. Fifty years of light scattering studies in gases, liquids and solids [abstract] (Discussion). *Trans.* A **293**, 213 (1979).

Stoker, M.G.P. The Leeuwenhoek Lecture, 1971. Tumour viruses and the sociology of fibroblasts.
Proc. B **181**, 1–17 (1972).

Stokes, B.E. High volume component manufacture (Discussion). *Trans.* A **275**, 391–400 (1973).

Stokes, M.J. & Ettles, C.M.M. A general evaluation method for the diabatic journal bearing.
Proc. A **336**, 307–325 (1974).

Stolp, H. Interactions between *Bdellovibrio* and its host cell (Discussion).
Proc. B **204**, 211–217 (1979).

Stone, A.J. Spin-orbit coupling and the intersection of potential energy surfaces in polyatomic molecules. Proc. A **351**, 141–150 (1976).
See also Curran, Macdonald, Stone & Thrush.

Stone, E.L. Effects of species on nutrient cycles and soil change (Discussion).
Trans. B **271**, 149–162 (1975).

Stone, F.S. & Vickerman, J.C. Hydrogen–deuterium exchange catalysed by chromia-alumina solid solutions. Proc. A **354**, 331–347 (1977).

Stone, N.J. *See* Allsop, Bleaney, Bowden, Nambudripad, Stone & Suzuki.

Storåkers, B. *See* Hayhurst (D.R.) & Storåkers.

Storey, P.D. *See* Marshall (R.M.), Purnell & Storey.

Stork, S.T. *See* Doyle (M.J.), Maranci, Orowan & Stork.

Størmer, L. Olaf Holtedahl. *Biogr. Mem.* 22, 193–205 (1976).

Stott, A.N.B. *See* Chamberlain, Clough, Heard, Newton, Stott & Wells.

Stott, F.H. *See* Wilson (J.E.), Stott & Wood.

Stow, C.D. *See* Bradley (S.G.) & Stow.

Stoward, P.J. *See* Christie (K.N.) & Stoward.

Strähle, J. *See* Anderson (J.S.), Bevan, Cheetham, Von Dreele and others.

Strange, W.E. & Hothem, L.D. Establishment of scale and orientation for satellite Doppler positions (Discussion). *Trans.* A 294, 335–340 (1980).

Strangway, D.W. & Olhoeft, G.R. Electrical properties of planetary surfaces (Discussion). *Trans.* A 285, 441–450 (1977).

Straughan, B. Uniqueness and stability for the conduction-diffusion solution to the Boussinesq equations backward in time. *Proc.* A 347, 435–446 (1976).

Strausfeld, N.J. *See* Hausen & Strausfeld.

Strausfeld, N.J. & Hausen, K. The resolution of neuronal assemblies after cobalt injection into neuro-pil. *Proc.* B 199, 463–476 (1977).

Streeck, R.E. *See* Steinmetz, Streeck & Zachau.

Street, K.N. *See* Kelly (A). & Street.

Streett, W.B. & Tildesley, D.J. Computer simulations of polyatomic molecules.
 I. Monte Carlo studies of hard diatomics. *Proc.* A 348, 485–510 (1976).
 II. Molecular dynamics studies of diatomic liquids with atom—atom and quadrupole—quadrupole potentials. *Proc.* A 355, 239–266 (1977).

Streilein, J.W. Neonatal tolerance of H-2 alloantigens. II. *I* region dependence of tolerance expressed to K and D antigens. *Proc.* B 207, 475–486 (1980).
 See also Zakarian, Streilein & Billingham.

Streilein, J.W. & Klein, J. Neonatal tolerance of H-2 alloantigens. I. *I* region modulation of tolerogenic potential of K and D antigens. *Proc.* B 207, 461–474 (1980).

Stride, A.H. *See* Lloyd (A.J.), Savage, Stride & Donovan.

Stride, A.H., Belderson, R.H. & Kenyon, N.H. Evolving miogeanticlines of the east Mediterranean (Hellenic, Calabrian and Cyprus Outer Ridges) (with an appendix by M.L. Somers). *Trans.* A 284, 255–285 (1977).

Stringer, J. *See* Sambrook, Botchan, Hu, Mitchison & Stringer; Whittle (D.P.) & Stringer.

Strominger, J.L. *See* Waxman (D.J.), Yocum and Strominger.

Strominger, J.L., Blumberg, P.M., Suginaka, H., Umbreit, J. & Wickus, G.G. How penicillin kills bacteria: progress and problems (Discussion). *Proc.* B 179, 369–383 (1971).

Stuart, A.J. The history of the mammal fauna during the Ipswichian/Last interglacial in England. *Trans.* B 276, 221–250 (1976).
 The vertebrates of the Last Cold Stage in Britain and Ireland (Discussion). *Trans.* B 280, 295–312 (1977).
 The vertebrate fauna from the interglacial deposits at Sugworth, near Oxford. *Trans.* B 289, 87–97 (1980).

Stuart, C.A. *See* Edmunds, Potter & Stuart.

Stuart, J.T. & DiPrima, R.C. The Eckhaus and Benjamin-Feir resonance mechanisms. *Proc.* A 362, 27–41 (1978).
 On the mathematics of Taylor-vortex flows in cylinders of finite length. *Proc.* A 372, 357–365 (1980).

Stubblefield, Sir James. William Sawney Bisat. *Biogr. Mem.* 20, 27–40 (1974).
 Oliver Meredith Boone Bulman. *Biogr. Mem.* 21, 175–195 (1975).
 George Hoole Mitchell. *Biogr. Mem.* 23, 367–383 (1977).

Stubbs, L.C. *See* Golding & Stubbs.

Stumpfl, E.F. Sediments, ores, and metamorphism: new aspects (Discussion). *Trans.* A 286, 507–525 (1977).

Sturiale, C. *See* Romano (R.) & Sturiale.

Styles, P. *See* Girdler, Brown, Noy & Styles.

Subramanian, J. *See* Fuhrhop & Subramanian.

Subramanyam, S.V. *See* Gopal, Sekhar, Ananthakrishna and others.

Sugden, T.M. The classification of pollutants and their pathways in the atmosphere (Discussion). *Trans.* A **290**, 469–476 (1979).

Suginaka, H. *See* Strominger, Blumberg, Suginaka, Umbreit & Wickus.

Sugiyama, M. *See* Suzuki (Y.), Tsuchiya, Kinoshita, Sugiyama & Inuzuka.

Sullivan, F.M. & Barlow, S.M. Congenital malformations and other reproductive hazards from environmental chemicals (Discussion). *Proc.* B **205**, 91–110 (1979).

Sullivan, J.P. *See* Widnall & Sullivan.

Sullivan, R.A.L. *See* Parkin (D.W.), Sullivan & Andrews.

Sulston, J.E. Post-embryonic development in the ventral cord of *Caenorhabditis elegans*. *Trans.* B **275**, 287–297 (1976).

Sümengen, Z. *See* Saunders (G.A.) & Sümengen.

Summerfield, W. Circular islands as resonators of long-wave energy. *Trans.* A **272**, 361–402 (1972).

Summers, C.J. *See* Dennis (R.B.), Smith & Summers.

Summers, D. An exact solution to the equation of propagation of magneto-acoustic waves. *Proc.* A **343**, 421–425 (1975).

On the effects of viscosity in one-fluid wind theory. *Proc.* A **349**, 53–62 (1976).

Sun, D.-C. On the effects of two-dimensional Reynolds roughness in hydrodynamic lubrication. *Proc.* A **364**, 89–106 (1978).

Sun, S.-S. Lead isotopic study of young volcanic rocks from mid-ocean ridges, ocean islands and island arcs (Discussion). *Trans.* A **297**, 409–445 (1980).

See also Nesbitt & Sun.

Sunderland, E. & Coope, Elizabeth. Biological studies of Yemenite and Kurdish Jews in Israel and other groups in southwest Asia. XII. Genetic studies in Jordan. *Trans.* B **266**, 207–220 (1973).

Sundvor, E. *See* Eldholm (O.) & Sundvor.

Sutcliffe, L.H. *See* Ledwith, Russell & Sutcliffe.

Sutcliffe, R.C. *See* Pasquill, Sheppard & Sutcliffe.

Sutcliffe, R.C. & Best, A.C. Ernest Gold. *Biogr. Mem.* **23**, 115–131 (1977).

Sutcliffe, R.C. & Pasquill, F. Percival Albert Sheppard. *Biogr. Mem.* **25**, 535–553 (1979).

Sutherland, Joan. *See* Manton (Irene), Sutherland & Leadbeater; Manton (Irene), Sutherland & Oates.

Sutherland, N.S. & Longuet-Higgins, H.C. Introduction to a Discussion. *Trans.* B **290**, 3–4 (1980).

Sutton, J. Introductory remarks to a Discussion. *Trans.* A **273**, 316 (1973).

Introductory remarks to a Discussion. *Trans.* A **280**, 399–403 (1976).

Antarctica, a key to the understanding of the evolution of Gondwanaland (Discussion). *Trans.* B **279**, 197–205 (1977).

Evolution of the continental crust (Discussion). *Trans.* A **291**, 257–268 (1979).

William Quarrier Kennedy. *Biogr. Mem.* **26**, 275–303 (1980).

Sutton, J., Sabine, P.A. & Skelhorn, R.R. Introductory remarks to a Discussion. *Trans.* A **271**, 103 (1972).

Sutton, L.E. Introductory remarks to a Discussion. *Trans.* B **272**, 3–4 (1975).

Suzuki, H. *See* Allsop, Bleaney, Bowden, Nambudripad, Stone & Suzuki; Gordon (M.), Leonis & Suzuki.

Suzuki, Y., Tsuchiya, Y., Kinoshita, K., Sugiyama, M. & Inuzuka, E. Recent developments in picosecond streak camera systems (Discussion). *Trans.* A **298**, 295–302 (1980).

Svanberg, S. Atomic spectroscopy by resonance scattering (Discussion). *Trans.* A **293**, 215–222 (1979).

Svestka, Z. Solar particle events (Discussion). *Trans.* A **270**, 157–165 (1971).

Optical observations of flares (Discussion). *Trans.* A **281**, 435–441 (1976).

Activated solar filaments and flares (Discussion). *Trans.* A **297**, 575–585 (1980).

Svoboda, J.A. *See* Thompson (M.J.), Svoboda, Kaplanis & Robbins.

Swain, C.J. *See* Banks (R.J.) & Swain.

Swain, M.V. Microfracture about scratches in brittle solids. *Proc.* A **366**, 575–597 (1979).

Swale, E.M.F. *See* Belcher & Swale.

Swallow, J.C. The *Aries* current measurements in the western North Atlantic (Discussion).
Trans. A 270, 451—460 (1971).

Swann, Sir Michael. Introductory remarks to a Discussion. *Proc.* A 345, 435—436 (1975).

Swanson, S.A.V. Design and testing of replacement prostheses (Discussion).
Proc. B 192, 173—189 (1976).

Swindale, N.V. A model for the formation of ocular dominance stripes. *Proc.* B 208, 243—264 (1980).

Swindale, N.V. & Benjamin, P.R. The anatomy of neurosecretory neurones in the pond snail *Lymnaea stagnalis* (L.). *Trans.* B 274, 169—202 (1976).

Swingland, I.R. *See* Coe (M.J.), Bourn & Swingland.

Swingland, I.R. & Coe, M.J. The natural regulation of giant tortoise populations on Aldabra Atoll: recruitment (Discussion). *Trans.* B 286, 177—188 (1979).

Swingler, D.L. *See* Bolton (H.C.), Grant, McWilliam, Nicholson & Swingler.

Swithinbank, C. Glaciological research in the Antarctic Peninsula (Discussion).
Trans. B 279, 161—183 (1977).

Switsur, V.R. Appendix to a paper by Hilary H. Birks. *Trans.* B 270, 224 (1975).
See also Hibbert (F.A.), Switsur & West.

Sydenham, P.H. Strain measurement in Australia with particular reference to the Cooney Observatory (Discussion). *Trans.* A 274, 323—330 (1973).

von Sydow, E. Flavour — a problem for the consumer or for the food producer? (Discussion).
Proc. B 191, 145—153 (1975).

Sykes, R.I. Stratification effects in boundary layer flow over hills. *Proc.* A 361, 225—243 (1978).
On three-dimensional boundary layer flow over surface irregularities. *Proc.* A 373, 311—329 (1980).

Sykes, R.M. *See* Ashkenazi, Dodson, Sykes, Dean & Blanchard; Ashkenazi, McLintock & Sykes; Ashkenazi & Sykes; Ashkenazi, Sykes, Gough & Williams.

Sykes, Sir Charles. Norman Percy Allen. *Biogr. Mem.* 19, 1—18 (1973).

Sylvester-Bradley, P.C. The search for protolife (Discussion). *Proc.* B 189, 213—233 (1975).

Syme, Gabrielle & Levin, R.J. Effect of altered thyroid status induced by thyroid hormones, goitrogens and diet on intestinal electrogenic valine transfer. *Proc.* B 194, 121—139 (1976).

Symons, M.C.R. Water structure and hydration (Discussion). *Trans.* B 272, 13—28 (1975).

Symons, M.C.R. & Petersen, R.L. Electron capture by oxyhaemoglobin: an e.s.r. study.
Proc. B 201, 285—300 (1978).

Synge, F.M. Records of sea levels during the Late Devensian (Discussion).
Trans. B 280, 211—228 (1977).

Synge, J.L. Geometrical approach to the Heisenberg uncertainty relation and its generalization.
Proc. A 325, 151—156 (1971).
Maximum number of collisions of elastic particles. *Proc.* A 331, 1—18 (1972).
Eamon de Valera. *Biogr. Mem.* 22, 635—653 (1976). Corrigendum. *Biogr. Mem.* 23, 643 (1977).
See also Florides & Synge.

Szentágothai, J. The Ferrier Lecture, 1977. The neuron network of the cerebral cortex: a functional interpretation. *Proc.* B 201, 219—248 (1978).

Szczepaniak, Anna C. *See* Fulton, Miledi & Szczepaniak; Miledi & Szczepaniak.

Szer, J. *See* Rogers (L.J.), Oettinger, Szer & Mark.

Szigeti, B. *See* Leigh, Szigeti & Tewary.

Szulejko, J.E. *See* Amaya, Brenton, Szulejko & Beynon.

Szulejko, J.E., Amaya, A. Mendez, Morgan, R.P., Brenton, A.G. & Beynon, J.H. A method for calculating the shapes of peaks resulting from fragmentations of metastable ions in a mass spectrometer. I. Peak shapes arising from single valued kinetic energy releases.
Proc. A 373, 1—11 (1980).

Szurszewski, J.H. *See* Bülbring & Szurszewski.

Szurszewski, J.H. & Bülbring, Edith. The stimulant action of acetylcholine and catecholamines on the uterus (Discussion). *Trans.* B 265, 149—156 (1973).

Szwarc, M. *See* Fisher (M.), Rämme, Claesson & Szwarc; Karasawa, Levin & Szwarc; Rainis, Tung & Szwarc; Rämme, Fisher, Claesson & Szwarc.

Tabor, D. *See* Barnes (P.), Tabor & Walker; Doyle (E.D.), Horne & Tabor; Fuller (K.N.G.) & Tabor; Gane, Pfaelzer & Tabor; Israelachvili & Tabor; Kendall & Tabor; Pooley (Christine M.) & Tabor; Roberts (A.D.) & Tabor.

Tabor, H. Non-convecting solar ponds (Discussion). *Trans.* A **295**, 423–433 (1980).

Tabor, M. *See* Berry (M.V.) & Tabor.

Taff-Jones, D.H. *See* Matus & Taff-Jones.

Tager, H.S. *See* Steiner (D.F.), Patzelt, Chan, Quinn, Tager, Nielsen and others.

Taig, I.C. Principles of design of a carbon fibre composite aircraft wing (Discussion). *Trans.* A **294**, 565–575 (1980).

Tait, E.A. *See* Reyment & Tait.

Tait, J.F. *See* McDougall, Williams, Hyatt, Bell, Tait & Tait.

Tait, J.F., Tait, Sylvia A.S., Gould, R.P. & Mee, M.S.R. The properties of adrenal zona glomerulosa cells after purification by gravitational sedimentation. *Proc.* B **185**, 375–407 (1974).

Tait, Sylvia A.S. *See* McDougall, Williams, Hyatt, Bell, Tait & Tait; Tait (J.F.), Tait, Gould & Mee.

Tait, W. *See* Charap & Tait.

Tajana, G.F. *See* Metafora, Felsani, Cotrufo, Tajana, Del Rio, De Prisco and others; Metafora, Felsani, Cotrufo, Tajana, Di Iorio, Del Rio and others.

Takach, P. *See* Robbins (N.), Olek, Kelly, Takach & Christopher.

Takahashi, Ellen S. *See* Oyster & Takahashi.

Takahashi, T. Intracellular recording from visually identified motoneurons in rat spinal cord slices. *Proc.* B **202**, 417–421 (1978).

See also Fulton, Miledi & Takahashi.

Takezawa, I. Development of the automated shipyard (Discussion). *Trans.* A **273**, 151–172 (1972).

Talbot, C.J. A plate tectonic model for the Archaean crust (Discussion). *Trans.* A **273**, 413–427 (1973).

Talbot, D.R.S. & Willis, J.R. The effective sink strength of a random array of voids in irradiated material. *Proc.* A **370**, 351–374 (1980).

Talboys, P.W. Resistance to vascular wilt fungi (Discussion). *Proc.* B **181**, 319–332 (1972).

Taliaferro, S. *See* Nachman & Taliaferro.

Tamaru, K. *See* Watanabe (K.), Kondow, Soma, Onishi & Tamaru.

Tamura, H. *See* Kihara, Kanazawa & Tamura.

Tan, C.L. & Fenner, R.T. Elastic fracture mechanics analysis by the boundary integral equation method. *Proc.* A **369**, 243–260 (1979).

Tan, K.L. & von Engel, A. Energy transfer from excited mercury atoms to electrons. *Proc.* A **324**, 183–200 (1971).

Tang, P.Y. *See* Naghdi & Tang.

Tang, T.B. & Chaudhri, M.M. The thermal decomposition of silver azide. *Proc.* A **369**, 83–104 (1979).

Tanguy, J.C. The 1971 Etna eruption: petrography of the lavas (Symposium). *Trans.* A **274**, 45–53 (1973).

Tanguy, J.C. & Wilson, R.L. Palaeomagnetism of Mount Etna (Symposium). *Trans.* A **274**, 163 (1973).

Tansey, E.M. Aminergic fluorescence in the cephalopod brain. *Trans.* B **291**, 127–145 (1980).

Tanswell, P. *See* Daniels (A.), Korda, Tanswell, Williams & Williams.

Tapp, R.L. *See* Hardy (R.N.), Hockaday & Tapp.

Tarney, J., Wood, D.A., Saunders, A.D., Cann, J.R. & Varet, J. Nature of mantle heterogeneity in the North Atlantic: evidence from deep sea drilling (Discussion). *Trans.* A **297**, 179–202 (1980).

Tarzwell, C.M. Bioassays to determine allowable waste concentrations in the aquatic environment (Discussion). *Proc.* B **177**, 279–285 (1971).

Tash, J.S. & Mann, T. Adenosine $3':5'$-cyclic monophosphate in relation to motility and senescence of spermatozoa. *Proc.* B **184**, 109–114 (1973).

Tassie, L.J. *See* Pope (R.L.) & Tassie.

Tate, A. *See* Billington & Tate.

Tatnall, A.R.L. *See* Bullough (K.), Denby, Gibbons, Hughes and others.

Tavener-Smith, R. & Williams, A. The secretion and structure of the skeleton of living and fossil

Bryozoa. *Trans.* B **264**, 97–159 (1972).

Taxi, J. Ultrastructural data on the cytology and cytochemistry of the autonomic nervous system (Discussion). *Trans.* B **261**, 311–312 (1971).

Tayler, A.B. *See* Ockendon & Tayler.

Taylor, A.C. & Brand, A.R. A comparative study of the respiratory responses of the bivalves *Arctica islandica* (L.) and *Mytilus edulis* L. to declining oxygen tension. *Proc.* B **190**, 443–456 (1975).

Taylor, A.R. *See* Elsworth, Taylor & James.

Taylor, D.B. & Crampin, S. Surface waves in anisotropic media: propagation in a homogeneous piezoelectric halfspace. *Proc.* A **364**, 161–179 (1978).

Taylor, D.G. The costs of arthritis and the benefits of joint replacement surgery (Discussion). *Proc.* B **192**, 145–155 (1976).

Taylor, D.L. Nutrition of algal-invertebrate symbiosis.

I. Utilization of soluble organic nutrients by symbiont-free hosts. *Proc.* B **186**, 357–368 (1974).

II. Effects of exogenous nitrogen sources on growth, photosynthesis and the rate of excretion by algal symbionts *in vivo* and *in vitro*. *Proc.* B **201**, 401–412 (1978).

See also Chalker & Taylor.

Taylor, F. Shipyards of the future: possibilities and prospects (Discussion). *Trans.* A **273**, 173–181 (1972).

Taylor, F.J.R. Symbionticism revisited: a discussion of the evolutionary impact of intracellular symbioses (Discussion). *Proc.* B **204**, 267–286 (1979).

Taylor, G. *See* Chapman (J.A.), Grant, Taylor, Mahmud, Sardar-ul-Mulk & Shahid.

Taylor, H.F. & Burden, R.S. Xanthoxin, a recently discovered plant growth inhibitor. *Proc.* B **180**, 317–346 (1972).

Taylor, J.B. *See* Connor, Hastie & Taylor; Edwards (S.F.) & Taylor.

Taylor, J.D. Intertidal zonation at Aldabra Atoll (Discussion). *Trans.* B **260**, 173–213 (1971).

See also Braithwaite (C.J.R.), Taylor & Kennedy; Stoddart (D.R.), Taylor, Fosberg & Farrow.

Taylor, J.D., Braithwaite, C.J.R., Peake, J.F. & Arnold, E.N. Terrestrial faunas and habitats of Aldabra during the late Pleistocene (Discussion). *Trans.* B **286**, 47–66 (1979).

Taylor, J.G. Quantizing super-Riemannian space. *Proc.* A **362**, 493–507 (1978).

See also Nouri-Moghadam & Taylor.

Taylor, J.R. *See* Adams (M.C.), Bradley, Sibbett & Taylor.

Taylor, J.R.B. Liverpool Civic and Social Centre: preliminary wind tunnel testing to determine environmental conditions (Discussion). *Trans.* A **269**, 487–491 (1971).

Taylor, K. A transformation of the acoustic equation with implications for wind-tunnel and low-speed flight tests. *Proc.* A **363**, 271–281 (1978).

Taylor, M.J. *See* Fröhlich (A.) & Taylor.

Taylor, N. *See* De Sanctis, Grdenić, Taylor & Hodgkin; Guil, Hayward & Taylor.

Taylor, R.B. & Iverson, G.M. Hapten competition and the nature of cell-cooperation in the antibody response (Discussion). *Proc.* B **176**, 393–418 (1971). Corrigenda. **178**, 477 (1971).

Taylor, R.E. *See* Keynes, Bezanilla, Rojas & Taylor.

Taylor, R. Eatock. Analysis of the flexural vibrations of variable density spheroids immersed in an ideal fluid, with application to ship structural dynamics. *Trans.* A **277**, 623–646 (1975).

See also Bishop (R.E.D.) & Taylor.

Tazieff, H. Structural implications of the 1971 Mount Etna eruption (Symposium). *Trans.* A **274**, 97–82 (1973).

Tchalenko, J.S. Seismicity and structure of the Kopet Dagh (Iran, U.S.S.R.). *Trans.* A **278**, 1–28 (1975).

Tchalenko, J.S. & Braud, J. Seismicity and structure of the Zagros (Iran): the Main Recent Fault between 33 and 35° N. *Trans.* A **277**, 1–25 (1974).

Tedder, J.M. *See* Izod & Tedder.

Tedford, D.J. *See* Chalmers, Duffy & Tedford.

Teesdale, P. *See* Lehmann (H.), Ala, Hedeyat, Montazemi and others.

Telfer, D.J. *See* Geake, Walker, Telfer & Mills.

Telfer, D.J. & Fielder, G. Optical excitation spectroscopy of the Luna 24 sample 24125. *Trans.* A **297**, 23–25 (1980).

Telford, N.R. *See* Hayhurst (A.N.) & Telford.

Tel-Or, E. & Stewart, W.D.P. Photosynthetic components and activities of nitrogen-fixing isolated heterocysts of *Anabaena cylindrica*. *Proc.* B **198**, 61–86 (1977).

Temperley, H.N.V. On the possible transition of the Gaussian model of an imperfect gas. *Proc.* A **357**, 345–353 (1977).
See also Baxter, Temperley & Ashley.

Temperley, H.N.V. & Ashley, Susan E. Some exact results for the Ashkin–Teller model. *Proc.* A **365**, 371–380 (1979).

Temperley, H.N.V. & Lieb, E.H. Relations between the 'percolation' and 'colouring' problem and other graph-theoretical problems associated with regular planar lattices: some exact results for the 'percolation' problem. *Proc.* A **322**, 251–280 (1971).

Temperley, H.N.V. & Trevena, D.H. Metastability of phase transitions and the tensile strength of liquids. *Proc.* A **357**, 395–402 (1977).

Tenyi, I. *See* Logan (A.G.), Tenyi, Peart, Breathnach & Martin.

Teodorovich, E.V. On the contribution of macroscopic van der Waals interactions to frictional force. *Proc.* A **362**, 71–77 (1978).

Teoh, S.B. & Rees, H. B chromosomes in White Spruce. *Proc.* B **198**, 325–344 (1977).

Tera, F. *See* Wasserburg, Papanastassiou, Tera & Huneke.

Terry, P.D. A complex ray tracing study of ion cyclotron whistlers in the ionosphere. *Proc.* A **363**, 425–443 (1978).
See also Budden & Terry.

Terwilliger, D.T., Beynon, J.H. & Cooke, R.G. Kinetic energy distributions from the shapes of metastable peaks. *Proc.* A **341**, 135–146 (1974).

Tescari, M. *See* Clarke (T.A.), Mason & Tescari.

Teufel, L.W. *See* Friedman (M.), Teufel & Morse.

Tevaarwerk, J.L. *See* Johnson (K.L.) & Tevaarwerk.

Tevesz, M.J.S. *See* Carter (J.G.) & Tevesz.

Tevlin, M.P. *See* Moriarty (D.J.W.), Darlington, Dunn, Moriarty & Tevlin.

Tewari, J.P. *See* Malhotra (S.K.) & Tewari.

Tewary, V.K. *See* Leigh, Szigeti & Tewary.

Textor, M. *See* Gay (I.D.), Textor, Mason & Iwasawa; Mason (R.) & Textor; Mason (R.), Textor, Iwasawa & Gay.

Textor, M., Gay, I.D. & Mason, R. Photoelectron spectroscopy of the α-Fe(111)-carbon monoxide surface. *Proc.* A **356**, 37–45 (1977).

Thake, Brenda. *See* Wymer & Thake.

Thewlis, G. *See* Llewellyn (D.T.), Marriott, Naylor & Thewlis.

Thiede, J. Palaeo-oceanography, margin stratigraphy and palaeophysiography of the Tertiary North Atlantic and Norwegian–Greenland Seas (Discussion). *Trans.* A **294**, 177–185 (1980).

Thieffry, M., Bruner, J. & Personne, P. Effects of high calcium solutions on glutamate sensitivity of crayfish muscle fibres. *Proc.* B **209**, 415–429 (1980).

Thimm, K. *See* Connerade, Mansfield, Newsom, Tracy, Baig & Thimm; Connerade, Mansfield & Thimm.

Thirunamachandran, T. Theory of laser-induced optical activity. *Proc.* A **365**, 327–343 (1979).
See also Andrews (D.L.) & Thirunamachandran; Babiker, Power & Thirunamachandran; Craig (D.P.), Power & Thirunamachandran; Power (E.A.) & Thirunamachandran.

Thoday, J.M. Review Lecture. Disruptive selection. *Proc.* B **182**, 109–143 (1972).

Thom, A. Astronomical significance of prehistoric monuments in Western Europe (Discussion). *Trans.* A **276**, 149–156 (1974).

Thom, B.G. *See* Polach, McLean, Caldwell & Thom; Stoddart (D.R.), McLean, Scoffin, Thom & Hopley.

Thom, B.G. & Chappell, J. Holocene sea level change: an interpretation (Discussion). *Trans.* A **291**, 187–194 (1978).

Thom, B.G., Orme, G.R. & Polach, H.A. Drilling investigation of Bewick and Stapleton islands (Discussion). *Trans.* A **291**, 37–54 (1978).

Tho

Thom, N.S. & Agg, A.R. The breakdown of synthetic organic compounds in biological processes (Discussion). *Proc.* B **189**, 347–357 (1975).

Thomas, A.R. & Paton, Sir Angus. Claude Cavendish Inglis. *Biogr. Mem.* **21**, 367–388 (1975).

Thomas, G.B. & Gibbons, T.B. The influence of trace elements on the creep and stress-rupture properties of Nimonic 105 [abstract] (Discussion). *Trans.* A **295**, 296 (1980).

Thomas, G.P. *See* Peregrine & Thomas.

Thomas, H.H.B.M. & Küchemann, D. Sidney Barrington Gates. *Biogr. Mem.* **20**, 181–212 (1974).

Thomas, Jennifer A. *See* Sanderson, Hall & Thomas; Sanderson & Thomas.

Thomas, J.M. Review Lecture. Topography and topology in solid-state chemistry. *Trans.* A **277**, 251–286 (1974).

See also Cohen (M.D.), Ludmer, Thomas & Williams; Evans (S.), Adams & Thomas; Evans (S.) & Thomas; Jefferson (D.A.) & Thomas; Mallinson (L.G.), Jefferson, Thomas & Hutchison; Nakanishi (H.), Jones, Thomas, Hasegawa & Rees.

Thomas, J.M., Evans, E.L. & Williams, J.O. Microscopic studies of enhanced reactivity at structural faults in solids (Discussion). *Proc.* A **331**, 417–427 (1972).

Thomas, J.O. Experiments with plasma waves (Discussion). *Trans.* A **280**, 193–224 (1975).

Thomas, L. The composition of the mesosphere and lower thermosphere (Discussion). *Trans.* A **296**, 243–260 (1980).

See also Dickinson (P.H.G.), Bain, Thomas, Williams, Jenkins & Twiddy.

Thomas, R.D.K. Limits to opportunism in the evolution of the Arcoida (Bivalvia) (Discussion). *Trans.* B **284**, 335–344 (1978).

Thomas, R.G. High temperature mechanical properties of AISI 316 weld metal [abstract] (Discussion). *Trans.* A **295**, 292 (1980).

Thomas, R.G.O. & Thrush, B.A. Energy transfer in the quenching of singlet molecular oxygen.
I. Kinetics of quenching of singlet oxygen. *Proc.* A **356**, 287–294 (1977).
II. The rates of formation and quenching of vibrationally excited molecules. *Proc.* A **356**, 295–306 (1977).
III. Application of statistical theory. *Proc.* A **356**, 307–314 (1977).

Thomas, R.K. Hydrogen bonding in the gas phase: the thermodynamic properties of hydrogen fluoride–ether complexes and their far infrared spectra. *Proc.* A **322**, 137–146 (1971).
Hydrogen bonding in the gas phase: the infrared spectra of complexes of hydrogen fluoride with hydrogen cyanide and methyl cyanide. *Proc.* A **325**, 133–149 (1971).
Photoelectron spectroscopy of hydrogen-bonded systems: spectra of monomers, dimers and mixed complexes of carboxylic acids. *Proc.* A **331**, 249–261 (1972).
Hydrogen bonding in the vapour phase between water and hydrogen fluoride: the infrared spectrum of the 1:1 complex. *Proc.* A **344**, 579–592 (1975).

See also Bomchil, Hüller, Rayment, Roser, Smalley, Thomas & White; Mines & Thomas; Mines, Thomas & Thompson.

Thomas, R.K., Leisegang, E.C. & Thompson, Sir Harold. Vibration–rotation bands of methyl isocyanide and its d_3-derivative. *Proc.* A **330**, 15–28 (1972).

Thomas, R.K. & Thompson, Sir Harold. Photoelectron spectra of carbonyl halides and related compounds. *Proc.* A **327**, 13–22 (1972).
The photoelectron spectra of allene, deuteroallenes and tetrafluoroallene. *Proc.* A **339**, 29–36 (1974).

Thomason, P.F. The mechanics of fracture in metal-matrix composites. *Proc.* A **348**, 265–284 (1976).

Thomaz, M.F. & Davies, G. The decay time of N3 luminescence in natural diamond. *Proc.* A **362**, 405–419 (1978).

Thomée, V. Some convergence results in elliptic difference equations (Discussion). *Proc.* A **323**, 191–199 (1971).

Thompson, A.W. Automotive drive shafts (Discussion). *Trans.* A **294**, 577–582 (1980).

Thompson, D.O. & Thompson, R.B. Quantitative ultrasonics (Discussion). *Trans.* A **292**, 233–250 (1979).

Thompson, E.W. *See* Boulter, Ramshaw, Thompson, Richardson & Brown.

Thompson, G. *See* Melson & Thompson.

Thompson, G.B. On the relation between information technology and socio-economic systems (Discussion). *Trans.* A **289**, 207–212 (1978).

Thompson, G.F. *See* Dainton, May, Morrow, Salmon & Thompson; Dainton, Morrow, Salmon & Thompson.

Thompson, G.I. *See* Nandy, Thompson, Jamar, Monfils & Wilson.

Thompson, J.E.S. Maya astronomy (Discussion). *Trans.* A **276**, 83–98 (1974).

Thompson, J.M.T. Stability predictions through a succession of folds. *Trans.* A **292**, 1–23 (1979).

Thompson, K. *See* Day (M.J.), Dixon-Lewis & Thompson.

Thompson, M.J., Svoboda, J.A., Kaplanis, J.N. & Robbins, W.E. Metabolic pathways of steroids in insects (Discussion). *Proc.* B **180**, 203–221 (1972).

Thompson, R.B. *See* Thompson (D.O.) & Thompson.

Thompson, R.C. *See* Malvern, Pinder, Stacey & Thompson.

Thompson, R.N. *See* Morrison (M. Ann), Thompson, Gibson & Marriner.

Thompson, R.W. *See* Roberts (D.G.), Montadert, Thompson, Auffret and others.

Thompson, Sir Harold. Cyril Norman Hinshelwood. *Biogr. Mem.* **19**, 375–431 (1973).
See also Massey & Thompson; Mines & Thompson; Mines, Thomas & Thompson; Thomas (R.K.), Leisegang & Thompson; Thomas (R.K.) & Thompson.

Thompson, S.M. *See* Leng, Rowlinson & Thompson.

Thompson, T.L. *See* Roberts (D.G.), Montadert, Thompson, Auffret and others.

Thompson, W. *See* Kuffler (D.P.), Thompson & Jansen.

Thomson, J.N. *See* Albertson & Thomson; White (J.G.), Southgate, Thomson & Brenner.

Thomson, R.E. *See* Stewart (R.W.) & Thomson.

Thomson, Sir George & Hall, Sir Arnold. William Scott Farren. *Biogr. Mem.* **17**, 215–241 (1971).

Thong, K.C. & Weinberg, F.J. Electrical control of the combustion of solid and liquid particulate suspensions. *Proc.* A **324**, 201–215 (1971).

Thorne, K.S. Sources of gravitational waves [abstract] (Discussion). *Proc.* A **368**, 9 (1979).

Thornhill, R.A. The development of the labyrinth of the lamprey (*Lampetra fluviatilis* Linn. 1758). *Proc.* B **181**, 175–198 (1972).

Thornley, Margaret J., Glauert, Audrey M. & Sleytr, Uwe B. Structure and assembly of bacterial surface layers composed of regular arrays of subunits (Discussion). *Trans.* B **268**, 147–153 (1974).

Thornton, I. & Webb, J.S. Geochemistry and health in the United Kingdom (Discussion). *Trans.* B **288**, 151–168 (1979).

Thornton, J. *See* Hensens, Hill, Thornton, Turner & Williams.

Thorpe, R.S. *See* Gass, Chapman, Pollack & Thorpe.

Thorpe, S.A. Turbulence and mixing in a Scottish Loch. *Trans.* A **286**, 125–181 (1977).

Thorpe, W.H. William Robin Thompson. *Biogr. Mem.* **19**, 655–678 (1973).
David Lambert Lack. *Biogr. Mem.* **20**, 271–293 (1974).

Thrush, B.A. Aspects of the chemistry of ozone depletion (Discussion).
Trans. A **290**, 505–514 (1979).
The chemistry of the stratosphere (Discussion). *Trans.* A **296**, 149–160 (1980).
See also Burrows (J.P.), Cliff, Harris, Thrush & Wilkinson; Clough (P.N.), Curran & Thrush; Curran, Macdonald, Stone & Thrush; Davies (P.B.), Russell, Thrush & Radford; Gartner & Thrush; Golde & Thrush; Orchard (S.W.) & Thrush; Thomas (R.G.O.) & Thrush.

Thurrell, R.C. *See* Harris (P.M.), Thurrell, Healing & Archer.

Thurston, E.L. *See* Ito (S.), Thurston & Nicol; Wang (R.T.), Nicol, Thurston & McCants; Watson (M.), Thurston & Nicol.

Tiang, K.M. *See* Meyer-Rochow & Tiang.

Tideman, T.N. *See* Good & Tideman.

Tijssen, S.B. *See* van Bennekom, Gieskes & Tijssen.

Tildesley, D.J. *See* Rowlinson & Tildesley; Streett & Tildesley.

Till, S.M., Freedman, P.A., Tuckett, R.P. & Jones, W.J. A single-double pass Sisam spectrometer for the near infrared. *Proc.* A **353**, 421–430 (1977).

Till, Susan M., Jones, W.J. & Shotton, K.C. A double-pass Sisam spectrometer for the near infrared. *Proc.* A **346**, 395–412 (1975).

Tilley, C.E. Concluding remarks to a Discussion. *Trans.* A **268**, 745 (1971).

Tilley, R.J.D. *See* Iguchi & Tilley.

Tills, D. *See* Godber, Kopeć, Mourant, Tills & Lehmann; Lehmann (H.), Ala, Hedeyat, Montazemi and others; Mourant, Godber, Kopeć, Tills & Woodhead.

Tilly, J.F. Submarine systems (Discussion). *Trans.* A **289**, 151—158 (1978).

Tilton, G.R. *See* Jahn, Vidal & Tilton.

Timmins, P.A. *See* Finch (J.T.), Lewit-Bentley, Bentley, Roth & Timmins.

Timmons, C.J. *See* Cehelnik, Cundall, Timmons & Bowley.

Timofeev, P.P. *See* Roberts (D.G.), Montadert, Thompson, Auffret and others.

Tinbergen, N. The Croonian Lecture, 1972. Functional ethology and the human sciences. *Proc.* B **182**, 385—410 (1972).

Tindall, D.A. *See* Pippard, Shepherd & Tindall.

Ting, S.C.C. Search for new particles (Discussion). *Proc.* A **355**, 493—513 (1977).

Tinker, M.H. *See* Butts, Tinker & Kernahan.

Tinker, P.B. Transport of water to plant roots in soil (Discussion). *Trans.* B **273**, 445—461 (1976).

Tinsley, Beatrice M. The detectability of young galaxies (Discussion). *Trans.* A **296**, 303—308 (1980).

Tipler, H.R. The influence of purity on the strength and ductility in creep of CrMoV steels of varied microstructures (Discussion). *Trans.* A **295**, 213—233 (1980).

Tipnis, U. *See* Malhotra (S.K.) & Tipnis.

Tissières, A. *See* Moran, Mirault, Arrigo, Goldschmidt-Clermont & Tissières.

Tissot, B. Organic geochemistry of margin sediments and its significance for hydrocarbon generation [abstract] (Discussion). *Trans.* A **294**, 187 (1980).

Tittmann, B.R. Lunar rock Q in 3000—5000 range achieved in laboratory (Discussion). *Trans.* A **285**, 475—479 (1977).

Titulaer, C. *See* Fielder, Fryer, Titulaer, Herring & Wise.

To, L.P. *See* Margulis, Chase & To.

Tocher, D. *See* King (C-Y.), Nason & Tocher.

Tocher, K.D. Planning systems (Discussion). *Trans.* A **287**, 425—441 (1971).

Tod, K.P. *See* Hansen (R.O.), Newman, Penrose & Tod.

Tod, K.P. & Ward, R.S. Self-dual metrics with self-dual Killing vectors. *Proc.* A **368**, 411—427 (1979).

Todaro, G.J., Callahan, R., Rapp, U.R. & De Larco, J.E. Genetic transmission of retroviral genes and cellular oncogenes (Discussion). *Proc.* B **210**, 367—385 (1980).

Todd, J.J. *See* Carnochan, Dworetsky, Todd, Willis & Wilson.

Todd, Lord. Anniversary Addresses:

    1976 *Proc.* A **352**, 451—462 (1977); *also Proc.* B **196**, 1—12 (1977).

    1977 *Proc.* A **359**, v-xiv (1978); *also Proc.* B **200**, v-xiv (1978).

    1978 *Proc.* A **365**, v-xviii (1979); *also Proc.* B **204**, 1—14 (1979).

    1979 *Proc.* A **369**, 295—306 (1980); *also Proc.* B **206**, 369—380 (1980).

    1980 *Proc.* B **211**, 1—13 (1980).

    Robert Arthur James Gascoyne-Cecil, 5th Marquess of Salisbury. *Biogr. Mem.* **19**, 621—627 (1973).

    James Kenner. *Biogr. Mem.* **21**, 389—405 (1975).

    Introductory remarks at the afternoon session, 24 February 1977 (Discussion). *Trans.* A **288**, 97 (1978).

    George Wallace Kenner. *Biogr. Mem.* **25**, 391—420 (1979).

Todd, Lord & Cornforth, J.W. Robert Robinson. *Biogr. Mem.* **22**, 415—527 (1976).

Tohline, J.E. Star formation in the early Universe (Discussion). *Trans.* A **296**, 309—311 (1980).

Tokuno, H. *See* Tomita, Tokuno & Usune.

Toland, J.F. On the existence of a wave of greatest height and Stokes's conjecture. *Proc.* A **363**, 469—485 (1978).

Tomasz, A. On the mechanism of the irreversible antimicrobial effects of β-lactams (Discussion). *Trans.* B **289**, 303—308 (1980).

Tomblin, J.F. *See* Briden, Rex, Faller & Tomblin.

Tomita, T. *See* Bülbring & Tomita.

Tomita, T., Tokuno, H. & Usune, S. Confirmation of conductance increase by adrenalin in the

guinea-pig taenia coli (α-action). *Proc.* B **198**, 473–477 (1977).

Tomita, T. & Watanabe, H. Factors controlling myogenic activity in smooth muscle (Discussion). *Trans.* B **265**, 73–85 (1973).

Tomkins, F.S. *See* Garton, Reeves & Tomkins; Garton, Reeves, Tomkins & Ercoli; Garton, Tomkins & Crosswhite; Lu, Tomkins & Garton; Parkinson (W.H.), Reeves & Tomkins; Smith (G.) & Tomkins.

Tomlinson, J.R. *See* Harding (G.L.), Pippard & Tomlinson.

Tomlinson, R.C. Introductory remarks to a Discussion. *Trans.* A **287**, 353 (1977).
Operational research and systems analysis: from practice to precept (Discussion). *Trans.* A **287**, 355–371 (1977).

Tommerup, I.C. *See* Sargent (J.A.), Ingram & Tommerup.

Tommerup, Inez C. *See* Ingram (D.S.) & Tommerup.

Tompkins, F.C. *See* Duš & Tompkins.

Tompkins, F.C. & Goodeve, Sir Charles. Edward Armand Guggenheim. *Biogr. Mem.* **17**, 303–326 (1971).

Tong, E.Y. *See* Schmidt (H.) & Tong.

Tonn-Ehlers, Margrit. *See* Martin (H.H.), Tonn-Ehlers & Schilf.

Torre, V. & Poggio, T. A synaptic mechanism possibly underlying directional selectivity to motion. *Proc.* B **202**, 409–416 (1978).

Torrence, M.H. *See* Kolenkiewicz, Smith, Rubincam, Dunn & Torrence; Smith (D.E.), Kolenkiewicz, Wyatt, Dunn & Torrence.

Torza, S., Cox, R.G. & Mason, S.G. Electrohydrodynamic deformation and burst of liquid drops. *Trans.* A **269**, 295–319 (1971).

Tosi, M.P. *See* March & Tosi; Parrinello, Tosi & March.

Tosney, T. & Hoyle, G. Computer-controlled learning in a simple system. *Proc.* B **195**, 365–393 (1977).

Tournier, P. *See* Cassingena & Tournier.

Tourtelot, H.A. Geochemical surveys in the United States in relation to health (Discussion). *Trans.* B **288**, 113–125 (1979).

Tousey, R. Observations of the extreme ultraviolet solar spectrum (Discussion). *Trans.* A **270**, 59–70 (1971).
Eruptive prominences recorded by the X u.v. spectroheliograph on Skylab (Discussion). *Trans.* A **281**, 359–364 (1976).

Tracy, D.H. Dynamics of windowless vapour containment systems for absorption spectroscopy. *Proc.* A **344**, 563–577 (1975).
Photoabsorption structure in lanthanides: 5p subshell spectra of Sm I, Eu I, Dy I, Ho I, Er I, Tm I, and Yb I. *Proc.* A **357**, 485–498 (1977).
See also Connerade, Mansfield, Newsom, Tracy, Baig & Thimm.

Trainor, L.E.H. *See* Ihrig, Rosensteel, Chow & Trainor; Rosensteel, Ihrig & Trainor; Wise (M.B.) & Trainor.

Tranter, R.L. *See* Bell (R.P.) & Tranter.

Traut, W. *See* Clarke (Sir Cyril), Mittwoch & Traut.

Trebst, A. Plastoquinones in photosynthesis (Discussion). *Trans.* B **284**, 591–599 (1978).

Tredre, Barbara E. *See* Edholm (O.G.), Humphrey, Lourie, Tredre & Brotherhood.

Tregear, R.T. *See* Holmes (K.C.), Tregear & Leigh.

Treherne, J.E. *See* Birch (M.C.), Cheng & Treherne; Foster (W.A.) & Treherne.

Treloar, L.R.G. The mechanics of rubber elasticity (Discussion). *Proc.* A **351**, 301–330 (1976).

Treloar, L.R.G. & Riding, G. A non-Gaussian theory for rubber in biaxial strain.
I. Mechanical properties. *Proc.* A **369**, 261–280 (1979).
II. Optical properties. *Proc.* A **369**, 281–293 (1979).

Trench, R.K. The physiology and biochemistry of zooxanthellae symbiotic with marine coelenterates.
I. The assimilation of photosynthetic products of zooxanthellae by two marine coelenterates. *Proc.* B **177**, 225–235 (1971).
II. Liberation of fixed ^{14}C by zooxanthellae *in vitro*. *Proc.* B **177**, 237–250 (1971).

III. The effect of homogenates of host tissues on the excretion of photosynthetic products *in vitro* by zooxanthellae from two marine coelenterates. *Proc.* B **177**, 251–264 (1971).
   See also Schoenberg & Trench; Siebens & Trench.
Trench, R.K., Boyle, J. Elizabeth & Smith, D.C. The association between chloroplasts of *Codium fragile* and the mollusc *Elysia viridis*.
I. Characteristics of isolated *Codium* chloroplasts. *Proc.* B **184**, 51–61 (1973).
II. Chloroplast ultrastructure and photosynthetic carbon fixation in *E. viridis.*
   Proc. B **184**, 63–81 (1973).
III. Movement of photosynthetically fixed ^{14}C in tissues of intact living *E. viridis* and in *Tridachia crispata*. *Proc.* B **185**, 453–464 (1974).
Trench, R.K., Pool, R.R., Jr, Logan, M. & Engelland, A. Aspects of the relation between *Cyanophora paradoxa* (Korschikoff) and its endosymbiotic cyanelles *Cyanocyta korschikoffiana* (Hall & Claus). I. Growth, ultrastructure, photosynthesis and the obligate nature of the association. *Proc.* B **202**, 423–443 (1978).
Trench, R.K. & Ronzio, G.S. Aspects of the relation between *Cyanophora paradoxa* (Korschikoff) and its endosymbiotic cyanelles *Cyanocyta korschikoffiana* (Hall & Claus). II. The photosynthetic pigments. *Proc.* B **202**, 445–462 (1978).
Trench, R.K. & Siebens, H.C. Aspects of the relation between *Cyanophora paradoxa* (Korschikoff) and its endosymbiotic cyanelles *Cyanocyta korschikoffiana* (Hall & Claus). IV. The effects of rifampicin, chloramphenical and cycloheximide on the synthesis of ribosomal ribonucleic acids and chlorophyll. *Proc.* B **202**, 473–482 (1978).
Treuil, M. *See* Allègre, Shimizu & Treuil; Bougault, Joron & Treuil.
Trevena, D.H. *See* Temperley & Trevena.
Tribe, D.E. The conservation and improvement of resources: the grazing animal (Discussion).
   Trans. B **278**, 565–582 (1977).
Trice, R. *See* Warren (N.) & Trice.
Triller, A. *See* Korn, Triller & Faber.
Trimm, D.L. *See* Brown (D.M.) & Trimm; Cullis, Keene & Trimm.
Tripathi, R.S. *See* Lukes & Tripathi.
Trivedi, V.K. & Kumar, I.J. On a Mellin transform technique for the asymptotic solution of a nonlinear Volterra integral equation. *Proc.* A **352**, 339–349 (1977).
Truckle, P.H. *See* Norry, Truckle, Lippard, Hawkesworth, Weaver & Marriner.
Trudgill, S.T. Surface lowering and landform evolution on Aldabra (Discussion).
   Trans. B **286**, 35–45 (1979).
   The soils of Aldabra (Discussion). *Trans.* B **286**, 67–77 (1979).
Truter, Mary R. Recognition of metal cations by biological systems (Discussion).
   Trans. B **272**, 29–41 (1975).
Tsai, C.-C. *See* Sobell, Tsai, Jain & Sakore.
Tsai, C.-C., Jain, S.C. & Sobell, H.M. Drug—nucleic acid interaction: X-ray crystallographic determination of an ethidium—dinucleoside monophosphate crystalline complex, ethidium:5-iodouridylyl (3′–5′)adenosine (Discussion). *Trans.* B **272**, 137–146 (1975).
Tsai, Chon-Yin. *See* Widnall & Tsai.
Tschesche, R. Biosynthesis of cardenolides, bufadienolides and steroid sapogenins (Discussion).
   Proc. B **180**, 187–202 (1972).
Tse, J.S. *See* Gupta (R.P.), Tse & Bancroft.
Tsien, R.Y., Green, D.P.L., Levinson, S.R., Rudy, B. & Sanders, J.K.M. A pharmacologically active derivative of tetrodotoxin. *Proc.* B **191**, 555–559 (1975).
Tsuchiya, Y. *See* Suzuki (Y.), Tsuchiya, Kinoshita, Sugiyama & Inuzuka.
Tsukagoshi, N. *See* Franklin (R.M.), Hinnen, Schäefer & Tsukagoshi.
Tsukahara, Y. *See* Horridge, Mimura & Tsukahara.
Tubbs, E.F. *See* Huber, Sandeman & Tubbs.
Tuck, A.F. Numerical model studies of the effect of injected nitrogen oxides on stratospheric ozone.
   Proc. A **355**, 267–299 (1977).
   A comparison of one-, two- and three-dimensional model representations of stratospheric gases

(Discussion). *Trans.* A **290**, 477–494 (1979).

Tuck, E.O. *See* Vanden Broeck, Schwartz & Tuck.

Tucker, E.M., Ellory, J.C., Wooding, F.B.P., Morgan, G. & Herbert, J. The number and specificity of L antigen sites on low potassium type sheep red cells. *Proc.* B **194**, 271–277 (1976).

Tucker, M.O. & Turnbull, J.A. The morphology of interlinked porosity in nuclear fuels. *Proc.* A **343**, 299–314 (1975).

Tuckett, R.P. *See* Till (S.M.), Freedman, Tuckett & Jones.

Tuffrey, M. *See* Ford (C.E.), Evans, Burtenshaw, Clegg, Tuffrey & Barnes.

Tulunay, Y. Kabasakal. *See* Goodall, Hopkins, Tulunay & D'Arcy.

Tung, R. *See* Rainis, Tung & Szwarc.

Turcotte, D.L. *See* Spence & Turcotte.

Turcotte, D.L. & Oxburgh, E.R. Intra-plate volcanism (Discussion). *Trans.* A **288**, 561–579 (1978).

Turcsanyi, B. *See* Bernstein (J.), Regev, Herbstein, Main and others.

Turekian, K.K. *See* Clark (S.P.) & Turekian.

Turnbull, J.A. *See* Tucker (M.O.) & Turnbull.

Turner, A.M. *See* Hensens, Hill, Thornton, Turner & Williams.

Turner, C. *See* Kerney, Preece & Turner.

Turner, D. *See* Franks (A.), Lindsey, Bennett, Speer, Turner & Hunt.

Turner, D.W. *See* Ames & Turner.

Turner, G. The early chronology of the Moon: evidence for the early collisional history of the solar system (Discussion). *Trans.* A **285**, 97–103 (1977).

See also Cadogan & Turner; Hennessy (J.) & Turner.

Turner, Jill F. *See* Allen (J.A.) & Turner.

Turner, J.R.G. *See* Brown (K.S.), Sheppard & Turner.

Turner, Judith, Hewetson, Valerie P., Hibbert, F.A., Lowry, Katharine H. & Chambers, C. The history of the vegetation and flora of Widdybank Fell and the Cow Green reservoir basin, Upper Teesdale. *Trans.* B **265**, 327–408 (1973).

Turner, R.S. *See* Burger, Turner, Kuhns & Weinbaum.

Turner, S.C. *See* Coles (E.C.), Beilin, Bulpitt, Dollery, Johnson and others.

Turver, K.E. *See* Dixon (H.E.), Earnshaw, Hook, Hough, Smith, Stephenson & Turver; Dixon (H.E.) & Turver; Dixon (H.E.), Turver & Waddington.

Twiddy, N.D. *See* Dickinson (P.H.G.), Bain, Thomas, Williams, Jenkins & Twiddy.

Tyndall, A.M. Appendix to the Biographical Memoir of Cecil Frank Powell. *Biogr. Mem.* **17**, 555–557 (1971).

Tyrrell, D.A.J. Introductory remarks to a Discussion. *Proc.* B **184**, 349 (1973).

Introduction to a Discussion. *Proc.* B **199**, 5–8 (1977).

Some health hazards associated with agricultural improvements (Discussion). *Proc.* B **199**, 33–35 (1977).

Influenza vaccines (Discussion). *Trans.* B **288**, 449–460 (1980).

Tyrrell, D.A.J. & Elliott, Katherine. Introduction to a Discussion. *Proc.* B **209**, 5–6 (1980).

Tyrrell, D.A.J. & Pereira, H.G. Preface to a Discussion. *Trans.* B **288**, 289 (1980).

Ubatuba, F.B. *See* Beddell, Clark, Hardy, Lowe, Ubatuba, Vane & Wilkinson.

Ubbelohde, A.R. Electronic anomalies in dilute synthetic metals. *Proc.* A **321**, 445–460 (1971).

Electrical anisotropy of synthetic metals based on graphite. *Proc.* A **327**, 289–303 (1972).

See also Bach & Ubbelohde; Drummond (I.) & Ubbelohde; Duruz, Michels & Ubbelohde; Duruz & Ubbelohde; Michels & Ubbelohde.

Uchitel, O.D. *See* Cull-Candy, Miledi, Nakajima & Uchitel.

Udintsev, G.B. *See* Dmitriev, Vinogradov & Udintsev; Vinogradov, Dmitriev & Udintsev.

Udintsev, G.B., Dmitriev, L.V. & Vinogradov, A.P. The tectonics of the Mid-Indian Ocean Ridge and the petrography of the solid rocks of its rift zones (Discussion). *Trans.* A **268**, 653–659 (1971).

Corrigendum. *Trans.* A **269**, 555, 647 (1971).

Udovich, Nancy. *See* Greeley, Iversen, Pollack, Udovich & White.

Ueda, H. *See* Mizushina, Ogino, Ueda & Komori.

Ueda, M. *See* Araki, Yano, Ueda & Noda.

Ueda, Mirthes. *See* Faulk (W. Page), Yeager, McIntyre & Ueda.

Ulbricht, W. & Wagner, H.-H. The reaction between tetrodotoxin and membrane sites at the node of Ranvier: its kinetics and dependence on pH (Discussion). *Trans.* B **270**, 353–363 (1975).

Ullman, S. The interpretation of structure from motion. *Proc.* B **203**, 405–426 (1979).

Umbreit, J. *See* Strominger, Blumberg, Suginaka, Umbreit & Wickus.

Underhill, Anne B. The ultraviolet spectrum of B-type supergiants (Discussion). *Trans.* A **279**, 429–442 (1975).

Underwood, E.J. Trace elements and health: an overview (Discussion). *Trans.* B **288**, 5–14 (1979). Concluding remarks to a Discussion. *Trans.* B **288**, 215–216 (1979).

Unruh, W.G. Alternative Fock quantization of neutrinos in flat space-time. *Proc.* A **338**, 517–525 (1974).

Self force on charged particles. *Proc.* A **348**, 447–465 (1976).

See also Davies (P.C.W.) & Unruh.

Unsöld, A. Abundance of iron in the photosphere (Discussion). *Trans.* A **270**, 23–28 (1971).

Unsworth, J.F. *See* Kenner, Rimmer, Smith & Unsworth.

Unsworth, P.J. *See* Newton (G.), Andrews & Unsworth.

Unwin, P.N.T. Phase contrast and interference microscopy with the electron microscope (Discussion). *Trans.* B **261**, 95–104 (1971).

Electron microscopy of biological specimens by means of an electrostatic phase plate. *Proc.* A **329**, 327–359 (1972).

Unwin, S.D. & Critchley, R. Atomic ground state energy in multiply connected universes. *Proc.* A **372**, 297–306 (1980).

Upadhyaya, J.C. On the volume-dependent contribution to the elastic constants of cubic metals. *Proc.* A **356**, 345–350 (1977).

Upstill, C. Light caustics from rippling water. *Proc.* A **365**, 95–104 (1979).

Upton, G.J. & Wadsworth, W.J. Aspects of magmatic evolution on Réunion Island (Discussion). *Trans.* A **271**, 105–130 (1972).

Urbański, T. Degradation of amber and formation of free radicals by mechanical action. *Proc.* A **325**, 377–381 (1971).

Urey, H.C. *See* O'Keefe (J.A.) & Urey.

Uribe, R. *See* Woodgate (B.E.), Knight, Uribe, Sheather, Bowles & Nettleship.

Ursell, F. *See* Cartwright (D.E.) & Ursell.

Uscinski, B.J. The propagation and broadening of pulses in weakly irregular media. *Proc.* A **336**, 379–392 (1974).

Parabolic moment equations and acoustic propagation through internal waves. *Proc.* A **372**, 117–148 (1980).

See also Budden & Uscinski.

Ussing, H.H. Introductory remarks to a Discussion. *Trans.* B **262**, 85–90 (1971).

Usune, S. *See* Tomita, Tokuno & Usune.

Uttenthal, L.O. *See* Livett, Uttenthal & Hope.

Uttenthal, L.O. & Hope, D.B. Neurophysins and posterior pituitary hormones in the Suiformes. *Proc.* B **182**, 73–87 (1972).

Uttenthal, L.O., Livett, B.G. & Hope, D.B. Release of neurophysin together with vasopressin by a Ca^{2+} dependent mechanism (Discussion). *Trans.* B **261**, 379–380 (1971).

Uvarov, D.B. *See* Andreeva, Katasyev & Uvarov; Andreeva, Katasyev, Uvarov, Nesterov & Chasovitin.

Vaadia, Y. Plant hormones and water stress (Discussion). *Trans.* B **273**, 513–522 (1976).

Vaiana, G.S. The X-ray corona from Skylab (Discussion). *Trans.* A **281**, 365–374 (1976).

Vail, J.R. Outline of the geochronology and tectonic units of the basement complex of northeast Africa. *Proc.* A **350**, 127–141 (1976).

Vail, P.R., Mitchum, R.M., Jr, Shipley, T.H. & Buffler, R.T. Unconformities of the North Atlantic (Discussion). *Trans.* A **294**, 137–155 (1980).

Vajravelu, K. & Sastri, K.S. Correction to 'Free convection effect on the oscillatory flow past an infinite, vertical, porous plate with constant suction. I.'. (Soundalgekar, V.M. 1973). *Proc.* A **353**, 221–223 (1977).

Valasinas, Aldonia. *See* Frydman (B.), Frydman, Valasinas, Levy & Feinstein.

Valeri, C.R. *See* Costa (J.L.), Dobson, Kirk, Poulsen, Valeri & Vecchione.

Valliant, H.D. A technique for the precise calibration of continuously recording gravimeters (Discussion). *Trans.* A **274**, 227–230 (1973).

Vanden Broeck, J.-M., Schwartz, L.W. & Tuck, E.O. Divergent low-Froude-number series expansion of nonlinear free-surface flow problems. *Proc.* A **361**, 207–244 (1978).

Van der Loos, H. Structural changes in the cerebral cortex upon modification of the periphery: barrels in somatosensory cortex (Discussion). *Trans.* B **278**, 373–376 (1977).

Vane, J.R. *See* Beddell, Clark, Hardy, Lowe, Ubatuba, Vane & Wilkinson.

Van Montagu, M. *See* Schell, Van Montagu, De Beuckeleer, De Block and others.

Van Montagu, M., Holsters, M., Zambryski, P., Hernalsteens, J.P., Depicker, A., De Beuckeleer, M., Engler, G., Lemmers, M., Willmitzer, L. & Schell, J. The interaction of *Agrobacterium* Ti-plasmid DNA and plant cells (Discussion). *Proc.* B **210**, 351–365 (1980).

Van Schmus, W.R. Early and Middle Proterozoic history of the Great Lakes area, North America (Discussion). *Trans.* A **280**, 605–628 (1976).

Van Vleck, J.H. *See* Foglio, Sekerka & Van Vleck; Foglio & Van Vleck.

Van Vliet, F. *See* Schell, Van Montagu, De Beuckeleer, De Block and others.

Varet, J. *See* Tarney, Wood, Saunders, Cann & Varet.

Varghese, J.N. & Mason, R. Electron densities in transition metal complexes: population analysis of polarized neutron diffraction data. *Proc.* A **372**, 1–7 (1980).

Varley, E. *See* Cekirge & Varley; Kazakia & Varley.

Varley, E., Kazakia, J.Y. & Blythe, P.A. The interaction of large amplitude barotropic waves with an ambient shear flow: critical flows. *Trans.* A **287**, 189–236 (1977).

Varshavsky, A.J., Bakayev, V.V., Bakayeva, T.G., Chumackov, P.M., Shmatchenko, V.V. & Georgiev, G.P. On the structure of cellular and viral chromatin (Discussion). *Trans.* B **283**, 275–285 (1978).

Vasileff, R.T. *See* Hatfield, Fisher, Dunigan, Burchfield and others.

Vassie, J.M. *See* Cartwright (D.E.), Edden, Spencer & Vassie.

Vasudevan, S. *See* Rao (C.N.R.), Sarma, Vasudevan & Hegde.

Vaughan, D.J. & Burns, R.G. Electronic absorption spectra of lunar minerals (Discussion). *Trans.* A **285**, 249–258 (1977).

Vaughan, H. The finite compression of elastic solid cylinders in the presence of gravity. *Proc.* A **321**, 381–396 (1971).

Vaughan, J.M. *See* Randall (Sir John) & Vaughan.

Vaughan, P.R. The deformations of the Empingham valley slope (appendix to a paper by Horswill & Horton) (Discussion). *Trans.* A **283**, 451–461 (1976).

Vecchione, J.J. *See* Costa (J.L.), Dobson, Kirk, Poulsen, Valeri & Vecchione.

van der Veen, J.H. Development of steels for offshore structures (Symposium). *Trans.* A **282**, 319–328 (1976).

Veldhuis, K.H. The use of operational research and systems analysis in decision making in Unilever (Discussion). *Trans.* A **287**, 487–492 (1977).

Venables, J.A. *See* English (C.A.) & Venables; English (C.A.), Venables & Salahub; Niebel & Venables.

Venables, J.A. & Ball, D.J. Nucleation and growth of rare-gas crystals. *Proc.* A **322**, 331–354 (1971).

Venkatesan, K., Dale, D., Hodgkin, Dorothy Crowfoot, Nockolds, C.E., Moore, F.H. & O'Connor, B.H. The structure of vitamin B_{12}. IX. The crystal structure of cobyric acid, factor V $1a$ (with appendixes by N. Waters & Joyce Waters and Eleanor Dodson). *Proc.* A **323**, 455–487 (1971).

Vered, M. *See* Singh (S.J.), Ben-Menahem & Vered.

Verhulst, F. Discrete symmetry dynamical systems at the main resonances with applications to axisymmetric galaxies. *Trans.* A **290**, 435–465 (1979).

Verma, S.C., Malik, Renuka & Dhir, Indra. Genetics of the incompatibility system in the crucifer

Eruca sativa L. *Proc.* B **196**, 131–159 (1977).

Vermeulen, L.A. *See* Walker (J.), Vermeulen & Clark.

Vermilion, J. *See* Schroepfer, Lutsky, Martin, Huntoon and others.

Vernazza, J.E. *See* Reeves (E.M.), Vernazza & Withbroe.

Verney, R.B. A future forest policy for Britain (Discussion). *Trans.* B **271**, 219–232 (1975).

Vernon, K.R. Hydro (including tidal) energy (Discussion). *Trans.* A **276**, 485–493 (1974).

Veron, J.E.N. Deltaic and dissected reefs of the far Northern Region (Discussion).
Trans. B **284**, 23–37 (1978).
   Evolution of the far northern barrier reefs (Discussion). *Trans.* B **284**, 123–127 (1978).

Veron, J.E.N. & Hudson, R.C.L. Ribbon reefs of the Northern Region (Discussion).
Trans. B **284**, 3–21 (1978).

Verrall, R.A. *See* Ashby (M.F.) & Verrall.

Vesseur, H.J.A. Results of E-layer drift measurements at De Bilt (Discussion).
Trans. A **271**, 485–497 (1972).

Vessey, M.P. & Doll, Sir Richard. Is 'the pill' safe enough to continue using? (Discussion).
Proc. B **195**, 69–80 (1976).

Vest, C.M. *See* Debler & Vest.

Vibert, P.J. *See* Lowy, Vibert, Haselgrove & Poulsen.

Vickerman, J.C. *See* Stone (F.S.) & Vickerman.

Vickers, D.G. & Bastin, J.A. The interaction of lunar rock and far infrared radiation (Discussion).
Trans. A **285**, 319–324 (1977).

Vidal, P. *See* Jahn, Vidal & Tilton.

Vietmeyer, N.D. Underexploited village resources (Discussion). *Proc.* B **209**, 47–58 (1980).

Villarroel, R. *See* Schell, Van Montagu, De Beuckeleer, De Block and others.

Vincent, Angela. *See* Green (D.P.L.), Ito, Miledi & Vincent; Green (D.P.L.), Miledi, de la Mora &
   Vincent; Green (D.P.L.), Miledi & Vincent; Ito (Y.), Miledi & Vincent; Ito (Y.), Miledi, Molenaar,
   Vincent, Polak and others.

Vincent, E.A. Introduction to a Symposium. *Trans.* A **274**, 3 (1973).

Vincent, J.F.V. Locust oviposition: stress softening of the extensible intersegmental membranes.
Proc. B **188**, 189–201 (1975).

Vincent, R. A theoretical analysis and computer simulation of the growth of epitaxial films.
Proc. A **321**, 53–68 (1971).

Vine, F.J. *See* Moores (E.M.) & Vine.

Vinen, W.F. Light scattering by liquid helium [abstract] (Discussion). *Trans.* A **293**, 377–390 (1979).

Viner, A.B. Responses of a mixed phytoplankton population to nutrient enrichments of ammonia
   and phosphate, and some associated ecological implications. *Proc.* B **183**, 351–370 (1973).
   See also Ganf & Viner.

Viner, A.B. & Smith, I.R. Geographical, historical and physical aspects of Lake George (Discussion).
Proc. B **184**, 235–270 (1973).

Viney, I.V.F. *See* Carey (R.), Coleman & Viney.

Vinogradov, A.P. *See* Dmitriev, Vinogradov & Udintsev; Udintsev, Dmitriev & Vinogradov.

Vinogradov, A.P., Dmitriev, L.V. & Udintsev, G.B. Distribution of trace elements in crystalline rocks
   of rift zones (Discussion). *Trans.* A **268**, 487–491 (1971).
   Corrigendum. *Trans.* A **269**, 555, 647 (1971).

Vinogradov, A.P., Yaroshevsky, A.A. & Ilyin, N.P. A physico-chemical model of element separation
   in the differentiation of the mantle material (Discussion). *Trans.* A **268**, 409–421 (1971).
   Corrigendum. *Trans.* A **269**, 555, 647 (1971).

Vinson, G.P. *See* Bell (Janet B.G.), Vinson & Lacy.

Viñuela, E., Camacho, A., Jiménez, F., Carrascosa, J.L., Ramírez, G. & Salas, Margarita. Structure and
   assembly of phage ϕ29 (Discussion). *Trans.* B **276**, 29–35 (1976).

Virden, R. *See* Carrey, Mitchinson, Pain & Virden.

Vishnu-Mittre. *See* Godwin & Vishnu-Mittre.

Vítek, V. Computer simulation of the screw dislocation motion in b.c.c. metals under the effect of
   the external shear and uniaxial stresses. *Proc.* A **352**, 109–124 (1976).

See also Duesbery, Vítek & Bowen; Pond & Vítek.

Viton, M. *See* Courtès, Laget, Sivan, Viton and others.

Vizi, E.S. *See* Garamvölgyi, Vizi & Knoll.

Voge, J.P. & Arifon, P. Rationalization for a better management of the radio frequency space allocated to radiocommunications between specified fixed points and mainly to point to point microwave links (Discussion). *Trans.* A **289**, 103—112 (1978).

Vogel, F. 'Our load of mutation': reappraisal of an old problem (Discussion).
 Proc. B **205**, 77—90 (1979).

Vogel, K.A. & Anderson, A.J. An improved servo-controlled tiltmeter system and latest measurements in Sweden (Discussion). *Trans.* A **274**, 305—309 (1973).

Vogt, Marthe L. Geoffrey Wingfield Harris. *Biogr. Mem.* **18**, 309—329 (1972).

Voigt-Martin, I.G. *See* Andrews (E.H.) & Voigt-Martin.

Volkman, J.K. *See* Eglinton, HajIbrahim, Maxwell, Quirke, Shaw and others.

Volkmann, D. *See* Sievers & Volkmann.

Vollrath, L. *See* Knowles (Sir Francis), Vollrath & Meurling.

Vonbun, F.O. Goddard laser systems and their accuracies (Discussion).
 Trans. A **284**, 443—450 (1977).
 Probing the Earth's gravity field by means of satellite-to-satellite tracking (Discussion).
 Trans. A **284**, 475—483 (1977).
 The N.A.S.A. Earth and ocean dynamics programme (Discussion). *Trans.* A **284**, 607—619 (1977).

Von Dreele, R.B. *See* Anderson (J.S.), Bevan, Cheetham, Von Dreele and others.

Von Dreele, R.B. & Cheetham, A.K. The structures of some titanium—niobium oxides by powder neutron diffraction. *Proc.* A **338**, 311—326 (1974).

Voordouw, G. *See* Eisenberg (H.), Borochov, Kam & Voordouw.

van de Vooren, A.I. A numerical investigation of the rolling-up of vortex sheets.
 Proc. A **373**, 67—91 (1980).

Vuillemin, A. *See* Courtès, Laget, Sivan, Viton and others.

Wachtel, E.J. *See* Marvin & Wachtel.

Waddington, C.H. The recognition of alien life at the level of macroscopic morphology (Discussion).
 Proc. B **189**, 155—159 (1975).
 The Bernal Lecture, 1975. The New Atlantis revisited. *Proc.* B **190**, 301—314 (1975).

Waddington, C.J. *See* Dixon (H.E.), Turver & Waddington.

Wadsworth, R.M. *See* Elston, Karamanos, Kassam & Wadsworth.

Wadsworth, W.J. *See* Upton & Wadsworth.

Waechter, R.T. *See* Stewartson & Waechter.

Wagner, A. *See* Bartel, Duinker, Heintze, Heinzelmann and others.

Wagner, H.-H. *See* Ulbricht & Wagner.

Wagner, J. & Laemmli, U.K. Studies on the maturation of the head of bacteriophage T4 (Discussion).
 Trans. B **276**, 15—26 (1976).

Wagner, L.E. *See* Hattangadi, Wagner & Seth.

Wagner, S. Heterojunction solar cells (Discussion). *Trans.* A **295**, 445—451 (1980).

Wain, R.L. Review Lecture. Some developments in research on plant growth inhibitors.
 Proc. B **191**, 335—352 (1975).

Waite, P.M.E. & Cragg, B.G. The effect of destroying the whisker follicles in mice on the sensory nerve, the thalamocortical radiation and cortical barrel development.
 Proc. B **204**, 41—55 (1979).

Wake, G.C. *See* Boddington, Gray & Wake; Carter (M.R.), Druce & Wake.

Wakefield, J. *See* Coward, Graham, James & Wakefield.

Wakeham, W.A. *See* Pratt (K.C.) & Wakeham.

Wakelin, L.P.G. *See* Dean (P.M.) & Wakelin.

Wakely, Jennifer. *See* Clint (Jane M.), Wakely & Ockleford; Fisher (R.F.) & Wakely.

Wakely, Jennifer & England, Marjorie A. Scanning electron microscopical and histochemical study of the structure and function of basement membranes in the early chick embryo.
Proc. B **206**, 329–352 (1979).

Wakerley, J.B. *See* Poulain, Wakerley & Dyball.

Walcott, B. & Horridge, G.A. The compound eye of *Archichauliodes* (Megaloptera).
Proc. B **179**, 65–72 (1971).

Wald, R.M. Construction of metric and vector potential perturbations of a Reissner—Nordström black hole. *Proc.* A **369**, 67–81 (1979).

Waldichuk, M. Review of the problems (Discussion). *Trans.* B **286**, 399–424 (1979).

Waldram, J.R. Chemical potential and boundary resistance at normal-superconducting interfaces.
Proc. A **345**, 231–249 (1975).
See also Hook (J.R.) & Waldram.

Wale, M.J. *See* Drummond (J.R.), Houghton, Peskett, Rodgers, Wale and others.

Walenta, A.H. *See* Bartel, Duinker, Heintze, Heinzelmann and others.

Waley, S.G. *See* Hill (H.A.O.), Sammes & Waley.

Walkden, F. & Caine, P. Surface pressures on a wing moving with supersonic speed.
Proc. A **341**, 177–193 (1974).

Walker, A.D.M. The propagation of very low-frequency radio waves in ducts in the magnetosphere.
Proc. A **321**, 69–93 (1971).
The propagation of very low-frequency waves in ducts in the magnetosphere. II.
Proc. A **329**, 219–231 (1972).
Excitation of the Earth—ionosphere waveguide by downgoing whistlers.
·I. Isotopic model. *Proc.* A **340**, 367–374 (1974).
II. Propagation in the magnetic meridian. *Proc.* A **340**, 375–393 (1974).

Walker, B.J. *See* Andrews (E.H.) & Walker.

Walker, D. & Flenley, J.R. Late Quaternary vegetational history of the Enga Province of upland Papua New Guinea. *Trans.* B **286**, 265–344 (1979).

Walker, Doreen M.C. Cosmos 462 (1971–106A): orbit determination and analysis.
Trans. A **292**, 473–512 (1979).
See also King-Hele & Walker.

Walker, E.B. Non-conventional hydrocarbons and future trends in oil utilization in North America and their effect on world supplies (Discussion). *Trans.* A **276**, 541–546 (1974).

Walker, E.F. *See* May (M.J.), Gladman & Walker.

Walker, G. A study of the oviducal glands and ovisacs of *Balanus balanoides* (L.), together with comparative observations on the ovisacs of *Balanus hameri* (Ascanius) and the reproductive biology of the two species. *Proc.* B **291**, 147–162 (1980).
See also Geake, Walker, Telfer & Mills.

Walker, G.P.L. Viscosity control of ocean floor volcanics (Discussion). *Trans.* A **268**, 727–729 (1971).
Lengths of lava flows (Symposium). *Trans.* A **274**, 107–118 (1973).
A brief account of the 1971 eruption of Mount Etna (Symposium). *Trans.* A **274**, 177–179 (1973).
See also Booth (B.), Croasdale & Walker; Booth (B.) & Walker.

Walker, I.K. *See* Boddington, Gray & Walker.

Walker, Isobel C. *See* Dance & Walker.

Walker, J. *See* Clark (C.D.) & Walker.

Walker, J., Vermeulen, L.A. & Clark, C.D. Electronic transitions at the diamond vacancy.
Proc. A **341**, 253–266 (1974).

Walker, J.C.F. *See* Barnes (P.), Tabor & Walker.

Walker, J.D.A. The boundary layer due to rectilinear vortex. *Proc.* A **359**, 167–188 (1978).

Walker, M. *See* Perry (V.H.) & Walker; Stewart (J.M.) & Walker.

Walker, P. Opening address to a Discussion. *Trans.* A **276**, 407–412 (1974).

Walker, P.M.B. Introductory remarks: DNA and genes (Discussion). *Trans.* B **283**, 305–307 (1978).

Walker, R.M. *See* Crozaz, Poupeau, Walker, Zinner & Morrison.

Walker, S. *See* Calderwood, Coffey, Morita & Walker.

Walker, S.M. *See* Block (H.), Ions, Powell, Singh & Walker.

Walker, T.E.H. *See* Hinkley, Walker & Richards.
Wall, C.T.C. Nets of quadrics, and theta-characteristics of singular curves.
　Trans. A **289**, 229–269 (1978).
Wall, D.J.N. *See* Bates (R.H.T.) & Wall.
Wall, D.N. *See* Franklin (R.N.), MacKinlay, Edgley & Wall.
Wall, J.V. Evidence for the cosmological evolution of active galaxies (Discussion).
　Trans. A **296**, 367–383 (1980).
Wall, P.D. The presence of ineffective synapses and the circumstances which unmask them (Discussion). *Trans.* B **278**, 361–372 (1977).
Wallace, B.G., Adal, M.N. & Nicholls, J.G. Regeneration of synaptic connections by sensory neurons in leech ganglia maintained in culture. *Proc.* B **199**, 567–585 (1977).
Wallace, J.S. *See* Biscoe (P.V.), Cohen & Wallace.
Wallace, L. *See* Richmond (M.H.), Bennett, Choi, Brown, Brunton and others.
Waller, I. Memories of my early work on lattice dynamics and X-ray diffraction (Symposium).
　Proc. A **371**, 120–124 (1980).
Waller, S. *See* Breuer, Ryan & Waller.
Waller, T.R. Morphology, morphoclines and a new classification of the Pteriomorphia (Mollusca: Bivalvia) (Discussion). *Trans.* B **284**, 345–365 (1978).
Wallis, Jenifer. *See* Richards (W.G.) & Wallis.
Wallis, R.H. *See* Mott (Sir Nevill), Pepper, Pollitt, Wallis & Adkins.
Wallis, V. *See* Davies (A.J.S.), Leuchars, Wallis & Doenhoff.
Wallwork, S.C. *See* Ashwell (G.J.), Eley, Wallwork & Willis.
Walmsley, R. *See* Dixon-Lewis, Isles & Walmsley.
Walmsley, S.H. *See* Schipper & Walmsley.
Waloff, Zena. *See* Hemming (C.F.), Popov, Roffey & Waloff.
Walpole, L.J. Iterative solution of linear problems. *Proc.* A **334**, 119–133 (1973).
Walsby, A.E. The pressure relationships of gas vacuoles. *Proc.* B **178**, 301–326 (1971).
　The water relations of gas-vacuolate prokaryotes. *Proc.* B **208**, 73–102 (1980).
Walsh, D., Hayes, A.P. & Harrison, V.A.W. Observations of radio frequency noise from Ariel 4 (Discussion). *Proc.* A **343**, 227–240 (1975).
Walsh, D.J. *See* Allen (G.), Burgess, Edwards & Walsh.
Walsh, F.S. *See* Crumpton, Snary, Walsh, Barnstable and others.
Walsh, J.F. *See* Le Berre, Garms, Davies, Walsh & Philippon.
Walsh, Joan. Finite-difference and finite-element methods of approximation (Discussion).
　Proc. A **323**, 155–165 (1971).
Walsh, R.J. Geographical, historical and social background of the peoples studied in the I.B.P. (Discussion). *Trans.* B **268**, 223–228 (1974).
Walsh, R.P.D. *See* Stoddart (D.R.) & Walsh.
Walshaw, C.D. *See* Houghton & Walshaw.
Walston, Lord. Farming structure in Britain (Discussion). *Trans.* B **267**, 23–36 (1973).
Walters, A.G. Non-symmetric flow in Laval type nozzles. *Trans.* A **273**, 185–235 (1972).
Walton, A. *See* Baxter (M.S.) & Walton.
Walton, D.W.H. *See* Callaghan, Smith & Walton.
Walton, I.C. Viscous shear layers in an oscillating rotating fluid. *Proc.* A **344**, 101–110 (1975).
　The transition to Taylor vortices in a closed rapidly rotating cylindrical annulus.
　Proc. A **372**, 201–218 (1980).
　See also Hall (P.) & Walton; Stewartson & Walton.
Wampler, D.L. *See* Kennard, Isaacs, Motherwell, Coppola and others.
Wan, T. *See* Baldwin (J.E.), Jung, Singh, Wan, Haber and others.
Wanas, M.I. *See* Mikhail & Wanas.
Wang, K.C. Boundary layer over a blunt body at high incidence with an open-type of separation.
　Proc. A **340**, 33–55 (1974).
　Boundary layer over spinning blunt body of revolution at incidence including Magnus forces.
　Proc. A **363**, 357–380 (1978).

Wang, R.T., Nicol, J.A.C., Thurston, E.L. & McCants, M. Studies on the eyes of bigeyes (Teleostei Priacanthidae) with special reference to the tapetum lucidum. *Proc.* B **210**, 499—512 (1980).

Wänke, H. *See* Dreibus, Spettel & Wänke; Palme & Wänke.

Wänke, H., Palme, H., Baddenhausen, H., Dreibus, G., Kruse, H. & Spettel, B. Element correlations and the bulk composition of the Moon (Discussion). *Trans.* A **285**, 41—48 (1977).

Ward, A.J.I. *See* Eley, Hey & Ward.

Ward, C.W. *See* Laver, Air, Webster, Gerhard, Ward & Dopheide.

Ward, H. *See* Drever (R.W.P.), Hough, Pugh, Edelstein and others.

Ward, I.M. Ultra-high modulus polyolefins (Discussion). *Trans.* A **294**, 473—482 (1980).

Ward, Margaret E. *See* Flowers, Ward & Hall.

Ward, P. Rational strategies for the control of queleas and other migrant bird pests in Africa (Discussion). *Trans.* B **287**, 289—300 (1979).

Ward, P. & Kendall, Marion D. Morphological changes in the thymus of young and adult red-billed queleas *Quelea quelea* (Aves). *Trans.* B **273**, 55—64 (1975).

Ward, P.F.V. *See* Peters, Shorthouse, Ward & McDowell.

Ward, R.D. *See* Killick-Kendrick & Ward; Lainson, Ward & Shaw.

Ward, R.S. A class of self-dual solutions of Einstein's equations. *Proc.* A **363**, 289—295 (1978).
See also Tod & Ward.

Ward, W.H. & Burland, J.B. The use of ground strain measurements in civil engineering (Discussion). *Trans.* A **274**, 421—428 (1973).

Wardroper, A.M.K. *See* Eglinton, Hajlbrahim, Maxwell, Quirke, Shaw and others.

Wareing, P.F. Abscisic acid as a natural growth regulator (Discussion). *Trans.* B **284**, 483—498 (1978).

Wareing, P.F. & Allen, E.J. Physiological aspects of crop choice (Discussion). *Trans.* B **281**, 107—119 (1977).

Warman, H.R. Hydrocarbon potential of deep water (Discussion). *Trans.* A **290**, 33—42 (1978).

Warner, G. *See* Leslie (M.), Jenkin, Hayter, White, Cox & Warner.

Warner, Sir Frederick. Sources and extent of pollution (Discussion). *Proc.* B **205**, 5—15 (1979).
See also Cook (Sir James) & Warner.

Warren, F.W. On the method of Hermitian forms and its application to some problems of hydrodynamic stability. *Proc.* A **350**, 213—237 (1976).
Restrictions upon instabilities of plane and helical gas flows. *Proc.* A **368**, 225—237 (1979).

Warren, G.B. *See* Lee (A.G.), Birdsall, Metcalfe, Warren & Roberts.

Warren, N. & Trice, R. Structure in the upper lunar crust (Discussion).
Trans. A **285**, 469—473 (1977).

Warren, R.C. *See* Hicks (R. Marian), Ketterer & Warren.

Warrington, D.M. *See* Baird (P.E.G.), Brambley, Burnett, Stacey, Warrington & Woodgate.

Warwick, R.S. Extragalactic X-ray sources and the diffuse X-ray background (Discussion).
Proc. A **366**, 391—402 (1979).

Wass, Suzanne Y., Henderson, P. & Elliott, C.J. Chemical heterogeneity and metasomatism in the upper mantle: evidence from rare earth and other elements in apatite-rich xenoliths in basaltic rocks from eastern Australia (Discussion). *Trans.* A **297**, 333—346 (1980).

Wasserburg, G.J., Papanastassiou, D.A., Tera, F. & Huneke, J.C. Outline of a lunar chronology (Discussion). *Trans.* A **285**, 7—22 (1977).

Wässle, H. *See* Boycott, Peichl & Wässle.

Wässle, H., Boycott, B.B. & Peichl, L. Receptor contacts of horizontal cells in the retina of the domestic cat. *Proc.* B **203**, 247—267 (1978).

Wässle, H., Peichl, L. & Boycott, B.B. Topography of horizontal cells in the retina of the domestic cat. *Proc.* B **203**, 269—291 (1978).

Wässle, H. & Riemann, H.J. The mosaic of nerve cells in the mammalian retina.
Proc. B **200**, 441—461 (1978).

Watanabe, H. *See* Tomita & Watanabe.

Watanabe, K., Kondow, T., Soma, M., Onishi, T. & Tamaru, K. Molecular-sieve type sorption on alkali graphite intercalation compounds. *Proc.* A **333**, 51—67 (1973).

Waterfield, M.D. *See* Skehel, Waterfield, McCauley, Elder & Wiley.

Waters, Joyce. *See* Waters (N.) & Waters.

Waters, N. & Waters, Joyce. Appendix to a paper by Venkatesan, Dale, Hodgkin, Nockolds, Moore & O'Connor. *Proc.* A **323**, 480–484 (1971).

Waters, R.T., Allibone, T.E., Dring, D. & Allen, N.L. The structure of the impulse corona in a rod/plane gap. II. The negative corona: propagation and streamer/anode ineraction. *Proc.* A **367**, 321–342 (1979).

Watkins, N.D. & Paster, T.P. The magnetic properties of igneous rocks from the ocean floor (Discussion). *Trans.* A **268**, 507–550 (1971).

Watkins, R.D. *See* Heddle, Keesing & Watkins.

Watkins, Winifred M. Genetics and biochemistry of some human blood groups (Discussion). *Proc.* B **202**, 31–53 (1978).

Watkinson, I.A. *See* Akhtar, Wilton, Watkinson & Rahimtula.

Watson, D.G. *See* Kennard, Isaacs, Motherwell, Coppola and others.

Watson, D.M.S. Pterodactyls past and present. *Trans.* B **267**, 583–585 (1974).

Watson, E. The periglacial environment of Great Britain during the Devensian (Discussion). *Trans.* B **280**, 183–198 (1977).

Watson, Janet. *See* Bridgwater, Watson & Windley.

Watson, Janet V. Effects of reworking on high-grade gneiss complexes (Discussion). *Trans.* A **273**, 443–455 (1973).

Vertical movements in Proterozoic structural provinces (Discussion). *Trans.* A **280**, 629–640 (1976). Precambrian thermal régimes (Discussion). *Trans.* A **288**, 431–440 (1978). Review Lecture. Ore-deposition through geological time. *Proc.* A **362**, 305–328 (1978). Metallogenesis in relation to mantle heterogeneity (Discussion). *Trans.* A **297**, 347–352 (1980).

Watson, Janet V. & Plant, Jane. Regional geochemistry of uranium as a guide to deposit formation (Discussion). *Trans.* A **291**, 321–338 (1979).

Watson, L.M. *See* Padalia, Lang, Norris, Watson & Fabian.

Watson, M., Thurston, E.L. & Nicol, J.A.C. Reflectors in the light organ of *Anomalops* (Anomalopidae, Teleostei). *Proc.* B **202**, 339–351 (1978).

Watson, M.G. Galactic X-ray sources (Discussion). *Proc.* A **366**, 329–344 (1979).

Watt, W. Introduction to a Discussion. *Trans.* A **294**, 409 (1980).

Wattam, D.G. *See* Whittam, Hallam & Wattam.

Watters, W.A. & Fleming, C.A. Contributions to the geology and palaeontology of Chiloe Island, southern Chile. *Trans.* B **263**, 369–408 (1972).

Watterson, J. *See* Bridgwater, Escher & Watterson; Escher, Jack & Watterson.

Watts, M. *See* Miller (D.S.), Baker, Bowden, Evans, Holt and others.

Watts, M.E. *See* Mark & Watts.

Watts, M.E. & Mark, R.F. Drug inhibiton of memory formation in chickens. II. Short-term memory. *Proc.* B **178**, 455–464 (1971).

Watts, W.A. The Late Devensian vegetation of Ireland (Discussion). *Trans.* B **280**, 273–293 (1977).

Waugh, A.R., Mills, P.F. & Southon, M.J. Imaging atom probe microscopy for segregation studies [abstract] (Discussion). *Trans.* A **295**, 133 (1980).

Waxman, D.J., Yocum, R.R. & Strominger, J.L. Penicillins and cephalosporins are active site-directed acylating agents: evidence in support of the substrate analogue hypothesis (Discussion). *Trans.* B **289**, 257–271 (1980).

Waxman, S.G., Bradley, W.G. & Hartwieg, E.A. Organization of the axolemma in amyelinated axons: a cytochemical study in dy/dy dystrophic mice. *Proc.* B **201**, 301–308 (1978).

Waxman, S.G. & Foster, R.E. Development of the axon membrane during differentiation of myelinated fibres in spinal nerve roots. *Proc.* B **209**, 441–446 (1980).

Waylen, P.C. On the degree of sharpness in solutions of Einstein's field equations. *Proc.* A **321**, 397–408 (1978).

Green's functions in the early universe. *Proc.* A **362**, 233–244 (1978). Gravitational waves in an expanding universe. *Proc.* A **362**, 245–250 (1978).

Wayne, R.P. *See* Giachardi & Wayne; Jones (I.T.N.) & Wayne.

Waynick, A.H. The early history of ionospheric investigations in the United States (Discussion). *Trans.* A **280**, 11–25 (1975).

Wdowczyk, J. Contribution from pulsars (Discussion). *Trans.* A **277**, 443–451 (1974).

Weatherall, D.J. & Clegg, J.B. The α-chain-termination mutants and their relation to the α-thalassae-mias. *Trans.* B **271**, 411–455 (1975).

Weatherley, Marie-Louise P.M. *See* Craxford & Weatherley.

Weatherley, P.E. Introduction: water movement through plants (Discussion).
Trans. B **273**, 435–444 (1976).

Weaver, S.D. *See* Norry, Truckle, Lippard, Hawkesworth, Weaver & Marriner.

Webb, A.C. The effects of changing levels of arousal on the spontaneous activity of cortical neurones.
I. Sleep and wakefulness. *Proc.* B **194**, 225–237 (1976).
II. Relaxation and alarm. *Proc.* B **194**, 239–251 (1976).
See also Burns (B. Delisle), Stean & Webb; Burns (B. Delisle) & Webb.

Webb, C.J. Systematics of the *Pomatoschistus minutus* complex (Teleostei: Gobioidei).
Trans. B **291**, 201–241 (1980).

Webb, G. *See* Robertson (J.) & Webb.

Webb, J.P. *See* Orme, Webb, Kelland & Sargent.

Webb, J.R.L. *See* Edmunds & Webb.

Webb, J.S. *See* Thornton (I.) & Webb.

Webb, J.S. & Howarth, R.J. Regional geochemical mapping (Discussion).
Trans. B **288**, 81–93 (1979).

Webb, K.R. *See* Carrington (A.), Hills & Webb.

Webb, W.L. *See* Thornton (I.) & Webb.

Webber, J.A. *See* Hatfield, Fisher, Dunigan, Burchfield and others.

Webster, F. *See* Fofonoff & Webster.

Webster, R.G. *See* Laver, Air, Webster, Gerhard, Ward & Dopheide.

Webster, R.G., Hinshaw, V.S., Bean, W.J. & Sriram, G. Influenza viruses: transmission between species (Discussion). *Trans.* B **288**, 439–447 (1980).

Weck, R. Review Lecture. The need for new welding processes. *Proc.* A **356**, 1–23 (1977).

Wedgwood, F.A. Electron-spin transfer due to covalent bonding in the $(CrF_6)^{3-}$ group of K_2NaCrF_6.
Proc. A **349**, 447–465 (1976).

Weeks, A.G. See Skempton & Weeks.

Weertman, J. Creep laws for the mantle of the Earth (Discussion). *Trans.* A **288**, 9–26 (1978).

Weese, G.M. *See* Beddard, Porter & Weese.

Wegner, Ch. *See* Franck, Rowold, Wegner & Eckert.

Weigel, W. *See* Goldflam, Hinz, Weigel & Wissmann.

Weightman, J.A. Geoid float techniques in satellite geodesy (Discussion).
Trans. A **294**, 299–305 (1980).

Weihs, D. A hydrodynamical analysis of fish turning manoeuvres. *Proc.* B **182**, 59–72 (1972).

Weiler, R. *See* Marchiafava & Weiler.

Weinbaum, G. *See* Burger, Turner, Kuhns & Weinbaum.

Weinbaum, S. *See* Leichtberg, Weinbaum, Pfeffer & Gluckman.

Weinberg, F.J. *See* Colver & Weinberg; Cox (J.B.), Jones & Weinberg; Jones (A.R.), Lloyd & Weinberg;
Harrison (A.J.) & Weinberg; Hong, Jones & Weinberg; Jones (A.R.), Schwar & Weinberg; Thong & Weinberg.

Weinberg, F.J. & Wilson, J.R. A preliminary investigation of the use of focused laser beams for minimum ignition energy studies. *Proc.* A **321**, 41–52 (1971).

Weinberg, F.J. & Wong, W.W.Y. A laser optical method for the direct measurement of normal propagation velocities of phase objects. *Proc.* A **345**, 379–385 (1975).

Weiner, J.S. Work capacity, thermal responses and lung function: United Kingdom studies in the I.B.P. (Discussion). *Trans.* B **274**, 457–472 (1976).

Weinhold, F. Electric polarizabilities of two-electron atoms by a lower-bound procedure.
Proc. A **327**, 209–227 (1972).

Weinstein, M. Wavy vortices in the flow between two long eccentric rotating cylinders.
I. Linear theory. *Proc.* A **354**, 441–457 (1977).
II. Nonlinear theory. *Proc.* A **354**, 459–489 (1977).

Weir, A.H. *See* Mitchell (G.F.), Catt, Weir, McMillan, Margerel & Whatley; West (R.G.), Dickson, Catt, Weir & Sparks.

Weir, G.J. & Kerr, R.P. Diverging type-D metrics. *Proc.* A 355, 31–52 (1977).

Weir, R.D. *See* Callanan, Weir & Staveley.

Weiskrantz, L. Review Lecture. Behavioural analysis of the monkey's visual nervous system. *Proc.* B 182, 427–455 (1972).

Weiss, H. Materials behaviour in low gravity environment (Discussion). *Proc.* A 361, 157–164 (1978).

Weissman, M.A. Nonlinear wave packets in the Kelvin—Helmholtz instability. *Trans.* A 290, 639–681 (1979).

Welander, P. The thermocline problem (Discussion). *Trans.* A 270, 415–421 (1971).

Welberry, T.R. Order and disorder in acenaphthylene. *Proc.* A 334, 19–48 (1973).
Solution of crystal growth disorder models by imposition of symmetry. *Proc.* A 353, 363–367 (1977).

Welberry, T.R., Miller, G.H. & Pickard, D.K. A 3-D crystal growth-disorder model with cubic symmetry. *Proc.* A 367, 175–192 (1979).

Wellburn, A.R. *See* Mansfield (T.A.), Wellburn & Moreira.

Wells, A.A. Keynote speech (Symposium). *Trans.* A 282, 64–68 (1976).
Strength as a design parameter (Symposium). *Trans.* A 282, 131–148 (1976).

Wells, A.C. *See* Chamberlain, Clough, Heard, Newton, Stott & Wells.

Wells, B. *See* La Cour & Wells.

Wells, G.P. Lancelot Thomas Hogben. *Biogr. Mem.* 24, 183–221 (1978).

Wells, J.D. Appendix to a paper by Ogston, Preston & Wells. *Proc.* A 333, 314–316 (1973).
Salt activity and osmotic pressure in connective tissue. I. A study of solutions of dextran sulphate as a model system. *Proc.* B 183, 399–419 (1973).
See also Ogston, Preston & Wells.

Wells, M.G. *See* King (D.A.) & Wells.

Wells, M.R. *See* Bleaney, Robinson & Wells; Bleaney & Wells.

Wells, P.N.T. Ultrasonics in medical diagnosis (Discussion). *Trans.* A 292, 187–199 (1979).

Welsh, H.L. *See* McKellar (A.R.W.) & Welsh.

Wenner, M.L. *See* Green (A.E.), Naghdi & Wenner.

Wenzel, K.-P. The scientific objectives of the International Solar Polar Mission (Discussion). *Trans.* A 297, 565–573 (1980).

West, E.A. *See* Horai, Winkler, Keihm, Langseth, Fountain & West.

West, J.B. & Marr, G.V. The absolute photoionization cross sections of helium, neon, argon and krypton in the extreme vacuum ultraviolet region of the spectrum. *Proc.* A 349, 397–421 (1976).

West, M.A. *See* Formosinho, Porter & West.

West, R.G. Relative land—sea-level changes in southeastern England during the Pleistocene (Discussion). *Trans.* A 272, 87–98 (1972).
Early and Middle Devensian flora and vegetation (Discussion). *Trans.* B 280, 229–246 (1977).
See also Funnell, Norton & West; Hibbert (F.A.), Switsur & West.

West, R.G., Dickson, Camilla A., Catt, J.A., Weir, A.H. & Sparks, B.W. Late Pleistocene deposits at Wretten, Norfolk. II. Devensian deposits (with appendixes by G.R. Coope and J.H. Dickson). *Trans.* B 267, 337–420 (1974).

West, R.G. & Norton, P.E.P. The Icenian Crag of southeast Suffolk. *Trans.* B 269, 1–28 (1974).

Westbrook, D.R. *See* Chee-Seng, Majumdar & Westbrook.

Westbrook, J.H. Problems with residual and additive elements and their control through specifications (Discussion). *Trans.* A 295, 25–43 (1980).

Westcott, J.H. Review Lecture. The application of microprocessors. *Proc.* A 367, 451–484 (1979).

Westcott, M. The random record model. *Proc.* A 356, 529–547 (1977).
See also Gates, O'Connor & Westcott.

Westgaard, R.H. *See* Lømo, Westgaard & Dahl.

Westheimer, F.H. *See* Kistiakowsky & Westheimer.

Westoll, T.S. Introductory remarks to a Discussion. *Trans.* B 260, 3–4 (1971).
See also Parrington & Westoll.

Westoll, T.S. & Parrington, F.R. Alfred Sherwood Romer. *Biogr. Mem.* **21**, 497–516 (1975).

Wetherell, A.M. Experiments on proton–proton interactions at the Institute of High-Energy Physics, Serpukhov, U.S.S.R. (Discussion). *Proc.* A **335**, 421–430 (1973).

von Wettstein, D. The assembly of the synaptinemal complex (Discussion). *Trans.* B **277**, 235–243 (1977).

Weymann, R.J. Uncondensed matter in the Universe: optical evidence from quasar absorption lines (Discussion). *Trans.* A **296**, 399–405 (1980).

Whalley, E. *See* Kell, McLaurin & Whalley.

Whalley, W.B. *See* Haworth & Whalley.

Whan, D.A. *See* Dowie, Kemball, Kempling & Whan.

Whangbo, Myung-Hwan, Hoffmann, Roald & Woodward, R.B. Conjugated one and two dimensional polymers. *Proc.* A **366**, 23–46 (1979).

Whatley, F.R. Recycling through higher plants (Discussion). *Proc.* B **179**, 193–200 (1971).
  See also Whatley (Jean M.), John & Whatley.

Whatley, Jean M., John, P. & Whatley, F.R. From extracellular to intracellular: the establishment of mitochondria and chloroplasts (Discussion). *Proc.* B **204**, 165–187 (1979).

Whatley, R.C. *See* Mitchell (G.F.), Catt, Weir, McMillan, Margerel & Whatley.

Wheeler, A.G. *See* Baird (D.T.), Land, Scaramuzzi & Wheeler.

Wheeler, M.W.L. *See* Ross (G.), Fiddy, Nieto-Vesperinas & Wheeler.

Wheeler, Sir Mortimer. King Gustaf VI Adolf of Sweden, K.G. *Biogr. Mem.* **20**, 213–216 (1974).

Whelan, M.J. *See* Saldin, Stathopoulos & Whelan; Saldin & Whelan.

Wherrett, B.S. *See* Dennis (R.B.), Pidgeon, Smith, Wherrett & Wood.

Whitaker, M.J. *See* Baker (P.F.), Knight & Whitaker.

White, A. *See* Edwards (R.N.), Law & White.

White, A.I. *See* Dwek, Jones, Marsh, McLaughlin, Press and others.

White, B. *See* Greeley, Iversen, Pollack, Udovich & White.

White, C.L., Clausing, R.E. & Heatherly, L. The effect of trace element additions on the grain boundary composition of Ir + 0.3 % W alloys [abstract] (Discussion). *Trans.* A **295**, 303 (1980).

White, D.J. Prospects for greater efficiency in the use of different energy sources (Discussion). *Trans.* B **281**, 261–275 (1977).

White, G.K. & Collins, J.G. The thermal expansion of alkali halides at low temperatures. II. Sodium, rubidium and caesium halides. *Proc.* A **333**, 237–259 (1973).

White, G.W.T. & Simmons, M.D. Analysis of complex systems (Discussion). *Trans.* A **287**, 405–423 (1977).

White, J.G., Southgate, Eileen, Thomson, J.N. & Brenner, S. The structure of the ventral nerve cord of *Caenorhabditis elegans. Trans.* B **275**, 327–348 (1976).

White, J.W. Neutron diffraction, inelastic scattering and the structure of amphiphiles and macro-molecules in solution (Discussion). *Proc.* A **345**, 119–144 (1975).
  Erratum. *Proc.* A **349**, 577 (1976).
  See also Bomchil, Hüller, Rayment, Roser, Smalley, Thomas & White; Leslie (M.), Jenkin, Hayter, White, Cox & Warner.

White, P.W. Finite-difference methods in numerical weather prediction (Discussion). *Proc.* A **323**, 285–292 (1971).

White, S. The effects of strain on the microstructures, fabrics, and deformation mechanisms in quartzites (Discussion). *Trans.* A **283**, 69–86 (1976). Corrigendum. *Trans.* A **284**, 621 (1977).

White, S.D.M. Tidal interactions and the merging of galaxies (Discussion). *Trans.* A **296**, 347–349 (1980).

White, Sir Frederick. Early work in Australia, New Zealand and at the Halley Stewart Laboratory, London (Discussion). *Trans.* A **280**, 35–46 (1975).
  Robert Gordon Menzies. *Biogr. Mem.* **25**, 445–476 (1979).

Whitehead, R.G. Some quantitative considerations of importance to the improvement of the nutritional status of rural children (Discussion). *Proc.* B **199**, 49–60 (1977).
  The better use of food resources for infants and mothers (Discussion). *Proc.* B **209**, 59–69 (1980).

Whitehead, Valerie L. *See* Woodruff (Sir Michael) & Whitehead.

Whitehouse, D.J. & Phillips, M.J. Discrete properties of random surfaces.
　　Trans. A **290**, 267—298 (1978).
Whitelaw, J.H. *See* Durão & Whitelaw; Durst & Whitelaw; Ribeiro & Whitelaw.
Whiteman, J.R. Finite-difference techniques for a harmonic mixed boundary problem having a
　　reentrant boundary (Discussion). *Proc.* A **323**, 271—276 (1971).
Whitfield, G.R. *See* Bramwell (Cherrie D.) & Whitfield.
Whitham, G.B. *See* Fornberg & Whitham; Jimenez (J.) & Whitham.
Whiting, J.S.S. *See* Ahmad, Prutton & Whiting.
Whitney, J. *See* Drummond (J.R.), Houghton, Peskett, Rodgers, Wale and others.
Whittaker, J.M. Dynamics with variable masses. *Proc.* A **372**, 485—487 (1980).
Whittam, R. *See* Chipperfield & Whittam; Hallam (C.) & Whittam.
Whittam, R., Hallam, C. & Wattam, D.G. Observations on ouabain binding and membrane phos-
　　phorylation by the sodium pump. *Proc.* B **193**, 217—234 (1976).
Whittembury, G. *See* Giebisch, Boulpaep & Whittembury.
Whittembury, G., de Martínez, Clara Verde, Linares, H. & Paz-Aliaga, A. Solvent drag of large solutes
　　indicates paracellular water flow in leaky epithelia. *Proc.* B **211**, 63—81 (1980).
Whitteridge, D. *See* Clarke (P.G.H.), Ramachandran & Whitteridge; Donaldson (I.M.L.) & Whitteridge.
Whitteridge, Gweneth. The Wilkins Lecture, 1979. Of the local movement of animals.
　　Proc. B **206**, 1—13 (1979).
Whittingham, C.P. The potentialities of chloroplast fragments (Discussion).
　　Proc. B **179**, 237—246 (1971).
Whittington, H.B. The enigmatic animal *Opabinia regalis*, Middle Cambrian, Burgess Shale, British
　　Columbia. *Trans.* B **271**, 1—43 (1975).
　　The Middle Cambrian trilobite *Naraoia*, Burgess Shale, British Columbia.
　　Trans. B **280**, 409—443 (1977).
　　The lobopod animal *Aysheaia pedunculata* Walcott, Middle Cambrian, Burgess Shale, British
　　Columbia. *Trans.* B **284**, 165—197 (1978). Corrigendum. *Trans.* B **285**, 408 (1979).
Whittington, H.B. & Hughes, C.P. Ordovician geography and faunal provinces deduced from trilobite
　　distribution. *Trans.* B **263**, 235—278 (1972).
Whittle, D.P. *See* Bastow (B.D.), Whittle & Wood.
Whittle, D.P. & Stringer, J. Improvements in high temperature oxidation resistance by additions of
　　reactive elements or oxide dispersions (Discussion). *Trans.* A **295**, 309—329 (1980).
Whittle, M.J. *See* Coffey (J.M.), Oates & Whittle.
Whittle, P.D.J., Clarke, Sir Cyril, Sheppard, P.M. & Bishop, J.A. Further studies on the industrial
　　melanic moth *Biston betularia* (L.) in the northwest of the British Isles.
　　Proc. B **194**, 467—480 (1976).
Whitton, B.A. Terrestrial and freshwater algae of Aldabra (Discussion).
　　Trans. B **260**, 249—255 (1971).
　　See also Potts (M.) & Whitton.
Wibberley, G. Land use and rural planning (Discussion). *Trans.* B **271**, 213—218 (1975).
Wick, U. *See* Gerisch, Hülser, Malchow & Wick.
Wickens, G.E. Speculations on seed dispersal and the flora of the Aldabra archipelago (Discussion).
　　Trans. B **286**, 85—97 (1979).
Wickus, G.G. *See* Strominger, Blumberg, Suginaka, Umbreit & Wickus.
Widdel, H.U. *See* Rose (G.), Widdel, Azcárraga & Sanchez.
Widdows, J. *See* Bayne, Moore, Widdows, Livingstone & Salkeld.
Widdowson, Elsie M. *See* McCance, El Neil, El Din, Widdowson, Southgate and others; McCance &
　　Widdowson.
Wider de Xifra, E.A., Sandy, J.D., Davies, R.C. & Neuberger, A. Control of 5-aminolaevulinate
　　synthetase activity in *Rhodopseudomonas spheroides* (Discussion). *Trans.* B **273**, 79—98 (1976).
Widlund, O.B. Some recent applications of asymptotic error expansions to finite-difference schemes
　　(Discussion). *Proc.* A **323**, 167—177 (1971).
Widnall, Sheila E. & Sullivan, J.P. On the stability of vortex rings. *Proc.* A **332**, 335—353 (1973).
Widnall, Sheila E. & Tsai, Chon-Yin. The instability of the thin vortex ring of constant vorticity.
　　Trans. A **287**, 273—305 (1977).

Wiebkin, O.W. & Muir, Helen. The effect of hyaluronic acid on proteoglycan synthesis and secretion by chondrocytes of adult cartilage (Discussion). *Trans.* B **271**, 283—291 (1975).

de Wied, D. Behavioural actions of neurophypophysial peptides (Discussion).
Proc. B **210**, 183—194 (1980).

Wiesel, T.N. *See* Hubel & Wiesel; Hubel, Wiesel & LeVay.

Wiesenfeld, J.M. *See* Ippen, Shank, Wiesenfeld & Migus.

Wieslander, L. *See* Daneholt, Case, Lamb, Nelson & Wieslander.

Wiffen, R.D. *See* Chamberlain, Heard, Little & Wiffen.

Wigglesworth, Sir Vincent. Boris Petrovitch Uvarov. *Biogr. Mem.* **17**, 713—740 (1971).

Wigner, E.P. Restriction of irreducible representations of groups to a subgroup.
Proc. A **322**, 181—189 (1971).

Wigner, E.P. & Hodgkin, R.A. Michael Polanyi. *Biogr. Mem.* **23**, 413—448 (1977).

Wild, A.E. Role of the cell surface in selection during transport of proteins from mother to foetus and newly born (Discussion). *Trans.* B **271**, 395—410 (1975).

Wiley, D.C. *See* Skehel, Waterfield, McCauley, Elder & Wiley.

Wilkes, M.V. Appendix to a paper by W.G.J. Beynon. *Trans.* A **280**, 54—55 (1976).

Wilkie, D.R. *See* Dawson (M. Joan) & Wilkie; Kretzschmar & Wilkie.

Wilkie, N.M., Eglin, R.P., Sanders, P.G. & Clements, J.R. The association of herpes simplex virus with squamous carcinoma of the cervix, and studies of the virus thymidine kinase gene (Discussion). *Proc.* B **210**, 411—421 (1980).

Wilkins, G.A. An introductory review of ephemerides for lunar laser ranging (Discussion). *Trans.* A **284**, 461—466 (1977).

Wilkins, G.A. & Sinclair, A.T. The dynamics of the planets and their satellites (Symposium).
Proc. A **336**, 85—104 (1974).

Wilkins, M.B. Light- and gravity-sensing guidance systems in plants (Discussion).
Proc. B **199**, 513—524 (1977).

Wilkins, S.W. X-ray Bragg reflexion from perfect crystals — dependence on geometrical factors with particular regard to the zero-level-of-interaction limit. *Proc.* A **364**, 569—589 (1978).

Wilkinson, F. *See* Ellison, Salmon & Wilkinson.

Wilkinson, G.R. *See* Mead & Wilkinson.

Wilkinson, J.H. Introductory remarks to a Discussion. *Proc.* A **323**, 153 (1971).

Wilkinson, J.P.T. *See* Burrows (J.P.), Cliff, Harris, Thrush & Wilkinson.

Wilkinson, P. *See* Rees (H.), Shaw & Wilkinson.

Wilkinson, R. *See* Huxley (C.R.) & Wilkinson.

Wilkinson, R.T. *See* McCance, El Neil, El Din, Widdowson, Southgate and others.

Wilkinson, S. *See* Beddell, Clark, Hardy, Lowe, Ubatuba, Vane & Wilkinson.

Will, C.M. Experimental tests of General Relativity [abstract] (Discussion).
Proc. A **368**, 5—8 (1979).

Willetts, D.V. *See* Butcher, Willetts & Jones.

Williams, A. The secretion and structural evolution of the shell of thecideidine brachiopods.
Trans. B **264**, 439—478 (1973).
William John Pugh. *Biogr. Mem.* **21**, 485—495 (1975).
See also Daniels (A.), Korda, Tanswell, Williams & Williams; Tavener-Smith & Williams.

Williams, A. & Mackay, S. Secretion and ultrastructure of the periostracum of some terebratulide brachiopods. *Proc.* B **202**, 191—209 (1978).

Williams, A. & Hewitt, R.A. The delthyrial covers of some living brachiopods.
Proc. B **197**, 105—129 (1977).

Williams, A.D. *See* Cullen (A.L.), Nagenthiram & Williams.

Williams, B.C. *See* McDougall, Williams, Hyatt, Bell, Tait & Tait.

Williams, C.A. *See* Sibuet, Ryan, Arthur, Barnes, Blechsmidt and others.

Williams, D.E. *See* Mackie (G.O.), Paul, Singla, Sleigh & Williams.

Williams, D.H. Potential energy profiles for unimolecular reactions of organic ions (Discussion).
Trans. A **293**, 117—124 (1979).

Williams, D.P. *See* Chellone & Williams.

Williams, E.R. *See* Dickinson (P.H.G.), Bain, Thomas, Williams, Jenkins & Twiddy.

Williams, E.W. *See* Hemmings (W.A.) & Williams.

Williams, E.W. & Hemmings, W.A. Intestinal uptake and transport of proteins in the adult rat.
Proc. B **203**, 177–189 (1978).

Williams, G.C. The question of adaptive sex ratio in outcrossed vertebrates (Discussion).
Proc. B **205**, 567–580 (1979).

Williams, J.E.Ffowcs. *See* Dowling (A.P.), Williams & Goldstein; Howe & Williams.

Williams, J.G. Lunar ranging experiment ephemerides and the reduction of observations [abstract]
(Discussion). *Trans.* A **284**, 467 (1977).

Results from lunar laser ranging [abstract] (Discussion). *Trans.* A **284**, 587 (1977).

Williams, J.G. & Marshall, G.P. Environmental crack and craze growth phenomena in polymers.
Proc. A **342**, 55–77 (1975).

Williams, J.O. *See* Cohen (M.D.), Ludmer, Thomas & Williams; Thomas (J.M.), Evans & Williams.

Williams, J.T. *See* Katayama, North & Williams.

Williams, J.W. *See* Ashkenazi, Crane, Williams & Dean; Ashkenazi, Sykes, Gough & Williams.

Williams, M.B. International standards for telecommunications (Discussion).
Trans. A **289**, 185–205 (1978).

Williams, Melissa. *See* Hoyle (G.) & Williams.

Williams, M.M.R. A stochastic theory of particle transport. *Proc.* A **358**, 105–120 (1977).

Williams, Monica J. *See* Pau, Brunet & Williams.

Williams, P.G. *See* Michael & Williams.

Williams, P.R. *See* Beaven (P.A.), Miller, Williams, Delargy & Smith.

Williams, R.B.G. *See* Sparks, Williams & Bell.

Williams, R.J.P. Review Lecture. Energy states of proteins. enzymes and membranes.
Proc. B **200**, 353–389 (1978).

Introductory remarks to a Discussion. *Trans.* B **289**, 381–394 (1980).

See also Barry (C.D.), Hill, Sadler & Williams; Campbell (I.D.), Dobson & Williams; Daniels (A.),
Korda, Tanswell, Williams & Williams; Hensens, Hill, Thornton, Turner & Williams; Levine (B.A.)
& Williams.

Williams, R.T. *See* James (Margaret O.), Smith, Williams & Reidenberg.

Williams, T.A. *See* Potts (A.W.), Williams & Price.

Williamson, D.T.N. Review Lecture. The anachronistic factory. *Proc.* A **331**, 139–160 (1972).

Williamson, E.J. *See* Drummond (J.R.), Houghton, Peskett, Rodgers, Wale and others; Ellis (P.),
Holah, Houghton, Jones, Peckham and others.

Williamson, I.P. The broadening of pulses due to multipath propagation of radiation.
Proc. A **342**, 131–147 (1975).

Williamson, J.B.P. *See* Pullen & Williamson.

Williamson, J.B.P. & Hunt, R.T. Asperity persistence and the real area of contact between rough
surfaces. *Proc.* A **327**, 147–157 (1972).

Williamson, R.M. & Roberts, B.L. The timing of motoneuronal activity in the swimming spinal dogfish. *Proc.* B **211**, 119–133 (1980).

Willis, A.J. *See* Carnochan, Dworetsky, Todd, Willis & Wilson.

Willis, F.R. *See* Briscoe, Scruton & Willis.

Willis, J.R. Self-similar problems in elastodynamics. *Trans.* A **274**, 435–491 (1973).

See also Talbot (D.R.S.) & Willis.

Willis, J.R.; Hayns, M.R. & Bullough, R. The dislocation void interaction.
Proc. A **329**, 121–136 (1972).

Willis, M.R. *See* Ashwell (G.J.), Eley, Wallwork & Willis.

Willmer, E.N. Emmanuel Fauré-Fremiet. *Biogr. Mem.* **18**, 187–221 (1972).

Willmitzer, L. *See* Van Montagu, Holsters, Zambryski, Hernalsteens, Depicker and others.

Willmore, A.P. Measurement of the positions of X-ray sources (Discussion).
Proc. A **340**, 439–446 (1974).

Transient X-ray sources (Discussion). *Proc.* A **350**, 463–479 (1976).

Introductory remarks to a Discussion. *Proc.* A **366**, 279–280 (1979).

Willmore, P.L. *See* Crampin (S.) & Willmore.

Willshaw, D.J. A simple network capable of inductive generalization. *Proc.* B **182**, 233—247 (1972).
See also Prestige & Willshaw.

Willshaw, D.J. & von der Malsburg, C. How patterned neural connections can be set up by self-organization. *Proc.* B **194**, 431—445 (1976).
A marker induction mechanism for the establishment of ordered neural mappings: its application to the retinotectal problem. *Trans.* B **287**, 203—243 (1979).

Wilson, A.L. Trace metals in waters (Discussion). *Trans.* B **288**, 25—39 (1979).

Wilson, A.M. Positions of galactic X-ray sources (Discussion). *Proc.* A **366**, 367—373 (1979).

Wilson, A.S. X-ray galaxies (Discussion). *Proc.* A **366**, 461—489 (1979).

Wilson, C.L. Review Lecture. World energy prospects to the year 2000.
Proc. A **358**, 121—136 (1977).

Wilson, H.J. *See* Steward, Israel, Mott, Wilson & Krikorian.

Wilson, I.D.L. *See* Gerratt & Wilson.

Wilson, J.E., Stott, F.H. & Wood, G.C. The development of wear-protective oxides and their influence on sliding friction. *Proc.* A **369**, 557—574 (1980).

Wilson, J.F. The Rhodesian Archaean craton — an essay in cratonic evolution (Discussion).
Trans. A **273**, 389—411 (1973).

Wilson, J.F., Goos, H.J.T. & Dodd, J.M. An investigation of the neural mechanisms controlling the colour change responses of the dogfish, *Scyliorhinus canicula* L. by mesencephalic and diencephalic lesions. *Proc.* B **187**, 171—190 (1974).

Wilson, J.M. *See* Kennett, Lee & Wilson.

Wilson, Joan F. *See* Pigott (C.D.) & Wilson.

Wilson, J.R. *See* Weinberg & Wilson.

Wilson, J. Warren. The position of regenerating cambia: auxin/sucrose ratio and the gradient induction hypothesis. *Proc.* B **203**, 153—176 (1978).

Wilson, K.W. The laboratory estimation of the biological effects of organic pollutants (Discussion).
Proc. B **189**, 459—477 (1975).

Wilson, M. *See* Horridge, Giddings & Wilson.

Wilson, M.R. *See* Adamek & Wilson.

Wilson, Penelope A. *See* Daniel, Pratt & Wilson.

Wilson, P.N. Livestock physiology and nutrition (Discussion). *Trans.* B **267**, 101—112 (1973).

Wilson, R. *See* Bates (B.), Bradley, McBride and others; Boland (B.C.), Jones, Wilson, Engstrom & Noci; Burton (W.M.), Jordan, Ridgeley & Wilson; Carnochan, Dworetsky, Todd, Willis & Wilson; Evans (R.G.), Nandy & Wilson; Gondhalekar & Wilson; McWhirter & Wilson; Nandy, Thompson, Jamar, Monfils & Wilson.

Wilson, R.L. *See* Tanguy & Wilson.

Wilson, S., Silver, D.M. & Farrell, R.A. Special invariance properties of the $[N+1/N]$ Padé approximants in Rayleigh—Schrödinger perturbation theory. *Proc.* A **356**, 363—374 (1977).

Wilson, Sir Alan. Solid state physics 1925—33: opportunities missed and opportunities seized (Symposium). *Proc.* A **371**, 39—48 (1980).

Wilson, Sir Graham. *See* Gladstone, Knight & Wilson.

Wilson, Sir Graham & Zinnemann, K.S. James Walter McLeod. *Biogr. Mem.* **25**, 421—444 (1979).

Wilson, T. *See* Sheppard (C.J.R.) & Wilson.

Wilton, D.C. *See* Akhtar, Wilton, Watkinson & Rahimtula.

Wind, R.A. *See* Bovée, Creyghton, Getreuer, Korbee, Lobregt and others.

Windley, B.F. Crustal development in the Precambrian (Discussion). *Trans.* A **273**, 321—341 (1973).
Appendix to a paper by P.C. Sylvester-Bradley (Discussion). *Proc.* B **189**, 230 (1975).
See also Bridgwater, Watson & Windley.

Windsor, M.W. & Holten, D. Picosecond studies of primary charge separation in bacterial photsynthesis (Discussion). *Trans.* A **298**, 335—349 (1980).

Winfield, D.A., Hiorns, R.W. & Powell, T.P.S. A quantitative electron-microscopical study of the postnatal development of the lateral geniculate nucleus in normal kittens and in kittens with eyelid suture. *Proc.* B **210**, 211—234 (1980).

Winfield, D.A. & Powell, T.P.S. An electron-microscopical study of the postnatal development of the lateral geniculate nucleus in the normal kitten and after eyelid suture.
Proc. B **210**, 197–210 (1980).

Winkler, H. The membrane of the chromaffin granule (Discussion). *Trans.* B **261**, 293–303 (1971).

Winkler, J.L., Jr. *See* Horai, Winkler, Keihm, Langseth, Fountain & West.

Winter, R.E. *See* Hutchings, Winter & Field.

Winter, R.E. & Field, J.E. The role of localized plastic flow in the impact initiation of explosives.
Proc. A **343**, 399–413 (1975).

Winterer, E.L. Sedimentary facies on the rises and slopes of passive continental margins (Discussion).
Trans. A **294**, 169–176 (1980).

Wise, A.F.E. Effects due to groups of buildings (Discussion). *Trans.* A **269**, 469–488 (1971).

Wise, B. *See* Fielder, Fryer, Titulaer, Herring & Wise.

Wise, M.B. & Trainor, L.E.H. Group theory and many-body diagrams. III. Many-particle Green diagrams. *Proc.* A **359**, 111–119 (1978).

Wissmann, G. *See* Goldflam, Hinz, Weigel & Wissmann.

Withbroe, G.L. *See* Reeves (E.M.), Vernazza & Withbroe.

Witkovsky, P. *See* Roberts (B.L.) & Witkovsky.

Witkovsky, P. & Roberts, B.L. The light microscopical structure of the mesencephalic nucleus of the fifth nerve in the selachian brain. *Proc.* B **190**, 457–471 (1975).

Witz, J. *See* Jacrot, Pfeiffer & Witz; Jonard, Briand, Bouley, Witz & Hirth.

Wixson, B.G. Geochemistry and pollution (Discussion). *Trans.* B **288**, 179–184 (1979).

Wofsy, S.C. *See* Logan (Jennifer A.), Prather, Wofsy & McElroy; McElroy, Wofsy & Yung.

Wohn, J. *See* Pearlman, Lanham, Lehr & Wohn.

Wolde-Gabriel, Y. *See* Miller (D.S.), Baker, Bowden, Evans, Holt and others.

Wolde-Gabriel, Z. *See* Miller (D.S.), Baker, Bowden, Evans, Holt and others.

Wolfe, H.G. *See* Grüneberg, Cattanach, McLaren, Wolfe & Bowman.

Wolfendale, A.W. Explanations of the spectral shape in the energy range 10^{14}–10^{20} eV (Discussion).
Trans. A **277**, 429–442 (1974).
See also Krishnaswamy (M.R.), Menon, Narasimham, Hinotani, Ito and others; Rochester & Wolfendale.

Wolff, H.S. Biological experiments in *Spacelab* (Discussion). *Proc.* B **199**, 479–483 (1978).

Wolstenholme, A.J. *See* Mahy, Barrett, Briedis, Brownson & Wolstenholme.

Wong, W.H. & Burns, G. Dynamics of collisional dissociation: I_2 in Ar and Xe.
Proc. A **341**, 105–119 (1974).

Wong, W.W.Y. *See* Weinberg & Wong.

Wong-Ng, W. *See* Nyburg & Wong-Ng.

Wood, A.D. *See* Paris & Wood.

Wood, B.J. The activities of components in clinopyroxene and garnet solid solutions and their application to rocks (Discussion). *Trans.* A **286**, 331–342 (1977).
See also Carmichael, Nicholls, Spera, Wood & Nelson.

Wood, C. *See* Hemmings (C.), Hemmings, Patey & Wood.

Wood, C.D. & Perry, G.E. The Russian satellite navigation system (Discussion).
Trans. A **294**, 307–315 (1980).

Wood, D.A. *See* Tarney, Wood, Saunders, Cann & Varet.

Wood, D.M. *See* Allen (R.V.), Wood & Mortensen.

Wood, D.R. *See* Morgan (D.V.) & Wood.

Wood, D.S. Patterns and magnitudes of natural strain in rocks (Discussion).
Trans. A **274**, 373–382 (1973).
See also Donath & Wood.

Wood, D.S., Oertel, G., Singh, J. & Bennett, H.F. Strain and anisotropy in rocks (Discussion).
Trans. A **283**, 27–42 (1976).

Wood, G.C. *See* Bastow (B.D.), Whittle & Wood; Wilson (J.E.), Stott & Wood.

Wood, I. *See* Steinberg & Wood; Steinberg, Wood & Hogan.

Wood, M.D., Allen, R.V. & Allen, S.S. Methods for prediction and evaluation of tidal tilt data from

borehole and observatory sites near active faults (Discussion). *Trans.* A **274**, 245–252 (1973).

Wood, M.H. *See* Clement & Wood; Guest (M.F.), Hillier, Saunders & Wood; Hayns & Wood; Little (E.A.), Bullough & Wood.

Wood, P.C. *See* Preston (A.) & Wood.

Wood, P.H.N. Arthritis calling for surgical treatment – aetiology and epidemiology (Discussion). *Proc.* B **192**, 131–143 (1976).

Wood, R.A. *See* Dennis (R.B.), Pidgeon, Smith, Wherrett & Wood.

Wood, R.K.S. Introduction: disease resistance in plants (Discussion). *Proc.* B **181**, 213–232 (1972).

Wood, W.W. Inertial oscillations in a rigid axisymmetric container. *Proc.* A **358**, 17–30 (1977).

Wood, W.W. & Head, A.K. The motion of dislocations. *Proc.* A **336**, 191–209 (1974).

Woodcock, L.V. Some quantitative aspects of ionic melt microstructure. *Proc.* A **328**, 83–95 (1972). Corresponding states theory for the fusion of ionic crystals. *Proc.* A **348**, 187–202 (1976).

Woodell, S.R.J. The rôle of unspecialized pollinators in the reproductive success of Aldabran plants (Discussion). *Trans.* B **286**, 99–108 (1979).

Woodgate, B.E., Knight, D.E., Uribe, R., Sheather, P., Bowles, J. & Nettleship, R. Extreme ultraviolet line intensities from the Sun. *Proc.* A **332**, 291–309 (1973).

Woodgate, G.K. *See* Angel, Sandars & Woodgate; Baird (P.E.G.), Brambley, Burnett, Stacey, Warrington & Woodgate.

Woodhead, Bridget G. *See* Mourant, Godber, Kopeć, Tills & Woodhead.

Woodhead, D.S. The biological effects of radioactive waste. (Discussion). *Proc.* B **177**, 423–437 (1971).

Woodhead-Galloway, J. Structure of the collagen fibril: some variations on a theme of tetragonally packed dimers. *Proc.* B **209**, 275–297 (1980).
See also Doyle (B.B.), Hulmes, Miller, Parry, Piez & Woodhead-Galloway.

Woodhead-Galloway, J. & Knight, D.P. Some observations on the fine structure of elastoidin. *Proc.* B **195**, 355–364 (1977).

Woodhouse, M.A. *See* Raymont, Krishnaswamy, Woodhouse & Griffin.

Woodhull, Ann M. *See* Hille, Woodhull & Shapiro.

Wooding, F.B.P. *See* Tucker (E.M.), Ellory, Wooding, Morgan & Herbert.

Woodruff, Pamela R. & Marr, G.V. The photoelectron spectrum of N_2, and partial cross sections as a function of photon energy from 16 to 40 eV. *Proc.* A **358**, 87–103 (1977).

Woodruff, Sir Michael. Review Lecture. Cancer – the elusive enemy. *Proc.* B **183**, 87–104 (1973).

Woodruff, Sir Michael, Dunbar, Noreen & Ghaffar, A. The growth of tumours in T-cell deprived mice and their response to treatment with *Corynebacterium parvum*. *Proc.* B **184**, 97–102 (1973).

Woodruff, Sir Michael & Speedy, Gillian. Inhibition of chemical carcinogenesis by *Corynebacterium parvum*. *Proc.* B **201**, 209–215 (1978).

Woodruff, Sir Michael & Whitehead, Valerie L. Mechanism of inhibition of immunization with irradiated tumour cells by a large dose of *Corynebacterium parvum*. *Proc.* B **197**, 505–514 (1977).

Woods, P.T. *See* Blaney, Bradley, Edwards, Jolliffe and others.

Woodward, A.C. *See* Johnson (K.L.), O'Connor & Woodward.

Woodward, Patricia M. *See* Fox (R.H.), Even-Paz, Woodward & Jack; Fox (R.H.), Budd, Woodward, Hackett & Hendrie.

Woodward, R.B. Penems and related substances (Discussion). *Trans.* B **289**, 239–250 (1980).
See also Whangbo, Hoffmann & Woodward.

Wooley, J.C. *See* Richards (B.M.), Pardon, Lilley, Cotter *et al.*

Woolfson, M.M. Star formation in a galactic cluster. *Trans.* A **291**, 219–252 (1979).
See also Dormand & Woolfson.

Woolhouse, G.R. & Ipohorski, M. On the interaction between radiation damage and coherent precipitates. *Proc.* A **324**, 415–431 (1971).

Woollacott, M. & Hoyle, G. Neural events underlying learning in insects: changes in pacemaker. *Proc.* B **195**, 395–415 (1977).

Woolley, M.L. On the isometric motion of a charged fluid in the Einstein–Maxwell theory. *Proc.* A **336**, 273–284 (1974).

A tetrad approach to charged fluid motion in the Einstein—Maxwell theory.
Proc. A **342**, 79–91 (1975).

Woolley, R.G. Molecular quantum electrodynamics. *Proc.* A **321**, 557–572 (1971).
See also Boyd (P.D.W.), Gerloch, Harding & Woolley.

Woolley, Sir Richard. Charles Rundle Davidson. *Biogr. Mem.* **17**, 193–194 (1971).
See also Griffin (R.F.) & Woolley.

Worcel, A. & Benyajati, C. Molecular architecture of *Drosophila* chromatin (Discussion).
Trans. B **283**, 407–409 (1978).

Worcester, D.L. *See* Richards (B.M.), Pardon, Lilley, Cotter *et al.*

Wrigglesworth, J.M. Electrostatic interactions at the plasma membrane (Discussion).
Trans. B **271**, 273–275 (1975).

Wright, C.A. *Bulinus* on Aldabra and the subfamily Bulininae in the Indian Ocean area (Discussion).
Trans. B **260**, 299–313 (1971).

Wright, C.E. *See* Hamlin & Wright.

Wright, C.J. *See* Mason (D.P.), McIlroy & Wright.

Wright, E. *See* Petch & Wright.

Wright, F.J. *See* Berry (M.V.), Nye & Wright.

Wright, M. *See* Osborne & Wright.

Wright, P.G. *See* Denton, Gilpin-Brown & Wright.

Wu, T.Y. *See* Chwang & Wu.

Wunsch, L. *See* Metz, Howard, Wunsch, Neusser & Schlag.

Wurtz, M. *See* Hohn (T.), Wurtz & Hohn.

Wuytack, F. *See* van Breemen, Farinas, Casteels, Gerba, Wuytack & Deth.

Wyatt, A.F.G. *See* King (P.J.) & Wyatt.

Wyatt, F. *See* Berger & Wyatt.

Wyatt, G.H. *See* Smith (D.E.), Kolenkiewicz, Wyatt, Dunn & Torrence.

Wyatt, R.J. *See* Chandler (R.J.), Kellaway, Skempton & Wyatt.

Wybourne, B.G. *See* Haskell & Wybourne.

Wyllie, A.H. *See* Gurdon, Wyllie & De Robertis.

Wymer, P.E.O. & Thake, Brenda. The importance of phosphorus in microalgal growth and species composition in mixed populations: experiments and simulations. *Proc.* B **209**, 333–353 (1980).

Wyss, M. Derivation of rupture area and stress-drop from body wave displacement spectra and the relative material strength in deep seismic zones (Discussion). *Trans.* A **274**, 361–368 (1973).

Xanthopoulos, B.C. A technique for generating solutions of Einstein's equation.
Proc. A **365**, 381–394 (1979).
See also Chandrasekhar (S.) & Xanthopoulos.

Yagoda, H.P. *See* Boley & Yagoda.

Yajima, S. Development of ceramics, especially silicon carbide fibres, from organosilicon polymers by heat treatment (Discussion). *Trans.* A **294**, 419–426 (1980).

Yale, B. *See* Proctor & Yale.

Yallop, B.D. *See* Murray (C.A.) & Yallop.

Yamaguchi, S., Kobayashi, H., Matsumiya, T. & Hayami, S. The effect of minor elements on the hot-workability of nickel-based superalloys [abstract] (Discussion). *Trans.* A **295**, 122 (1980).

Yamanaka, C. High-power neodymium glass laser systems for fusion research (Discussion).
Trans. A **298**, 393–405 (1980).

Yano, T. *See* Araki, Yano, Ueda & Noda.

Yaroshevsky, A.A. *See* Vinogradov, Yaroshevsky & Ilyin.

Yarwood, G. *See* O'Hara & Yarwood.

Yates, F. Appendix to the Biographical Memoir of William Ogilvy Kermack.
Biogr. Mem. **17**, 416–420 (1971).

Yates, J.T., Jr. *See* King (D.A.), Goymour & Yates.

Yau, A.W. & Pritchard, H.O. Ionization in a dense hydrogen plasma: analytic solution of the master equation. *Proc.* A **362**, 113—127 (1978).

Yeager, Carol. *See* Faulk (W. Page), Yeager, McIntyre & Ueda.

Yeo, D. The choice and management of research (Discussion). *Trans.* B **287**, 471—473 (1979).

Yocum, R.R. *See* Waxman (D.J.), Yocum & Strominger.

Yoneda, M. *See* Morita (K.), Nomura, Numata, Ochiai & Yoneda.

Yonge, C.M. Functional morphology with particular reference to hinge and ligament in *Spondylus* and *Plicatula* and a discussion on relations within the superfamily Pectinacea (Mollusca: Bivalvia). *Trans.* B **267**, 173—208 (1973).

Form and evolution in the Anomiacea (Mollusca:Bivalvia) — *Pododesmus, Anomia, Patro, Enigmonia* (Anomiidae): *Placunanomia, Placuna* (Placunidae Fam. Nov.) *Trans.* B **276**, 453—526 (1977).

Significance of the ligament in the classification of the Bivalvia. *Proc.* B **202**, 231—248 (1978).

Introductory remarks to a Discussion. *Trans.* A **291**, 3—4 (1978).

Introductory remarks to a Discussion. *Trans.* B **284**, 1—2 (1978).

Introductory remarks to a Discussion. *Trans.* B **284**, 201 (1978).

Significance of the ligament in the classification of the Bivalvia [abstract] (Discussion). *Trans.* B **284**, 375 (1978).

See also Yonge, Sir Maurice.

Yonge, Sir Maurice. Preface to a Discussion. *Trans.* B **272**, 268 (1975).

See also Yonge, C.M.

York, D.G. Physical state of interstellar atoms (Discussion). *Proc.* A **340**, 447—455 (1974).

See also Gull, York, Snow & Henize.

Yoshida, M. *See* Baldwin (J.E.), Jung, Singh, Wan, Haber and others.

Yoshida, T. Structural requirements for antibacterial activity and β-lactamase stability of 7β-arylmalonylamino-7α-methoxy-1-oxacephems (Discussion). *Trans.* B **289**, 231—237 (1980).

Yoshimura, T. *See* Hunt (G.W.), Reay & Yoshimura.

Young, A.D., Pankhurst, R.C. & Schultz, D.L. Douglas William Holder. *Biogr. Mem.* **24**, 223—244 (1978).

Young, B.R. *See* Sabine.

Young, C.A.J. The chemical and petrochemical industries (Discussion). *Trans.* A **275**, 329—356 (1973).

Young, Catherine M. *See* Cartwright (D.E.) & Young.

Young, C.Y. *See* Benedek (G.B.), Clark, Serrallach, Young and others.

Young, D.H. *See* Mansfield (E.H.) & Young.

Young, J. *See* Palese, Racaniello, Desselberger, Young & Baez.

Young, J.D., Jones, Siân E.M. & Ellory, J.C. Amino acid transport in human and in sheep erythrocytes. *Proc.* B **209**, 355—375 (1980).

Young, J.G. Why we need non-destructive testing of welded constructions (Discussion). *Trans.* A **292**, 201—206 (1979).

Young, J.Z. The organization of a cephalopod ganglion. *Trans.* B **263**, 409—429 (1972).

The central nervous system of *Loligo*. I. The optic lobe. *Trans.* B **267**, 263—302 (1974).

The nervous system of *Loligo*.

II. Suboesophageal centres. *Trans.* B **274**, 101—167 (1976).

III. Higher motor centres: the basal supraoesophageal lobes. *Trans.* B **276**, 351—398 (1977).

Concluding remarks to a Discussion. *Trans.* B **278**, 435—436 (1977).

The nervous system of *Loligo*. V. The vertical lobe complex. *Trans.* B **285**, 311—354 (1979).

See also Sanders (G.D.) & Young.

Young, L. *See* Cornish & Young.

Young, N.A. *See* Bates (D.R.), Malaviya & Young.

Young, R.E. *See* Schubert, Young & Cassen.

Young, Sir Frank & Foglia, V.G. Bernardo Alberto Houssay. *Biogr. Mem.* **20**, 247—270 (1974).

Yu, J. *See* McMahon (C.J.) & Yu.

Yu, P.K. *See* Cullen (A.L.) & Yu.

Yudkin, J.S. Drug control: a prerequisite for health (Discussion). *Proc.* B **209**, 159—163 (1980).

Yung, Y.L. *See* McElroy, Wofsy & Yung.

Zachau, H.G. *See* Steinmetz, Streeck & Zachau.

Zagalsky, P.F. & Herring, P.J. Studies of the blue astaxanthin-proteins of *Velella velella* (Coelenterata: Chondrophora). *Trans.* B **279**, 289—326 (1977).

Zakarian, S., Streilein, J.W. & Billingham, R.E. Studies on transplantation antigen extracts in Syrian hamsters. *Proc.* B **180**, 1—20 (1972).

Zala, C. *See* Jones (M.N.) & Zala.

Zaman, Z. *See* Akhtar, Abboud, Barnard, Jordan & Zaman.

Zambryski, P. *See* Van Montagu, Holsters, Zambryski, Hernalsteens, Depicker and others.

Zamorano, N.A. *See* Matzner & Zamorano.

Zannoni, C. *See* Luckhurst & Zannoni.

Zarnecki, J.C. Some recent observations of supernova remnants emitting X-rays (Discussion). *Proc.* A **366**, 311—327 (1979).

van Zeist, W. On macroscopic traces of food plants in southwestern Asia (with some reference to pollen data) (Discussion). *Trans.* B **275**, 27—41 (1976).

Zeki, S. The response properties of cells in the middle temporal area (area MT) of owl monkey visual cortex. *Proc.* B **207**, 239—248 (1980).

A direct projection from area VI to area V3A of rhesus monkey visual cortex.
Proc. B **207**, 499—506 (1980).

See also Zeki, S.M.

Zeki, S. & Fries, W. A function of the corpus callosum in the Siamese cat.
Proc. B **207**, 249—258 (1980).

Zeki, S.M. The projections to the superior temporal sulcus from areas 17 and 18 in the rhesus monkey. *Proc.* B **193**, 199—207 (1976).

Simultaneous anatomical demonstration of the representation of the vertical and horizontal meridians in areas V2 and V3 of rhesus monkey visual cortex. *Proc.* B **195**, 517—523 (1977).

Colour coding in the superior temporal sulcus of rhesus monkey visual cortex.
Proc. B **197**, 195—223 (1977). Erratum. *Proc.* B **199**, 588 (1977).

See also Zeki, S.

Functional specialization and binocular interaction in the visual areas of rhesus monkey prestriate cortex (Discussion). *Proc.* B **204**, 379—397 (1979). Erratum, p. 513.

Zeki, S.M. & Sandeman, D.R. Combined anatomical and electrophysiological studies on the boundary between the second and third visual areas of rhesus monkey cortex.
Proc. B **194**, 555—562 (1976).

Zel'dovich, B.Ya. *See* Baranova & Zel'dovich.

Zembala, M. *See* Asherson & Zembala.

Zentgraf, H., Keller, W. & Müller, Ulrike. The structure of SV40 chromatin (Discussion).
Trans. B **283**, 299—303 (1978).

Zerahn, K. Active transport of the alkali metals by the isolated mid-gut of *Hyalophora cecropia* (Discussion). *Trans.* B **262**, 315—321 (1971).

Zeuthen, J. *See* Bak & Zeuthen.

Zeuthen, T. & Monge, C. Intra- and extracellular gradients of electrical potential and ion activities of the epithelial cells of the rabbit ileum *in vivo* recorded by microelectrodes (Discussion).
Trans. B **271**, 277—281 (1975).

Zhen-Hong, Mai. *See* Lang & Zhen-Hong.

Ziegler, R.E. Improved accuracy from Doppler satellite positioning (Discussion).
Trans. A **294**, 353—356 (1980).

Zienkiewicz, O.C. The finite element method and the solution of some geophysical problems (Discussion). *Trans.* A **283**, 139—151 (1976).

Zimmern, D. The region of tobacco mosaic virus RNA involved in the nucleation of assembly (Discussion). *Trans.* B **276**, 189—204 (1976).

Zinnemann, K.S. *See* Wilson (Sir Graham) & Zinnemann.

Zinner, E. *See* Crozaz, Poupeau, Walker, Zinner & Morrison; Morrison (D.A.) & Zinner.

Zirin, H. The manifold structure of the chromosphere and corona (Discussion). *Trans.* A **270**, 77—80 (1971).

Observations of stellar chromospheres (Discussion). *Trans.* A **270**, 183—188 (1971).

Zobel, F.G.R. *See* Mingins, Zobel, Pethica & Smart.

Zonsveld, J.J. *See* Hannant & Zonsveld.

Zoro, J.A. *See* Eglinton, Simoneit & Zoro.

Zucker, U.F. *See* Dainton, Salmon & Zucker.

Zuckerman, Lord. Wilfrid Edward Le Gros Clark. *Biogr. Mem.* **19**, 217—233 (1973).

Scientific advice during and since World War II (Discussion). *Proc.* A **342**, 465—480 (1975).

Zyznar, E.S., Cross, F.B. & Nicol, J.A.C. Uric acid in the tapetum lucidum of mooneyes *Hiodon* (Hiodontidae Teleostei). *Proc.* B **201**, 1—6 (1978).

Zyznar, E.S. & Nicol, J.A. Reflecting materials in the eyes of three teleosts, *Orthopristes chrysopterus, Dorosoma cepedianum* and *Anchoa mitchilli. Proc.* B **184**, 15—27 (1973).

APPENDIX I. ANNIVERSARY ADDRESSES

1970 By **Lord Blackett**. *Proc.* A **321**, 1–14 (1971); *Proc.* B **177**, 1–14 (1971).
1971 By **A.L. Hodgkin**. *Proc.* A **326**, v–xx (1971); *Proc.* B **180**, v–xx (1972).
1972 By **Sir Alan Hodgkin**. *Proc.* A **331**, 285–303 (1972); *Proc.* B **183**, 1–19 (1973).
1973 By **Sir Alan Hodgkin**. *Proc.* A **336**, v–xx (1974); *Proc.* B **185**, v–xx (1974).
1974 By **Sir Alan Hodgkin**. *Proc.* A **342**, 1–17 (1975); *Proc.* B **188**, 103–119 (1975).
1975 By **Sir Alan Hodgkin**. *Proc.* A **348**, 153–173 (1976); *Proc.* B **192**, 371–391 (1976).
1976 By **Lord Todd**. *Proc.* A **352**, 451–462 (1977); *Proc.* B **196**, 1–12 (1977).
1977 By **Lord Todd**. *Proc.* A **359**, v–xiv (1978); *Proc.* B **200**, v–xiv (1978).
1978 By **Lord Todd**. *Proc.* A **365**, v–xviii (1979); *Proc.* B **204**, 1–14 (1979).
1979 By **Lord Todd**. *Proc.* A **369**, 295–306 (1980); *Proc.* B **206**, 369–380 (1980).
1980 By **Lord Todd**. *Proc.* B **211**, 1–13 (1980).

APPENDIX II. LECTURES

Bakerian Lectures

1971 **B.J. Mason**. The physics of the thunderstorm. *Proc.* A **327**, 433–466 (1972).
1972 **Dorothy Crowfoot Hodgkin**. Insulin, its chemistry and biochemistry.
 Proc. B **186**, 191–215 (1974); *Proc.* A **338**, 251–275 (1974).
1974 **D.G. King-Hele**. A view of Earth and air. *Trans.* A **278**, 67–109 (1975).
1975 **M.F. Atiyah**. Global geometry. *Proc.* A **347**, 291–299 (1976).
1976 **G.W. Kenner**. Towards synthesis of proteins. *Proc.* A **353**, 441–457 (1977);
 Proc. B **197**, 237–253 (1977).
1977 **Sir George Porter**. *In vitro* models for photosynthesis. *Proc.* A **362**, 281–303 (1978).
 [Abstract.] *Proc.* B **202**, 539 (1978).
1978 **R.L.F. Boyd**. Cosmic exploration by X-rays. *Proc.* A **366**, 1–21 (1979).

Bernal Lectures

1971 **Sir Eric Ashby**. Science and antiscience. *Proc.* B **178**, 29–42 (1971).
1975 **C.H. Waddington**. The New Atlantis revisited. *Proc.* B **190**, 301–314 (1975).
1976 **P.L. Kapitza**. Scientific and social approaches for the solution of global problems.
 Proc. A **357**, 1–14 (1977). [Extract from the introduction.] *Proc.* B **199**, 327–328 (1977).

Blackett Memorial Lecture

1978 **B.V. Sreekantan**. Fundamental research in India in the area of the physical sciences.
 Proc. A **365**, 145–160 (1979).

Clifford Paterson Lectures

1976 **Sir Eric Eastwood**. Radar: new techniques and applications. *Proc.* A **354**, 137–155 (1977).
1977 **G.H. Rawcliffe**. Induction motors: old and new. *Proc.* A **362**, 145–178 (1978).
1979 **G.G. Scarrott**. From computing slave to knowledgeable servant: the evolution of computers.
 Proc. A **369**, 1–30 (1979).

Lectures

Croonian Lectures

1967 **A.F. Huxley**. The activation of striated muscle and its mechanical response.
  Proc. B **178**, 1–27 (1971).
1970 **H.E. Huxley**. The structural basis of muscular contraction. *Proc.* B **178**, 131–149 (1971).
1971 **Henry Harris**. Cell fusion and the analysis of malignancy. *Proc.* B **179**, 1–20 (1971).
1972 **N. Tinbergen**. Functional ethology and the human sciences. *Proc.* B **182**, 385–410 (1972).
1973 **E.J. Denton**. On buoyancy and the lives of modern and fossil cephalopods.
  Proc. B **185**, 273–299 (1974).
1974 **J. Heslop-Harrison**. The physiology of the pollen grain surface. *Proc.* B **190**, 275–299 (1975).
1975 **F. Sanger**. Nucleotide sequences in DNA. *Proc.* B **191**, 317–333 (1975).
1976 **J.B. Gurdon**. Egg cytoplasm and gene control in development. *Proc.* B **198**, 211–247 (1977).
1977 **J.W.S. Pringle**. Stretch activation of muscle: function and mechanism.
  Proc. B **201**, 107–130 (1978).
1978 **M. Abercrombie**. The crawling movement of metazoan cells. *Proc.* B **207**, 129–147 (1980).
1979 **S. Ebashi**. Regulation of muscle contraction. *Proc.* B **207**, 259–286 (1980).
1980 **R.R. Porter**. The complex proteases of the complement system. *Proc.* B **210**, 477–498 (1980).

Ferrier Lectures

1974 **W. Feldberg**. Body temperature and fever: changes in our views during the last decade.
  Proc. B **191**, 199–229 (1975).
1972 **D.H. Hubel & T.N. Wiesel**. Functional architecture of macaque monkey visual cortex.
  Proc. B **198**, 1–59 (1977).
1977 **J. Szentágothai**. The neuron network of the cerebral cortex: a functional interpretation.
  Proc. B **201**, 219–248 (1978).

Leeuwenhoek Lectures

1970 **P.H. Gregory**. Airborne microbes: their significance and distribution.
  Proc. B **177**, 469–483 (1971).
1971 **M.G.P. Stoker**. Tumour viruses and the sociology of fibroblasts. *Proc.* B **181**, 1–17 (1972).
1972 **H.L. Kornberg**. Carbohydrate transport by micro-organisms. *Proc.* B **183**, 105–123 (1973).
1974 **R. Dulbecco**. The control of cell growth regulation by tumour-inducing viruses: a challenging
  problem. *Proc.* B **189**, 1–14 (1975).
1975 **J. Mandelstam**. Bacterial sporulation: a problem in the biochemistry and genetics of a primitive
  developmental system. *Proc.* B **193**, 89–106 (1976).
1976 **G.H. Beale**. Protozoa and genetics. *Proc.* B **196**, 13–27 (1977).
1977 **F. Jacob**. Mouse teratocarcinoma and mouse embryo. *Proc.* B **201**, 249–270 (1978).
1979 **Patricia H. Clarke**. Experiments in microbial evolution: new enzymes, new metabolic activities.
  Proc. B **207**, 385–404 (1980).
1978 **John Cairns**. Bacteria as proper subjects for cancer research. *Proc.* B **208**, 121–133 (1980).

Royal Society Technology Lectures

1970 **Lord Rothschild**. Petrol and pollution. *Proc.* A **322**, 147–163 (1971).
1972 **D.S. Davies**. Discontinuities in chemistry and chemical technology.
  Proc. A **330**, 149–172 (1972).
1973 **H.M. Finniston**. Nuclear energy for the steel industry. *Proc.* A **340**, 129–146 (1974).
1975 **Sir Ieuan Maddock**. Science, technology, and industry. *Proc.* A **345**, 295–326 (1975).
1976 **Sir Angus Paton**. Dams and their interfaces. *Proc.* A **351**, 1–17 (1976).

Lectures

Review Lectures

1969 **Sir William Hudson**. The Snowy Moutains hydroelectric and irrigation scheme (Australia). *Proc.* A **326**, 23–37 (1971).

1970 **P. Schagen**. Electronic aids to night vision. *Trans.* A **269**, 233–263 (1971).
  R.D. Martin. Adaptive radiation and behaviour of the Malagasy lemurs. *Trans.* B **264**, 295–352 (1972).

1971 **R.A. Smith**. Lasers and light scattering. *Proc.* A **323**, 305–320 (1971).
  J.F. Coales. The control of industrial processes. *Proc.* A **325**, 291–311 (1971).
  H.M. Finniston. Steel: an industry with a future. *Proc.* A **326**, 1–22 (1971).
  J.L. Brachet. Nucleocytoplasmic interactions in morphogenesis. *Proc.* B **178**, 227–243 (1971).
  J.P. Elliott. Calculations of nuclear structure. *Proc.* A **326**, 199–213 (1972).
  H.G. Callan. Replication of DNA in the chromosomes of eukaryotes. *Proc.* B **181**, 19–41 (1972).

1972 **Alan H. Cook**. The dynamical properties and internal structures of the Earth, the Moon and the planets. *Proc.* A **328**, 301–336 (1972).
  D.T.N. Williamson. The anachronistic factory. *Proc.* A **331**, 139–160 (1972).
  J.M. Thoday. Disruptive selection. *Proc.* B **182**, 109–143 (1972).
  H.C. Longuet-Higgins. The algorithmic description of natural language. *Proc.* B **182**, 255–276 (1972).
  E.W. Horton. The prostaglandins. *Proc.* B **182**, 411–426 (1972).
  L. Weiskrantz. Behavioural analysis of the monkey's visual nervous system. *Proc.* B **182**, 427–455 (1972).
  J.H. Chesters. The prevention of metal breakouts through refractories. *Proc.* A **333**, 133–148 (1973).
  Sir Michael Woodruff. Cancer – the elusive enemy. *Proc.* B **183**, 87–104 (1973).
  K.L. Blaxter. The nutrition of ruminant animals in relation to intensive methods of agriculture. *Proc.* B **183**, 321–336 (1973).

1973 **F.T. Bacon & T.M. Fry**. The development and practical application of fuel cells. *Proc.* A **334**, 427–452 (1973).
  Sir George Edwards. The technical aspects of supersonic civil transport aircraft. *Trans.* A **275**, 529–565 (1974).
  V.E. Cosslett. Perspectives in high voltage electron microscopy. *Proc.* A **338**, 1–16 (1974).
  R.H. Pritchard. On the growth and form of a bacterial cell. *Trans.* B **267**, 303–336 (1974).
  R.A. McCance & Elsie M. Widdowson. The determinants of growth and form. *Proc.* B **185**, 1–17 (1974).

1974 **J.M. Thomas**. Topography and topology in solid-state chemistry. *Trans.* A **277**, 251–286 (1974).
  A.R. Burkin. The winning of non-ferrous metals, 1974. *Proc.* A **338**, 419–437 (1974).
  Mary F. Lyon. Mechanisms and evolutionary origins of variable X-chromosome activity in mammals. *Proc.* B **187**, 243–268 (1974).
  Sir Ernst Gombrich. Mirror and map: theories of pictorial representation. *Trans.* B **270**, 119–149 (1975).
  T.P. Hoar. Corrosion of metals: its cost and control. *Proc.* A **348**, 1–18 (1976).

1975 **D.G. Kendall**. The recovery of structure from fragmentary information. *Trans.* A **279**, 547–582 (1975).
  M. Hart. Ten years of X-ray interferometry. *Proc.* A **346**, 1–22 (1975).
  R.L. Wain. Some developments in research on plant growth inhibitors. *Proc.* B **191**, 335–352 (1975).
  D.V. Glass. Recent and prospective trends in fertility in developed countries. *Trans.* B **274**, 1–52 (1976).
  T. Mann. Relevance of physiological and biochemical research to problems in animal fertility. *Proc.* B **193**, 1–15 (1976).

1976 **A.W. Fletcher**. Metal recycling from scrap and waste materials. *Proc.* A **351**, 151–178 (1976).
  G.V. Groves. Rocket studies of atmospheric tides. *Proc.* A **351**, 437–469 (1976).

1977 **R. Weck.** The need for new welding processes. *Proc.* A **356**, 1–23 (1977).
    A.S. Curry. Forensic science. *Proc.* B **199**, 189–198 (1977).
    C.L. Wilson. (Arranged by the Fellowship of Engineering in conjunction with The Royal Society.) World energy prospects to the year 2000 (with a vote of thanks by Sir William Hawthorne). *Proc.* A **358**, 121–139 (1978).
    P.W. Brian. Hormones in healthy and diseased plants. *Proc.* B **200**, 231–243 (1978).
    R.J.P. Williams. Energy states of proteins, enzymes and membranes.
    Proc. B **200**, 353–389 (1978).
1978 **Janet V. Watson.** Ore-deposition through geological time. *Proc.* A **362**, 305–328 (1978).
    B.J. Mason. Recent advances in the numerical prediction of weather and climate.
    Proc. A **363**, 297–333 (1978).
    J.L. Harley. Ectomycorrhizas as nutrient absorbing organs. *Proc.* B **203**, 1–21 (1978).
    Sir John Cornforth. The imitation of enzymic catalysis. *Proc.* B **203**, 101–117 (1978).
    S. Cohen. Immunity to malaria. *Proc.* B **203**, 323–345 (1979).
1979 **A. Carrington.** Spectroscopy of molecular ion beams. *Proc.* A **367**, 433–449 (1979).
    J.H. Westcott. The application of microprocessors. *Proc.* A **367**, 451–484 (1979).
    H.C. Longuet-Higgins. The perception of music. *Proc.* B **205**, 307–322 (1979).
    G.A. Horridge. Apposition eyes of large diurnal insects as organs adapted to seeing.
    Proc. B **207**, 287–309 (1980).
    N.A. Locket. Some advances in coelacanth biology. *Proc.* B **208**, 265–307 (1980).
1980 **K.A. Browning.** Local weather forecasting. *Proc.* A **371**, 179–211 (1980).
    R. Freeman. Nuclear magnetic resonance spectroscopy in two frequency dimensions.
    Proc. A **373**, 149–178 (1980).
    M.F. Perutz. Stereochemical mechanism of oxygen transport by haemoglobin.
    Proc. B **208**, 135–162 (1980).

Rutherford Memorial Lectures

1971 **P.H. Fowler.** Evolution of the elements. *Proc.* A **329**, 1–16 (1972).
1975 **P.B. Moon.** Yarns and spinners: recollections of Rutherford and applications of swift rotation.
    Proc. A **360**, 303–315 (1978).
1977 **N. Feather.** Some episodes of the α-particle story, 1903–1977. *Proc.* A **357**, 117–129 (1977).

Wilkins Lectures

1973 **A. Rupert Hall.** Newton and his editors. *Proc.* A **338**, 397–417 (1974).
1979 **Gweneth Whitteridge.** Of the local movement of animals. *Proc.* B **206**, 1–13 (1979).

APPENDIX III. DISCUSSIONS AND SYMPOSIA

1969 The petrology of igneous and metamorphic rocks from the ocean floor (organized by Sir Edward Bullard, J.R. Cann & D.H. Matthews). *Trans.* A **268**, 381–745 (1971).
    The results of the Royal Society Expedition to Aldabra 1967–68 (organized by T.S. Westoll & D.R. Stoddart). *Trans.* B **260**, 1–654 (1971).
1970 Computer aids in mechanical engineering design and manufacture (organized by Sir William Hawthorne & Sir George Edwards). *Proc.* A **321**, 143–248 (1971).
    Numerical analysis of partial differential equations (organized by J.H. Wilkinson).
    Proc. A **323**, 151–304 (1971).
    Cooperation between lymphocytes in the immune response (organized by J.L. Gowans, J.H. Humphrey & N.A. Mitchison). *Proc.* B **176**, 367–426 (1971).
    Animal viruses as genetic modifiers of the cell (organized by M.G.P. Stoker).
    Proc. B **177**, 15–108 (1971).
    Biological effects of pollution in the sea (organized by H.A. Cole).
    Proc. B **177**, 275–468 (1971).

Architectural aerodynamics (organized by M.J. Lighthill & A. Silverleaf).
Trans. A **269**, 321–554 (1971).

Solar studies with special reference to space observations (arranged by the British National Committee on Space Research under the leadership of Sir Harrie Massey, C.W. Allen, A.H. Gabriel, B.E.J. Pagel & R. Wilson). *Trans.* A **270**, 1–195 (1971).

Ocean currents and their dynamics (organized by G.E.R. Deacon).
Trans. A **270**, 349–465 (1971).

Volcanism and the structure of the earth (organized by J. Sutton, P.A. Sabine & R.R. Skelhorn).
Trans. A **271**, 101–323 (1972).

New developments in electron microscopy with special emphasis on their application in biology (organized by H.E. Huxley & A. Klug). *Trans.* B **261**, 1–230 (1971).

Subcellular and macromolecular aspects of synaptic transmission (organized by H.K.F. Blaschko & A.D. Smith). *Trans.* B **261**, 273–437 (1971).

Active transport of salts and water in living tissues (organized by R.D. Keynes).
Trans. B **262**, 83–342 (1971).

1971 Optimal conditions for photosynthesis in a wholly artificial environment (organized by N.W. Pirie). *Proc.* B **179**, 171–246 (1971).

Penicillin and related antibiotics – past, present and future (organized by Sir Ernst Chain & Joan Stokes). *Proc.* B **179**, 291–432 (1971).

Biosynthesis of sterols and related compounds (organized by T.W. Goodwin).
Proc. B **180**, 113–246 (1972).

Freshwater and estuarine studies of the effects of industry (organized by Sir Frederick Russell & H.C. Gilson). *Proc.* B **180**, 363–536 (1972).

Disease resistance in plants (organized by P.W. Brian & S.D. Garrett).
Proc. B **181**, 213–351 (1972).

D and E region winds over Europe (arranged by the British National Committee on Space Research under the leadership of Sir Harrie Massey & G.V. Groves).
Trans. A **271**, 455–629 (1972).

Problems associated with the subsidence of southeastern England (organized by K.C. Dunham & D.A. Gray). *Trans.* A **272**, 79–274 (1972).

Building technology in the 1980s (organized by Sir James Lighthill, K.W. Pepper & P.L. Bakke).
Trans. A **272**, 493–661 (1972).

1972 The physics and chemistry of surfaces (organized by J.W. Linnett).
Proc. A **331**, 305–443 (1972).

Ship technology in the 1980s (organized by Sir James Lighthill, F.B. Bolton & R. Hurst).
Trans. A **273**, 1–184 (1972).

The evolution of the Precambrian crust (organized by J. Sutton & B.F. Windley).
Trans. A **273**, 315–581 (1973).

Mount Etna and the 1971 eruption (edited by J.E. Guest & R.R. Skelhorn).
Trans. A **274**, 1–179 (1973).

The measurement and interpretation of changes of strain in the Earth (organized by Alan H. Cook, R.V. Jones, R.G. Bilham & G.C.P. King). *Trans.* A **274**, 181–433 (1973).

Manufacturing technology in the 1980s (organized by G.B.R. Feilden & D.T.N. Williamson).
Trans. A **275**, 309–424 (1973).

The place of astronomy in the ancient world (organized jointly for the Royal Society and the British Academy by D.G. Kendall, S. Piggott, D.G. King-Hele & I.E.S. Edwards).
Trans. A **276**, 1–276 (1974).

Recent developments in vertebrate smooth muscle physiology (organized by Edith Bülbring & Dorothy M. Needham). *Trans.* B **265**, 1–231 (1973).

1973 Proton–proton scattering at very high energies (organized by J.M. Cassels & A.M. Wetherell).
Proc. A **335**, 407–507 (1973).

Planetary science (in celebration of the quincentenary of Nicolaus Copernicus 1473–1543) (organized by Sir Harrie Massey & W.H. McCrea). *Proc.* A **336**, 1–114 (1974).

The exploitation of British mineral resources (other than coal and hydrocarbons) in relation to

countryside conservation (organized by Sir Kingsley Dunham & M.E.D. Poore).
Proc. A **339**, 271–416 (1974).
Scientific results from the Copernicus satellite (arranged by the British National Committee on Space Research under the leadership of Sir Harrie Massey and R.L.F. Boyd).
Proc. A **340**, 397–469 (1974).
The biology of an equatorial lake: Lake George, Uganda (organized by P.H. Greenwood & J.W.G. Lund). *Proc.* B **184**, 227–346 (1973).
The value of automation in the health service (organized by D.A.J. Tyrrell).
Proc. B **184**, 347–476 (1973).
The assessment of the environmental impact of chemical substances (organized by Sir James Cook & Sir Frederick Warner). *Proc.* B **185**, 123–224 (1974).
Recent advances in heavy electrical plant (organized by J.S. Forrest).
Trans. A **275**, 33–253 (1973).
Energy in the 1980s (organized by Sir Peter Kent). *Trans.* A **276**, 405–615 (1974).
Agricultural productivity in the 1980s (organized by G.D.H. Bell, C.J. Moss & O.G. Williams).
Trans. B **267**, 1–172 (1973).
The electron microscopy and composition of biological membranes and envelopes (organized by R. Markham, R.W. Horne & R. Marian Hicks). *Trans,* B **268**, 1–159 (1974).
Human adaptability in a tropical ecosystem: an I.B.P. human biological investigation of two New Guinea communities (organized by G.A. Harrison & R.J. Walsh).
Trans. B **268**, 221–400 (1974).

1974 The effects of the two World Wars on the organization and development of science in the United Kingdom (organized by R.V. Jones). *Proc.* A **342**, 439–591 (1975).
The scientific results from the Prospero and Ariel 4 satellites (arranged by the British National Committee on Space Research under the leadership of Sir Harrie Massey, R. Dalziel & D.G. King-Hele). *Proc.* A **343**, 159–287 (1975).
The determination of structures and conformation of molecules in solution (organized by R.E. Richards & R.J.P. Williams). *Proc.* A **345**, 1–144 (1975).
The introduction of satellites into education systems (organized by Sir James Lighthill, Sir Harrie Massey, C.C. Butler & G.K.C. Pardoe). *Proc.* A **345**, 431–605 (1975).
Incompatibility in flowering plants (organized by J. Heslop-Harrison & D. Lewis).
Proc. B **188**, 233–375 (1975).
The recognition of alien life (organized by N.W. Pirie). *Proc.* B **189**, 137–274 (1975).
Organic pollutants in the sea: their origin, distribution, degradation and ultimate fate (organized by H.A. Cole & J.E. Smith). *Proc.* B **189**, 275–483 (1975).
Food technology in the 1980s (organized by Sir Ernst Chain, S.A. Goldblith, L. Rey & A. Spicer). *Proc.* B **191**, 1–198 (1975).
The origin of the cosmic radiation (organized by G.D. Rochester & W.A. Wolfendale).
Trans. A **277**, 317–501 (1975).
The geology of the English Channel (organized by Sir Kingsley Dunham & A.J. Smith for the Royal Society and the Marine Studies Group of the Geological Society).
Trans. A **279**, 1–295 (1975).
Astronomy in the ultraviolet (arranged by the British National Committee on Space Research under the leadership of Sir Harrie Massey & R. Wilson). *Trans.* A **279**, 297–485 (1975).
(1) The early days of ionospheric research; (2) the theory of electric and magnetic waves in the ionosphere and magnetosphere (organized by W.J.G. Beynon & J.A. Ratcliffe).
Trans. A **280**, 1–224 (1975).
Excitable membranes (organized by R.D. Keynes). *Trans.* B **270**, 295–559 (1975).
Forests and forestry in Britain (organized by G.D. Holmes, P.G. Wareing & J.L. Harley).
Trans. B **271**, 45–232 (1975).
The pericellular environment and its regulation in vertebrate tissues (organized by Dame Honor Fell & J.T. Dingle). *Trans.* B **271**, 233–410 (1975).
The physics and chemistry of biological recognition (organized by D.C. Phillips & G.K. Radda).
Trans. B **272**, 1–198 (1975).

Discussions and Symposia

The results of the 1971 Royal Society—Percy Sladen Expedition to the New Hebrides (organized by E.J.H. Corner & K.E. Lee). *Trans.* B **272**, 267–486 (1975).

1975 Scientific results from the Ariel 6 satellite (arranged by the British National Committee on Space Research under the leadership of Sir Harrie Massey, K.A. Pounds and A.P. Willmore). *Proc.* A **350**, 419–545 (1976).

Rubber elasticity (organized by G. Gee, G. Allen & C. Price). *Proc.* A **351**, 295–406 (1976).

The treatment of arthritis by joint replacement (organized by Sir Hugh Ford, S.A.V. Swanson & D.A.J. Tyrrell). *Proc.* B **192**, 129–219 (1976).

Human adaptability in Ethiopia (organized in cooperation with the Society for the Study of Human Biology by E.J. Clegg). *Proc.* B **194**, 1–98 (1976).

Recent developments in medical endoscopy and related fields (organized by H.H. Hopkins & D.A.J. Tyrrell). *Proc.* B **195**, 225–306 (1977).

Global tectonics in Proterozoic times (organized by J. Sutton, R.M. Shackleton & J.C. Briden). *Trans.* A **280**, 397–663 (1976).

The physics of the solar atmosphere (arranged by the British National Committee on Space Research under the leadership of Sir Harrie Massey, P.A. Sweet & A.H. Gabriel). *Trans.* A **281**, 293–513 (1976).

Rosenhain Centenary Conference. The contribution of physical metallurgy to engineering practice (a joint symposium of The Metals Society, The National Physical Laboratory and The Royal Society). *Trans.* A **282**, 1–483 (1976).

Natural strain and geological structure (organized by J.G. Ramsay & D.S. Wood). *Trans.* A **283**, 1–344 (1976).

Valley slopes and cliffs in southern England: morphology, mechanics and Quaternary history (organized by A.W. Skempton & J.N. Hutchinson). *Trans.* A **283**, 421–631 (1976).

The Moon — a new appraisal from space missions and laboratory analyses (arranged by the British National Committee on Space Research under the leadership of Sir Harrie Massey, G.M. Brown, G. Eglinton, S.K. Runcorn & H.C. Urey). *Trans.* A **285**, 1–606 (1977).

The biosynthesis of porphyrins, chlorophyll and vitamin B_{12} (organized by A. Neuberger and G.W. Kenner). *Trans.* B **273**, 75–357 (1976).

Water relations of plants (organized by J.L. Monteith & P.E. Weatherley). *Trans.* B **273**, 433–613 (1976).

A review of the United Kingdom contribution to the International Biological Programme (organized by A.R. Clapham, C.E. Lucas & N.W. Pirie). *Trans.* B **274**, 275–554 (1976).

The early history of agriculture (organized jointly for the Royal Society and the British Academy by Sir Joseph Hutchinson, Grahame Clark, E.M. Jope & R. Riley). *Trans.* B **275**, 1–213 (1976).

The assembly of regular viruses (organized by A. Klug & P.J.G. Butler). *Trans.* B **276**, 1–204 (1976).

The meiotic process (organized by R. Riley, M.D. Bennett & R.B. Flavell). *Trans.* B **277**, 183–376 (1977).

Water structure and transport in biology (organized by R.E. Richards & F. Franks). *Trans.* B **278**, 1–205 (1977).

Structural and functional aspects of plasticity in the nervous system (organized by H.B. Barlow & R.M. Gaze). *Trans.* B **278**, 241–436 (1977).

1976 New particles and new quantum numbers (organized by R.H. Dalitz, B. Richter, B. Wiik & W.T. Toner). *Proc.* A **355**, 441–631 (1977).

Materials behaviour in low gravity conditions (arranged by the British National Committee on Space Research under the leadership of Sir Harrie Massey, E.G.C. Burt, J.A. Champion, J.W. Christian and R.F. Rissoné). *Proc.* A **361**, 129–179 (1978).

Contraceptives of the future (organized by R.V. Short & D.T. Baird). *Proc.* B **195**, 1–224 (1976).

The biology of chemical carcinogenesis [abstract] (organized by R. Dulbecco). *Proc.* B **196**, 117–130 (1977).

The scientific aspects of nature conservation in Great Britain (organized by J.E. Smith, A.R. Clapham & D.A. Ratcliffe). *Proc.* B **197**, 1–103 (1977).

253

Technologies for rural health (organized by D.A.J. Tyrrell, D.P. Burkitt & Sir William Henderson). *Proc.* B **199**, 1–187 (1977).

Turning points in zoological science (organized by a committee under the chairmanship of R.J. Harrison, as part of the 150th anniversary celebrations of the Zoological Society of London). *Proc.* B **199**, 335–443 (1977).

Gravity and biological systems (organized by the British National Committee on Space Research under the leadership of Sir Harrie Massey, N.W. Pirie, Daphne J. Osborne & H.S. Wolff). *Proc.* B **199**, 477–566 (1978).

Methods and applications of ranging to artificial satellites and the Moon (organized by Alan H. Cook, D.G. King-Hele, S.A. Ramsden & A.R. Robbins). *Trans.* A **284**, 419–619 (1977).

Mineralogy: towards the twenty-first century (organized by J.E.T. Horne & Sir Kingsley Dunham). *Trans.* A **286**, 231–638 (1977).

The use of operational research and systems analysis in decision making (organized by J.F. Coales & R.C. Tomlinson). *Trans.* A **287**, 351–544 (1977).

Resource development in semi-arid lands (organized by Sir Joseph Hutchinson, A.H. Bunting, A.R.Jolly & H.C. Pereira). *Trans.* B **278**, 437–614 (1977).

Scientific research in Antarctica (organized by Sir Vivian Fuchs & R.M. Laws). *Trans.* B **279**, 1–288 (1977).

The changing environmental conditions in Great Britain and Ireland during the Devensian (Last) cold stage (organized jointly for the Royal Society and the Royal Irish Academy by G.F. Mitchell & R.G. West). *Trans.* B **280**, 103–374 (1977).

The management of inputs for yet greater agricultural yield and efficiency (organized by G.W. Cooke, N.W. Pirie & G.D.H. Bell). *Trans.* B **281**, 73–301 (1977).

The northern Great Barrier Reef (organized by D.R. Stoddart & Sir Maurice Yonge). (Part A.) *Trans.* A **291**, 1–197 (1978). (Part B.) *Trans.* B **284**, 1–163 (1978).

1977 Scientific aspects of the 1975–76 drought in England and Wales (organized by Sir Charles Pereira, H.L. Penman, O. Gibb & R.A.S. Ratcliffe on behalf of the British National Committee on Hydrological Sciences). *Proc.* A **363**, 1–133 (1978).

Genetics of the cell surface (organized by W.F. Bodmer). *Proc.* B **202**, 1–189 (1978).

Long-term hazards to man from man-made chemicals in the environment (organized by Sir Richard Doll & A.E.M. McLean). *Proc.* B **205**, 1–197 (1979).

Creep of engineering materials and of the Earth (organized by A. Kelly, Alan H. Cook & G.W. Greenwood). *Trans.* A **288**, 1–236 (1978).

Terrestrial heat and the generation of magmas (organized by G.M. Brown, M.J. O'Hara & E.R. Oxburgh). *Trans.* A **288**, 383–643 (1978).

Telecommunications in the 1980s and after (organized by Sir James Lighthill, Sir Eric Eastwood, C.A. May & K.W. Cattermole). *Trans.* A **289**, 1–228 (1978).

Sea floor development: moving into deep water (organized by Sir Angus Paton, Sir Peter Kent, Sir George Deacon, Sir Kenneth Hutchison & M.B.F. Ranken in collaboration with the British National Committee on Ocean Engineering of the Council of Engineering Institutions). *Trans.* A **290**, 1–185 (1978).

Pathways of pollutants in the atmosphere (organized by T.M. Sugden). *Trans.* A **290**, 467–637 (1979).

Theoretical and practical aspects of uranium geology (organized by S.H.U. Bowie, W.S. Fyfe, D. Ostle, Jane Plant & P.R. Simpson). *Trans.* A **291**, 253–452 (1979).

The evolution of passive continental margins in the light of recent deep drilling results (organized by Sir Peter Kent, A.S. Laughton, D.G. Roberts & E.J.W. Jones on behalf of the National Committee for Geodynamics of The Royal Society, the Marine Studies Group of the Geological Society and the Joint Oceanographic Institutions Deep Earth Sampling (JOIDES). *Trans.* A **294**, 1–208 (1980).

Structure of eukaryotic chromosomes and chromatin (organized for the Royal Society and the Malacological Society of London by H.G. Callan & A. Klug). *Trans.* B **283**, 231–416 (1978).

Evolutionary systematics of bivalve molluscs (organized by Sir Maurice Yonge & T.E. Thompson). *Trans.* B **284**, 199–436 (1978).

The terrestrial ecology of Aldabra (organized by D.R. Stoddart & T.S. Westoll).
Trans. B 286, 1–263, I–XI (1979).
Strategy and tactics of control of migrant pests (organized by D.L. Gunn & R.C. Rainey).
Trans. B 287, 245–488 (1979).
1978 Some recent results in X-ray astronomy (arranged by The British National Committee on Space Research, under the leadership of Sir Harrie Massey, R.L.F. Boyd & A.P. Willmore).
Proc. A 366, 277–489 (1979).
Recent developments in General Relativity (extended abstracts of contributions) (organized by S.W. Hawking & R. Penrose). *Proc.* A 368, 1–36 (1979).
The cell as habitat (organized by M.H. Richmond & D.C. Smith). *Proc.* B 204, 113–286 (1979).
Stereoscopic vision (organized by D. Whitteridge & P.G.H. Clarke).
Proc. B 204, 377–512 (1979).
The evolution of adaptation by natural selection (organized by J. Maynard Smith & R. Holliday). *Proc.* B 205, 433–604 (1979).
Ultrasonic and radiological methods in engineering and medical diagnosis: limitations and future prospects (organized by Sir James Menter, R. Weck, P. Duncumb, D.W. Pashley, G.N. Hounsfield & J.G. Young). *Trans.* A 292, 135–305 (1979).
Mass spectrometry in organic and biological chemistry (organized by A.W. Johnson & J.H. Beynon). *Trans.* A 293, 1–168 (1979).
Light scattering in physics, chemistry and biology (organized by G.W. Series, E.R. Pike & J.G. Powles). *Trans.* A 293, 209–471 (1979).
Satellite Doppler tracking and its geodetic applications (organized by A.R. Robbins, V. Ashkenazi & D.G. King-Hele for the British National Committee for Geodesy and Geophysics). *Trans.* A 294, 209–406 (1980).
New fibres and their composites (organized by W. Watt, B. Harris & A.C. Ham).
Trans. A 294, 407–597 (1980).
Residuals, additives and materials properties (organized as a joint symposium of the Metals Society, the National Physical Laboratory and the Royal Society).
Trans. A 295, 1–341 (1980).
Solar energy (organized by Sir George Porter & Sir William Hawthorne).
Trans. A 295, 343–511 (1980).
The middle atmosphere as observed from balloons, rockets and satellites (arranged by the British National Committee on Space Research & Solar–Terrestrial Physics, under the leadership of Sir Harrie Massey, Sir Granville Beynon, J.T. Houghton & L. Thomas).
Trans. A 296, 1–268 (1980).
The evidence for chemical heterogeneity in the Earth's mantle (organized by D.K. Bailey, J. Tarney & Sir Kingsley Dunham on behalf of the Geochemistry Group of the Mineralogical Society). *Trans.* A 297, 135–493 (1980).
The biochemical functions of terpenoids in plants (organized by T.W. Goodwin).
Trans. B 284, 437–599 (19878).
The assessment of sublethal effects of pollutants in the sea (organized by H.A. Cole in collaboration with the Marine Pollution Subcommittee of the British National Committee on Oceanic Research). *Trans.* B 286, 397–633 (1979).
Environmental geochemistry and health (organized by S.H.U. Bowie & J.S. Webb).
Trans. B 288, 1–216 (1979).
1979 The beginnings of solid state physics (organized by Sir Nevill Mott).
Proc. A 371, 1–177 (1980).
More technologies for rural health (organized by D.A.J. Tyrrell, Sir William Henderson & Katherine Elliott). *Proc.* B 209, 1–186 (1980).
The origin and early evolution of the galaxies (organized by W.H. McCrea & M.J. Rees).
Trans. A 296, 269–435 (1980).
The Sun and the heliosphere (arranged by the British National Committee on Space Research under the leadership of Sir Harrie Massey, A.H. Gabriel & H. Elliot).
Trans. A 297, 519–640 (1980).

Ultra-short laser pulses (organized by D.J. Bradley, Sir George Porter & M.H. Key). *Trans.* A **298**, 209–414 (1980).

Influenza (organized by D.A.J. Tyrrell & H.G. Pereira). *Trans.* B **288**, 287–460 (1980).

Penicillin fifty years after Fleming (organized by Sir James Baddiley & E.P. Abraham). *Trans.* B **289**, 165–378 (1980).

Nuclear magnetic resonance of intact biological systems (organized by R.J.P. Williams, E.R. Andrew & G.K. Radda). *Trans.* B **289**, 379–553 (1980).

The psychology of vision (organized by H.C. Longuet-Higgins & N.S. Sutherland). *Trans.* B **290**, 1–218 (1980).

New horizons in industrial microbiology (organized by S. Brenner, B.S. Hartley & P.J. Rodgers). *Trans.* B **290**, 277–430 (1980).

Neutron scattering in biology, chemistry and physics (organized by Sir Ronald Mason, E.W.J. Mitchell & J.W. White). *Trans.* B **290**, 498–681 (1980).

1980 Neuroactive peptides (organized by Sir Arnold Burgen, H.W. Kosterlitz & L.L. Iversen). *Proc.* B **210**, 1–195 (1980).

Interactions between virus and host molecules (organized by M.G.P. Stoker, L.V. Crawford & S. Brenner). *Proc.* B **210**, 317–476 (1980).

APPENDIX IV. BIOGRAPHICAL MEMOIRS

Abercrombie, M. By Sir Peter Medawar. *Biogr. Mem.* **26**, 1–15 (1980).

Adam, N.K. By A. Carrington, G.J. Hills & K.R. Webb. *Biogr. Mem.* **20**, 1–26 (1974).

Adrian, E.D. (Baron Adrian of Cambridge). By Sir Alan Hodgkin. *Biogr. Mem.* **25**, 1–73 (1979).

Allen, N.P. By Sir Charles Sykes. *Biogr. Mem.* **19**, 1–18 (1973).

Andrade, N.E. da C. By Sir Alan Cottrell. *Biogr. Mem.* **18**, 1–20 (1972).

Barber, H.N. By C.D. Darlington. *Biogr. Mem.* **18**, 21–33 (1972).

Barker, J. By T.A. Bennet-Clark. *Biogr. Mem.* **18**, 35–42 (1972).

Bawden, F.C. By N.W. Pirie. *Biogr. Mem.* **19**, 19–63 (1973).

de Beer, G.R. By E.J.W. Barrington. *Biogr. Mem.* **19**, 65–93 (1973).

Bennet-Clark, T.A. By R. Brown. *Biogr. Mem.* **23**, 1–18 (1977).

Bernal, J.D. By Dorothy M.C. Hodgkin. *Biogr. Mem.* **26**, 17–84 (1980).

Besicovitch, A.S. By J.C. Burkill. *Biogr. Mem.* **17**, 1–16 (1971).

Bisat, W.S. By Sir James Stubblefield. *Biogr. Mem.* **20**, 27–40 (1974).

Blackett, P.M.S. (Baron Blackett). By Sir Bernard Lovell. *Biogr. Mem.* **21**, 1–115 (1975).

Born, M. By N. Kemmer & R. Schlapp. *Biogr. Mem.* **17**, 17–52 (1971).

Bose, S.N. By J. Mehra. *Biogr. Mem.* **21**, 117–154 (1975).

Boys, S.F. By C.A. Coulson. *Biogr. Mem.* **19**, 95–115 (1973).

Bradley, A.J. By H. Lipson. *Biogr. Mem.* **19**, 117–128 (1973).

Bragg, W.L. By Sir David Phillips. *Biogr. Mem.* **25**, 75–143 (1979).
  Corrigendum. *Biogr. Mem.* **26**, 543 (1980).

Brambell, F.W.R. By C.L. Oakley. *Biogr. Mem.* **19**, 129–171 (1973).

Bronk, D.W. By Lord Adrian. *Biogr. Mem.* **22**, 1–9 (1976).

Brown, G.L. By F.C. MacIntosh & W.D.M. Paton. *Biogr. Mem.* **20**, 41–73 (1974).

Brown, W. By S.D. Garrett. *Biogr. Mem.* **21**, 155–174 (1975).

Browning, C.H. By C.L. Oakley. *Biogr. Mem.* **19**, 173–215 (1973).

Bullen, K.E. By Sir Harold Jeffreys. *Biogr. Mem.* **23**, 19–39 (1977).

Bullerwell, W. By Sir Kingsley Dunham. *Biogr. Mem.* **24**, 1–13 (1978).

Bulman, O.M.B. By Sir James Stubblefield. *Biogr. Mem.* **21**, 175–195 (1975).

Butler, J.A.V. By W.V. Mayneord. *Biogr. Mem.* **25**, 145–178 (1979).
  Corrigendum. *Biogr. Mem.* **26**, 543 (1980).

Chadwick, J. By Sir Harrie Massey & N. Feather. *Biogr. Mem.* **22**, 11–70 (1976).

Chapman, S. By T.G. Cowling. *Biogr. Mem.* **17**, 53–89 (1971).

Christophers, S.R. By H.E. Shortt & P.C.C. Garnham. *Biogr. Mem.* **25**, 179–207 (1979).
Clark, W.E.Le G. By Lord Zuckerman. *Biogr. Mem.* **19**, 217–233 (1973).
Colebrook, L. By C.L. Oakley. *Biogr. Mem.* **17**, 91–138 (1971).
Collingwood, E.F. By Dame Mary Cartwright & W.K. Hayman. *Biogr. Mem.* **17**, 139–192 (1971).
Collip, J.B. By M.L. Barr & R.J. Rossiter. *Biogr. Mem.* **19**, 235–267 (1973).
Conant, J.B. By G.B. Kistiakowsky & F.H. Westheimer. *Biogr. Mem.* **25**, 209–232 (1979).
Constant, H. By Sir William Hawthorne, H. Cohen & A.R. Howell. *Biogr. Mem.* **19**, 269–279 (1973).
Cook, J.W. By J.M. Robertson. *Biogr. Mem.* **22**, 71–103 (1976).
Coulson, C.A. By S.L. Altmann & E.J. Bowen. *Biogr. Mem.* **20**, 75–134 (1974).
Craigie, J. By Sir Christopher Andrewes. *Biogr. Mem.* **25**, 233–240 (1979).
Crew, F.A.E. By L. Hogben. *Biogr. Mem.* **20**, 135–153 (1974).
Daly, I. de B. By H. Barcroft. *Biogr. Mem.* **21**, 197–226 (1975).
Davenport, H. By C.A. Rogers, D.A. Burgess, H. Halberstam & B.J. Birch.
    Biogr. Mem. **17**, 159–192 (1971).
Davidson, C.R. By Sir Richard Woolley. *Biogr. Mem.* **17**, 193–194 (1971).
Davidson, J.N. By A. Neuberger. *Biogr. Mem.* **19**, 281–303 (1973).
Dent, C.E. By A. Neuberger. *Biogr. Mem.* **24**, 15–31 (1978).
Dent, F.J. By Sir Kenneth Hutchison & D. Hebden. *Biogr. Mem.* **20**, 155–180 (1974).
Dobson, G.M.B. By J.T. Houghton & C.D. Walshaw. *Biogr. Mem.* **23**, 41–57 (1977).
Dobzhansky, T.G. By E.B. Ford. *Biogr. Mem.* **23**, 59–89 (1977).
Dodds, E.C. By F. Dickens. *Biogr. Mem.* **21**, 227–267 (1975).
Dorey, S.F. By D.M. Smith. *Biogr. Mem.* **19**, 305–316 (1973).
Duke-Elder, W.S. By T.K. Lyle, Sir Stephen Miller & N.H. Ashton. *Biogr. Mem.* **26**, 85–105 (1980).
Eccles, W.H. By J.A. Ratcliffe. *Biogr. Mem.* **17**, 195–214 (1972).
Edman, P.V. By S. Miles Partridge & Birger Blombäck. *Biogr. Mem.* **25**, 241–265 (1979).
Erdélyi, A. By D.S. Jones. *Biogr. Mem.* **25**, 267–286 (1979).
Evans, H.McL. By E.C. Amoroso & G.W. Corner. *Biogr. Mem.* **18**, 83–186 (1972).
Ewing, W.M. By Sir Edward Bullard. *Biogr. Mem.* **21**, 269–311 (1975).
Fage, A. By A.R. Collar. *Biogr. Mem.* **24**, 33–53 (1978).
Farren, W.S. By Sir George Thomson & Sir Arnold Hall. *Biogr. Mem.* **17**, 215–241 (1971).
Fauré-Fremiet, E. By E.N. Willmer. *Biogr. Mem.* **18**, 187–221 (1972).
Fildes, P.G. By G.P. Gladstone, B.C.J.G. Knight & Sir Graham Wilson.
    Biogr. Mem. **19**, 317–347 (1973).
Finch, G.I. By M. Blackman. *Biogr. Mem.* **18**, 223–239 (1972).
Fleck, A. (Baron Fleck of Saltcoats). By Sir Ronald Holroyd. *Biogr. Mem.* **17**, 243–254 (1971).
Florey, H.W. (Baron Florey of Adelaide and Marston). By E.P. Abraham.
    Biogr. Mem. **17**, 255–302 (1971).
Folley, S.J. By Sir Alan Parkes. *Biogr. Mem.* **18**, 241–265 (1972).
    Corrigendum. *Biogr. Mem.* **19**, 695 (1973).
Fraser, F.C. By N.B. Marshall. *Biogr. Mem.* **25**, 287–317 (1979).
Freeth, F.A. By Sir Peter Allen. *Biogr. Mem.* **22**, 105–118 (1976).
Gabor, D. By T.E. Allibone. *Biogr. Mem.* **26**, 107–147 (1980).
Gascoyne-Cecil, R.A.J. *See* Salisbury, 5th Marquess.
Gates, S.B. By H.H.B.M. Thomas & D. Küchemann. *Biogr. Mem.* **20**, 181–212 (1974).
Glanville, W.H. By Lord Baker. *Biogr. Mem.* **23**, 91–113 (1977).
Gödel, K. By G. Kreisel. *Biogr. Mem.* **26**, 149–224 (1980).
Gold, E. By R.C. Sutcliffe & A.C. Best. *Biogr. Mem.* **23**, 115–131 (1977).
Gray, J. By H.W. Lissmann. *Biogr. Mem.* **24**, 55–70 (1978).
Guggenheim, E.A. By F.C. Tompkins & Sir Charles Goodeve. *Biogr. Mem.* **17**, 303–326 (1971).
Gustaf VI Adolf, King of Sweden. By Sir Mortimer Wheeler. *Biogr. Mem.* **20**, 213–216 (1974).
Haddow, A. By F. Bergel. *Biogr. Mem.* **23**, 133–191 (1977).
Haddow, A.J. By P.C.C. Garnham. *Biogr. Mem.* **26**, 225–254 (1980).
Hanson, E. Jean. By Sir John Randall. *Biogr. Mem.* **21**, 313–344 (1975).
Harington, C.R. By Sir Harold Himsworth & Rosalind Pitt-Rivers. *Biogr. Mem.* **18**, 267–308 (1972).

Harris, G.W. By Marthe L. Vogt. *Biogr. Mem.* **18**, 309–329 (1972).
Hartley, H.B. By A.G. Ogston. *Biogr. Mem.* **19**, 349–373 (1973).
    Corrigendum. *Biogr. Mem.* **19**, 695 (1973).
Hartridge, H. By W.A.H. Rushton. *Biogr. Mem.* **23**, 193–211 (1977).
Harvey, H.W. By L.H.N. Cooper. *Biogr. Mem.* **18**, 331–347 (1972).
Havelock, T.H. By A.M. Binnie & P.H. Roberts. *Biogr. Mem.* **17**, 327–377 (1971).
Heilbronn, H.A. By J.W.S. Cassels & A. Fröhlich. *Biogr. Mem.* **22**, 119–135 (1976).
Heisenberg, W. By Sir Nevill Mott & Sir Rudolf Peierls. *Biogr. Mem.* **23**, 213–251 (1977).
Hill, A.V. By Sir Bernard Katz. *Biogr. Mem.* **24**, 71–149 (1978).
Hindle, E. By P.C.C. Garnham. *Biogr. Mem.* **20**, 217–234 (1974).
Hinshelwood, C.N. By Sir Harold Thompson. *Biogr. Mem.* **19**, 375–431 (1973).
Hinton, H.E. By G. Salt. *Biogr. Mem.* **24**, 151–182 (1978).
Hirst, E.L. By M. Stacey & Elizabeth Percival. *Biogr. Mem.* **22**, 137–168 (1976).
Hodge, W.V.D. By M.F. Atiyah. *Biogr. Mem.* **22**, 169–192 (1976).
Hogben, L.T. By G.P. Wells. *Biogr. Mem.* **24**, 183–221 (1978).
Holder, D.W. By A.D. Young, R.C. Pankhurst & D.L. Schultz. *Biogr. Mem.* **24**, 223–244 (1978).
Holroyd, R. By J.D. Rose. *Biogr. Mem.* **20**, 235–245 (1974).
Holtedahl, O. By L. Størmer. *Biogr. Mem.* **22**, 193–205 (1976).
Houssay, B.A. By Sir Frank Young & V.G. Foglia. *Biogr. Mem.* **20**, 247–270 (1974).
Hudson, W. By Sir Angus Paton. *Biogr. Mem.* **25**, 319–335 (1979).
Hutchinson, J. By C.E. Hubbard. *Biogr. Mem.* **21**, 345–365 (1975).
Huxley, J.S. By J.R. Baker. *Biogr. Mem.* **22**, 207–238 (1976).
Ing, H.R. By H.O. Schild & F.L. Rose. *Biogr. Mem.* **22**, 239–255 (1976).
Inglis, C.C. By A.R. Thomas & Sir Angus Paton. *Biogr. Mem.* **21**, 367–388 (1975).
Ingold, C.K. By C.W. Shoppee. *Biogr. Mem.* **18**, 349–411 (1972).
    Corrigenda. *Biogr. Mem.* **19**, 695 (1973); **24**, 605 (1978).
Jackson, W. (Baron Jackson of Burnley). By D. Gabor & J. Brown.
    Biogr. Mem. **17**, 379–398 (1971).
James, W.O. By A.R. Clapham & J.L. Harley. *Biogr. Mem.* **25**, 337–364 (1979).
    Corrigendum. *Biogr. Mem.* **26**, 543 (1980).
Jones, B.M. By Sir Arnold Hall & Sir Morien Morgan. *Biogr. Mem.* **23**, 253–282 (1977).
Jones, J.K.N. By M. Stacey. *Biogr. Mem.* **25**, 365–389 (1979).
Karrer, P. By O. Isler. *Biogr. Mem.* **24**, 245–321 (1978).
Kay, H.D. By Sir Kenneth Blaxter. *Biogr. Mem.* **23**, 283–310 (1977).
Kendall, J.P. By N. Campbell & C. Kemball. *Biogr. Mem.* **26**, 255–273 (1980).
Kennedy, W.Q. By J. Sutton. *Biogr. Mem.* **26**, 275–303 (1980).
Kenner, G.W. By Lord Todd. *Biogr. Mem.* **25**, 391–420 (1979).
Kenner, J. By Lord Todd. *Biogr. Mem.* **21**, 389–405 (1975).
Kermack, W.O. By J.N. Davidson with appendixes by F. Yates & W.H. McCrea.
    Biogr. Mem. **17**, 399–429 (1971).
Kidd, F. By H.B.S. Montgomery & A.F. Posnette. *Biogr. Mem.* **21**, 407–430 (1975).
Kleczkowski, A.A.P. By Sir Frederick Bawden. *Biogr. Mem.* **17**, 431–440 (1971).
Knowles, F.G.W. By E.J.W. Barrington. *Biogr. Mem.* **21**, 431–446 (1975).
Kronberger, H. By L. Rotherham. *Biogr. Mem.* **18**, 413–426 (1972).
Küchemann, D. By P.R. Owen & E.C. Maskell. *Biogr. Mem.* **26**, 305–326 (1980).
Lack, D.L. By W.H. Thorpe. *Biogr. Mem.* **20**, 271–293 (1974).
Lefschetz, S. By Sir William Hodge. *Biogr. Mem.* **19**, 433–453 (1973).
Lemberg, M.R. By C. Rimington & C.H. Gray. *Biogr. Mem.* **22**, 257–294 (1976).
Linnett, J.W. By A.D. Buckingham. *Biogr. Mem.* **23**, 311–343 (1977).
Littlewood, J.E. By J.C. Burkill. *Biogr. Mem.* **24**, 323–367 (1978).
London, H. By D. Shoenberg. *Biogr. Mem.* **17**, 441–461 (1971).
Lonsdale, Kathleen. By Dorothy M.C. Hodgkin. *Biogr. Mem.* **21**, 447–484 (1975).
McLeod, J.W. By Sir Graham Wilson & K.S. Zinnemann. *Biogr. Mem.* **25**, 421–444 (1979).
Mahalanobis, P.C. By C.R. Rao. *Biogr. Mem.* **19**, 455–492 (1973).
    Corrigenda. *Biogr. Mem.* **20**, 505 (1974); **21**, 585 (1975).

Maizels, M. By R.G. Macfarlane. *Biogr. Mem.* **23**, 345–366 (1977).
Manton, S.M. By G. Fryer. *Biogr. Mem.* **26**, 327–356 (1980).
Mapson, L.W. By Robert Hill. *Biogr. Mem.* **18**, 427–444 (1972).
Marion, L.E. By R.U. Lemieux & O.E. Edwards. *Biogr. Mem.* **26**, 357–370 (1980).
Marsden, E. By C.A. Fleming. *Biogr. Mem.* **17**, 463–496 (1971).
Marshall, S.M. By Sir Frederick Russell. *Biogr. Mem.* **24**, 369–389 (1978).
Martin, D.C. By Sir Harrie Massey & Sir Harold Thompson. *Biogr. Mem.* **24**, 391–407 (1978).
Martyn, D.F. By Sir Harrie Massey. *Biogr. Mem.* **17**, 497–510 (1971).
Menzies, R.G. By Sir Frederick White. *Biogr. Mem.* **25**, 445–476 (1979).
    Corrigendum. *Biogr. Mem.* **26**, 543 (1980).
Mitchell, G.H. By Sir James Stubblefield. *Biogr. Mem.* **23**, 367–383 (1977).
Monod, J.L. By A.M. Lwoff. *Biogr. Mem.* **23**, 385–412 (1977).
Mordell, L.J. By J.W.S. Cassels. *Biogr. Mem.* **19**, 493–520 (1973).
Morgan, M.B. By E.G. Broadbent. *Biogr. Mem.* **26**, 371–410 (1980).
Morton, R.A. By J. Glover, J.F. Pennock, G.A.J. Pitt & T.W. Goodwin.
    Biogr. Mem. **24**, 409–442 (1978).
Neish, A.C. By J.K.N. Jones. *Biogr. Mem.* **20**, 295–315 (1974).
Nyholm, R.S. By D.P. Craig. *Biogr. Mem.* **18**, 445–475 (1972).
Oakley, C.L. By D.G. Evans. *Biogr. Mem.* **22**, 295–305 (1976).
Onsager, L. By H.C. Longuet-Higgins & M.E. Fisher. *Biogr. Mem.* **24**, 443–471 (1978).
Orr, J.B. (Baron Boyd Orr of Brechin Mearns). By H.D. Kay. *Biogr. Mem.* **18**, 43–81 (1972).
Pearsall, W.H. By A.R. Clapham. *Biogr. Mem.* **17**, 511–540 (1971).
Penfield, W.G. By Sir John Eccles & W. Feindel. *Biogr. Mem.* **24**, 473–513 (1978).
Penrose, L.S. By Harry Harris. *Biogr. Mem.* **19**, 521–561 (1973).
Pfeil, L.B. By N.P. Allen. *Biogr. Mem.* **18**, 477–487 (1972).
Polanyi, M. By E.P. Wigner & R.A. Hodgkin. *Biogr. Mem.* **23**, 413–448 (1977).
Powell, A.R. By G.V. Raynor. *Biogr. Mem.* **22**, 307–318 (1976).
Powell, C.F. By F.C. Frank & D.H. Perkins, with an appendix by A.M. Tyndall.
    Biogr. Mem. **17**, 541–563 (1971).
Proudman, J. By D.E. Cartwright & F. Ursell. *Biogr. Mem.* **22**, 319–333 (1976).
Pugh, W.J. By A. Williams. *Biogr. Mem.* **21**, 485–495 (1975).
Raistrick, H. By J.H. Birkinshaw. *Biogr. Mem.* **18**, 489–509 (1972).
Raman, C.V. By S. Bhagavantam. *Biogr. Mem.* **17**, 565–592 (1971).
Redman, R.O. By R.F. Griffin & Sir Richard Woolley. *Biogr. Mem.* **22**, 335–357 (1976).
Relf, E.F. By A.R. Collar. *Biogr. Mem.* **17**, 593–616 (1971).
Ricardo, H.R. By Sir William Hawthorne. *Biogr. Mem.* **22**, 359–380 (1976).
Rideal, E.K. By D.D. Eley. *Biogr. Mem.* **22**, 381–413 (1976).
    Corrigendum. *Biogr. Mem.* **25**, 575 (1979).
Roberts, G. By O.A. Kerensky. *Biogr. Mem.* **25**, 477–503 (1979).
Robertson, Alexander. By R.D. Haworth & W.B. Whalley. *Biogr. Mem.* **17**, 617–642 (1971).
Robertson, Andrew. By Sir Alfred Pugsley. *Biogr. Mem.* **24**, 515–528 (1978).
Robertson, Muriel. By Ann Bishop & Sir Ashley Miles. *Biogr. Mem.* **20**, 317–347 (1974).
Robinson, R. By Lord Todd & J.W. Cornforth. *Biogr. Mem.* **22**, 415–527 (1976).
Romer, A.S. By T.S. Westoll & F.R. Parrington. *Biogr. Mem.* **21**, 497–516 (1975).
Rose, J.D. By A.W. Johnson. *Biogr. Mem.* **23**, 449–463 (1977).
Rosenheim, M.L. (Baron Rosenheim of Camden). By Sir George Pickering.
    Biogr. Mem. **20**, 349–358 (1974).
Roughton, F.J.W. By Q.H. Gibson. *Biogr. Mem.* **19**, 563–582 (1973).
Rous, F.P. By Sir Christopher Andrewes. *Biogr. Mem.* **17**, 643–662 (1971).
Russell, B.A.W., Earl Russell. By G. Kreisel. *Biogr. Mem.* **19**, 583–620 (1973).
Ruzicka, L. By V. Prelog & O. Jeger. *Biogr. Mem.* **26**, 411–501 (1980).
Salisbury, E.J. By A.R. Clapham. *Biogr. Mem.* **26**, 503–541 (1980).
Salisbury, 5th Marquess (Gascoyne-Cecil, R.A.J.). By Lord Todd. *Biogr. Mem.* **19**, 621–627 (1973).
Schonland, B.F.J. By T.E. Allibone. *Biogr. Mem.* **19**, 629–653 (1973).

Scott, D.A. By C.J. Best & A.M. Fisher. *Biogr. Mem.* **18**, 511–524 (1972).
Seshadri, T.R. By Wilson Baker & S. Rangaswami. *Biogr. Mem.* **25**, 505–533 (1979).
Shaw, T.I. By E.J. Denton. *Biogr. Mem.* **20**, 359–380 (1974).
Sheppard, P.A. By R.C. Sutcliffe & F. Pasquill. *Biogr. Mem.* **25**, 535–553 (1979).
Sheppard, P.M. By Sir Cyril Clarke. *Biogr. Mem.* **23**, 465–500 (1977).
Slater, W.K. By H.D. Kay. *Biogr. Mem.* **17**, 663–680 (1971).
Smith, F.E. By Sir Charles Goodeve. *Biogr. Mem.* **18**, 525–548 (1972).
Smith, T. By K.J. Habell. *Biogr. Mem.* **17**, 681–687 (1971).
Southwell, R.V. By Sir Derman Christopherson. *Biogr. Mem.* **18**, 549–565 (1972).
Spence, R. By E. Glueckauf. *Biogr. Mem.* **23**, 501–528 (1977).
Stedman, E. By H.J. Cruft. *Biogr. Mem.* **22**, 529–553 (1976).
Stoll, A. By L. Ruzicka. *Biogr. Mem.* **18**, 567–593 (1972).
Stoneley, R. By Sir Harold Jeffreys. *Biogr. Mem.* **22**, 555–564 (1976).
Sutton, O.G. By F. Pasquill, P.A. Sheppard & R.C. Sutcliffe. *Biogr. Mem.* **24**, 529–546 (1978).
Svedberg, T. By S. Claesson & K.O. Pedersen. *Biogr. Mem.* **18**, 595–627 (1972).
Taylor, G.I. By G.K. Batchelor. *Biogr. Mem.* **22**, 565–633 (1976).
Taylor, H.S. By C. Kemball. *Biogr. Mem.* **21**, 517–547 (1975).
Thomas, M. By Helen K. Porter & S.L. Ranson. *Biogr. Mem.* **24**, 547–568 (1978).
Thompson, W.R. By W.H. Thorpe. *Biogr. Mem.* **19**, 655–678 (1973).
Thomson, G.P. By P.B. Moon. *Biogr. Mem.* **23**, 529–556 (1977).
    Corrigenda. *Biogr. Mem.* **24**, 605 (1978).
Thornton, H.G. By P.S. Nutman. *Biogr. Mem.* **23**, 557–574 (1977).
Tilley, C.E. By W.A. Deer & S.R. Nockolds. *Biogr. Mem.* **20**, 381–400 (1974).
Timoshenko, S.P. By E.H. Mansfield & D.H. Young. *Biogr. Mem.* **19**, 679–694 (1973).
Tiselius, A. By R.A. Kekwick & Kai O. Pedersen. *Biogr. Mem.* **20**, 401–428 (1974).
Tolansky, S. By R.W. Ditchburn & G.D. Rochester. *Biogr. Mem.* **20**, 429–455 (1974).
Turrill, W.B. By C.E. Hubbard. *Biogr. Mem.* **17**, 689–712 (1971).
Uvarov, B.P. By Sir Vincent Wigglesworth. *Biogr. Mem.* **17**, 713–740 (1971).
de Valera, E. By J.L. Synge. *Biogr. Mem.* **22**, 635–653 (1976).
    Erratum. *Biogr. Mem.* **23**, 643 (1977).
Waddington, C.H. By Alan Robertson. *Biogr. Mem.* **23**, 575–622 (1977).
Walsh, A.D. By W.C. Price. *Biogr. Mem.* **24**, 569–582 (1978).
Walshe, F.M.R. By C.G. Phillips. *Biogr. Mem.* **20**, 457–481 (1974).
Warburg, O.H. By Sir Hans Krebs. *Biogr. Mem.* **18**, 629–699 (1972).
Watson, D.M.S. By F.R. Parrington & T.S. Westoll. *Biogr. Mem.* **20**, 483–504 (1974).
Watson-Watt. R.A. By J.A. Ratcliffe. *Biogr. Mem.* **21**, 549–568 (1975).
Wheeler, R.E.M. By S. Piggott. *Biogr. Mem.* **23**, 623–642 (1977).
Whiddington, R. By N. Feather. *Biogr. Mem.* **17**, 741–756 (1971).
Williams, F.C. By T. Kilburn & L.S. Piggott. *Biogr. Mem.* **24**, 583–604 (1978).
    Corrigenda. *Biogr. Mem.* **25**, 575 (1979).
Young, C.A.J. By R.L. Day & A. Spinks. *Biogr. Mem.* **25**, 555–573 (1979).
Ziegler, K. By C.E.H. Bawn. *Biogr. Mem.* **21**, 569–584 (1975).